"νοῦs ἄν εἴη τῶν ἀρχῶν, νοῦs ἄν εἴη ἐπιστήμηs ἀρχή καί ἀρχή τῆs ἀρχῆν..."

— Aristotle's Posterior Analytics

Nonlinear Mechanics

DEMETER G. FERTIS

CRC Press
Boca Raton Ann Arbor London Tokyo

Library of Congress Cataloging-in-Publication Data

Fertis, Demeter G.
Nonlinear mechanics / Demeter G. Fertis
p. cm.
Includes bibliographical references and index.
ISBN 0-8493-8933-X
1. Structural analysis (Engineering). 2. Nonlinear mechanics. I. Title.
TA646.F47 1993
624.1′71—dc2 93-12192
CIP

Direct all inquiries to CRC Press, Inc., 2000 Corporate Blvd., N.W., Boca Raton, Florida 33431.

International Standard Book Number 0-8493-8933-X
Library of Congress Card Number 93-12192
Printed in the United States of America 1 2 3 4 5 6 7 8 9 0
Printed on acid-free paper

Demeter G. Fertis is a professor of civil engineering at the University of Akron. He was previously an associate professor at the University of Iowa and at Wayne State University, and a research engineer in the Michigan State Department of Transportation. Dr. Fertis was also visiting professor at the National Technical University in Athens, Greece, where he received a Doctor of Engineering degree. He has also received a B.S. and an M.S. degree at the Michigan State University, and Diploma of engineering at the National Technical University of Athens. Dr. Fertis has consulted for NASA, Ford Motor Company, the Atomic Power Development Associates, General Motors, Boeing Aircraft Company, Lockheed California Company, Goodyear Aerospace, and Department of the Navy. He has developed patents, where one of his patents on *Airfoil design* was ranked in the top 2% of the country. He is the author of several books and numerous articles in professional journals and proceedings. Some of his books are *Dynamics and Vibration of Structures*, *Dynamics of Structural Systems*, and co-author of *Transverse Vibration Theory*. He is a member of ASCE, American Academy of Mechanics, Who's Who in America, Who's Who in the World, and other organizations, and served as professional journal editor and as a member in many national and international technical committees.

PREFACE

This book is written to be used as a text for a graduate course, in nonlinear mechanics, which is presently offered in many civil, mechanical, aeronautical, and engineering mechanics departments. It could also be used as a text for an introductory senior/first year graduate course on nonlinear mechanics with proper selection of chapters and chapter sections. It is written in a way that progressively exposes the student from the simpler to the most challenging material that makes it easy for the instructor to adjust the course to a level that better fits the background of its students. It also may be used as a professional book for practicing civil, mechanical, aeronautical, and polymer engineers, and as a research tool for engineers that are actively engaged in research in this area.

Over the past years, nonlinear mechanics has gained an important position in our modern technology, and its need and importance to our future technology grows at a much faster rate. In order to understand material and structural behavior and determine the nonlinear response of structures and machines, we must develop new methodologies and mathematical modeling that adequately represent such types of problems. In other words, we need to develop more sophisticated tools for our students and practicing engineers that will help them meet present needs, and also prepare them for the forthcoming challenges of our 21st century technology.

The purpose of this book is to establish a basis on which the field of nonlinear mechanics could develop in a way that meets both present and future needs. It is not intended to provide a complete and comprehensive treatment in all areas of nonlinear mechanics, but it does provide new concepts, theories, and methods that provide convenient and easy solutions to very complex nonlinear mechanics problems. Large deflections and inelastic analysis of structural components and machine elements can be examined in detail by using the theory and method of the equivalent systems. The structural or machine element can have any stiffness variation and loading along its length; it can be subjected to large deformations; it can vibrate from its large static equilibrium configuration; it can be subjected to cyclic loadings; and it can be permitted to be stressed well into the inelastic range and all the way to failure. The method of the equivalent systems as developed by the author and his collaborators, converts the nonlinear problem into an equivalent pseudolinear problem that can be handled conveniently with known methods of linear analysis. Since the derived equivalent system is exact, the equations representing this system may be solved exactly, or to a desired degree of accuracy. In other words, exact, as well as approximate methods of analysis, can be used.

The first chapter of the book introduces the reader to the theory of "elastica", which defines the exact shape of the deflection curve of a flexible member. This includes brief historical aspects of the large deformation problem and existing methods of solution. In the same chapter, the basics of the theory and method of the equivalent systems for flexible members are established. Chapters 2 and 3 provide a comprehensive treatment of statically determinate and stati cally indeterminate flexible members with various loading conditions and stiffness variations along their length. Exact as well as very accurate approximate methods of analysis are discussed.

Chapters 4 and 5 examine in detail the vibration of flexible members that are subjected to large static deformations. They formulate the exact theory regarding the vibration response of such flexible members, and its application to very challenging problems. Chapters 6 and 7 are dealing with the inelastic response of uniform and variable thickness members. They formulate the general theory based on small deformations by including the effect of axial restraints such as tensile and compressive loads. Chapter 8 provides an extensive treatment regarding the elastic and inelastic analysis of uniform and variable thickness rectangular and circular plates. It establishes the theory and method of the equivalent systems for two-dimensional problems and its application to plate problems with various boundary and loading conditions.

The subject of inelastic analysis of flexible members with uniform and variable thickness along their length, is examined in detail in Chapters 9 and 10. The flexible member is permitted to be subjected to large deformations, it can be stressed well beyond the elastic limit of its material, it can have any stiffness variation along its length, and it can be subjected to axial restraints. The derivation of hysteretic models for continuous systems, and the analysis of uniform and variable stiffness members subjected to cyclic loadings, is treated in detail in Chapter 11. In the final chapter, the nonlinear vibration of elastically supported beams with or without axial restraints, and the static and flutter instabilities that are developed during vibration, are investigated in detail. This includes transversed restraints in the form of initial displacements, as well as axial restraints produced by axial boundary displacements or axial tensile and compressive forces.

The material in this book is the result of many years of research and teaching of the author at Wayne State University, the University of Iowa, and the University of Akron, and the author's active collaboration and experience with industry and government. Most of the material in the book is taught at the University of Akron in the graduate course on nonlinear mechanics as part of the graduate curriculum of the department of Civil Engineering. I wish to express my thanks and appreciation to my students for their overwhelming response regarding the material in the book and their fine suggestions.

My appreciation and thanks go to my graduated Ph.D. students who spend endless hours with me during the development of the various theories involved, and the solution of the many challenging problems. In particular I wish to thank, in alphabetical order, Drs. Alexander O. Afonta, Paul A. Bosela, and Chin T. Lee. My gratitude and thanks go to Ms. Anna Kapetanaki for her outstanding work regarding the typing of the manuscript, and her excellent suggestions and encouragements. Last but not least, I wish to thank CRC Press for making my work available to the academic and professional audience.

—DEMETER G. FERTIS
Akron, Ohio
February, 1993

TABLE OF CONTENTS

Chapter 1

INTRODUCTION TO NONLINEAR MECHANICS

1.1 INTRODUCTION

The minimum weight criteria in the design of aircraft and aerospace vehicles, coupled with the ever growing use of light polymer materials that can undergo large displacements without exceeding their specified elastic limit, prompted a renewed interest in the analysis of flexible structures that are subjected to static and dynamic loads. Due to the geometry of their deformation, the behavior of such structures is highly nonlinear and the solution of such problems becomes very complex. The solution complexity becomes immense when flexible structural components are of variable cross sections. Such members are often used to improve strength and weight requirements, and in some cases, architects and planners are using variable cross section members to improve the architectural aesthetics and design of the structure.

In this chapter, the well known theory of elastica is discussed, as well as the methods used to solve the elastica. In addition, the solution of flexible members of variable cross section is developed in detail. This solution utilizes equivalent pseudolinear systems of constant cross section, as well as equivalent simplified nonlinear systems of constant cross section. This approach simplifies the solution of such complex problems (see, e.g., Fertis and Afonta[1]).

1.2 THE THEORY OF ELASTICA

The exact shape of the deflection curve of a flexible member is called the "elastica". The problem of the elastica was first investigated by Bernoulli, Lagrange, Euler, and Plana,[2–4] and mathematical solutions of some simple elastica problems have been obtained. Solutions of the elastica problem were also obtained by Frisch-Fay.[5] However, the most popular elastica is the solution of the flexible uniform cantilever beam loaded with a concentrated load P at the free end, as shown in Figure 1.1a.

The large deformation configuration of the cantilever beam caused by the load P is also shown in Figure 1.1a. Note that the end point B moved to point B′ during the large deformation of the member. The symbol Δ is used to denote the large horizontal displacement of point B. The beam is assumed to be inextensible, and thus the arc length AB′ of the deflection curve is equal to the initial length AB.

The expression for the bending moment M_x at any $0 \leq x \leq L_0$ may be obtained by using the free body diagram in Figure 1.1b and applying statics, i.e.,

$$M_x = -Px \tag{1.1}$$

In rectangular coordinates, the Euler-Bernoulli equation is given by the expression

$$\frac{y''}{\left[1+(y')^2\right]^{3/2}} = -\frac{M_x}{E_x I_x} \tag{1.2}$$

where E_x is the modulus of elasticity along the length of the member and I_x is the moment of inertia. By substituting Equation 1.1 into Equation 1.2 and assuming that E and I are uniform, we obtain

$$\frac{y''}{\left[1+(y')^2\right]^{3/2}} = \frac{Px}{EI} \tag{1.3}$$

Equation 1.2 may be also written in terms of the arc length x_o as

$$E_{x_o} I_{x_o} \frac{d\theta}{dx_o} = -M_x \tag{1.4}$$

By using Equation 1.1 and assuming uniform E and I, we have

$$\frac{d\theta}{dx_o} = \frac{Px}{EI} \tag{1.5}$$

By differentiating Equation 1.5 once with respect to x_o, we obtain

$$\frac{d^2\theta}{dx_o^2} = \frac{P}{EI}\cos\theta \tag{1.6}$$

By assuming that

$$E_x I_x = E_1 I_1\, g(x_o)\, f(x_o) \tag{1.7}$$

where $g(x_o)$ represents the variation of E_x with respect to a reference value E_1 and $f(x_o)$ represents the variation of I_x with respect to a reference value I_1, we can differentiate Equation 1.4 once to obtain

$$\frac{d}{dx_o}\left\{E_1 I_1\, g(x_o)\, f(x_o) \frac{d\theta}{dx_o}\right\} = -V_{x_o} \cos\theta \tag{1.8}$$

For members of uniform cross section and of linearly elastic material, we have $g(x_o) = f(x_o) = 1.0$.

Equations 1.3 and 1.6 are nonlinear second order differential equations, and exact solutions of these two equations are not presently available. Elliptic integral solutions are often used by investigators, e.g., Frisch-Fay,[4] but they are very complicated. This problem is discussed in detail in later sections of this chapter, where convenient methods of analysis are developed to simplify the solution of such problems.

By integrating Equation 1.3 once, we obtain

$$\frac{y'}{\left[1+(y')^2\right]^{1/2}} = \frac{Px^2}{2EI} + C \tag{1.9}$$

where C is the constant of integration that can be determined by applying the boundary condition of zero rotation y' at $x = L_0 = (L - \Delta)$. This yields

$$\frac{y'}{\left[1+(y')^2\right]^{1/2}} = G(x) \tag{1.10}$$

where

$$G(x) = \frac{P}{2EI}\left[x^2 - (L-\Delta)^2\right] \tag{1.11}$$

Thus, by solving Equation 1.10, we obtain

$$y'(x) = \frac{G(x)}{\left\{1-[G(x)]^2\right\}^{1/2}} \tag{1.12}$$

The large deflection y at any $0 \le x \le L_o$ may now be obtained by integrating Equation 1.12 once and satisfying the boundary condition of zero deflection at $x = L_o$ for the evaluation of the integration constant. It should be noted, however, that G(x) in Equation 1.11 is a function of the unknown horizontal displacement Δ of the free end of the beam. The value of Δ may be determined from the equation

$$L = \int_0^{L_o} \left[1+(y')^2\right]^{1/2} dx \tag{1.13}$$

by a trial and error procedure. That is, assume a value of Δ in Equation 1.12 and then carry out the integration in Equation 1.13 to determine the length L.

The procedure may be repeated for various values of Δ until the correct length L is obtained.

The integration in Equation 1.13 becomes more convenient if we introduce the variable

$$\xi = \frac{x}{L - \Delta} \tag{1.14}$$

$$d\xi = \frac{dx}{L - \Delta} \tag{1.15}$$

On this basis, Equation 1.13 becomes

$$L = (L - \Delta \int_0^1 \left\{1 + \left[y'(\xi)\right]^2\right\}^{1/2} d\xi \tag{1.16}$$

or, by using Equations 1.11 and 1.12 and the variable ξ,

$$L = (L - \Delta) \int_0^1 \frac{1}{\left\{1 - \left[G(\xi)\right]^2\right\}^{1/2}} d\xi \tag{1.17}$$

where

$$G(\xi) = \frac{P(L - \Delta)^2}{2EI}\left(\xi^2 - 1\right) \tag{1.18}$$

It would be appropriate to emphasize again here that Equations 1.2 and 1.4 are second order nonlinear differential equations which provide the exact shape of the deflection curve of the beam. In conventional applications these equations are linearized by neglecting the square of the slope (y') in Equation 1.2, in comparison with unity. This assumption is permissible, provided that the deflections are very small compared to the length of the beam. For flexible bars, however, where deflections are large compared to the length of the member, this assumption is not permissible, and Equation 1.2 or Equation 1.4 must be used in its entirety. This means that the deflections are no longer a linear function of the bending moment, or of the load, and consequently the principle of superposition is not applicable. Therefore, every case involving large deformations has to be solved separately, since combinations of load types already solved cannot be superimposed. The situation yields much greater consequences when the stiffness EI of the member is variable.

1.3 METHODS OFTEN USED FOR THE SOLUTION OF THE ELASTICA

A trial and error procedure may be used to solve Equation 1.13 or Equation 1.16 for the value of the horizontal displacement Δ. For more complicated problems, the utilization of an equivalent system will simplify the solution of the problem (see, e.g., Fertis,[6,7] Fertis and Afonta,[5] and Fertis and Lee.[8,9] This problem is examined in detail in the later parts of this chapter.

Newton's method, the nonlinear least squares method, and the iterative method may be also used in place of the trial and error procedure. In the application of these three methods care must be taken in choosing the starting deformation parameters. The utilization of equivalent systems facilitates the application of these methods.

The trial and error procedure is used here to solve the cantilever beam problem in Figure 1.1a by assuming that L = 1000 in. (25.4 m), EI = 180,000 kip-in.^{2} (516,541 Nm^2), and values of the concentrated load P are between 0 and 10 kips (44.48 kN). The trial and error procedure is used to determine the value of the horizontal displacement Δ from Equations 1.17 and 1.18. The required integrations are carried out by using Simpson's rule. With known Δ, the values of y' at any $0 \leq x \leq L_0$, and consequently the slope $\theta = \tan^{-1}(y')$ of the flexible member, may be determined from Equation 1.12. Integration of Equation 1.12 yields the deflections y at any $0 \leq x \leq L_0$. The constant of integration may be determined by applying the boundary condition of zero deflection at $x = L_0$.

Table 1.1 shows the values obtained for the free end horizontal and vertical displacements Δ_B and δ_B, respectively, and for the rotation θ_B at the same end for various values of the load P. In the same table, values of δ_B that are obtained by using linear analysis are also shown in the second column for comparison purposes. The linear analysis gives erroneous results for values of $P > 0.4$ kips (1.779 kN). Figure 1.2 shows the deflection curves of the cantilever beam for values of P = 1 kip (4.448 kN), P = 3 kips (13.344 kN), and P = 10 kips (44.48 kN). For the last value of P, the cantilever beam is practically hanging in the downward direction, and the concentrated load P becomes almost an axially applied load since the rotation θ_B at the free end is 89.61°.

1.4 HISTORICAL ASPECTS OF THE LARGE DEFORMATION PROBLEM

The static analysis of flexible members involves primarily the solution of elastica problems by using simplified closed form solutions and finite element approaches. The theory of elastica which examines the elastic deformation of the axis of straight bars that are subjected to bending is a classical one, and it was developed in the 18th century by Jacob Bernoulli, his younger brother Johann Bernoulli, and Leonhard Euler. The first known published works

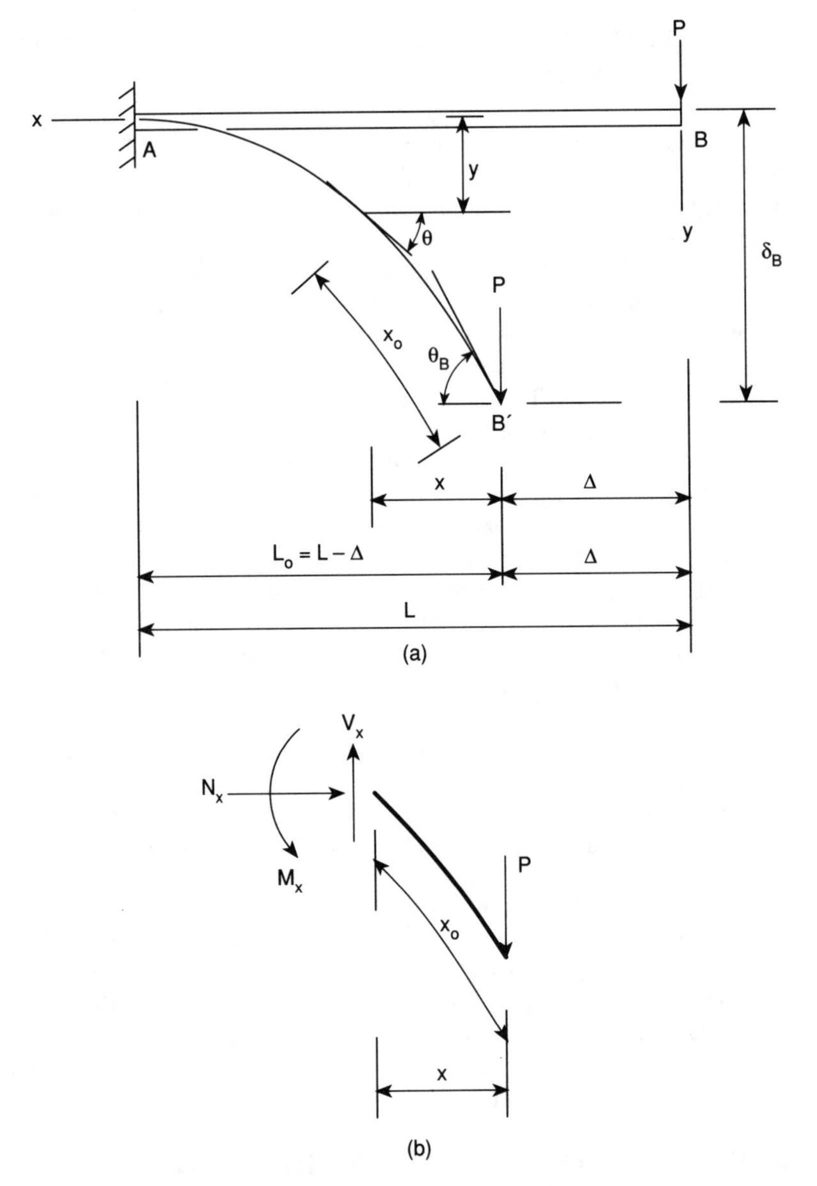

FIGURE 1.1. (a) Large deformation of a cantilever beam of uniform cross section; (b) free body diagram of a beam element.

regarding the large deflection of flexible members was written by Euler[2] in 1744. In the appendix of his book *De Curvis Elastics,* Euler stated that the slope dy/dx cannot be neglected in the expression of the curvature unless the deflections are small. This problem was also analyzed by Lagrange,[1] but, as pointed out later by Plana,[3] his solution provided erroneous results.

The extensively used Euler-Bernoulli law states that the bending moment is proportional to the change in the curvature produced by the action of the load, i.e.,

TABLE 1.1
Values of δ_B, Δ_B, and θ_B for Various Values of P and Comparisons with Linear Theory

	Linear analysis	Nonlinear analysis		
Load P (kips)	**d_B (in.)**	**d_B (in.)**	**D_B (in.)**	**q_B (°)**
0.2	370.37	328.61	67.36	28.90
0.4	740.74	523.27	183.10	47.86
0.6	1111.11	629.00	281.29	59.42
0.8	1481.48	691.58	356.71	66.87
1.0	—	732.14	414.95	71.95
1.4	—	821.37	577.22	83.23
2.5	—	841.64	621.12	85.48
3.0	—	856.20	653.84	86.92
5.0	—	888.86	731.69	89.00
10.0	—	921.40	810.27	89.61

Note: 1 in. = 0.0254 m, 1 kip = 4.448 kN.

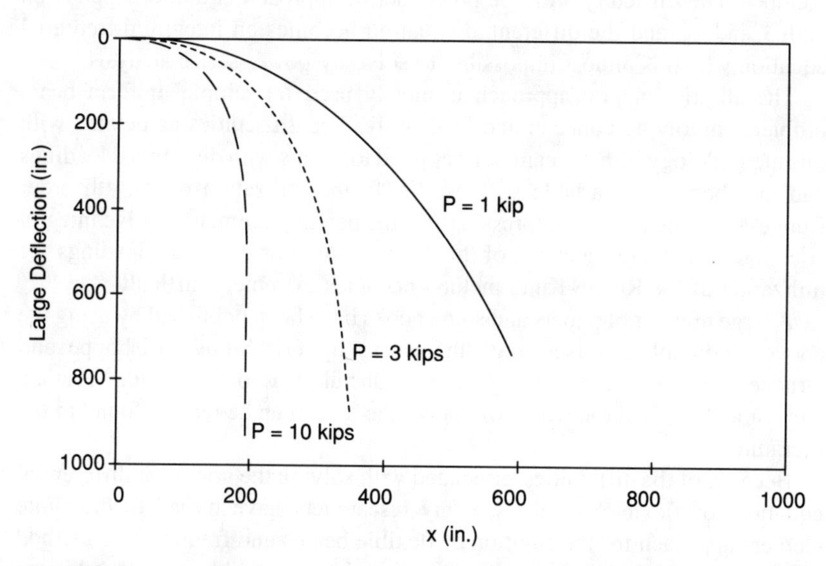

FIGURE 1.2. Large deflection curves for P = 1, 3, and 10 kips (1 kip = 4.448 kN, 1 in. = 0.0254 m).

$$\frac{1}{r} = \frac{d\theta}{dx_o} = \frac{M}{EI} \tag{1.19}$$

where r is the radius of curvature and θ is the slope at any point x_o, where x_o is measured along the length of the arc as shown in Figure 1.1a. In rectangular x,y coordinates the equation is written as

$$\frac{1}{r} = \frac{d^2y/dx^2}{\left[1+(dy/dx)^2\right]^{3/2}} = -\frac{M}{EI} \tag{1.20}$$

which is similar to Equation 1.2 used in the preceding section.

Commonly used methodologies for the solution of Equation 1.20 involve the utilization of power series, complete and incomplete elliptic integrals, and numerical procedures using the Runge-Kutta method. In the power series approach, the rotation θ is expressed as a function of x_o by using Maclaurin's series, i.e.,

$$\theta(x_o) = \theta(c) + (x_o - c)\,\theta'(c) + \frac{(x_o - c)^2}{2!}\theta''(c) + \frac{(x_o - c)^3}{3!}\theta'''(c) + \ldots\ldots \tag{1.21}$$

where c is an arbitrary point taken along the arc length of the deformed member. The difficulty with the power series approach is that θ depends on both x and x_o, and the differential equation becomes an integral differential equation which is almost impossible to solve by power series analysis.

The elliptic integral approach is mostly used for simple uniform beam problems involving concentrated loads only. The difficulties associated with this methodology is that it cannot be applied to beams with distributed loadings and members with variable stiffness EI. In the utilization of the 4th order Runge-Kutta method, the expressions for the bending moment involve integral equations which are functions of the deformation. For multistate loadings the utilization of the Runge-Kutta method becomes extremely difficult.

A large number of papers and some books have been published by using the above methodologies (see, e.g., the works by Frisch-Fay,[4] Bishoppe and Drucker,[10] Lau,[11] Seide,[12] Wang, et al.,[13] Ohtsuki,[14] Rao,[77] Liebold,[15] Prathap and Varandar,[16] and the works of many others that can be readily found in the literature.

Because of the difficulties associated with solving the nonlinear differential equation for flexible members, many researchers have turned to the finite element approach for the solution of flexible beams and frames. This method would be particularly attractive to beam and frame problems with arbitrary geometry and complex loading conditions. However, difficulties are also associated with this method in representing rigid body motions of oriented bodies that undergo large displacements. The works by Hsiao et al.,[17] Tada et al.,[18] and Yang[19] are examples of such applications.

In recent years the large deflection problem of flexible members, by including both static and dynamic analysis, was extensively investigated by Fertis and Afonta[5,20–22] Fertis and Pallaki,[23] and Fertis and Lee.[8,9,24] They have utilized equivalent pseudolinear and simplified nonlinear equivalent systems of uniform stiffness to solve flexible members with complicated stiffness variations and loading conditions. Both elastic and inelastic analyses of such flexible members have been considered. The effects of axial compressive forces are considered by Fertis and Lee[25] for both elastic and inelastic ranges.

Considerable work regarding the nonlinear vibration of members with axial restraints and small deformations may be found in the works of Prathap,[26] Mei,[27] Bhashyam and Prathap,[28] and others listed in the references of this text. Fertis and Lee[29] have also investigated the nonlinear vibration of axially restrained elastically supported beams. A limited amount of work regarding large beam amplitude vibrations is found in the literature (see, e.g., the work of Fertis and Afonta[21,22]).

1.5 PSEUDOLINEAR EQUIVALENT SYSTEMS

In this section, the initial large deflection nonlinear problem will be solved by using an equivalent pseudolinear system that has a deflection curve identical to the initial nonlinear problem. In other words, the initial nonlinear problem is transformed into a pseudolinear equivalent system that can be solved by applying linear analysis. The initial nonlinear system can have an arbitrary moment of inertia variation and a variable modulus of elasticity along the length of the member, and it can be subjected to arbitrary loading conditions. The equivalent pseudolinear system will always be of uniform stiffness EI throughout its equivalent length, but its loading will be different. The utilization of equivalent pseudolinear systems greatly simplifies the solution of complicated large deflection problems as shown later in the text.

The derivation of pseudolinear equivalent systems of constant stiffness EI may be initiated by employing the Euler-Bernoulli law given by Equation 1.2. This equation is written again below:

$$\frac{y''}{\left[1+(y')^2\right]^{3/2}} = -\frac{M_x}{E_x I_x} \tag{1.22}$$

where the bending moment M_x, modulus of elasticity E_x, and moment of inertia I_x, are assumed to vary in any arbirtary manner.

The curvature of the member that is represented by the left-hand side of Equation 1.22 is geometrical in nature, and it requires that the parameters M_x, E_x, and I_x on the right-hand side of the equation also be associated with the deformed configuration of the member. When the loading on the member is distributed and/or the cross-sectional moment of inertia is variable, the expressions for these parameters are, in general, nonlinear integral equations of the deformation, and contain functions of horizontal displacement. That is, the bending moment M_x, depth h_x of the member, and moment of inertia I_x, are all

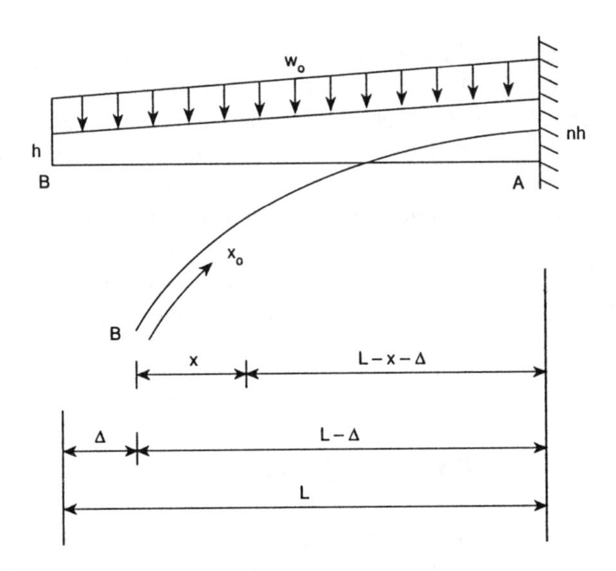

FIGURE 1.3. Tapered cantilever beam loaded with a uniformly distributed load w_o.

functions of both x and x_0. This is easily observed by examining the deformed configuration of the doubly tapered cantilever beam in Figure 1.3. Therefore, the bending moment M_x has to be defined with respect to the deformed segment. On the other hand, the total load acting on an undeformed segment of a member does not change after the segment is deformed. This subject will be addressed in greater detail in Section 1.7 of this chapter.

The variable stiffness E_xI_x may be expressed as

$$E_xI_x = E_1I_1g(x)f(x) \tag{1.23}$$

where g(x) represents the variation of E_x with respect to a reference value E_1 and f(x) represents the variation of I_x with respect to a reference value I_1. If the member has a constant E and I throughout its length, then g(x) = f(x) = 1.00 and $E_xI_x = E_1I_1 = EI$. In this case, if preferred, the constant stiffness EI may be taken as the reference stiffness value E_1I_1; however, this is not mandatory. By substituting Equation 1.23 into Equation 1.22, we obtain

$$\frac{y''}{\left[1+(y')^2\right]^{3/2}} = -\frac{1}{E_1I_1}\cdot\frac{M_x}{g(x)f(x)} \tag{1.24}$$

By integrating Equation 1.24 twice, the expression for the large transversed displacement y may be written as

$$y(x) = \frac{1}{E_1I_1}\int\left\{-\int\left[1+(y')^2\right]^{3/2}\frac{M_x}{g(x)f(x)}\right\}dx + C_1\int dx + C_2 \tag{1.25}$$

where C_1 and C_2 are the constants of integration that can be determined by using the boundary conditions of the member.

For a member of constant stiffness E_1I_1 and with length and reference system of axes identical to the one used for Equation 1.25, the expression for its large deflection y_e may be written as

$$y_e = \frac{1}{E_1I_1}\int\left\{-\int\left[1+(y_e')^2\right]^{3/2} M_e dx\right\}dx + C_1'\int dx + C_2' \tag{1.26}$$

where M_e is the bending moment at any cross section x and C_1' and C_2' are constants of integration.

The deflection curves y and y_e given by Equations 1.25 and 1.26, respectively, will be identical if

$$C_1 = C_1' \qquad C_2 = C_2' \tag{1.27}$$

$$\int\left\{-\int\left[1+(y')^2\right]^{3/2}\frac{M_x dx}{f(x)g(x)}\right\}dx = \int\left\{-\int\left[1+(y_e')^2\right]^{3/2} M_e dx\right\}dx \tag{1.28}$$

The conditions in Equation 1.27 are satisfied if the two members have the same length and boundary conditions. Equation 1.28 is satisfied if $y_e' = y'$ and

$$M_e = \frac{M_x}{f(x)g(x)} \tag{1.29}$$

On this basis, we should have

$$\left[1+(y_e')^2\right]^{3/2} M_e = \left[1+(y')^2\right]^{3/2}\frac{M_x}{f(x)g(x)} \tag{1.30}$$

For small deflection theory Equation 1.30 reduces to Equation 1.29, because $(y')^2$ and $(y_e')^2$ are small compared with unity and they can be neglected. Thus, for small deflections, the moment diagram M_e of the equivalent system of constant stiffness E_1I_1 can be obtained from Equation 1.29. Its equivalent shear force V_e and loading W_e can be obtained from Equation 1.29 by defferentiation, i.e.,

$$V_e = \frac{d}{dx}(M_e) = \frac{d}{dx}\left[\frac{M_x}{f(x)g(x)}\right] \tag{1.31}$$

$$W_e = -\frac{d}{dx}(V_e)\cos\theta = -\frac{d^2}{dx^2}\left[\frac{M_x}{f(x)g(x)}\right]\cos\theta \tag{1.32}$$

where $\cos\theta \approx 1$ for small rotations θ of the member. The equivalent constant stiffness system in this case is linear, and linear small deflection theory can be used for its solution.

When the deflections and rotations are large, $(y')^2$ and $(y'_e)^2$ cannot be neglected, and Equations 1.24 and 1.30 suggest that for an equivalent pseudolinear analysis the moment M_e' of the equivalent pseudolinear system of constant stiffness E_1I_1 should be obtained from the equation

$$M_e' = \left[1+(y')^2\right]^{3/2} M_e = \left[1+(y')^2\right]^{3/2} \frac{M_x}{f(x)g(x)} = \frac{Z_e}{f(x)g(x)} M_x \tag{1.33}$$

where

$$Z_e = \left[1+(y')^2\right]^{3/2} \tag{1.34}$$

and $\theta = \tan^{-1}(y')$ represents the slope of the initial nonlinear system. If $f(x) = g(x) = 1.00$, the initial nonlinear system will have a uniform stiffness EI.

The shear force V_e' and loading W_e' of the equivalent constant stiffness pseudolinear system may be determined from the expressions

$$V_e' = \frac{d}{dx}(M_e') = \frac{d}{dx}\left[1+(y')^2\right]^{3/2} M_e' = \frac{d}{dx}\left[\frac{Z_e}{f(x)g(x)}\right] M_x \tag{1.35}$$

$$W_e' = -\frac{d}{dx}(V_e')\cos\theta = -\frac{d^2}{dx^2}\left[1+(y')^2\right]^{3/2} M_e' \cos\theta$$
$$= -\frac{d^2}{dx^2}\left[\frac{Z_e}{f(x)g(x)}\right] M_x \cos\theta \tag{1.36}$$

When the equivalent constant stiffness pseudolinear system is obtained, elementary linear deflection theory can be used to solve the pseudolinear system. The deflections and rotations obtained in this manner will be identical to those of the original nonlinear variable stiffness member. This is appropriate because the equivalent moment diagram M_e in Equation 1.29, which also represents the moment M_e at any x of a nonlinear equivalent system of constant stiffness E_1I_1, and consequently the curvature of the initial variable stiffness member, is

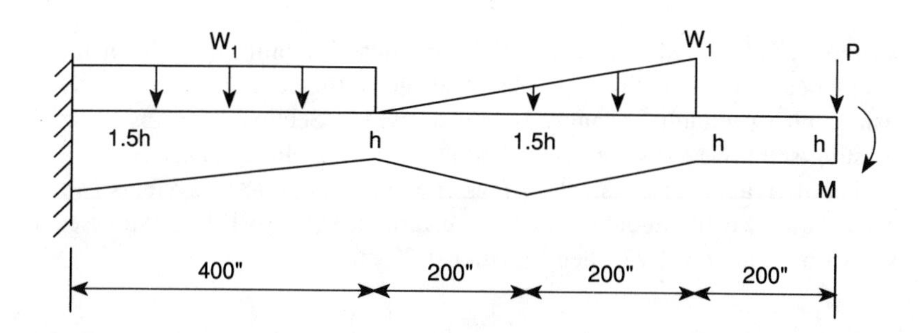

FIGURE 1.4. Variable stiffness flexible cantilever beam with elaborate loading condition (1 in. = 0.0254 m).

corrected by multiplying it by the expression $[1+(y')^2]^{3/2}$, as shown in Equation 1.33.

At this point it also should be emphasized that Equations 1.29, 1.31, and 1.32 represent, respectively, the moment M_e, shear V_e, and load W_e of a nonlinear equivalent system of constant stiffness E_1I_1, whose elastic curve is identical to that of the original variable stiffness member. Therefore, an alternative approach would be to use the constant stiffness nonlinear equivalent system for the computation of the large deflections and rotations of the initial variable stiffness member.

In order to simplify the mathematics regarding the computation of V_e' and W_e', or V_e and W_e, the slope of the moment diagram represented by Equation 1.33, or the one represented by Equation 1.29, may be approximated by a few straight lines judiciously selected. On this basis, the pseudolinear and equivalent nonlinear systems of constant stiffness E_1I_1 will always be loaded by a few concentrated loads. This approximation simplifies, to a large extent, the solution of the complex large deflection problem and yields very accurate results. It also provides a convenient way to solve large deflection problems where the stiffness EI and loading vary arbitrarily along the length of the member.

1.6 SIMPLIFIED NONLINEAR EQUIVALENT SYSTEMS

As stated earlier, see Section 1.2, the large deformations of flexible bars are no longer a linear function of the bending moment, or of the applied load, and consequently the principle of superposition is not applicable. This restriction creates enormous difficulties in the solution of beam problems with more elaborate loading conditions. The solution becomes even more complicated when the moment of inertia of a flexible member varies arbitrarily along its length. The flexible cantilever beam in Figure 1.4, loaded as shown, illustrates a case of an elaborate loading condition coupled with a variable depth along the length of the member. A reasonable solution to this problem can be obtained by finding a simpler equivalent mathematical model that accurately (or exactly) represents the initial complicated mathematical problem. This may be accomplished by reducing the initial complicated nonlinear problem into a simpler equivalent nonlinear problem that can be solved more conveniently by using either pseudolinear analysis as discussed in Section 1.5 or by utilizing existing solutions (or solution methodologies) of nonlinear analysis.

The derivation of constant stiffness nonlinear equivalent systems can be carried out by using Equation 1.22. If the variable stiffness E_xI_x is expressed as shown by Equation 1.23, then Equation 1.22 may be written as

$$\frac{y''}{\left[1+(y')^2\right]^{3/2}} = -E_1I_1\frac{M_x}{g(x)f(x)} \tag{1.37}$$

or

$$\frac{y''}{\left[1+(y')^2\right]^{3/2}} = -\frac{M_e}{E_1I_1} \tag{1.38}$$

where

$$M_e = \frac{M_x}{g(x)f(x)} \tag{1.39}$$

Equation 1.38 is the nonlinear differential equation of an equivalent system of constant stiffness E_1I_1 whose bending moment M_e at any cross section is given by Equation 1.39. Therefore, the variable stiffness system represented by Equation 1.37 and the one of constant stiffness represented by Equation 1.38 will have identical deflection curves. Thus, it may be concluded that Equation 1.38 may be used to solve the variable stiffness problem by applying nonlinear analysis. In order to make the solution easier, the shape of the M_e diagram represented by Equation 1.39 may be approximated with a few straight lines, which results in a constant stiffness equivalent system that is always loaded with a few concentrated loads.

In practical problems, such as the one in Figure 1.4 where I_x may have any arbitrary variation and the loading on the member is also arbitrary and discontinuous, the use of Equation 1.39 and the approximation of the shape of M_e with straight lines results in a constant stiffness equivalent system that yields a simple and accurate solution to the problem. The equivalent nonlinear system of constant stiffness E_1I_1 may be solved by applying known nonlinear methods of analysis, or it may be solved by using pseudolinear analysis as discussed in Section 1.5. Both procedures are discussed in detail in the following sections.

1.7 DEPENDENCE OF LOADING AND STIFFNESS ON THE GEOMETRY OF THE DEFORMATION

In order to apply the methodologies discussed in Sections 1.5 and 1.6, we have to realize that the expressions for the bending moment M_x and moment

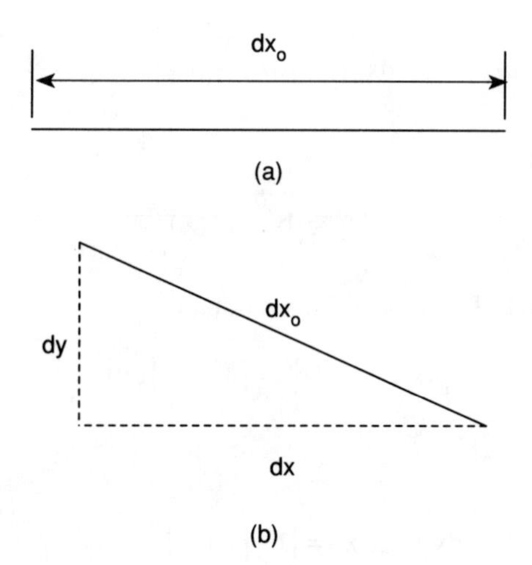

FIGURE 1.5. (a) Undeformed configuration of an arc length segment dx_o; (b) deformed configuration of dx_o.

of inertia I_x are generally nonlinear functions of the large deformation of the member, i.e.,

$$M_x = M(x, x_o) \tag{1.40}$$

$$I_x = I_1 f(x, x_o) \tag{1.41}$$

where x is the abscissa of center line points of the deformed configuration of the member, x_o is the arc length of the deformed segment, and I_1 is the reference moment of inertia. It should also be noted here that the equivalent bending moment M_e, or M'_e, should be defined with respect to the deformed configuration of the member, where the exact solutions for M_e, V_e, and W_e are functions of the horizontal displacement $\Delta(x)$ of the member.

In order to reduce the complexity of such problems, we express the arc length $x_o(x)$ in terms of the horizontal displacement $\Delta(x)$ of the member, where $0 \le x \le (L{-}x)$, i.e.,

$$x_o(x) = x + \Delta(x) \tag{1.42}$$

We also know that the expression for $x_o(x)$ is an integral function of the deformation, which can be expressed as

$$x_o(x) = \int_0^x \sqrt{1 + [y'(x)]^2}\, dx \tag{1.43}$$

The derivation of Equation 1.42 may be initiated by considering a segment dx_o before and after deformation, as shown in Figure 1.5. By applying the Pythagorean theorem we write

$$[dx_o]^2 = [dx]^2 + [dy]^2 \quad (1.44)$$

By assuming that

$$dx_o = dx + d\Delta(x) \quad (1.45)$$

and substituting into Equation 1.44, we obtain

$$[dx + d\Delta\,(x)]^2 = [dx]^2 + [dy]^2 \quad (1.46)$$

or

$$dx + d\Delta(x) = \left\{1 + [y'(x)]^2\right\}^{1/2} dx \quad (1.47)$$

Integration of Equation 1.47 with respect to x yields

$$x + \Delta(x) = \int_0^x \left\{1 + [y'(x)]^2\right\}^{1/2} dx \quad (1.48)$$

which gives the same results as Equations 1.42 and 1.43.

For beams where one of the end supports is permitted to move in the horizontal direction, such as cantilever beams, simply supported beams, etc., approximate expressions for the variation of $\Delta(x)$ may be used, which greatly facilitates the solution of the problem. The cases of $\Delta(x)$ that have been investigated (see Fertis and Afonta[1]), and are proven to provide accurate results as follows:

$$\Delta(x) = \text{constant} = \Delta \quad (1.49)$$

$$\Delta(x) = \Delta \frac{x}{L_o} \quad (1.50)$$

$$\Delta(x) = \Delta \sqrt{\frac{x}{L_o}} \quad (1.51)$$

$$\Delta(x) = \Delta \sin \frac{\pi x}{2L_o} \quad (1.52)$$

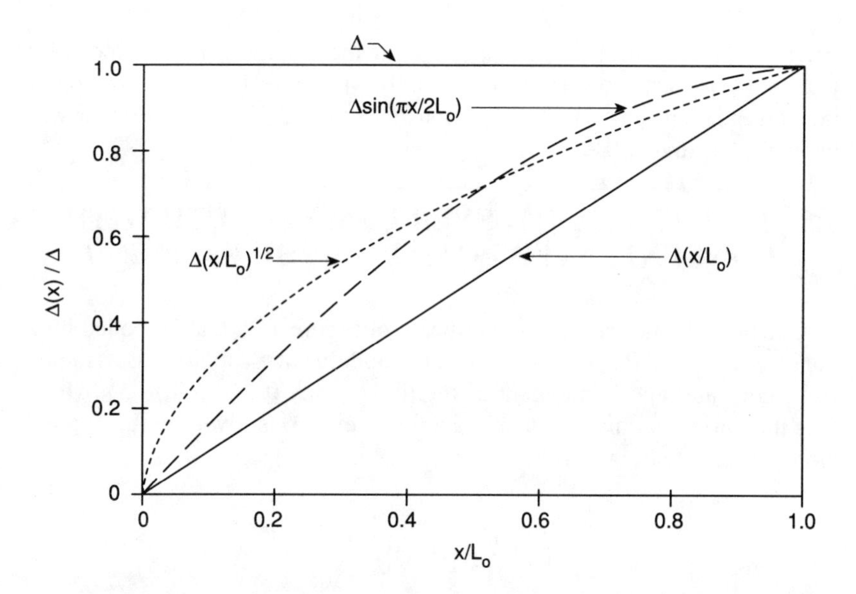

FIGURE 1.6. Graphs of various cases of Δ(x).

where Δ is the horizontal displacement of the movable end and $L_o = (L - \Delta)$. A plot of the variations of Δ(x) given by Equations 1.49 to 1.52 is shown in Figure 1.6.

The various cases examined in this text, as well as by Fertis and Afonta[5] and Fertis and Lee,[8] indicate that a reasonable solution with an error of about 3% or less, can be obtained by using Equation 1.49. This means that the variation of the bending moment M_x, and consequently the deformation of the member, are largely dependent upon the boundary condition of Δ(x) at the moving end of the member, and it is rather insensitive to the variation of Δ(x) between the ends of the member. This is particularly true when the deformations are very large. It should also be noted here that Equation 1.49 is an upper bound, as indicated by the graphs in Figure 1.6.

It was stated earlier that the variable moment of inertia I_x of a flexible member is also a nonlinear function of the deformation. For uniform and tapered members that are loaded with concentrated loads only, the variation of the height h(x) of the member may be approximated by the expression

$$h(x) = (n - 1)\left[\frac{1}{n - 1} + \frac{x}{L - \Delta}\right]h \tag{1.53}$$

where x is the abscissa of points of the centroidal axis of the member in its deformed configuration, n represents the taper, h is a reference height, and L is the undeformed length of the member. The error of 3% or less that is associated with the utilization of Equation 1.53 is considered to be small for practical applications. Under this assumption, the solution of flexible members that are loaded with concentrated loads only will not require the utilization of integral equations or the use of Equations 1.49 to 1.52. This point of view will be amply illustrated later.

1.8 TAPERED CANTILEVER BEAM LOADED WITH A CONCENTRATED LOAD P AT THE FREE END

Consider the tapered cantilever beam in Figure 1.7a that is loaded by a concentrated load P at the free end B. Its modulus of elasticity E is assumed to be constant, and its moment of inertia I_x at any $0 \le x \le (L - \Delta)$, where Δ is the horizontal displacement of the free end B, is given by the expression

$$
\begin{aligned}
I_x &= \frac{bh^3}{12}\left[f(x_o)\right] \\
&= I_B\left[1 + \frac{(n-1)}{L}x_o\right]^3 \\
&= I_B f(x)
\end{aligned}
\tag{1.54}
$$

where

$$I_B = \frac{bh^3}{12} \tag{1.55}$$

$$x_o = \int_0^x \left\{1 + \left[y'(\xi)\right]^2\right\}^{1/2} dx \tag{1.56}$$

$$f(x) = \left[1 + \frac{(n-1)}{L}x_o\right]^3 \tag{1.57}$$

b is the constant width of the member, and ξ is a dummy variable that is dependent of x.

From Figure 1.7a, the bending moment M_x at any x from the free end C is given by the expression

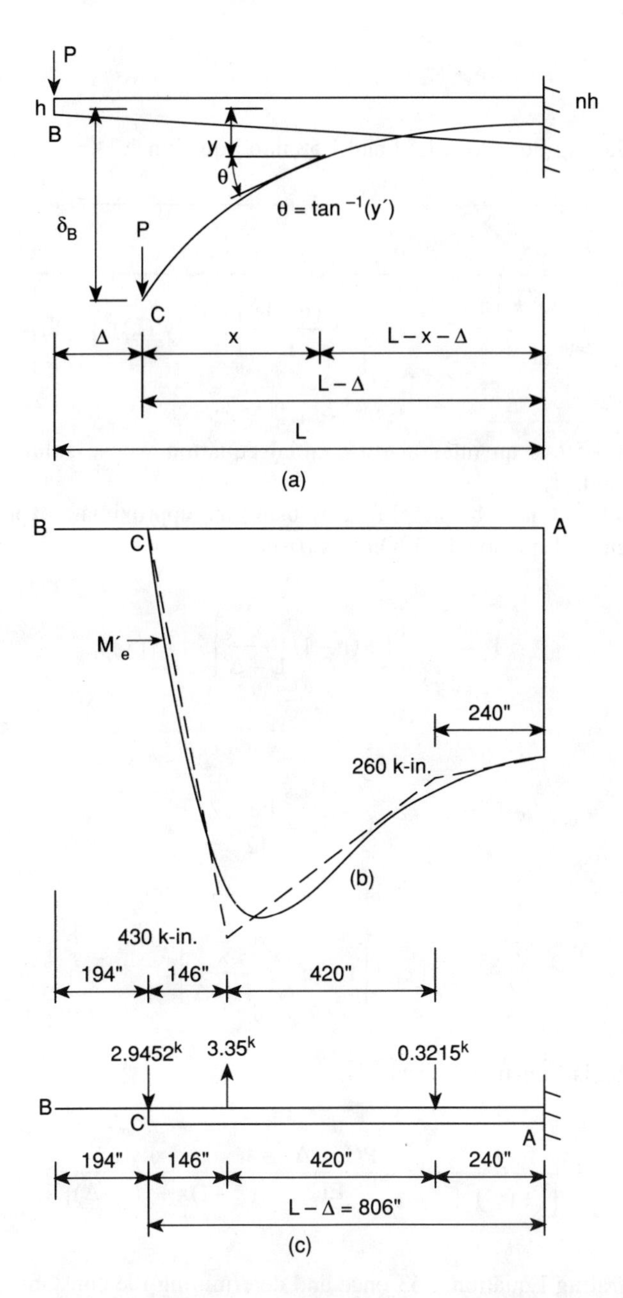

FIGURE 1.7. (a) Original variable stiffness member; (b) moment diagram M'_e of the pseudolinear system with its shape approximated with three straight lines; (c) equivalent pseudolinear system of constant stiffness (1 kip-in = 113 Nm, 1 kip = 4.448 kN, 1 in. = 0.0254 m).

$$M_x = Px \tag{1.58}$$

By substituting Equations 1.54 and 1.58 into Equation 1.24, with g(x) = 1, we obtain

$$\frac{y''}{\left[1+(y')^2\right]^{3/2}} = -\frac{P}{EI_B}\frac{x}{\left\{1+\frac{(n-1)}{L}\int_0^x\left[1+(y'(\xi))^2\right]^{1/2}d\xi\right\}^3} \tag{1.59}$$

Equation 1.59 is an integral differential equation whose solution is very complicated.

The solution may be simplified by using the approximate expression for h(x) given by Equation 1.53. On this basis

$$I_x = \frac{bh^3}{12}\left[1+(n-1)\frac{x}{L-\Delta}\right]^3 = I_1 f(x) \tag{1.60}$$

where

$$I_1 = I_B = \frac{bh^3}{12} \tag{1.61}$$

$$f(x) = \left[1+(n-1)\frac{x}{L-\Delta}\right]^3 \tag{1.62}$$

On this basis Equation 1.24 yields

$$\frac{y''}{\left[1+(y')^2\right]^{3/2}} = -\frac{P(L-\Delta)^3}{EI_B}\frac{x}{[(x-1)x+(L-\Delta)]^3} \tag{1.63}$$

By integrating Equation 1.63 once and determining the constant of integration by satisfying the boundary condition of zero rotation at x = (L – Δ), we find

$$y'(x) = \frac{Q(x)}{\left\{1-[Q(x)]^2\right\}^{1/2}} \tag{1.64}$$

TABLE 1.2
Calculated Values of the Moment Diagram M'_e of the Pseudolinear System

x(in.)	f(x)	y′(rad)	Z_e	M_x(kip-in.)	M'_e(kip-in.)
0	1.0000	1.3428	4.9631	0.00	0.0
100	1.0112	1.3423	4.6897	6.00	27.23
200	1.2105	1.2245	3.9514	106.00	346.01
300	1.4345	1.0064	2.8557	206.00	410.09
400	1.6894	0.7827	2.0598	306.00	374.20
500	1.9619	0.5927	1.5708	406.00	325.06
600	2.2682	0.4236	1.2809	506.00	285.75
700	2.6049	0.2728	1.1137	606.00	259.09
800	2.9733	0.1348	1.0274	706.00	243.95
806	3.3750	0.0000	1.0000	806.00	238.81

Note: 1 kip-in. = 113.0 Nm, 1 in. = 0.0254 m.

where

$$Q(x) =$$

$$\frac{P(L-\Delta)^3}{EI_B} \cdot \left[\frac{2(n-1)x + (L-\Delta)}{2(n-1)^2[(n-1)x + (L-\Delta)]^2} - \frac{(2n-1)}{2(n-1)^2 n^2 (L-\Delta)} \right] \tag{1.65}$$

Since the horizontal displacement Δ in Equation 1.65 is not known, the value of Δ can be determined from the equation

$$L = \int_0^{L-\Delta} \left[1 + (y')^2\right]^{1/2} dx \tag{1.66}$$

by a trial and error procedure, i.e., assume a value of Δ in Equation 1.64 and integrate Equation 1.66 to determine the length L. The procedure may be repeated for various values of Δ until the correct length L is obtained. With Δ known, the values of y′ at $0 \leq x \leq (L - \Delta)$ can be computed by using Equation 1.64. Thus, with known y′ the values at any x of the moment diagram M'_e of the pseudolinear system of constant stiffness EI_B, can be determined by using Equation 1.33.

For example, by assuming that L = 1000 in. (25.4 m), P = 1 kip (4.448 kN), EI_B = 180,000 kip-in.^{2} (516,541 Nm^2), and n = 1.5, the trial and error procedure for Equation 1.66 yields Δ = 194.0 in. (4.93 m). With known Δ, the values of y′, z_e, M_x, and M'_e at $0 \leq x \leq (L - \Delta)$ and at intervals of 100 in. (2.54 m) are determined from Equations 1.64, 1.34, 1.58, and 1.33, respectively, and they are shown in Table 1.2. Note in this table that (L – Δ) = 806.0 in. (20.47 m).

TABLE 1.3
Variation of Δ_B, θ_B and δ_B for Various Values of the Load P

	Equivalent pseudolinear systems			4th order Runge-Kutta method			
P (kips)	Δ_B (in.)	θ_B (°)	δ_B (in.)	Δ_B (in.)	θ_B (°)	δ_B (in.)	% Difference
0.5	71.00	32.19	327.10	68.87	31.65	322.70	1.36
1.0	194.00	53.31	520.00	181.82	51.42	507.11	2.54
1.5	298.00	65.86	622.75	274.90	62.99	606.27	2.71
2.0	376.80	75.53	682.30	345.71	70.22	665.61	2.51
2.5	436.59	78.44	720.82	400.33	75.02	704.88	2.26
3.0	483.11	81.70	747.33	443.61	78.37	732.93	1.96
3.5	520.25	84.00	768.05	478.83	80.80	754.13	1.84
4.0	550.56	85.61	783.49	508.13	82.61	770.85	1.64

Note: 1 kip = 4.448 kN, 1 in. = 0.0254 m.

The moment diagram M'_e of the pseudolinear system of constant stiffness EI_B is shown plotted in Figure 1.7b. The approximation of its shape with three straight line segments, as shown in the same figure, leads to the accurate pseudolinear system of constant stiffness EI_B shown in Figure 1.7c.

The large deflection at any $0 \le x \le (L - \Delta)$ of the original systems can be accurately determined by using the pseudolinear system in Figure 1.7c and applying elementary linear analysis. For example, the large deflection δ_B at the free end B in Figure 1.7a is equal to the deflection δ_C, at point C, of the pseudolinear system in Figure 1.7c. Thus, by applying the moment area method to the pseudolinear system in Figure 1.7c, we find $\delta_C = 517.8$ in. (13.15 m). The value obtained by solving Equation 1.63 directly and applying Simpson's rule to carry out the required integration is $\delta_B = 519.85$ in. (13.2042 m). The difference is only 0.4 %. In a similar manner, by using the pseudolinear system in Figure 1.7c and applying the moment area method, we find $y'_c = 1.6772$, and thus the rotation θ_c at C is $\theta_c = \tan^{-1}(y'_c) = 52.7°$. The solution of Equation 1.63 yielded $\theta_B = \theta_C = 53.31°$, a difference of only 0.11%. Note that the solution of the pseudolinear system in Figure 1.7c by using linear analysis yields the values of y', and consequently the rotation $\theta = \tan^{-1}(y')$.

In Table 1.3, the variation of the horizontal displacement Δ_B, rotation θ_B, and vertical displacement δ_B of the free end B in Figure 1.7a for various values of the vertical load P is shown. The variation of the same quantities for various values of the thickness parameter n in Figure 1.7a is shown in Table 1.4. In the same tables, for comparison purposes, the results obtained by using the 4th order Runge-Kutta method are shown. The last column in these tables shows the percentage difference in deflection between equivalent systems and Runge-Kutta method. Note that the maximum difference in deflection is less than 3%,

TABLE 1.4
Variation of Δ_B, θ_B, and δ_B for Various Values of the Parameter n in Figure 1.7c

	Equivalent pseudolinear systems			4th order Runge-Kutta method			
P (kips)	Δ_B (in.)	θ_B (°)	δ_B (in.)	Δ_B (in.)	θ_B (°)	δ_B (in.)	% Difference
1.0	414.94	71.96	732.38	414.98	71.97	732.04	0.04
1.2	314.88	64.88	649.16	304.25	63.67	641.81	1.14
1.4	229.90	57.19	562.75	216.91	55.37	550.80	2.16
1.5	194.00	53.31	519.85	181.82	51.42	507.11	2.51
1.6	162.80	49.51	478.39	151.93	47.65	465.43	2.78
1.8	113.26	42.39	401.23	105.65	40.65	389.76	2.94
2.0	78.42	36.14	334.47	73.57	34.89	325.35	2.80
2.2	54.60	30.00	278.81	51.67	29.93	272.03	2.49
2.5	32.48	24.58	214.00	31.16	24.01	209.89	1.95
3.0	14.82	17.43	142.73	14.48	17.17	140.96	1.25

Note: 1 in. = 0.0254 m.

which is reasonable for practical applications. Most of this difference, however, could be attributed to mathematical manipulations.

The horizontal displacement Δ_x of any point x along the length of the member in Figure 1.7a may be determined by using an expression similar to the one given by Equation 1.66. For example, the horizontal displacement Δ at x = L/2 may be determined from the equation

$$\frac{L}{2} = \int_0^{\frac{L}{2}-\Delta} \left[1+(y')^2\right]^{1/2} dx \qquad (1.67)$$

by using a trial and error procedure as it was suggested earlier, i.e., assume values of Δ in Equation 1.67, perform the integration, and repeat the procedure untill the equation is satisfied.

1.8.1 USING A SIMPLIFIED NONLINEAR EQUIVALENT SYSTEM

Another approach regarding the solution of the variable stiffness beam in Figure 1.7a would be the utilization of equivalent simplified nonlinear systems of constant stiffness EI_B. For example, by assuming P = 1000 lb (4.448 kN), L = 1000 in. (25.4 m), E = 30×10^3 ksi (206×10^6 kPa), and $I_B = 6.0$ in.^{4} (2.4972×10^{-6}m^4) (see Figure 1.8a) and applying Equation 1.39, the moment diagram M_e of the equivalent nonlinear system of constant stiffness EI_B is shown plotted in Figure 1.8b. The approximation of the shape of M_e with two straight line segments BC and CD and the application of elementary relationships between moment, shear, and load leads to the equivalent nonlinear system of constant stiffness EI_B shown in Figure 1.8c. The approximation of M_e with one straight line BE as shown in Figure 1.8b, yields the nonlinear equivalent system shown in Figure 1.8d.

The equivalent nonlinear system of constant stiffness shown in Figure 1.8c, or the one shown in Figure 1.8d, provides a simpler solution to the problem because the difficulty associated with the variable depth of the member is eliminated. The equivalent nonlinear systems in Figures 1.8c and 1.8d are of uniform depth, and consequently, of uniform stiffness. The applied load, however, is different in order to compensate for the change in stiffness. The nonlinear differential equation that can be used to solve the nonlinear equivalent systems of constant stiffness is given by Equation 1.38. If the exact expression for M_e is used and Equation 1.38 is solved exactly, the exact solution of the problem in Figure 1.8a is obtained. If the M_e from the problem in Figure 1.8c, or the one from Figure 1.8d, is used, then the solution of Equation 1.38 will provide an approximate solution. The accuracy of this approximate solution would depend upon how many straight lines are used to approximate the shape of the M_e diagram in Figure 1.8b. A reasonable solution that satisfies practical design requirements may be obtained by approximating the shape of M_e with only a few straight lines. Note that the juncture points between the straight lines can be adjusted above or below the M_e curve, as shown in Figure 1.8b, so that the areas added to the M_e diagram are approximately balanced by the areas subtracted from the M_e diagram. However, only reasonable care should be taken to do this, because the solution is not very sensitive to such approximations.

The solution of the simplified equivalent nonlinear system in Figure 1.8c, or the solution of the one in Figure 1.8d, can be carried out by using the pseudolinear system as discussed in Section 1.5. Other available methods of solution may also be used. The sole purpose in the utilization of equivalent nonlinear systems is to simplify the original problem so that the new mathematical model provides an easier solution. This approach becomes extremely useful when the geometry and loading of the initial system are complicated (see, i.e., the problem in Figure 1.4).

For comparison purposes only, the problem in Figure 1.8d is solved by using elliptic integrals. This solution yielded $\delta_B = 532.78$ in. (13.58 m), $\theta_B = 49.87°$, and $\Delta_B = 189.23$ in. (4.81 m). The pseudolinear analysis of the problem in Figure 1.8a (see Table 1.3) yields $\delta_B = 517.8$ in. (13.15 m), $\theta_B = 52.70°$, and $\Delta_B = 194.0$ in. (4.93 m). This indicates that very accurate results are obtained by approximating the shape of M_e in Figure 1.8b by only one straight line segment. However, the shape of the M_e diagram may be approximated by as many straight line segments as desired.

It should be pointed out here that the elliptic integral method does not apply to members where the stiffness E_xI_x or the moment of inertia I_x varies along the

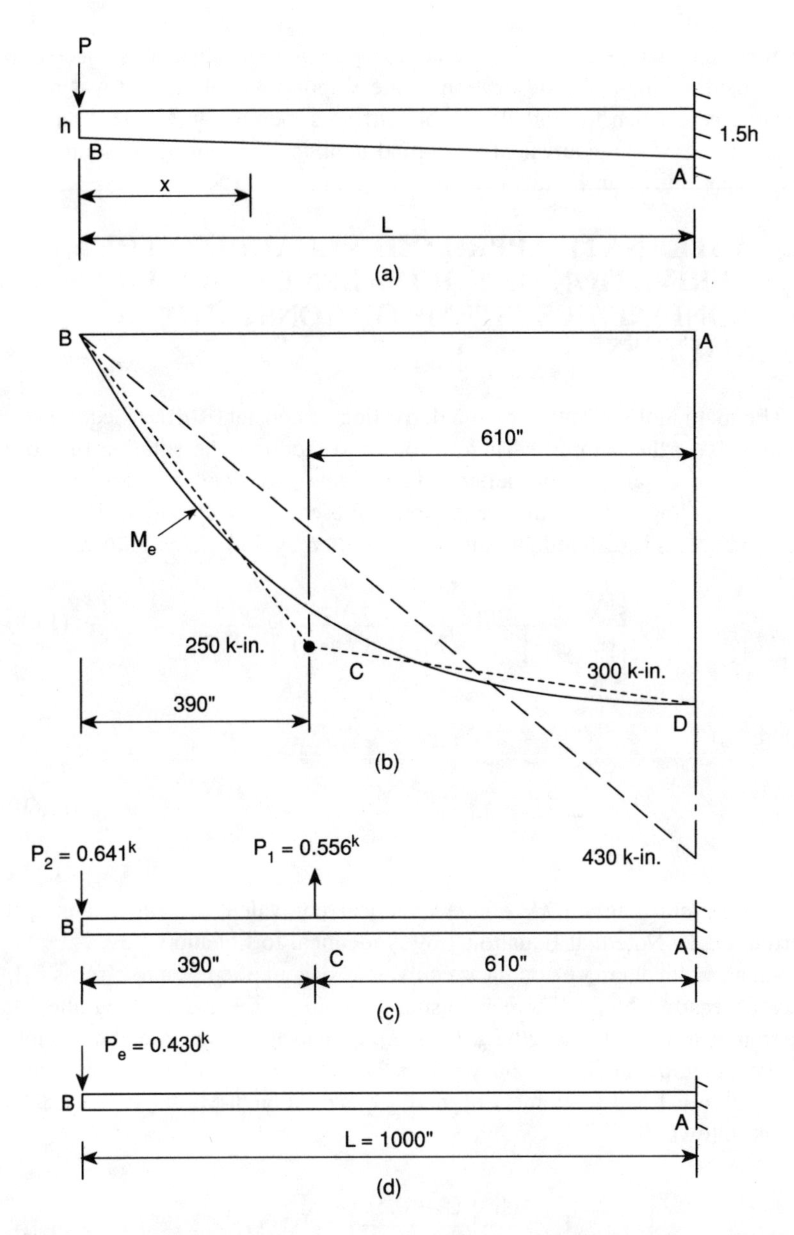

FIGURE 1.8. (a) Original variable stiffness member; (b) moment diagram M_e of the equivalent nonlinear system of constant stiffness; (c) constant stiffness equivalent nonlinear system loaded with two concentrated loads; (d) constant stiffness equivalent nonlinear system loaded with one concentrated load at the free end. (1 kip-in. = 113 Nm, 1 kip = 4.448 kN, 1 in. = 0.0254 m).

length of the member. Also, it does not apply to members that are subjected to distributed loadings. The utilization of the elliptic integral method was made possible here when the initial variable stiffness member in Figure 1.8a was transformed into an equivalent simplified nonlinear system of constant stiffness EI_B as shown in Figure 1.8d.

1.9 ALTERNATE APPROACH REGARDING THE DERIVATION OF EQUIVALENT SIMPLIFIED NONLINEAR SYSTEMS OF CONSTANT STIFFNESS

The methodology regarding the derivation of constant stiffness equivalent systems is further simplified here in order to facilitate the solution of more complicated cases of large deflection problems. Consider, for example, the general nonlinear differential equation represented by Equation 1.37. For convenience, this differential equation is written again here as follows:

$$\frac{y''}{\left[1+(y')^2\right]^{3/2}} = -\frac{1}{E_1 I_1} \cdot \frac{M_x}{g(x)f(x)} = -\frac{M_e}{E_1 I_1} \tag{1.68}$$

where

$$M_e = \frac{M_x}{g(x)f(x)} \tag{1.69}$$

is the moment diagram M_e of the nonlinear equivalent system of constant stiffness E_1I_1. Note that Equation 1.69 is identical to Equation 1.39.

An accurate alternate nonlinear equivalent system of constant stiffness E_1I_1 may be easily obtained here by using Equation 1.69 and plotting the M_e diagram in terms of the length $L_0 = (L - \Delta)$. Consider, for example, the variable stiffness member in Figure 1.9a, which is the same problem as the one in Figure 1.8a with n = 1.5. The moment of inertia I_x and moment M_x at any $0 \leq x \leq L_0$ are as follows:

$$I_x = \frac{bh^3}{12}\left[\frac{L_o + 0.5x}{L_o}\right]^3 = I_B f(x) \tag{1.70}$$

$$M_x = -Px = -x \tag{1.71}$$

where

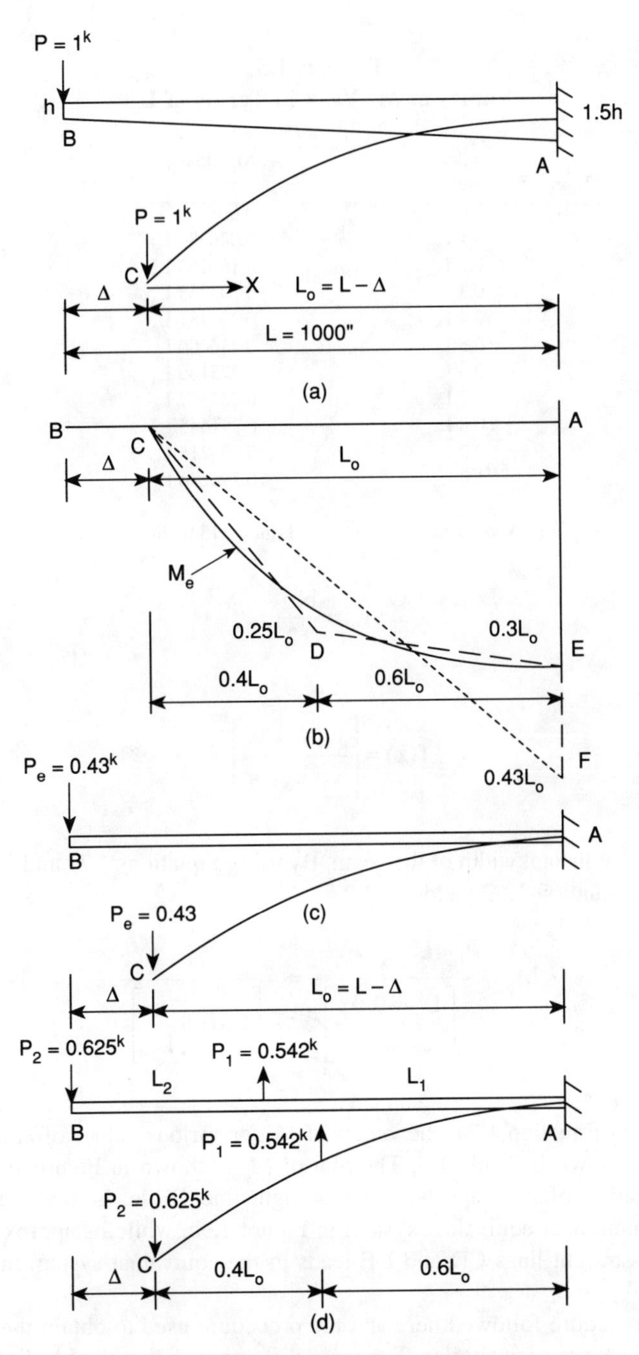

FIGURE 1.9. (a) Original variable stiffness member; (b) moment diagram M_e of the equivalent nonlinear system of constant stiffness; (c) constant stiffness equivalent nonlinear system loaded with one concentrated load; (d) constant stiffness equivalent nonlinear system loaded with two concentrated loads. (1 in. = 0.0254 m, 1 kip = 4.448 kN).

TABLE 1.5
Values of M_e Vs. x in Terms of L_o

x(in.)	M_e(kip-in.)
0.0	0.0
0.1 L_o	0.086384 L_o
0.2 L_o	0.150263 L_o
0.3 L_o	0.197255 L_o
0.4 L_o	0.231481 L_o
0.5 L_o	0.256000 L_o
0.6 L_o	0.273100 L_o
0.7 L_o	0.284509 L_o
0.8 L_o	0.291545 L_o
0.9 L_o	0.295215 L_o
1.0 L_o	0.296296 L_o

Note: 1 in. = 0.0254 m, 1 k-in. = 113.0 Nm.

$$I_B = \frac{bh^3}{12} \tag{1.72}$$

$$f(x) = \left[\frac{L_o + 0.5x}{L_o}\right]^3 \tag{1.73}$$

and b is the constant width of the beam. By using Equations 1.71 and 1.73, and $g(x) = 1$, Equation 1.69 yields

$$M_e = -\frac{x}{\left[\frac{L_o + 0.5x}{L_o}\right]^3} = -\frac{x}{\left[1 + 0.5\frac{x}{L_o}\right]^3} \tag{1.74}$$

By using Equation 1.74, the values of M_e for various values of x, in terms of L_o, are shown in Table 1.5. The plot of M_e is shown in Figure 1.9b. The approximation of its shape with one straight line CF leads to the constant stiffness nonlinear equivalent system in Figure 1.9c, while its approximation with two straight lines CD and DE leads to the equivalent system in Figure 1.9d.

The procedure followed here and the procedure used to obtain the results in Figure 1.8 are very similar. The main difference is that the M_e diagram in Figure 1.9b is plotted in terms of $L_o = (L - \Delta)$ along the length L_o. The one-line approximation of M_e yields identical nonlinear equivalent systems as shown by the results in Figures 1.8d and 1.9c. The two-line approximation, however, yields some important differences. In Figure 1.8c, the location of the equivalent loads P_1 and P_2 is given along the length L of the member and their position during the large deformation of the member, as well as L_o, must be determined. In Figure 1.9c, where the alternate approach is used, the position of P_1 and P_2 along the length L_o have been determined, while L_1, L_2 and L_o are unknowns. Both methods yield accurate results, but in the alternate approach, since the location of P_1 and P_2 along L_o is known, it only becomes necessary to determine L_o in order to be able to determine the values of y' along L_o. The approach used in Figure 1.8c will require, in addition, determining the location of P_1 and P_2 along L_o in order to obtain y'. The alternate approach would be even more advantageous when more than two straight lines are used to approximate the shape of the M_e diagram. The length L_o, and consequently Δ, for the system in Figure 1.9d may be determined from the equation

$$L = \int_{o}^{0.4L_o} \left[1+(y')^2\right]^{1/2} dx + \int_{0.4L_o}^{L_o} \left[1+(y')^2\right]^{1/2} dx \tag{1.75}$$

by a trial and error procedure, i.e., assume values of Δ and repeat the procedure until Equation 1.75 is satisfied, as it was done earlier for Equation 1.66.

The simplified nonlinear equivalent systems in Figures 1.8c and 1.9d, or the ones in Figures 1.8d and 1.9c, may be solved by using pseudolinear analysis as discussed earlier or, if preferred, by applying other methods of nonlinear analysis that are suitable for such types of problems.

PROBLEMS

1.1 The tapered cantilever beam in Figure 1.7a is loaded with a concentrated load P = 2.5 kips (11.12 kN) at the free end. By using a pseudolinear equivalent system of constant stiffness, determine the vertical and horizontal displacements δ_B and Δ_B, respectively, at the free end B of the beam, as well as the rotation θ_B at the same end. The length L = 1000 in. (25.4 m), EI_B = 180,000 kip-in^2. (516,541 Nm^2), and taper n = 1.5.
Answer: δ_B = 720.82 in. (18.31 m), Δ_B = 436.60 in. (11.10 m), and θ_B = 78.44°.

1.2 Solve Problem 1.1 by using a simplified nonlinear equivalent system of constant stiffness EI_B that is loaded with one equivalent concentrated load P_e at the free end B of the beam. Apply pseudolinear analysis to solve the simplified nonlinear equivalent system.

1.3 Solve Problem 1.1 with P = 1.5 kips (6.672 kN).
Answer: δ_B = 622.75 in. (15.82 m), Δ_B = 298.00 in. (7.57 m), and θ_B = 65.86°.

1.4 Solve Problem 1.1 with P = 1 kip (4.448 kN), and n = 1.80.
Answer: δ_B = 401.23 in. (10.19 m), Δ_B = 113.26 in. (2.88 m), and θ_B = 42.39°.

Chapter 2

FLEXIBLE BARS OF UNIFORM AND VARIABLE THICKNESS

2.1 INTRODUCTION

In the preceding chapter, basic aspects on nonlinear mechanics were discussed. The preliminary analysis centered on flexible bars of uniform or variable stiffness E_xI_x, and various methods of analysis were explored. The emphasis, however, was concentrated in the derivation of equivalent pseudolinear systems of constant stiffness E_1I_1 and in the derivation of simplified nonlinear equivalent systems of constant stiffness. Such methodologies, as stated in the preceding chapter, simplify the solution of complicated flexible beam problems. It should be pointed out, however, that any known method of analysis may be used to solve the equivalent pseudolinear system, or the simplified equivalent nonlinear systems. In fact, numerical methods of analysis such as the finite element and finite difference methods would be highly appropriate for the solution of such equivalent systems. This topic is extensively discussed in following chapters of the text.

In this chapter, various cases of statically determinate flexible bars with various loading conditions are examined, and solution methodologies are obtained. In all cases, the modulus of elasticity E is assumed to be constant, but the moment of inertia I_x can vary in any arbitrary manner along the length of the member. The inelastic analysis of flexible bars of either uniform or variable thickness is carried out in Chapters 6, 7, 9, 10, and 11.

2.2 UNIFORM FLEXIBLE CANTILEVER BEAMS SUBJECTED TO DISTRIBUTED AND COMBINED LOADINGS

In this section, pseudolinear equivalent systems will be used to solve the large deformations of cantilever beams that are subjected to distributed loadings. The methodology may be illustrated by considering a uniform flexible cantilever beam loaded with a uniformly distributed loading w_o, as shown in Figure 2.1a. Its large deformation configuration is shown in the same figure.

The bending moment M_x at any distance $0 \le x \le L_o$, where $L_o = (L - \Delta)$, is given by the expression

$$M_x = w_o x_o \frac{x}{2} = \frac{w_o x}{2} \int_o^x \left\{1 + [y'(\xi)]^2\right\}^{1/2} d\xi \tag{2.1}$$

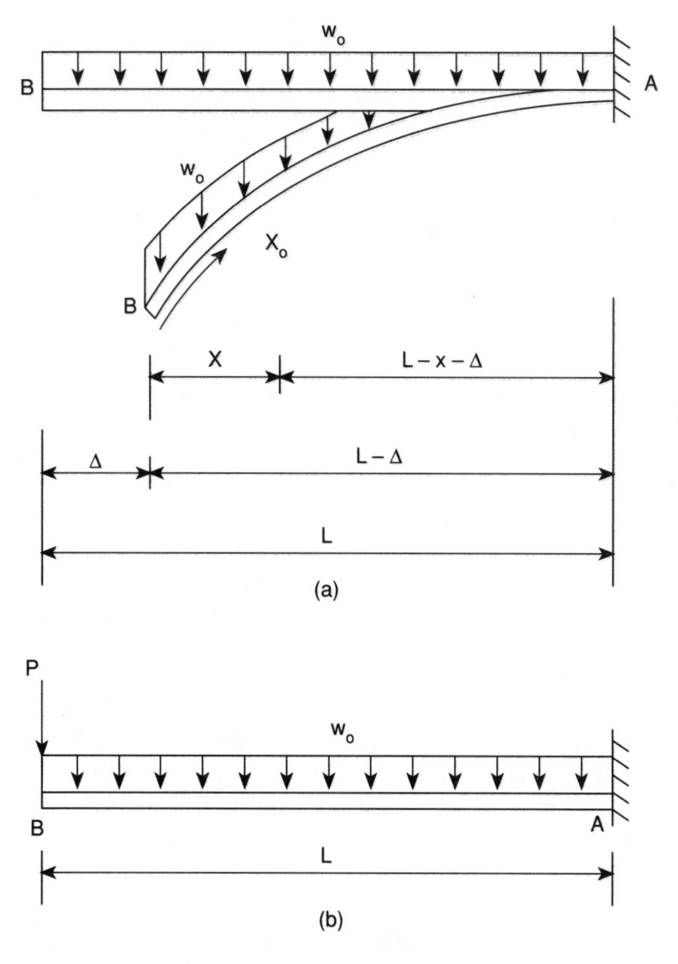

FIGURE 2.1. (a) Uniform cantilever beam loaded with a distributed loading w_0; (b) uniform cantilever beam loaded with a distributed load w_0 and a concentrated load P at the free end.

where ξ is a dummy variable that depends on x and x_0 is as shown in Equation 1.43. The stiffness EI of the member, in this case, is assumed to be constant.

By substituting Equation 2.1 into Equation 1.22 and carrying out the required manipulations, we obtain

$$\frac{y''}{\left[1+(y')^2\right]^{3/2}} = -\frac{w_0 x}{2EI}\int_0^x \left\{1+\left[y'(\xi)\right]^2\right\}^{1/2} d\xi \tag{2.2}$$

Equation 2.2 is a nonlinear integral differential equation that is very difficult to solve. A solution, however, may be obtained by using the expression for x_0 given by Equation 1.42, which requires knowledge of the function of the horizontal displacement $\Delta(x)$. A very accurate solution may be obtained by assuming that $\Delta(x)$ = constant = Δ, as shown in Equation 1.49, where Δ is the horizontal displacement of the free end of the member as shown in Figure 2.1. Any one of the expressions given by Equations 1.50 to 1.52 may be also used for this purpose. On this basis

$$M_x = \frac{w_o x}{2}(x + \Delta) \tag{2.3}$$

and Equation 2.1 may be simplified as follows:

$$\frac{y''}{\left[1+(y')^2\right]^{3/2}} = -\frac{w_o x}{2EI}(x + \Delta) \tag{2.4}$$

By integrating Equation 2.4 once and determining the constant of integration by satisfying the boundary condition of zero rotation at $x = (L - \Delta)$, we find

$$y'(x) = \frac{f(x)}{\left\{1-[f(x)]^2\right\}^{1/2}} \tag{2.5}$$

where

$$f(x) = -\frac{w_o}{12EI}\left[2x^3 + 3\Delta x^2 + (L - \Delta)^3 - 3L(L - \Delta)^2\right] \tag{2.6}$$

The horizontal displacement Δ in Equation 2.6 may be determined from Equation 1.66 by a trial and error procedure as discussed in Chapter 1.

In order to obtain numerical results we assume that the cantilever beam in Figure 2.1a has a stiffness EI = 180×10^3 kip-in.^{2} (516,541 Nm^2), and length L = 1000 in. (25.40 m). By utilizing pseudolinear equivalent systems and following the procedure discussed in Sections 1.5 and 1.8, the values of the rotation θ_B and vertical displacement δ_B at the free end B of the member are obtained for various values of the distributed load w_o, shown in Table 2.1. In the same table, the results obtained from the pseudolinear analysis are compared with the results obtained by using the power series method and the 4th order Runge-Kutta method. The maximum difference is about 3.2%, where a good part of it may be attributed to computational rounding up errors.

In Table 2.2 the values of the horizontal displacement Δ_B, rotation θ_B, and vertical displacement δ_B, at the free end B of the member are shown by assuming that $\Delta(x) = \Delta$, $\Delta x/L_o$, $\Delta(x/L_o)^{1/2}$, $\Delta\sin(\pi x/2L_o)$, and w_o = 1.5 lb/in. (262.69 N/m). The results are also compared with the ones obtained by using

TABLE 2.1
Variation of θ_B and δ_B for Various Values of the Distributed Load w_o

	Pseudolinear equivalent system method		Power series method		4th order Runge-Kutta method		
W_o (lb/in.)	θ_B (°)	δ_B (in.)	θ_B (°)	δ_B (in.)	θ_B (°)	δ_B (in.)	% Difference
1.0	43.41	522.42	43.14	531.48	43.01	529.53	−1.34
1.5	55.69	638.00	55.37	656.38	55.07	652.07	−2.15
2.0	66.30	705.72	63.62	731.04	63.23	725.34	−2.70
2.5	70.88	748.75	69.36	778.08	68.95	771.99	−3.01
3.0	73.88	777.86	73.49	809.39	73.12	803.57	−3.20

Note: 1 in. = 0.0254 m.

the 4th order Runge-Kutta method. Both methods provide reasonable results for practical applications. It should be also noted that reasonable agreement in results is obtained for the various cases of $\Delta(x)$ shown. A large error, however, is obtained if it is assumed in the analysis that $x_o = x$, as shown in Table 2.2. This is tantamount to assuming that the bending monent M_x of the beam is not a function of the deformation.

As a second example let it be assumed that a uniform cantilever beam is loaded by a combined loading consisting of a uniformly distributed load w_o and a concentrated load P applied at the free end of the member as shown in Figure 2.1b. This problem was also solved by Lau[30] using the power series method with about 12 coefficients. Peudolinear analysis will be used here to solve this problem and the results will be compared with the results obtained by Lau.

From the deformed configuration of the member, the bending moment M_x at any $0 \le x \le L_o$ is given by the expression

$$M_x = -\left[\frac{w_o x}{2} x_o + Px\right] \tag{2.7}$$

By substituting Equation 2.7 into Equation 1.22 we obtain

$$\frac{y''}{\left[1+(y')^2\right]^{3/2}} = \frac{w_o x}{2EI} x_o + \frac{Px}{EI} \tag{2.8}$$

Again here, an accurate solution may be obtained by assuming that $x_o = x + \Delta$, as it was done for the problem in Figure 2.1a. On this basis, Equation 2.8 yields

$$\frac{y''}{\left[1+(y')^2\right]^{3/2}} = \frac{P}{2EIL}\left[k(x^2 + \Delta x) + 2Lx\right] \tag{2.9}$$

TABLE 2.2
Values of Δ_B θ_B and δ_B for Various Assumed Cases of $\Delta(x)$ and with w_o = 1.5 lb/in.

Displacement	Assumed cases of $\Delta(x)$: $\Delta(x) = \Delta$	$\Delta(x) = \Delta x/L_o$	$\Delta(x) = \Delta(x/L_o)^{1/2}$	$\Delta(x) = \Delta \sin \pi x/2L_o$	$x_o = x$	Runge-Kutta Method
Δ_B (in.)	277.25	251.16	261.64	263.65	199.24	287.74
θ_B (°)	55.70	51.15	52.83	52.74	45.49	55.07
δ_B (in.)	637.96	617.55	626.27	628.64	558.46	652.06

Note: 1 in. = 0.0254 m, 1 lb/in. = 175.1268 N/m.

where

$$k = \frac{w_0 L}{P} \tag{2.9a}$$

By integrating Equation 2.9 once and satisfying the boundary condition of zero rotation at $x = L_0$, where $L_0 = L - \Delta$, the constant of integration may be determined. By making a few manipulations after the constant of integration is determined, the integration of Equation 2.9a) yields the expression

$$\frac{y'}{\left[1+(y')^2\right]^{1/2}} = \frac{P}{12EIL}\left[k(2x^3 + 3\Delta x^2 - 2L_o - 3\Delta L_o^2) + 6L(x^2 - L_o^2)\right] \tag{2.10}$$

In order to determine y' from Equation 2.10, the value of the horizontal displacement Δ of the free end of the cantilever beam must be known. This can be accomplished by using the equation

$$L = (L - \Delta)\int_0^1 \frac{1}{\left\{1 - \left[Q(\xi)\right]^2\right\}^{1/2}} d\xi \tag{2.11}$$

where

$$\xi = \frac{x}{L_o} \tag{2.12}$$

$$Q(\xi) = \frac{P}{12EIL}\left[kL_o^3\left(2\xi^3 + \frac{3\Delta}{L_o}\xi^2 - 2L_o^3 - 3\Delta L_o\right) + 6LL_o\left(\xi^2 - L_o^2\right)\right] \tag{2.13}$$

A trial and error procedure may be used here, as discussed in preceding sections, for the computation of Δ from Equation 2.11, i.e., assume a value of Δ and carry out the integration in Equation 2.11 in order to determine the length L of the member. If the first trial does not give you the correct value of L, then repeat the procedure with new values of Δ until the correct L is obtained.

This procedure accomplishes two important objectives. The first objective was to determine the horizontal displacement Δ of the free end of the beam. The second objective was the computation of y' at any $o \leq x \leq L_o$ from

TABLE 2.3
Comparison of Results Obtained for Θ_B, and δ_B by Using Pseudolinear Systems with the Ones Obtained by Lau

	Pseudolinear equivalent systems			Lau's solution		Relative difference
PL_2/EI	Θ_B (degrees)	δ_B (in.)	Δ_B (in.)	Θ_B (degrees)	δ_B (in.)	%
0.364	19.98	243.551	35.28	20.00	244.000	0.018
0.834	39.89	466.201	137.82	40.00	471.000	1.020
1.642	59.71	653.570	300.64	60.00	666.000	1.860
2.416	69.71	733.500	406.35	70.00	751.000	2.330
4.389	81.00	816.320	555.476	81.00	847.000	3.620

Note: 1 in. = 0.0254 m.

Equation 2.10, which requires Δ to be known. With known y', the pseudolinear analysis discussed in Sections 1.5 and 1.8 can be carried out in a very convenient manner. That is, the moment diagram M'_e of the equivalent pseudolinear system of constant stiffness EI can be determined from Equation 1.33, because z_e and y' can be determined from Equations 1.34 and 2.10, respectively. Note that $f(x) = g(x) = 1.0$ for this problem. The loading on the pseudolinear equivalent system may be obtained from Equation 1.36. If the shape of the M'_e diagram is approximated with straight lines as discussed in Section 1.8, then the loading on the pseudolinear equivalent system will consist of concentrated equivalent loads along the length L_o of the equivalent pseudolinear system. The pseudolinear system may be solved by applying simple linear methods of analysis such, as the moment area method, as discussed in Section 1.8.

Numerical results are shown in Table 2.3 for a beam length L = 1000 in. (25.4 m) and load factor k= 3.0. From Equation 2.9a, the factor k = 3.0 indicates that the concentrated load P is one third of the total distributed load w_0L of the member. Also the stiffness EI = 180,000 kip- in^2 (516,541 Nm^2). In Table 2.3 the values of the rotation $\theta_B = \tan^{-1}(y'_B)$, horizontal displacement Δ_B, and vertical displacement δ_B at the free end B of the member in Figure 2.1b, are obtained for various values of the parameter PL^2/EI, and the results are compared with the results obtained by Lau.[30] The maximum relative difference between pseudolinear systems and Lau's approach is 3.62% and it occurs when $PL^2/EI = 4.389$.

2.3 UNIFORM CANTILEVER BEAMS SUBJECTED TO ARBITRARY LOADING CONDITIONS

Consider a uniform stiffness cantilever beam loaded as shown in Figure 2.2. For such types of loading conditions, the utilization of equivalent

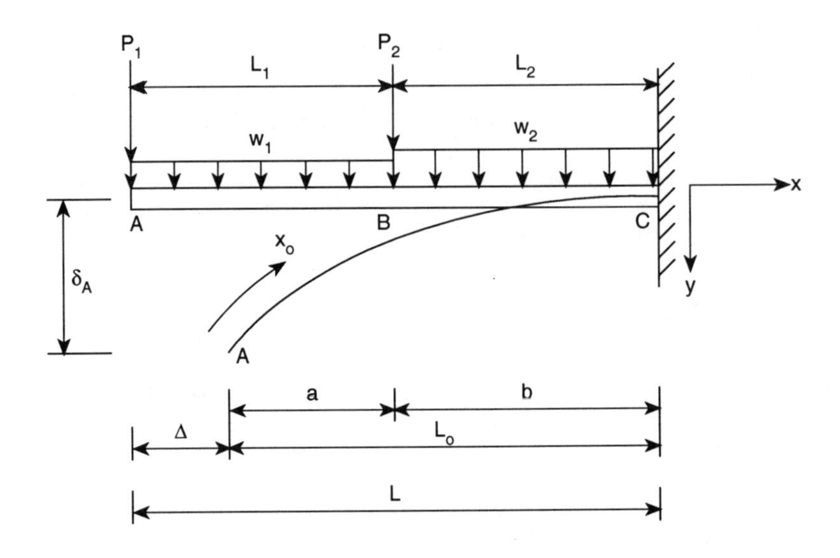

FIGURE 2.2. Uniform flexible cantilever beam loaded with an arbitrary loading condition.

pseudolinear systems will greatly facilitate the solution. The direct solution of this problem by using other known closed-form solutions will be difficult since the problem is nonlinear and the principle of superposition does not apply. Even the finite element method will encounter some difficulties for the same reason.

In the deformed configuration, the horizontal projection of the length L_1 in Figure 2.2 is represented by the length a, and the projection of the length L_2 is represented by the length b. For the given loading conditions, the expressions for the bending moments M_1 and M_2 for the intervals a and b, respectively, must be defined. In Figure 2.2 the coordinate x_{o1} is defined for the interval $0 \le x_{o1} \le L_1$, and x_{o2} is defined for the interval $L_1 \le x_{o2} \le L$. From the deformed configuration, the expressions for the bending moments M_1 and M_2 may be written as follows:

$$M_1 = -\left[\frac{w_1 x}{2} x_{o1} + P_1 x\right] \qquad (2.14)$$

$$0 \le x \le a \qquad 0 \le x_{o1} \le L_1$$

$$M_2 = -\left[\frac{w_1 L_1}{2}(2x - a) + P_1 x + \frac{w_2 x_{o2}}{2}(x - a) + P_2(x - a)\right] \qquad (2.15)$$

$$a \le x \le (a + b) \qquad L_1 \le x_{o2} \le L$$

It should be noted at this point that $x_{o1}(x)$ and $x_{o2}(x)$ are integral equations that are defined by the expressions

$$x_{o1}(x) = \int_0^x \left\{1 + [y_1'(\xi)]^2\right\}^{1/2} d\xi$$
$$0 \le x \le a \quad 0 \le x_{o1} \le L_1 \tag{2.16}$$

$$x_{o2}(x) = \int_0^x \left\{1 + [y_2'(\xi)]^2\right\}^{1/2} d\xi$$
$$a \le x \le (a+b) \quad L_1 \le x_{o2} \le L \tag{2.17}$$

where ξ is a dummy variable that depends on x.

A reasonable solution, as it was pointed out earlier, may be obtained by assuming that $x_{o1}(x)$ and $x_{o2}(x)$ are given by the expressions

$$x_{o1}(x) = x + \Delta_1 \tag{2.18}$$

$$x_{02}(x) = x + \Delta_2 \tag{2.19}$$

where Δ_1 is the horizontal movement of length L_1 and Δ_2 is the horizontal movement of length L_2. This implies that $\Delta = (\Delta_1 + \Delta_2)$. On this basis, the expressions for the bending moments M_1 and M_2 become

$$M_1 = -\left[\frac{w_1 x}{2}(x + \Delta_1) + P_1 x\right] \tag{2.20}$$

$$M_2 = -\left[\frac{w_1 L_1}{2}(2x - a) + P_1 x + \frac{w_2(x + \Delta_2)}{2}(x - a) + P_2(x - a)\right] \tag{2.21}$$

By using the notation

$$k_1 = \frac{w_1 L_1}{P_1} \qquad k_2 = \frac{w_2 L_2}{P_2} \tag{2.22}$$

the expressions of M_1 and M_2 become

$$M_1 = -\left[x^2\left(\frac{k_1 P_1}{2L_1}\right) + x\left(\frac{k_1 P_1 \Delta_1 + 2L_1 P_1}{2L_1}\right)\right] \tag{2.23}$$

$$M_2 = -\left[x^2\left(\frac{k_2P_2}{2L_2}\right) + x\left(\frac{2k_1P_1L_2 + 2L_2P_1 - k_2P_2a + k_2P_2\Delta_2 + 2P_2L_2}{2L_2}\right)\right.$$
$$\left. - \left(\frac{k_1P_1L_2a + k_2\Delta_2a + 2P_2L_2a}{2L_1}\right)\right] \tag{2.24}$$

On this basis, the Euler-Bernoulli equation yields

$$\frac{y_1''}{\left[1 + (y_1')^2\right]^{3/2}} = G_1x^2 + G_2x \qquad 0 \le x \le a \tag{2.25}$$

$$\frac{y_2''}{\left[1 + (y_2')^2\right]^{3/2}} = G_3x^2 + G_4x - G_5 \qquad a \le x \le (a + b) \tag{2.26}$$

where

$$G_1 = \frac{k_1P_1}{2L_1} \tag{2.27}$$

$$G_2 = \frac{k_1P_1\Delta_1 + 2L_1P_1}{2L_1} \tag{2.28}$$

$$G_3 = \frac{k_2P_2}{2L_2} \tag{2.29}$$

$$G_4 = \frac{2k_1P_1L_2 + 2L_2P_1 - k_2P_2a + k_2P_2\Delta_2 + 2P_2L_2}{2L_2} \tag{2.30}$$

$$G_5 = \frac{k_1P_1L_2a + k_2\Delta_2a + 2P_2L_2a}{2L_1} \tag{2.31}$$

By integrating Equations 2.25 and 2.26 once and making the required manipulations, the following expressions are obtained:

$$\frac{y_1'}{\left[1+(y_1')^2\right]^{1/2}} = Q_1x^3 + Q_2x^2 + C_1 \qquad 0 \le x \le a \tag{2.32}$$

$$\frac{y_2'}{\left[1+(y_2')^2\right]^{1/2}} = Q_3x^3 + Q_4x^2Q_5x + C_2 \qquad a \le x \le (a+b) \tag{2.33}$$

In Equations 2.32 and 2.33, the parameters Q_1, Q_2, Q_3, Q_4, and Q_5 are known quantities that are grouped in the indicated convenient format and C_1 and C_2 are the constants of integration.

By applying the boundary condition $y_2'(L_o) = 0$, and the continuity condition $y_1'(a) = y_2'(a)$, the constants of integration C_1 and C_2 may be determined as follows:

$$C_1 = Q_3a^3 + Q_4a^2 + Q_5a - Q_1a^3 - Q_2a^2 + C_2 \tag{2.34}$$

$$C_2 = -Q_3L_o^3 - Q_4L_o^2 - Q_5L_o \tag{2.35}$$

It should be pointed out here that the parameters Q_1, Q_2,…,Q_5 in Equations 2.32 and 2.33 contain the unknown horizontal displacements Δ_1 and Δ_2. These displacements can be determined from the expressions

$$L_1 = \int_0^a \left[1+(y_1')^2\right]^{1/2} dx \tag{2.36}$$

$$L_2 = \int_a^{(a+b)} \left[1+(y_2')^2\right]^{1/2} dx \tag{2.37}$$

by using a trial and error procedure as stated earlier. With known Δ_1 and Δ_2, the values of y_1', and y_2' at any $0 \le x \le a$ and $a \le x \le (a + b)$, respectively, may be determined from the respective Equations 2.32 and 2.33. With known y_1' and y_2', the pseudolinear system, and consequently the vertical displacements $y_1(x)$ and $y_2(x)$ of the member, may be determined as discussed in the preceding section.

If we assume in Figure 2.2 that EI = 180,000 kip-in.^{2} (516,541 Nm^2), L = 1000 in. (25.4 m), P_1 = 0.5 kips (2224 N), P_2 = 1.0 kip (4448 N), and $k_1 = k_2$ = 1.0, the above procedure yields

$$
\begin{aligned}
a &= 135.548 \text{ in. } (3.4429 \text{ m})\\
b &= 327.072 \text{ in. } (8.3076 \text{ m})\\
L_o &= (a + b) = 462.62 \text{ in. } (11.7505 \text{ m})\\
\Delta_1 &= 364.452 \text{ in. } (9.2571 \text{ m})\\
\Delta_2 &= 172.928 \text{ in. } (4.3924 \text{ m})\\
\Delta &= (\Delta_1 + \Delta_2) = 537.38 \text{ in. } (13.6495 \text{ m})\\
\delta_A &= 820.\ 375 \text{ in. } (20.8375 \text{ m})\\
\theta_A &= 76.562^\circ
\end{aligned}
$$

A simpler solution for the problem in Figure 2.2 could be obtained by using a simplified nonlinear equivalent system as discussed in Sections 1.6, 1.8, and 1.9, and then applying pseudolinear analysis to solve the simplified nonlinear system. This problem is addressed in detail in later sections of this chapter.

2.4 VARIABLE THICKNESS CANTILEVER BEAM LOADED WITH A UNIFORMLY DISTRIBUTED LOADING

Consider the doubly tapered cantilever beam in Figure 1.3, which is loaded by a uniformly distributed load w_o as shown. Its deformed configuration is also shown in the same figure. The modulus of elasticity E is assumed to be constant, and its moment of inertia I_x at any $0 \le x \le (L - \Delta)$, where Δ is the horizontal displacement of the free end B of the member, is given by the equation

$$I_X = I_B\left[1 + \frac{(n-1)}{L} x_o\right]^3 \tag{2.38}$$

where

$$I_B = \frac{bh^3}{12} \tag{2.39}$$

$$x_o = \int_0^x \left\{1 + [y'(\xi)]^2\right\}^{1/2} d(\xi) \tag{2.40}$$

and b is the constant width of the member.

The bending moment M_x at any $0 \le x \le (L - \Delta)$ is given by the expression

$$M_x = \frac{w_o x}{2} x_o \tag{2.41}$$

where x_o is given by Equation 2.40. By using Equations 2.38, 2.40, and 2.41 and substituting into the Euler-Bernoulli equation, we obtain the following nonlinear integral differential equation, which is again very difficult to solve:

$$\frac{y''}{\left[1+(y')^2\right]^{3/2}} = -\frac{w_o x}{2EI_B} \frac{\int_0^x \left\{1+\left[y'(\xi)\right]^2\right\}^{1/2} d\xi}{\left\{1+\frac{(n-1)}{L}\int_0^x \left[1+(y'(\xi))^2\right]^{1/2} d\xi\right\}^3} \tag{2.42}$$

In order to simplify the solution of Equation 2.42, a reasonable approximation for the variation of I_x at any $0 \le x \le (L - \Delta)$ (see also Fertis and Afonta[5]) may be written as

$$I_x = I_B\left[1+(n-1)\frac{x}{L-\Delta}\right]^3 \tag{2.43}$$

which is the same as Equation 1.60. If it is also assumed that $x_o = (x + \Delta)$, the simplified expression for the bending moment M_x at any $0 \le x \le (L - \Delta)$ becomes

$$M_x = \frac{w_o x}{2}(x+\Delta) \tag{2.44}$$

Based on Equations 2.43 and 2.44, the Euler-Bernoulli equation yields the following simplified nonlinear differential equation:

$$\frac{y''}{\left[1+(y')^2\right]^{3/2}} = -\frac{w_o(L-\Delta)^3}{2EI_B} \cdot \frac{(x^2+\Delta x)}{[(n-1)x+(L-\Delta)]^3} \tag{2.45}$$

By integrating Equation 2.45 once and determining the constant of integration by satisfying the boundary condition of zero rotation at $x = (L - \Delta)$, we find

TABLE 2.4
Variation of Δ_B, θ_B and δ_B for Various Values of the Load w_o

	Pseudolinear equivalent systems			4th order Runge-Kutta method			
w_o (lb/in.)	Δ_B (in.)	θ_B (°)	δ_B (in.)	Δ_B (in.)	θ_B (°)	δ_B (in.)	% Difference
2	121.20	37.99	435.41	118.27	36.66	432.74	0.61
5	363.42	68.38	691.35	354.17	63.80	697.14	–0.83
10	557.08	84.08	799.36	557.53	79.35	819.88	–2.50
15	645.08	88.13	837.99	654.68	84.79	862.43	–2.83

Note: 1 in. = 0.0254 m, 1 lb/in. = 175.1268 N/m.

$$y'(x) = \frac{g(x)}{\left\{1-[g(x)]^2\right\}^{1/2}} \tag{2.46}$$

where

$$\begin{aligned} g(x) = &-\frac{w_0(L-\Delta)^3}{2EI_B}\left\{\frac{-3(n-1)^2x^2 - 2(n-1)x[L-\Delta+\Delta(n-1)]}{2(n-1)^3[(n-1)x+(L-\Delta)]^2}\right. \\ &+\frac{-(L-\Delta)(n-1)\Delta + 2[(n-1)x+(L-\Delta)]^2\ln[(n-1)x+(L-\Delta)]}{2(n-1)^3[(n-1)x+(L-\Delta)]^2} \\ &-\frac{-(L-\Delta)^2\left[3(n-1)^2+2(n-1)\right]-\Delta(L-\Delta)\left[2(n-1)^2-(n-1)\right]}{2(n-1)^3n^2(L-\Delta)^2} \\ &\left.+\frac{2n^2(L-\Delta)^2\ln[n(L-\Delta)]}{2(n-1)^3n^2(L-\Delta)^2}\right\} \end{aligned} \tag{2.47}$$

The horizontal displacement Δ in Equation 2.47 may be determined from Equation 1.13 by a trial and error procedure as suggested in earlier sections. With known Δ, the values of y' can be determined from Equation 2.46 and the pseudolinear analysis can be carried out as stated in the preceding sections of this chapter, as well as in Sections 1.5 and 1.8.

In Table 2.4, the variation of Δ_B, θ_B, and δ_B, at the free end B of the member is shown for various values of the distributed load w_o. In obtaining these results, it was assumed that EI_B = 180,000 kip-in^2 (516,541 Nm^2), L = 1000 in. (25.4 m), and taper n = 1.5. The results are also compared with the results obtained by using the 4th order Runge-Kutta method.

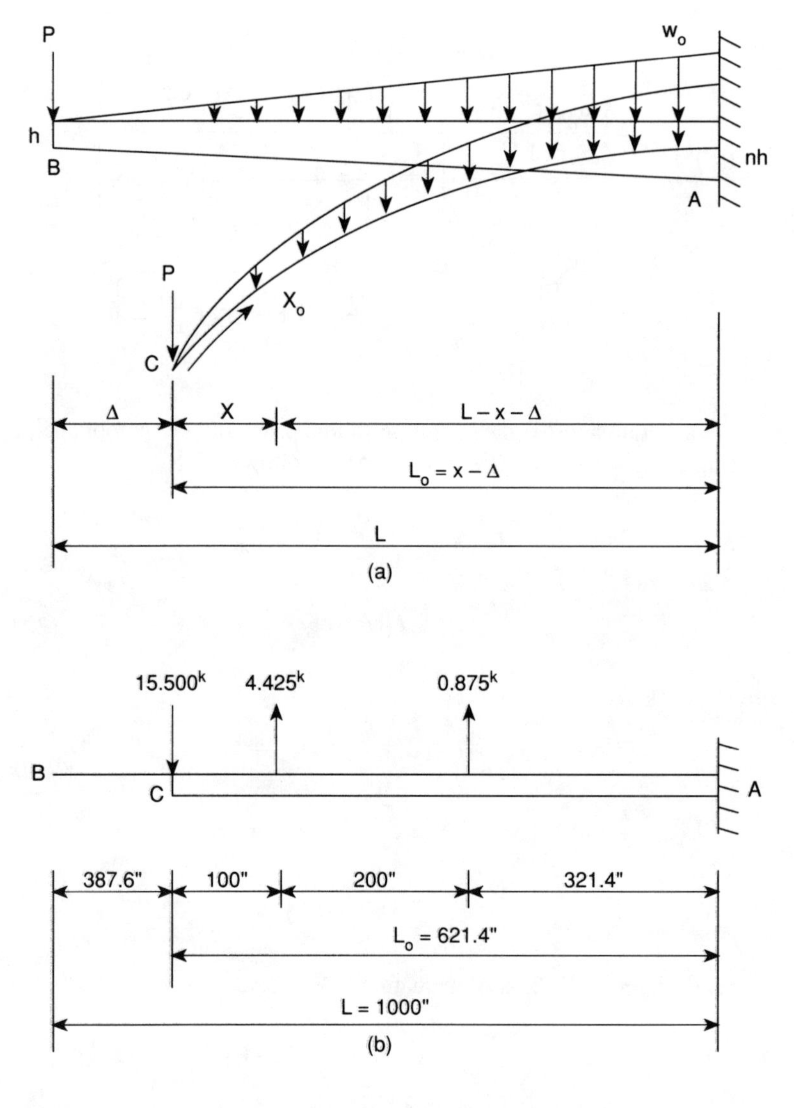

FIGURE 2.3. (a) Original tapered cantilever beam loaded as shown; (b) equivalent pseudolinear system of constant stiffness EI_B (1 kip = 4448 N, 1 in. = 0.0254 m).

2.5 VARIABLE THICKNESS CANTILEVER BEAM SUBJECTED TO COMBINED LOADINGS

In this section, the flexible tapered cantilever beam in Figure 2.3a is considered. The loading on the member consists of a concentrated load P at the free end of the member and a distributed triangluar loading that is applied as shown in the same figure. By proceeding as in the preceeding sections, the nonlinear integral equation that is extremely difficult to solve, takes the form

$$\frac{y''}{\left[1+(y')^2\right]^{3/2}} = -\frac{w_o x(L-\Delta)^3}{6LEI_B} \cdot \frac{\left[\int_o^x \left[1+(y'(\xi))^2\right]^{1/2} d\xi\right]^2}{\left[1+\frac{(n-1)}{L}\int_o^x \left[1+(y'(\xi))^2\right]^{1/2} d\xi\right]^3}$$
$$-\frac{Px(L-\Delta)^3}{EI_B} \cdot \frac{1}{\left[1+\frac{(n-1)}{L}\int_o^x \left[1+(y'(\xi))^2\right]^{1/2} d\xi\right]^3} \tag{2.48}$$

Again, an accurate approximate solution to this problem may be obtained by expressing M_x and I_x at any $0 \leq x \leq (L - \Delta)$ as follows:

$$M_x = -\frac{w_o x x_o^2}{6L} - Px \tag{2.49}$$

$$I_x = I_B f(x) \tag{2.50}$$

where

$$I_B = \frac{bh^3}{12} \tag{2.50a}$$

$$f(x) = \left[1 + \frac{(n-1)}{(L-\Delta)} x\right]^3 \tag{2.50b}$$

By replacing x_o with $(x + \Delta)$, Equation 2.49 yields

$$M_x = -\frac{w_o x}{6L}(x^2 + 2\Delta x + \Delta^2) - Px \tag{2.51}$$

By substituting Equations 2.50 and 2.51 into the Euler-Bernoulli equation, we obtain

$$\frac{y''}{\left[1+(y')^2\right]^{3/2}} = \frac{w_o(L-\Delta)^3}{6LEI_B}\left\{\frac{x^3 + 2\Delta x^2 + \Delta^2 x}{[(n-1)x + (L-\Delta)]^3}\right\}$$
$$+\frac{P(L-\Delta)^3}{EI_B}\left\{\frac{x}{[(n-1)x + (1-\Delta)]^3}\right\} \tag{2.52}$$

By integrating Equation 2.52 once, and satisfying the boundary condition of zero rotation at the fixed end, we find

$$y'(x) = \frac{H(x)}{\left\{1 - [H(x)]^2\right\}^{1/2}} \tag{2.53}$$

where

$$H(x) = -\frac{w_o(L-\Delta)^3}{6LEI_B} \cdot \frac{1}{2(n-1)^4[(n-1)x + (L-\Delta)]^2} \{-(n-1)^3 x^3$$

$$-3(n-1)^2 x^2 [(n-1)x + (L-\Delta)] + 6x(n-1)[(n-1)x + (L-\Delta)]^2$$

$$-6(L-\Delta)[(n-1)x + (L-\Delta)]^2 \ln[(n-1)x + (L-\Delta)]$$

$$+2\Delta\left[-3(n-1)^3 x^2 - 2(n-1)^2 x(L-\Delta)\right.$$

$$\left.+2(n-1)[(n-1)x + (L-\Delta]^2 \ln[(n-1)x + (L-\Delta)]\right]$$

$$\left.+\Delta^2\left[-2(n-1)^3 x - (L-\Delta)(n-1)^2\right]\right\}$$

$$+\frac{w_o(L-\Delta)^3}{6LEI_B} \cdot \frac{1}{2(n-1)^4 n^2 (L-\Delta)^2} \{-(n-1)^3 (L-\Delta)^3 \tag{2.54}$$

$$-3(n-1)^2 (L-\Delta)^3 n + 6(L-\Delta)^3 (n-1)n^2 - 6(L-\Delta)^3 n^2 \ln[n(L-\Delta)]$$

$$-2\Delta\left[-3(n-1)^3 (L-\Delta)^2 - 2(n-1)^2 (L-\Delta)^2\right.$$

$$\left.+2(n-1)n^2 (L-\Delta)^2 \ln[n(L-\Delta)]\right]$$

$$\left.-\Delta^2\left[-2(n-1)^3 (L-\Delta) - (n-1)^2 (L-\Delta)\right]\right\}$$

$$+\frac{P(L-\Delta)^3}{EI_B}\left\{\frac{2(n-1)^3 x + (L-\Delta)(n-1)^2}{2(n-1)^4[(n-1)x(L-\Delta)]^2}\right.$$

$$\left.-\frac{2(n-1)^3 (L-\Delta) - (n-1)^2 (L-\Delta)}{2(n-1)^4 n^2 (L-\Delta)^2}\right\}$$

The unknown value of Δ in Equation 2.54 may be determined from Equation 1.66 by applying a trial and error procedure similar to the one used in the preceding sections. With known Δ, the values of y' at any $0 \le x \le (L - \Delta)$ can be determined from Equation 2.53, and pseudoninear analysis can be applied as discussed in Sections 1.5 and 1.8.

TABLE 2.5
Calculated Values of the Moment Diagram M_e' of the Pseudolinear System

x (in.)	f(x)	y′ (rad)	Z_e	M_e (kip-in.)	$M_e' = M_e Z_e$ (kip-in.)
0	1.0	2.8174	26.7204	0.0	0.0
100	1.2613	2.2056	14.2031	109.548	1555.921
200	1.5646	1.4561	5.5118	199.146	1099.868
300	1.9130	0.9575	2.6540	277.175	735.622
400	2.3097	0.6050	1.5966	348.164	555.878
500	2.7576	0.3210	1.1584	414.585	480.255
600	3.2601	0.0577	1.0050	477.794	480.183
621.40	3.3750	0.0	1.0	490.983	490.983

Note: 1 in. = 0.0254 m, 1 k-in. = 113.0 Nm.

Numerical results are obtained here by assuming that the tapered cantilever beam in Figure 2.3a has a length L = 1000 in. (25.4 m), and taper n = 1.5. The stiffness EI_B at the free end B of the member is 180,000 kip-in.^{2} (516,541 Nm^2), the concentrated load P = 1000 lb (4.448 kN), and the distributed load $w_o = 10.0$ lb/in. (1751.27 N/m).

By using the above beam data, the value of the horizontal displacement Δ of the free end B of the member is obtained from Equation 1.66 by trial and error. The value obtained from this procedure is $\Delta = 378.6$ in. (9.6164 m). The values of f(x), y′, z_e, M_e, and M_e', at points $0 \leq x \leq (L - \Delta)$ are determined by using Equations 2.50b, 2.53, 1.34, 1.29, and 1.33, respectively. These values are shown in Table 2.5 for the indicated intervals of x. Note in this table that $L_o = L - \Delta = 1000 - 378.6 = 621.40$ in. (15.7836 m). The moment diagram M_e' of the pseudolinear system of constant stiffness $EI_B = 180{,}000$ kip-in^2 (561,541 Nm^2) may be plotted by using the values of M_e' in the last column of Table 2.5. By approximating the shape of the M_e' diagram with four straight line segments, the pseudolinear system of constant stiffness EI_B in Figure 2.3b is obtained, which is loaded with three concentrated loads as shown. Three to four straight line segments are usually sufficient for accurate results, but more straight line segments may be used.

By using this system and applying the moment area method, we find $\delta_B = \delta_C = 697.75$ in. (17.7076 m), which is the deflection at the free end of the beam. The rotation at the free end, by using the same method, is $\theta_C = \theta_B = 69.21°$. The value obtained for δ_B by solving Equation 2.52 by integration and using Simpson's rule to carry out the integrations involved is 701.00 in. (17.8054 m). The same problem was also solved by using the Runge-Kutta method. This method yielded $\theta_B = 67.54°$ and $\delta_B = 708.08$ in. (17.9852 m).

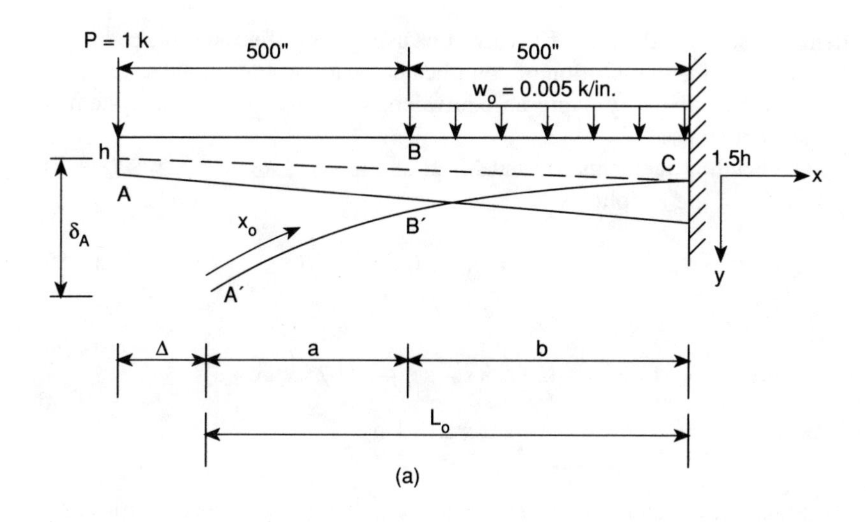

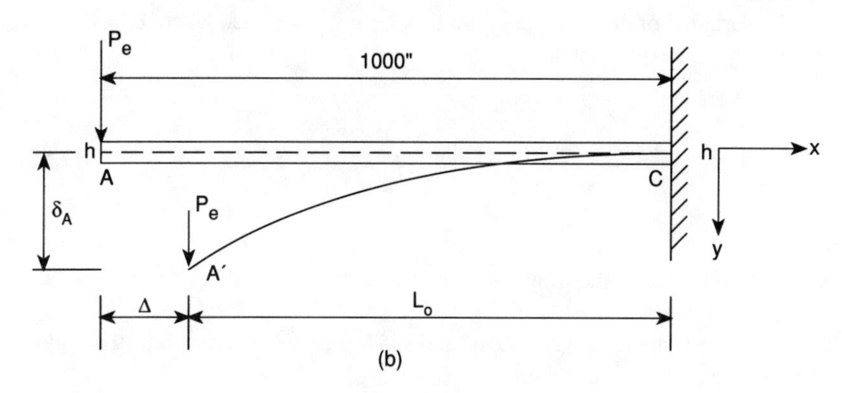

FIGURE 2.4. (a) Original variable stiffness member; (b) constant stiffness equivalent nonlinear system loaded with one concentrated load P_e at the free end (1 in. = 0.0254 m, 1 kip = 4448 N, 1 kip-in. = 175,127 N/m).

2.6 FLEXIBLE CANTILEVER BEAM LOADED WITH A PARTIALLY DISTRIBUTED LOAD AND A CONCENTRATED LOAD AT THE FREE END

Consider the tapered cantilever beam in Figure 2.4a which is loaded with a uniformly distributed load w_o = 0.005 kip/in. (875.635 N/m) over the half span of the member as shown, and with a concentrated load P = 1 kip (4448 N) at the free end. The problem in Figure 2.4a will be solved first by using pseudolinear systems as discussed in Sections 1.5 and 1.8. In order to demonstrate the advantages involved in using simplified nonlinear equivalent systems for the solution of complicated beam problems, the same problem will be solved by using an equivalent simplified nonlinear system of constant stiffness.

The bending moments M_1 and M_2 at any $0 \le x \le a$ and $a \le x \le (a + b)$, respectively, are as follows:

$$M_1 = -x \qquad 0 \le x \le a \tag{2.55}$$

$$M_2 = -x - 0.0025(x-a)^2 - 0.0025\Delta_2(x-a) \qquad a \le x \le (a+b) \tag{2.56}$$

The quantities Δ_1 and Δ_2 represent the horizontal displacements of portions AB and BC, respectively, of the beam in Figure 2.4a. It should be also noted that $\Delta = (\Delta_1 + \Delta_2)$, where Δ is the horizontal displacement of the free end of the member.

The variation of the moment of inertia I_x at any $0 \le x \le L_o$, where $L_o = (L - \Delta)$, is given by the expression

$$I_x = I_A \left[\frac{L_o + 0.5x}{L_o}\right]^3 \tag{2.57}$$

where $I_A = bh^3/12$ is the moment of inertia at the free end A of the beam, and b is its constant width.

By using Equations 2.55, 2.56, and 2.57, and substituting into the Euler-Bernoulli equation, we obtain

$$\frac{y_1''}{\left[1+(y_1')^2\right]^{3/2}} = \frac{L_o^3}{EI_A}\left[\frac{-x}{(L_o+0.5x)^3}\right] \qquad 0 \le x \le a \tag{2.58}$$

$$\frac{y_2''}{\left[1+(y_2')^2\right]^{3/2}} = \frac{L_o^3}{EI_A}\left[\frac{-x - 0.0025(x-a)^2 - 0.0025\Delta_2(x-a)}{(L_o+0.5x)^3}\right] \qquad a \le x \le (a+b) \tag{2.59}$$

By integrating Equations 2.58 and 2.59 once, we obtain

$$\frac{y_1'}{\left[1+(y_1')^2\right]^{1/2}} = \frac{0.0025L_o^3}{EI_A}\left[\frac{800(x+L_o)}{\left[L_o+0.5x\right]^2}\right]+C_1 \tag{2.60}$$

$$o \le x \le a$$

$$\frac{y_2'}{\left[1+(y_2')^2\right]^{1/2}} = \frac{0.0025L_o^3}{EI_A}\left[\frac{800(x+L_o)+(x-a)^2+4(x-a)(L-\Delta+0.5x)}{\left[L_o+0.5x\right]^2}\right.$$

$$\left.+\frac{(x-a)\Delta_2+2\Delta_2(L_o+0.5x)-8(L_o+0.5x)^2\ln\left[L_o+0.5x\right]}{\left[L_o+0.5x\right]^2}\right] \tag{2.61}$$

$$+C_2$$

$$a \le x \le (a+b)$$

where C_1 and C_2 are the constants of integration.

By applying the boundary conditions $y_1'(a) = y_2'(a)$, and $y_2'(L_o) = 0$, the constants of integration C_1 and C_2 are determined and they are as follows:

$$C_1 = \frac{0.0025L_o^3}{EI_A}\left[\frac{-8(L_o+0.5a)^2\ln\left[L_o+0.5a\right]+2\Delta_2(L_o+0.5a)}{\left[L_o+0.5a\right]^2}\right]+C_2 \tag{2.62}$$

$$C_2 = \frac{0.0025L_o^3}{EI_A}\left[\frac{1600L_o+(L_o-a)^2+4(L_o-a)(1.5L_o)}{2.25L_o^2}\right.$$

$$\left.+\frac{-18L_o^2\ln(1.5L_o)+(L_o-a)\Delta_2+3L_o\Delta_2}{2.25L_o^2}\right] \tag{2.63}$$

The parameters a, b, Δ_2, Δ_1, and Δ can be determined from the following equations by using a trial and error procedure as in the preceding section:

$$L = (L-\Delta)\left\{\int_0^{a_o}\frac{1}{\sqrt{1-\left[\Phi_1(\xi)\right]^2}}d\xi + \int_{a_o}^{1}\frac{1}{\sqrt{1-\left[\Phi_2(\xi)\right]^2}}d\xi\right\} \tag{2.64}$$

$$500 = (L - \Delta)\left\{\int_0^{a_o} \frac{1}{\sqrt{1 - [\Phi_1(\xi)]^2}} d\xi\right\} \tag{2.65}$$

$$500 = (L - \Delta)\left\{\int_{a_o}^{1} \frac{1}{\sqrt{1 - [\Phi_2(\xi)]^2}} d\xi\right\} \tag{2.66}$$

where

$$\Phi_1(\xi) = \frac{0.0025L_o^3}{EI_A}\left[\frac{800/L_o(\xi + 1)}{[1 + 0.5\xi]^2}\right] + C_1 \tag{2.67}$$

$$\Phi_2(\xi) = \frac{0.0025L_o^3}{EI_A}\left[\frac{800/L_o(\xi + 1) + (\xi - a_o)^2 + 4(\xi - a_o)(1 + 0.5\xi)}{[1 + 0.5\xi]^2}\right.$$
$$\left. + \frac{(\xi - a_o)\Delta_2/L_o + 2\Delta_2/L_o(1 + 0.5\xi) - 8(1 + 0.5\xi)^2 \ln(1 + 0.5\xi)}{(1 + 0.5\xi)^2}\right] + C_2 \tag{2.68}$$

In the above equations, $\xi = x/L_o$, $dx = L_o d\xi$, and $a_o = a/L_o$. The trial and error procedure for Equation 2.64 yields the value of Δ, while a trial and error procedure for Equations 2.65 and 2.66 yields the values of Δ_1 and Δ_2, respectively. With known values of Δ, Δ_1, and Δ_2, the values of y_1', and y_2', at any $0 \leq x \leq L_o$, may be determined using Equations 2.60 and 2.61, respectively, and pseudolinear analysis may be initiated in the usual way.

For the assumed length L = 1000 in. (25.4 m) and with EI_A = 180,000 kip-in^2 (561,541 Nm^2), the pseudolinear analysis yields the results

L_o = 744.19 in. (18.90 m)
a = 299.35 in. (7.6035 m)
b = 444.84 in. (11.2989 m)
Δ_1 = 200.65 in. (5.0965 m)
Δ_2 = 55.16 in. (1.4011 m)
Δ = $\Delta_1 + \Delta_2$ = 255.81 in. (6.4976 m)
δ_A = 600.096 in. (15.2424 m)
δ_B = 201.741 in. (5.1242 m)
θ_A = 58.65°
θ_B = 42.96°

TABLE 2.6
Values of M_x, f(x), and $M_e = M_x/f(x)$ for Various Values of $0 \leq x \leq 1000$ in.

x(in.)	M_x (kip-in.)	f(x)	$M_e = M_x/f(x)$ (kip-in.)
0	0	1.0	0.0
100	100	1.1576	86.3856
200	200	1.3331	150.0263
300	300	1.5209	197.2516
400	400	1.7280	231.4815
500	500	1.9531	256.0033
600	625	2.1970	284.4788
700	800	2.4604	325.1504
800	1025	2.7440	373.5423
900	1300	3.0486	426.4252
1000	1625	3.3750	481.4815

Note: 1 in. = 0.0254 m, 1 k-in = 112.9848 Nm.

2.6.1 SIMPLIFIED NONLINEAR EQUIVALENT SYSTEM OF CONSTANT STIFFNESS

The solution of the problem in Figure 2.4a can be made very convenient if a simplified nonlinear equivalent system of constant stiffness EI_A is used. Such a nonlinear equivalent system can be obtained by following the procedure in Sections 1.6 and 1.8. By using Equation 1.39 with g(x) = 1, and f(x) as shown by Equation 2.57, values of M_e at any $0 \leq x \leq L$ may be obtained as indicated in Table 2.6. The moment diagram M_e of the nonlinear equivalent system of constant stiffness EI_A is shown in Figure 2.5a. The approximation of the shape of M_e with one straight line AD as shown in Figure 2.5a leads to the simplified nonlinear equivalent system in Figure 2.5b, which is loaded with only one concentrated load $P_e = 0.52$ kips (2313 N) at the free end.

The solution of the nonlinear problem in Figure 2.5b is much simpler than the one followed earlier in this section. The problem in Figure 2.5b may be solved by following pseudolinear analysis or by using existing solutions that are available in undergraduate texts of mechanics of solids (see, e.g., Gere and Timoshenko[31]).

By following pseudolinear analysis, the results obtained are as follows:

$L_o =$ 755.03 in. (19.1778 m)
$\Delta =$ 244.97 in. (6.2222 m)
$\delta_A =$ 593.73 in. (15.0807 m)
$\theta_A =$ 55.43°

These results are, for all practical purposes, in close agreement with the ones obtained earlier in the section by using pseudolinear analysis to solve the nonlinear problem in Figure 2.4a.

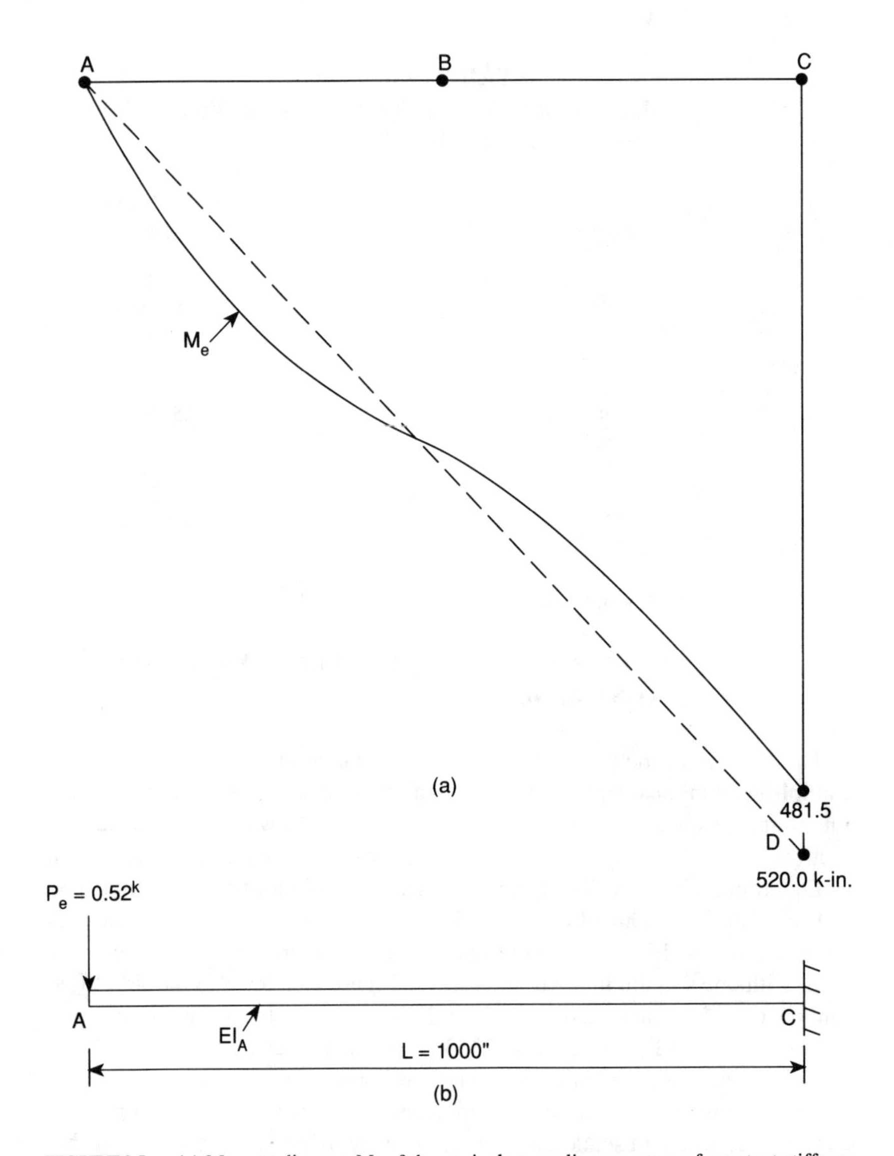

FIGURE 2.5. (a) Moment diagram M_e of the equivalent nonlinear system of constant stiffness EI_A approximated with one straight line AD; (b) simplified nonlinear equivalent system of constant stiffness EI_A (1 in. = 0.0254 m, 1 kip-in = 112.9848 Nm).

2.7 TAPERED FLEXIBLE CANTILEVER BEAM SUBJECTED TO A TRAPEZOIDAL LOADING

Consider the tapered contilever beam in Figure 2.6a, which is loaded with a trapezoidal load as shown. Its large deflection configuration is represented by the curve AB′. The factors n and m represent the variation of h and w_1, respectively. The bending moment M_x and the moment of inertia I_x at any distance $0 \leq x \leq L_o$, with $L_o = L - \Delta$, are as follows:

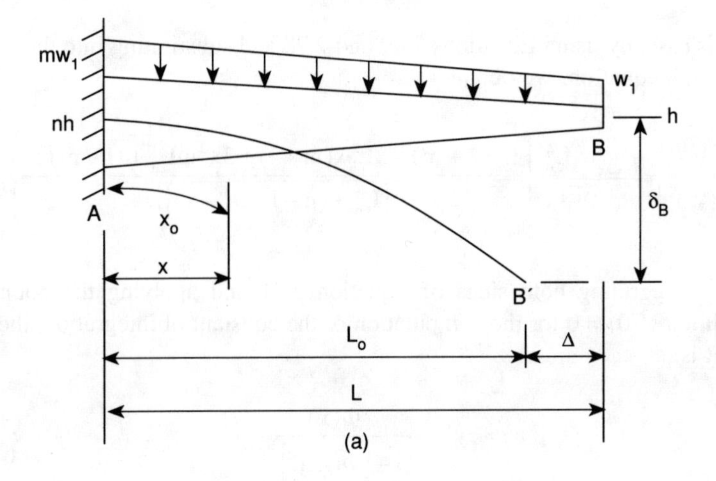

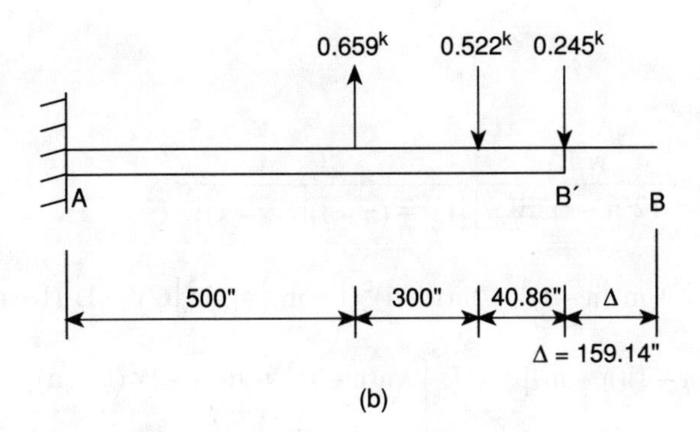

FIGURE 2.6. (a) Initial variable thickness member; (b) pseudolinear system of constant stiffness EI_B (1 in. = 0.0254 m, 1 kip = 4448.2 N).

$$M_x = -\frac{w_1}{6L_o}\left[L_o^3(2+m) - 3L_o^2x(m+1) + 3x^2mL_o - (m-1)x^3\right] \quad (2.69)$$

$$I_x = \frac{bh^3}{12}\left[\frac{L_o + (n-1)(L_o - x)}{L_o}\right]^3$$

$$I_x = I_B f(x) \quad (2.70)$$

where b is the constant width of the member, $I_B = bh^3/12$, and

$$f(x) = \left[\frac{L_o + (n-1)(L_o - x)}{L_o}\right]^3$$

In this case, by using Equations 2.69 and 2.70 and substituting into the Euler-Bernoulli equation, we obtain

$$\frac{y''}{\left[1+(y')^2\right]^{3/2}} = \frac{w_1 L_o^2}{6EI_B}\left\{\frac{L_o^3(2+m) - 3L_o^2 x(m+1) + 3x^2 mL_o + (1-m)x^3}{\left[L_o + (n-1)(L_o - x)\right]^3}\right\} \tag{2.71}$$

By integrating both sides of Equation 2.71 and applying the boundary condition $y'(0) = 0$ for the computation of the constant of integration, the end result is

$$y' = \frac{\Phi(x)}{\left\{1-[\Phi(x)]^2\right\}^{1/2}} \tag{2.72}$$

where

$$\begin{aligned}
\Phi(x) = {} & \frac{w_1 L_o^2}{12(n-1)^4 EI_B}\left\{\frac{1}{\left[L_o + (n-1)(L_o - x)\right]^2}\right. \\
& \left[L_o^3\left[(2+m)(n-1)^3 + 3n(n-1)^2(1+m)\right] - xL_o^2\left[6(n-1)^3(1+m)\right.\right. \\
& \left. +6n(n-1)(n-m)\right] + x^2 L_o\left[9m(n-1)^3 + 9n(n-1)^2(1-m)\right] \\
& +x^3\left[2(n-1)^3(m-1)\right] - 6L_o\left[L_o + (n-1)(L_o - x)\right]^2 \\
& \left.\ln\left[L_o + (n-1)(L_o - x)\right](n-m)\right] \\
& -\frac{1}{n^2}\left[L_o\left[(2+m)(n-1)^3 + 3n(n-1)^2(1+m)\right]\right. \\
& \left.\left. -6L_o n^2 \ln L_o n(n-m)\right]\right\}
\end{aligned} \tag{2.73}$$

The unknown horizontal displacement Δ in Equation 2.72 can be obtained from the equation

$$L = \int_0^{L_o = L-\Delta} \left[1+(y')^2\right]^{1/2} dx \tag{2.74}$$

TABLE 2.7
Calculated Values of the Moment Diagram M'_e of the Pseudolinear System of Constant Stiffness EI_B

x(in.)	f(x)	y′(rad)	Z_e	$M_e = M_x/f(x)$ (kip–in.)	M'_e (kip–in.)
0.0	8.0	0.0	1.0	294.60	294.60
100.0	6.6561	0.1580	1.0337	266.70	276.75
200.0	5.4718	0.3095	1.1471	235.32	269.93
300.0	4.4370	0.4586	1.3515	200.16	270.52
400.0	3.5417	0.6046	1.5957	161.16	257.18
500.0	2.7757	0.7408	1.9276	118.79	228.97
600.0	2.1290	0.8531	2.2711	74.63	169.49
700.0	1.5914	0.9239	2.5237	32.91	86.05
800.0	1.1530	0.9476	2.6146	3.68	9.62
840.86	1.000	0.9484	2.6176	0.0	0.0

Note: 1 in. = 0.0254 m, 1 k-in. = 112.9848 Nm.

by using a trial and error procedure, i.e., assume a value of Δ in Equation 2.72 and integrate to determine the length L as discussed earlier. The procedure may be repeated for various values of Δ until the correct length L is obtained. With Δ known, the values of y′ at any $0 \leq x \leq L_o$ may be computed by using Equation 2.72. With known y′, the values of the moment M'_e of the pseudolinear system of constant stiffness EI_B, at any $0 \leq x \leq L_o$, may be determined from Equation 1.33, and pseudolinear analysis can be carried out as in the earlier sections of the chapter.

Let it be assumed that L = 1000 in. (25.4 m), EI_B = 180,000 kip-in.^{2} (516,541 Nm^2), w_1 = 0.005 kips/in. (87.6 kN/m), and n = m = 2. On this basis, the trial and error procedure of Equation 2.74 yields Δ = 159.14 in. (4.0422 m). With known Δ, the values of y′, z_e, M_x, and M'_e, at any $0 \leq x \leq L_o$ and at the interval of 100 in. (2.54 m), are determined by using Equations 2.72, 1.34, 2.69, and 1.33, respectively, and they are shown in Table 2.7. Note in this table that L_o = 840.86 in. (21.3578 m). The values of M'_e are shown in the last column of the table. By using these values of M_e', the moment diagram M'_e of the pseudolinear system of constant stiffness EI_B can be plotted. The approximation of the shape of M'_e with three straight line segments leads to the constant stiffness pseudolinear equivalent system loaded as shown in Figure 2.6b.

The large deflections and rotations at any $0 \leq x \leq L_o$ of the original system in Figure 2.6a can be accurately determined by using the pseudolinear system in Figure 2.6b and applying elementary linear analysis. Thus, by applying the moment area method to the pseudolinear system in Figure 2.6b we obtain $\delta_B = \delta_{B'}$ = 493.82 in. (12.543 m). The value obtained by integrating Equation 2.72 directly, and using Simpson's rule to carry out the required integrations, gives δ_B = 492.30 in., yielding a difference of only 0.31%. In a similar manner, by

TABLE 2.8
Values of Δ_B, Θ_B, and δ_B for Various Values of the Taper Parameter n When m = 2 and w_1 = 0.005 kips/in.

n	Δ_B (in.)	θ_B (°)	δ (in.)
1.0	562.51	75.73	843.02
1.5	284.72	56.30	642.41
2.0	159.14	43.48	493.82
2.5	81.05	31.89	355.95
3.0	39.78	22.88	250.37

Note: 1 in = 0.0254 m, 1 k/in. = 175,126.8 N/m.

TABLE 2.9
Values of Δ_B, θ_B, and δ_B for Various Values of the Load Parameter m When n = 2 and w_1 = 0.005 kips/in.

m	Δ_B (in.)	θ_B (°)	δ_B (in.)
1.0	125.90	39.20	439.66
2.0	159.14	43.48	493.82
3.0	188.81	46.92	533.57
4.0	215.24	49.75	566.72

Note: 1 in. = 0.0254 m, 1 kip/in. = 175,126.8 N/m.

using the pseudolinear system in Figure 2.6b, we find $y'_B = y'_{B'} = 0.951$, and consequently $\theta_B = \theta_{B'} = 43.565°$. The solution of Equation 2.72 yields $y'_B = 0.948$, and $\theta_B = 43.48°$, a difference of only 0.20%

Table 2.8 shows the variations of the horizontal displacement Δ_B, rotation θ_B, and vertical displacement δ_B at the free end B of the member for various values of the taper parameter n and with load parameter m = 2. The variation of the same quantities for various values of the load parameter m and taper parameter n = 2 is shown in Table 2.9. The value of the load w_1 that is used for these results is 0.005 kips/in. (87.6 kN/m).

2.8 UNIFORM SIMPLY SUPPORTED BEAM SUBJECTED TO A CONCENTRATED LOAD P

Consider the uniform simply supported beam in Figure 2.7 that is loaded with a concentrated load P at any location along the length of the member. The large deformation configuration of the beam is shown in the same figure. In this case the Young's modulus of elasticity E and the moment of inertia I are constant along the length of the member, i.e., g(x) = f(x) = 1.00.

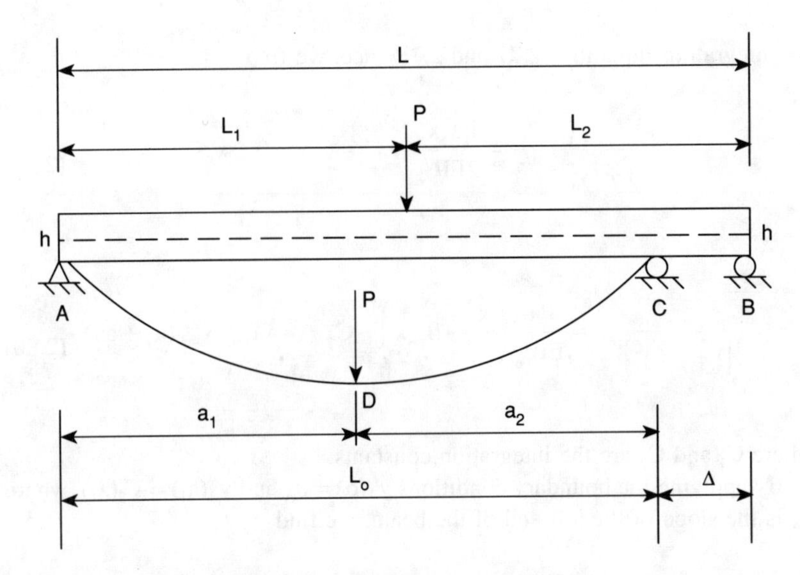

FIGURE 2.7. Uniform simply supported beam loaded with a concentrated load P at any point along its length.

The expressions for bending moment M_x for portions a_1 and a_2 are as follows:

$$M_1(x) = -\frac{Pa_2x}{L_o} \qquad 0 \le x \le a_1 \tag{2.75}$$

$$M_2(x) = -\frac{Pa_1}{L_o}(x - L_o) \qquad a_1 \le x \le (a_1 + a_2) \tag{2.76}$$

Note here that $L_o = (a_1 + a_2)$. By employing the Euler-Bernoulli equation and Equations 2.75 and 2.76, the following two expressions for the portions a_1 and a_2 of the member are obtained:

$$\frac{y_1''}{\left[1+(y_1')^2\right]^{3/2}} = \frac{Pa_2x}{EIL_o} \qquad 0 \le x \le a_1 \tag{2.77}$$

$$\frac{y_2''}{\left[1+(y_2')^2\right]^{3/2}} = \frac{Pa_1}{EIL_o}(x - L_o) \qquad a_1 \le x \le (a_1 + a_2) \tag{2.78}$$

By integrating Equations 2.77 and 2.78 once, we find

$$\frac{y_1'}{\left[1+(y_1')^2\right]^{1/2}} = \frac{Pa_2x^2}{2EIL_o} + C_1 \qquad 0 \le x \le a_1 \tag{2.79}$$

$$\frac{y_2'}{\left[1+(y_2')^2\right]^{1/2}} = \frac{Pa_1}{EIL_o}\left(\frac{x^2}{2} + L_o x\right) + C_2 \qquad a_1 \le x \le (a_1 + a_2) \tag{2.80}$$

where C_1 and C_2 are the integration constants.

By applying the boundary conditions $y_1'(o) = \theta_A$ and $y_1'(a_1) = y_2'(a_1)$, where θ_A is the slope of the left end of the beam, we find

$$C_1 = \frac{\theta_A}{\left[1+\theta_A^2\right]^{1/2}} \tag{2.81}$$

$$C_2 = \frac{Pa_1^2}{2EI} + \frac{\theta_A}{\left[1+\theta_A^2\right]^{1/2}} \tag{2.82}$$

It should be noted here that Equations 2.79 and 2.80 are functions of the horizontal displacement Δ of the right support B of the member. Thus, the two unknown quantities in these two equations are Δ and θ_A. The horizontal displacements Δ_1 and Δ_2, of the lengths L_1 and L_2, respectively, where $\Delta = (\Delta_1 + \Delta_2)$, are also unknown quantities.

The quantities Δ_1, Δ_2, and Δ may be determined by trial and error by using the following equations:

$$L_1 = \int_0^{a_1} \left[1+(y_1')^2\right]^{1/2} dx = \int_0^{a_1} \frac{1}{\left\{1-\left[Q_1(x)\right]^2\right\}^{1/2}} dx \tag{2.83}$$

$$L_2 = \int_{a_1}^{(a_1+a_2)} \left[1+(y_2')^2\right]^{1/2} dx = \int_{a_1}^{(a_1+a_2)} \frac{1}{\left\{1-\left[Q_2(x)\right]^2\right\}^{1/2}} dx \tag{2.84}$$

$$L = \int_0^{a_1} \frac{1}{\left\{1-\left[Q_1(x)\right]^2\right\}^{1/2}} dx + \int_{a_1}^{(a_1+a_2)} \frac{1}{\left\{1-\left[Q_2(x)\right]^2\right\}^{1/2}} dx \tag{2.85}$$

where

$$Q_1(x) = -\frac{Pa_2x^2}{2EIL_o} + \frac{\theta_A}{\left[1+\theta_A^2\right]^{1/2}} \tag{2.86}$$

$$Q_2(x) = \frac{Pa_1}{EIL_o}\left[\frac{x^2}{2} - L_o x\right] + \frac{Pa_1^2}{2EI} + \frac{\theta_A}{\left[1+\theta_A^2\right]^{1/2}} \tag{2.87}$$

In the above equations the rotation θ_A of the left end of the beam is also an unknown quantity. A trial and error procedure may be initiated by assuming a value for the slope θ_A. Then, we use Equations 2.83 and 2.84, and assume values of a_1 and a_2 until these two equations are satisfied. When this is done, we check if the condition of zero displacement at $x = L_o$ is satisfied. If this does not occur, we repeat the procedure with new values of θ_A until this displacement condition is satisfied. When this trial and error procedure is completed, then the values of θ_A, Δ_1, Δ_2, Δ, a_1, a_2, and $L_o = (a_1 + a_2)$ are all known. Therefore, the values of y' at any $0 \le x \le L_o$ may be determined from Equations 2.79 and 2.80. With known y', pseudolinear analysis can be carried out as in the preceding sections. If preferred, the displacements y at any $0 \le x \le L_o$ may be determined by integrating Equations 2.79 and 2.80 once, applying the displacement continuity conditions at point D in Figure 2.7, and applying the condition of zero displacement at support A in order to determine the two constants of integration. This approach, however, would be more tedious.

Let it be assumed in Figure 2.7 that EI = 312.50 lb in.^{2} (0.8968 Nm^2), $L_1 = 40.0$ in. (1.016 m), $L_2 = 64.60$ in. (1.6408 m), and $L = L_1 + L_2 = 104.60$ in. (2.6568 m). The applied load P = 0.50 lb (2.4482 N), depth h = 0.05 in. (0.00127 m), width b = 1.0 in. (0.0254 m), and modulus of elasticity E = 30×10^6 psi ($206{,}844 \times 10^6$ Pa). By applying the above methodology we find that $\theta_A = 49°$, $\Delta = 16.80$ in. (0.4267 m), and maximum vertical displacement $y_{max} = 25.48$ in. (0.6472 m). The same problem was also solved by Frisch-Fay,[5] by using the complicated method of elastic similarities His results are $\theta_A = 50°$, $\Delta = 16.10$ in. (0.4089 m), and $y_{max} = 23.80$ in. (0.6045 m). Frisch-Fay used an iterative procedure to evaluate the elliptic integrals involved in his analysis, and the difference in results may be attributed to the utilization of this procedure.

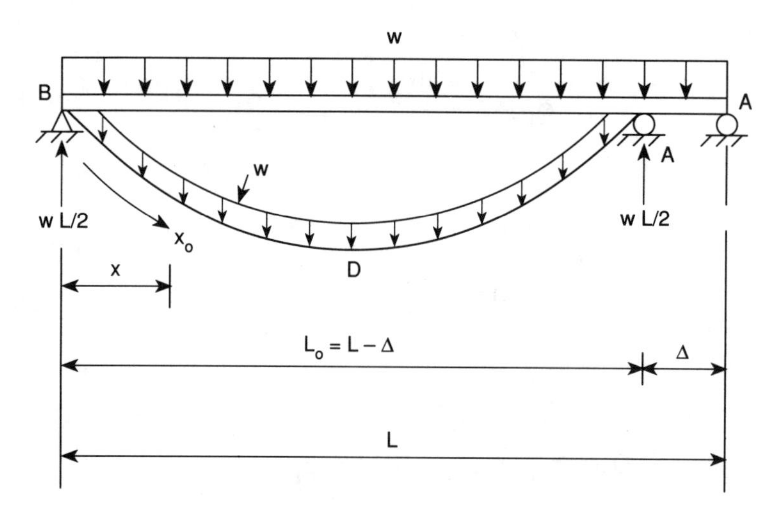

FIGURE 2.8. Straight and deflected configuration of a uniform simply supported beam loaded as shown.

2.9 UNIFORM SIMPLY SUPPORTED BEAMS LOADED WITH A UNIFORMLY DISTRIBUTED LOADING

Consider a simply supported beam of uniform cross section that is loaded by a uniformly distributed load w as shown in Figure 2.8. The bending moment M_x at any point $0 \le x \le L_o$, is

$$M_x = \frac{wLx}{2} - \frac{wx}{2}\int_0^x \left\{1+[y'(\xi)]^2\right\}^{1/2} d\xi \tag{2.88}$$

where ξ is a dummy variable. By substituting Equation 2.88 into the Euler-Bernoulli equation, we obtain

$$\frac{y''}{\left[1+(y')^2\right]^{3/2}} = \frac{wx}{2EI}\left\{-L + \int_0^x \left[1+(y'(\xi))^2\right]^{1/2} d\xi\right\} \tag{2.89}$$

Equation 2.89 is the exact nonlinear integral differential equation, which is very difficult to solve. A simpler solution may be obtained here again by assuming that x_o can be expressed as shown in Equation 1.42, where $\Delta(x)$ can take any of the forms given in Equations 1.49 through 1.52. Let it be assumed here that $x_o = x + \Delta$. On this basis, the bending moment M_x at any $0 \le x \le L_o$, can be expressed as

$$M_x = \frac{wLx}{2} - \frac{wx}{2}x_o = \frac{w(L-\Delta)x}{2} - \frac{wx^2}{2} \tag{2.90}$$

By substituting Equation 2.90 into the Euler-Bernoulli equation, we obtain

$$\frac{y''}{\left[1+(y')^2\right]^{3/2}} = \frac{w}{2EI}\left[-(L-\Delta)x + x^2\right] \tag{2.91}$$

By integrating Equation 2.91 once and satisfying the boundary condition of zero rotation at x = $L_o/2$, we obtain

$$y'(x) = \frac{G(x)}{\left\{1-[G(x)]^2\right\}^{1/2}} \tag{2.92}$$

where

$$G(x) = \frac{w}{24EI}\left[6(L-\Delta)x^2 - 4x^3 - (L-\Delta)^3\right] \tag{2.93}$$

The unknown horizontal displacement Δ of the end support A of the beam may be determined from the equation

$$L = \int_0^{L_o}\left[1+(y')^2\right]^{1/2} dx \tag{2.94}$$

as discussed in preceding sections by applying a trial and error procedure. With known Δ, y′ can be determined by Equation 2.92, and pseudolinear analysis can be carried out in the usual way.

By assuming L = 1000 in. (25.4 m) and EI = 75×10^3 kip-in.^{2} (215,224 Nm^2), the values of Δ_A at the end A, θ_B at the end B, and vertical displacement δ_B at x = $L_o/2$ are calculated for various values of the distributed load w by using the above procedure. The results are shown in Table 2.10. In the same table, the results are also compared by using the 4th order Runge-Kutta method.

2.10 TAPERED SIMPLY SUPPORTED BEAMS LOADED WITH A UNIFORMLY DISTRIBUTED LOADING

Consider now the tapered simply supported beam loaded as shown in Figure 2.9a. Its deformed configuaration is shown in Figure 2.9b where Δ is the horizontal displacement of support B. By using the expressions

$$I_x = \frac{bh^3}{12}\left\{1+\frac{(n-1)}{L}\int_0^x\left[1+(y'(\xi))^2\right]^{1/2} d\xi\right\}^3 \tag{2.95}$$

TABLE 2.10
Variation of Horizontal Displacement Δ_A, Rotation θ_B, and Vertical Displacement δ_D for Various Values of the Distributed Load w

	Pseudolinear equivalent systems			4th order Runge-Kutta method			
w (lb/in.)	Δ_A (in.)	θ_B (°)	δ_D (in.)	Δ_A (in.)	θ_B (°)	δ_D (in.)	% Difference
2	143.60	44.26	230.07	161.56	47.21	242.19	–5.00
5	311.43	65.08	321.25	358.65	70.59	337.72	–4.87
10	440.53	76.63	367.65	496.17	81.93	379.62	–3.15
15	508.55	81.54	389.87	562.53	85.81	396.42	–1.65
20	552.62	84.22	404.08	603.66	87.57	406.32	–0.55

Note: 1 in. = 0.0254 m, 1 lb/in. = 175.118 N/m.

$$M_x = \frac{wLx}{2} - \frac{wx}{2}\int_0^x \left[1 + (y'(\xi))^2\right]^{1/2} d\xi \tag{2.96}$$

and substituting into the Euler-Bernoulli equation, we get

$$\frac{y''}{\left[1+(y')^2\right]^{3/2}} = \frac{wx}{2EI_A}\left\{\frac{2L - \int_0^x \left[1+(y'(\xi))^2\right]^{1/2} d\xi}{\left\{1+\frac{n-1}{L}\int_0^x \left[1+(y'(\xi))^2\right]^{1/2} d\xi\right\}^3}\right\} \tag{2.97}$$

which is again a nonlinear integral differential equation.

We simplify again the complexity of the problem by assuming that $x_o = (x + \Delta)$. On this basis, the expressions for the bending moment M_x, and moment of inertia I_x, at any $0 \le x \le L_o$, are

$$M_x = \frac{wLx}{2} - \frac{wx}{2}x_o = \frac{w(L-\Delta)x}{2} - \frac{wx^2}{2} \tag{2.98}$$

$$I_x = \frac{bh^3}{12}\left[1 + (n-1)\frac{x}{L-\Delta}\right]^3 = I_A f(x) \tag{2.99}$$

where

$$I_A = \frac{bh^3}{12} \tag{2.100}$$

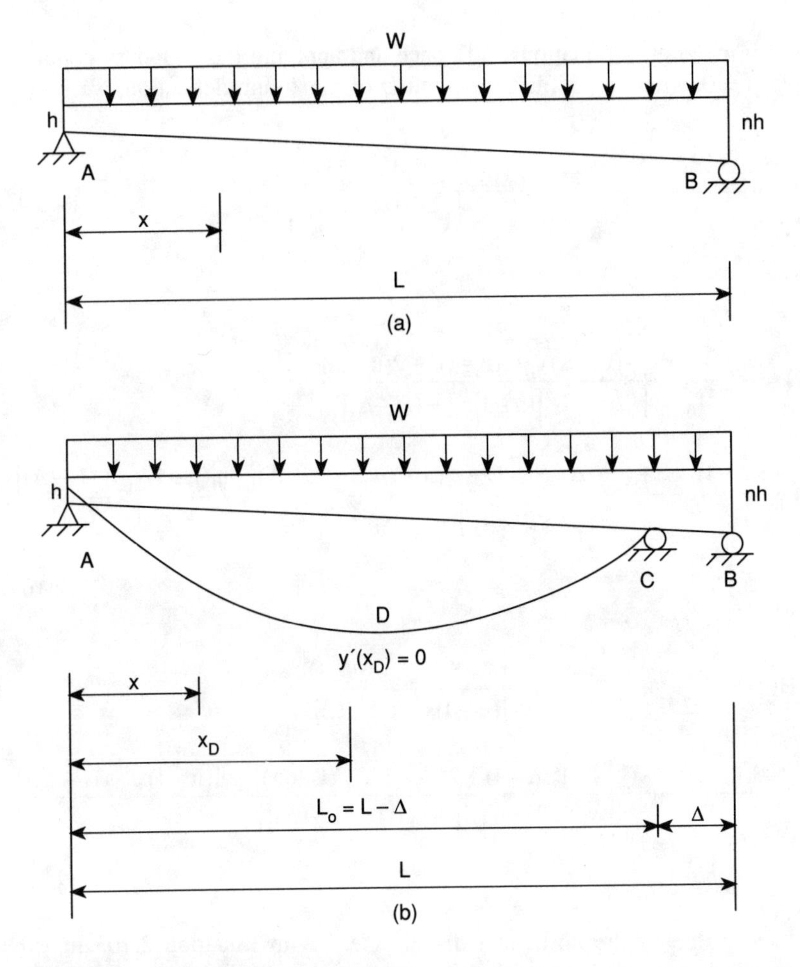

FIGURE 2.9. (a) Tapered simply supported beam loaded as shown; (b) deflected configuration of the tapered simply supported beam.

$$f(x) = \left[1 + (n-1)\frac{x}{L-\Delta}\right]^3 \tag{2.101}$$

and b is the constant width of the member. By substituting Equations 2.98 and 2.99 into the Euler-Bernoulli equation, we obtain

$$\frac{y''}{\left[1+(y')^2\right]^{3/2}} = \frac{w(L-\Delta)^3}{2EI_A}\left\{\frac{-(L-\Delta)x + x^2}{\left[(n-1)x + (L-\Delta)\right]^3}\right\} \tag{2.102}$$

which is a much simpler differential equation to solve compared to the one given by Equation 2.97.

By integrating Equation 2.102 once and applying the boundary condition $y'(x_D) = 0$, where x_D defines the point D of maximum deflection, we find

$$y'(x) = \frac{H(x)+C}{\left\{1-[H(x)+C]^2\right\}^{1/2}} \tag{2.103}$$

where

$$C = \frac{w(L-\Delta)^3}{2EI_A}\left\{\frac{-(L-\Delta)x_D(2(n-1)^2-2(n-1))}{2(n-1)^3[(n-1)x_D+(L-\Delta)]^2}\right.$$

$$\left.+\frac{-3(n-1)^2x_D^2+(L-\Delta)^2(n-1)+2[(n-1)x_D+(L-\Delta)]^2\ln[(n-1)x_D+(L-\Delta)]}{2(n-1)^3[(n-1)x_D+(L-\Delta)]^2}\right\} \tag{2.104}$$

$$H(x) = \frac{w(L-\Delta)^3}{2EI_A}\left\{\frac{(L-\Delta)x[2(n-1)^2-2(n-1)]}{2(n-1)^3[(n-1)x+(L-\Delta)]^2}\right.$$

$$\left.+\frac{-3(n-1)^2x+(L-\Delta)^2(n-1)+2[(n-1)x+(L-\Delta)]^2\ln[(n-1)x+(L-\Delta)]}{2(n-1)^3[(n-1)x+(L-\Delta)]^2}\right\} \tag{2.105}$$

The value of the horizontal displacement Δ in Equation 2.103 may be determined by a trial and error procedure, i.e., we may assume a value of x_D in Equation 2.103, where $x = x_D$ defines the position of the maximum deflection. Then, for the assumed x_D, we use Equation 2.94 to determine $L_o = (L - \Delta)$ by assuming values of Δ and applying a trial and error procedure as before. Simpson's rule may be used here to facilitate the integrations involved. When $L_o = (L - \Delta)$ is determined, we check if the displacement conditions at the ends C and A of the beam are satisfied. If not, the procedure is repeated for various values of x_D until these conditions are satisfied. The procedure converges fairly fast to the required value of Δ or L_o.

This procedure is applied to the problem in Figure 2.9b by assuming $EI_A = 75 \times 10^3$ kip-in.^{2} (215,224 Nm^2), L = 1000 in. (25.4 m), n = 1.5 and using various values of the uniformly distributed load w. When x_D and L_o or Δ are determined, the values of y′ at any $0 \leq x \leq L_o$ may be determined from Equation 2.103. The large deflection y at any $0 \leq x \leq L_o$ may be determined either by (1) integrating Equation 2.103 once and satisfying one displacement boundary condition for the constant of integration, or (2) by applying pseudolinear analysis as explained

TABLE 2.11
Variation of the Horizontal Displacement Δ_B, Rotation θ_A, and Vertical Displacement δ_D for Various Values of the Load w

	Pseudolinear equivalent system			4th order Runge-Kutta method			
w (lb/in.)	Δ_B (in.)	θ_A (°)	δ_D (in.)	Δ_B (in.)	θ_A (°)	δ_D (in.)	% Difference
2	62.99	33.05	155.48	67.36	32.15	145.17	7.10
5	194.21	57.73	262.04	219.35	58.14	251.59	4.15
10	326.59	73.49	326.98	368.78	74.79	315.15	3.75
15	403.10	80.17	358.55	452.27	81.41	342.81	4.59
20	454.32	83.60	377.10	504.30	84.68	359.31	4.95

Note: 1 in = 0.0254 m, 1 lb/in. = 175.118 N/m.

earlier. Both procedures yield accurate results, but pseudolinear analysis would be easier to apply. The results are shown in Table 2.11, where Δ_B is the horizontal displacement of the end B of the beam, θ_A is the rotation of end A, and δ_D is the maximum vertical deflection. In the same table, the results are compared to the results obtained by using the 4th order Runge-Kutta method. The maximum difference in deflection δ_D is 7.1% for w = 2 lb/in. (350,236 Nm) and reduces to 3.75% for w = 10 lb/in. (1751.18 N/m). A large portion of the difference may be attributed to mathematical manipulations.

2.11 EQUIVALENT SIMPLIFIED NONLINEAR SYSTEMS OF CONSTANT STIFFNESS FOR SIMPLY SUPPORTED BEAMS

The method discussed in Sections 1.6 through 1.11 will be used here to determine equivalent nonlinear systems of constant stiffness for simply supported beams. The procedure is particularly useful for practical cases where the loading on the member is a composition of various types of loadings and where the moment of inertia along the length of the member is arbitrary. A combination of concentrated and distributed loadings, coupled with an arbitrary variation of the depth of the member along its length, would be an example.

In order to compare results using this approach, a variable stiffness simply supported beam loaded as shown in Figure 2.10a is used. This problem is the same as the one in Figure 2.9a except that we measure x from the thick end of the member. From its deformed configuration and at any distance x from end A, $0 \leq x \leq (L-\Delta)$, we write

$$I_x = \frac{bh^3}{12}\left[\frac{(n-1)(L-\Delta-x)+(L-\Delta)}{(L-\Delta)}\right]^3 \tag{2.106}$$

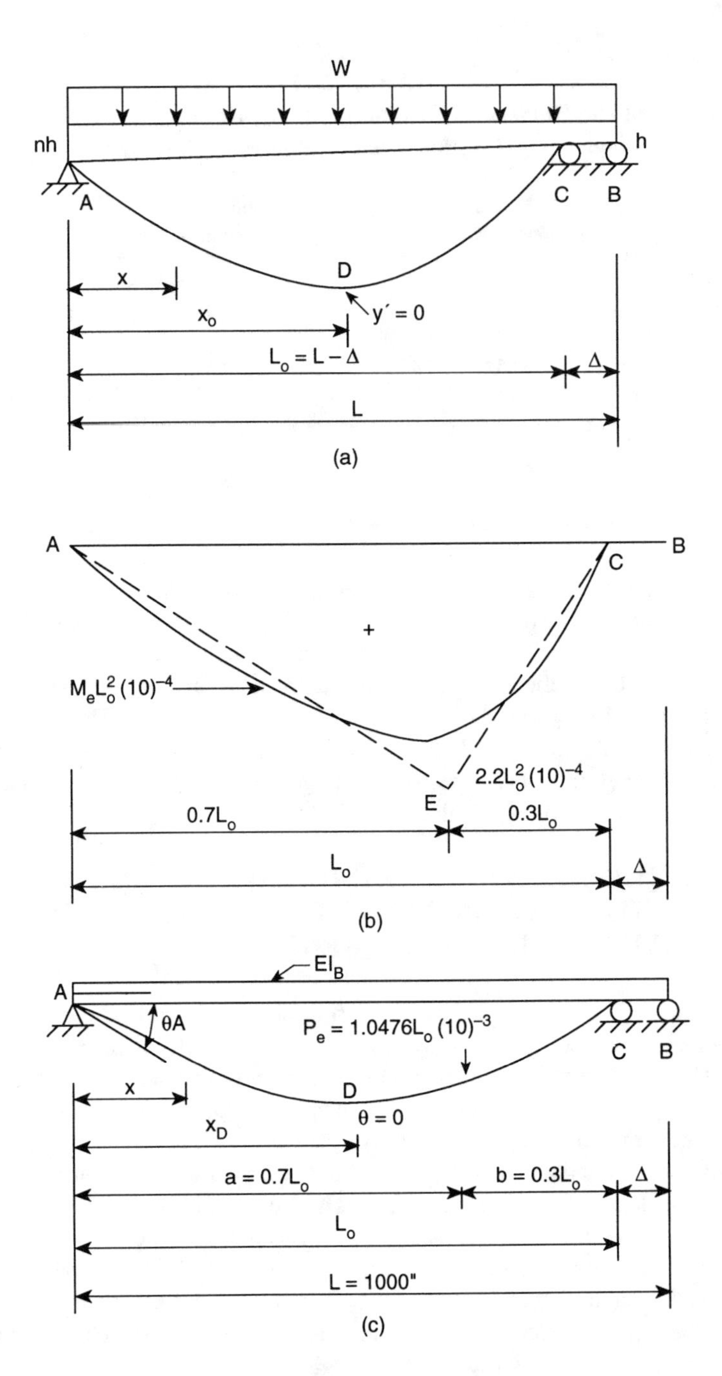

FIGURE 2.10. (a) Tapered simply supported beam loaded with a uniformly distributed load w as shown; (b) moment diagram M_e of the nonlinear equivalent system of constant stiffness with its shape approximated with two straight lines; (c) constant stiffness equivalent nonlinear system loaded with one concentrated load (1 in. = 0.0254 m).

$$M_x = \frac{w(L-\Delta)x}{2} - \frac{wx^2}{2} \tag{2.107}$$

$$f(x) = \left[\frac{(n-1)(L-\Delta-x)+(L-\Delta)}{(L-\Delta)}\right]^3 \tag{2.108}$$

If we assume that $n = 1.5$ and $L_o = (L - \Delta)$, we find

$$f(x) = \left[\frac{1.5L_o - 0.5x}{L_o}\right]^3 \tag{2.109}$$

$$M_x = \frac{wL_o x - wx^2}{2} \tag{2.110}$$

Thus, by substituting into Equation 1.39 with $g(x) = 1$, we find

$$M_e = \frac{wx(L_o - x)}{2\left(1.5 - 0.5\frac{x}{L_o}\right)^3} \tag{2.111}$$

By using Equation 2.111 with $w = 0.0025$ kips/in. (43.8 kN/m), the values of M_e for various values of x, in terms of L^2_o, are shown in Table 2.12. The plot of M_e is shown in Figure 2.10b. The approximation of its shape with two straight line segments AE and EC leads to the nonlinear equivalent system of constant stiffness EI_B shown in Figure 2.10c. The solution of the constant stiffness nonlinear system in Figure 2.10c is much simpler and it yields very accurate results.

This solution can be obtained by first writing the expressions for the moment due to P_e for the intervals $0 \leq x \leq 0.7L_o$ and $0.7L_o \leq x \leq L_o$. By utilizing the Euler-Bernoulli equation and the moment expressions stated above, the two nonlinear differential equations for these two intervals of x may be written. By integrating each equation once, the expression of y' for each interval may be derived. The two associated constants of integration and distance x_D to the point D of maximum deflection can be determined by utilizing the boundary conditions of zero deflection at the end A of the beam, y_D' at $x = x_D$ being zero, and that the rotations at $x = 0.7L_o$ for portions $0.7L_o$ and $0.3\ L_o$ are equal. The distance L_o can be determined from the equation

TABLE 2.12
Values of M_e Vs. x in Terms of L_o

xL_o (in.)	$M_eL_o^2 \times 10^{-4}$ (kip-in.)
0.0	0.0
0.10	0.3690
0.20	0.7289
0.30	1.0669
0.40	1.3655
0.50	1.6000
0.60	1.7361
0.70	1.7260
0.80	1.5026
0.90	0.9718
1.00	0.0

Note: 1 in. = 0.0254 m, 1 kip-in. = 113.0 N/m.

$$L = \int_0^{0.7L_o} \left[1+(y')^2\right]^{1/2} dx + \int_{0.7L_o}^{L_o} \left[1+(y')^2\right]^{1/2} dx \tag{2.112}$$

by a trial and error procedure as stated earlier, i.e., assume values of L_o in Equation 2.112 until the equation is satisfied.

When L_o is determined, the values of y' at any $0 \le x \le L_o$ may be determined from the equations of y' for portions 0.7 L_o and 0.3 L_o. With known y', the pseudolinear analysis may be initiated as stated earlier and can approximate the shape of M'_e with a few straight lines. By assuming that EI_B = 75,000 kip-in.^{2} (215,224 Nm^2), L = 1000 in. (25.4 m), w = 0.0025 kips/in. (43.8 kN/m) and applying the moment area method to the pseudolinear system, we find that the maximum deflection y_D at D is 186.31 in. (4.73 m) and the rotation θ_A at the support A is 30.03°. It is also found that x_D = 507.0 in. (12.88 m), L_o = 909.1 in. (23.09 m), and Δ = 90.95 in. (2.31 m). These results are in close agreement with the ones obtained by directly solving the nonlinear differential equation. The solution of the differential equation yielded θ_A = 30.54°, Δ = 87.22 in. (2.22 m), y_D = 181.38 in. (4.61 m), and x_D = 493.55 in. (12.54 m).

2.12 TAPERED SIMPLY SUPPORTED BEAMS LOADED WITH A TRAPEZOIDAL LOADING

Consider a doubly tapered simply supported beam that is loaded with a distributed trapezpoidal loading as shown in Figure 2.11a. The large deformation configuration of the member is shown in Figure 2.11b. The bending moment M_x and moment of inertia I_x at any point $0 \le x \le L_o$ are given by the expressions

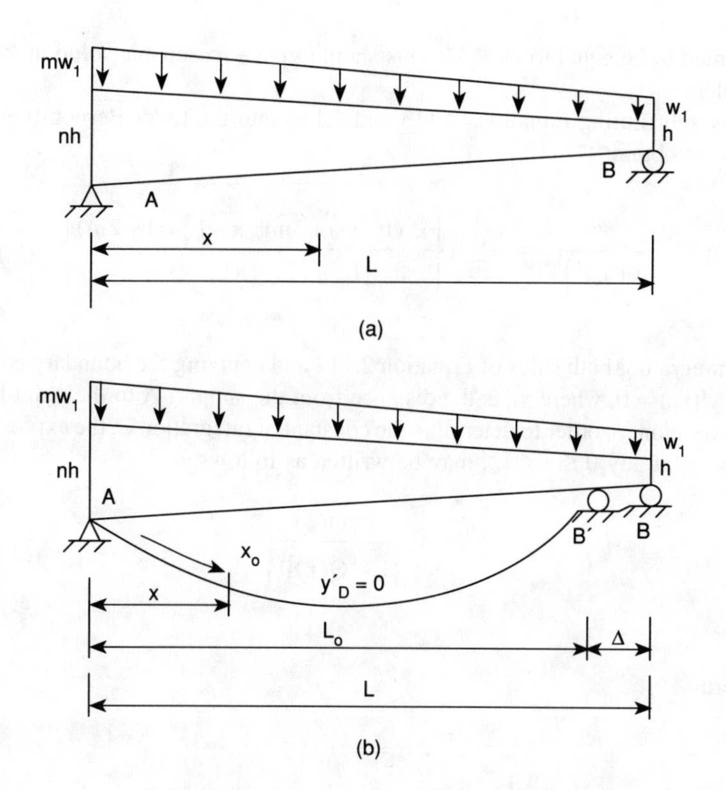

FIGURE 2.11. (a) Initial doubly tapered simply supported beam; (b) large deformation configuration of the member.

$$M_x = \frac{w_1}{6L_o}\left[x^3(1-m) + 3mL_o x^2 - L_o^2 x(1+2m)\right] \tag{2.113}$$

$$I_x = I_B f(x) \tag{2.114}$$

where

$$I_B = \frac{bh^3}{12} \tag{2.115}$$

$$f(x) = \left[\frac{L_o + (n-1)(L_o - x)}{L_o}\right]^3 \tag{2.116}$$

and b in Equation 2.115 is the constant width of the member. Since M_x and I_x are functions of the large deformation, x_o in Equations 2.113 and 2.114 was assumed to be equal to (x + Δ). This should give a reasonable solution to the problem.

By substituting Equations 2.113 and 2.114 into the Euler-Bernoulli equation, we obtain

$$\frac{y''}{\left[1+(y')^2\right]^{3/2}} = \frac{w_1 L_o^2}{6EI_B}\left\{\frac{x^3(1-m)+3mL_o x - L_o^2 x(1+2m)}{\left[L_o+(n-1)(L_o-x)\right]^3}\right\} \tag{2.117}$$

By integrating both sides of Equation 2.117 and applying the boundary condition $y'(x_D) = 0$, where x_D is the distance from the support A to the point D of zero rotation, in order to determine the constant of integration C, the expression for $y'(x)$ at any $0 \le x \le L_o$, may be written as follows:

$$y'(x) = \frac{Q(x)}{\left\{1-[Q(x)]^2\right\}^{1/2}} \tag{2.118}$$

where

$$\begin{aligned} Q(x) = {} & \frac{w_1 L_o^2}{12EI_B(n-1)^4\left[L_o+(n-1)(L_o-x)\right]^2}\Big\{L_o^3 n(n-1)^2(1+2m) \\ & -L_o^2 x\left[2(n-1)^3(1+2m)+6n(n-1)(n-m)\right] \\ & +L_o x^2\left[9m(n-1)^3+9n(n-1)^2(1-m)\right]-2x^3(n-1)^3(1-m) \\ & -6L_o\left[L_o+(n-1)(L_o-x)\right]^2 \ln\left[L_o+(n-1)(L_o-x)\right](n-m)\Big\}+C \end{aligned} \tag{2.119}$$

and

$$\begin{aligned} C = {} & \frac{w_1 L_o^2}{12EI_B(n-1)^4\left[L_o+(n-1)(L_o-x_D)\right]^2}\Big\{L_o^3 n(n-1)^2(1+2m) \\ & -L_o^2 x_D\left[2(n-1)^3(1+2m)+6n(n-1)(n-m)\right] \\ & +L_o x_D^2\left[9m(n-1)^3+9n(n-1)^2(1-m)\right]-2x_D^3(n-1)^3(1-m) \\ & -6L_o\left[L_o+(n-1)(L_o-x_D)\right]^2 \ln\left[L_o+(n-1)(L_o-x_D)\right](n-m)\Big\} \end{aligned} \tag{2.120}$$

TABLE 2.13
Variation of Δ, θ_A, θ_B, δ_D, and x_D for Various Values of w_1 and for m = n = 2

w_1 (kip/in.)	Δ (in.)	θ_A (°)	θ_B (°)	δ_D (in.)	x_D (in.)
0.0025	77.86	27.14	39.26	170.69	517.36
0.0050	181.20	41.63	59.38	252.10	468.63
0.0075	257.65	49.95	69.89	293.60	431.55
0.0100	313.85	55.49	76.10	318.60	403.42
0.0125	356.91	59.54	80.05	335.83	381.31

Note: 1 in. = 0.0254 m, 1 kip/in. = 175,126.8 N/m.

Note that n in the above equations is the taper parameter and m is the loading parameter, as shown in Figure 2.11.

The values of x_D and L_o in Equations 2.118, 2.119, and 2.120 are not known. However, they can be determined by using a trial and error procedure as in the preceding sections, i.e., assume a value for x_D and use Equation 2.94 to determine $L_o = (L - \Delta)$ by assuming values of Δ and applying a trial and error procedure until the correct length L of the member is obtained. Simpson's rule may be used to carry out the required integrations. When L_o is determined, check if the vertical displacement condition at one of the end supports of the member is satisfied. If not, the procedure may be repeated with new values of x_D until this condition is satisfied. With known x_D and L_o, the values of y' at any $0 \le x \le L_o$ may be determined from Equation 2.118.

The large vertical displacement y may now be determined by using one of two ways: (1) by integrating Equation 2.118 once and applying the condition of zero deflection at one of the end supports to determine the constant of integration and (2) by using pseudolinear analysis as discussed in the preceding sections. Both procedures yield accurate results, but pseudolinear analysis would be easier to use.

Numerical results are obtained here by assuming that m = n = 2, $w_1 = 0.005$ kips/in. (875.634 N/m), L = 1000 in. (25.4 m), and EI_B = 75,000 kip-in^2 (215,224 Nm^2). The above methodology yields $\theta_A = 41.63°$, $\Delta = 181.20$ in. (4.6025 m), and $y_{max} = \delta_D = 252.10$ in. (6.4033 m). In Table 2.13 the variation of θ_A, θ_B, Δ, and δ_D, for various values of the load w_1 and m = n = 2, is shown. The variation of the same quantities for various values of the load parameter n, when m = 2 and w_1 = 0.005 kips/in. (875.634 N/m), is shown in Table 2.14.

2.12.1 EQUIVALENT SIMPLIFIED NONLINEAR SYSTEMS OF CONSTANT STIFFNESS

An easier solution to the problem may be obtained by using simplified equivalent nonlinear systems of constant stiffness. The alternate approach discussed in Section 1.9 might be the most convenient to use. It should be noted here that equivalent nonlinear systems are extremely convenient to use for complicated loading conditions and arbitrary moments of inertia variations along the length of the member.

The procedure may be initiated by using Equation 1.69 and writing the expression for M_e for any point $0 \le x \le L_o$. Figure 2.11b may be used for this purpose. Values of M_e in terms of L_o at interval of $0.1L_o$ are shown in Table 2.15. It is assumed here that m = n = 2, L = 1000 in. (25.4 m), and $w_1 = 0.010$ kips/in. (1751.268 N/m). The simplified equivalent nonlinear system in Figure 2.12 is obtained by plotting the M_e diagram by using the values from Table 2.15 and approximating its shape with two straight line segments. The equivalent nonlinear system in Figure 2.12 has a constant stiffness EI_B, and it is loaded with a concentrated load $P_e = 4.386L_o \times 10^{-3}$ located as shown in the figure.

The solution of the simplified nonlinear system in Figure 2.12 may be obtained as discussed in Section 2.11, i.e., the expressions for the moment due to P_e at the intervals of $0 \le x \le 0.76L_o$, and $0.76L_o \le x \le L_o$ may be obtained by using statics. By using the Euler-Bernoulli equation, the two second order nonlinear differential equations for the two intervals may be formed. By integrating each nonlinear equation once, the expression of y' for each interval may be obtained. The two constants of integration may be determined by using the boundary condition of rotation $y' = y'_A$ at $x = 0$ and the continuity condition of the rotation at the point of application of the load P_e. On this basis, the equations for y' will have the unknowns y'_A and L_o. The value of y'_A may be determined by satisfying the boundary condition of zero vertical deflection at one of the end supports of the member, and L_o may be determined from the equation

$$L = \int_0^{0.76L_o} \left[1 + (y')^2\right]^{1/2} dx + \int_{0.76L_o}^{L_o} \left[1 + (y')^2\right]^{1/2} dx \tag{2.121}$$

by using a trial and error procedure as discussed earlier.

When the value of L_o is determined, y' at any $0 \le x \le L_o$ may be determined from the equations for y' at the intervals $0 \le x' \le 0.76L_o$ and $0.76L_o \le x \le L_o$. With known y', the pseudolinear analysis can be carried out in the usual way. The results obtained are $\theta_A = 54.87°$, $\theta_B = 72.09°$, $x_D = 410.31$ in. (10.4219 m), $L_o = 661.50$ in. (16.8021 m), $y_{max} = \delta_D = 317.85$ in. (8.0734 m), and $\Delta = 307.75$ in. (7.8169 m), which are in close agreement with the ones obtained in Table 2.13.

PROBLEMS

2.1 For the uniform cantilever beam loaded as shown in Figure 2.1a, determine the rotation θ_B and vertical displacement δ_B at the free end of the member by using a pseudolinear system. Assume L = 1000 in. (25.4 m),

TABLE 2.14
Variation of Δ, θ_A, θ_B, δ_D, and x_D for Various Values of the Taper Parameter n and with m = 2 and w_1 = 0.005 kips/in.

n	Δ (in.)	θ_A (°)	θ_B (°)	δ_D (in.)	x_D (in.)
1.0	338.56	73.52	71.08	350.49	298.62
1.5	267.63	55.67	65.59	301.55	398.60
2.0	181.20	41.63	59.38	252.10	468.63
3.0	77.62	24.11	44.51	167.16	553.67

Note: 1 in. = 0.0254 m, 1 kip/in. = 175, 126.8 N/m.

TABLE 2.15
Values of M_e in Terms of L_o

xL_o (in.)	f(x)	$M_eL_o^2(10)^{-5}$ (kip-in.)
0.0	8.0	0.0
0.1	6.859	10.716
0.2	5.832	21.948
0.3	4.913	33.483
0.4	4.096	44.922
0.5	3.375	55.556
0.6	2.744	64.140
0.7	2.197	68.503
0.8	1.728	64.815
0.9	1.331	46.206
1.0	1.0	0.0

Note: 1 in. = 0.0254 m, 1 kip-in. = 112.9848 Nm.

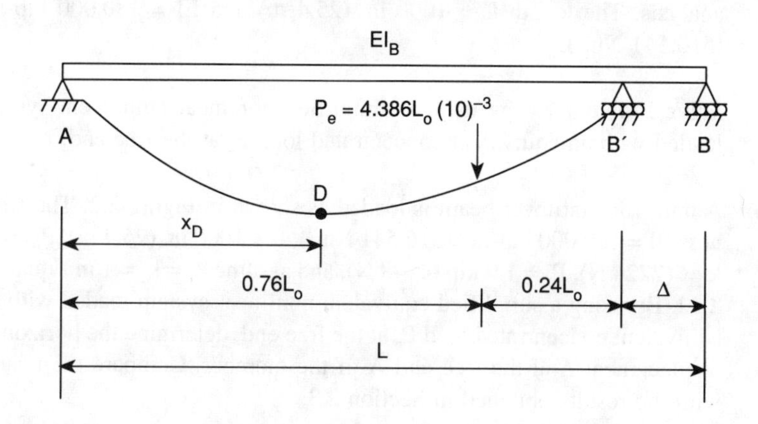

FIGURE 2.12. Simplified equivalent nonlinear system of constant stiffness EI_B loaded with one concentrated load.

$EI = 180 \times 10^3$ kip-in.^{2} (516,541 Nm^2), $w_o = 2$ lb/in. (350.2536 N/m), and $\Delta(x)$ = constant = Δ.
Answer: $\theta_B = 66.30°$, $\delta_B = 705.72$ in. (17.9253 m).

2.2 Solve Problem 2.1 by assuming $\Delta(x) = \Delta x/L_o$ and $w_o = 1.5$ lb/in. (262.6902 N/m). Determine also the horizontal displacement Δ and vertical displacement y of the member at $x = L_o/2$.
Answer: $\theta_B = 51.15°$, $\delta_B = 617.55$ in. (15.6858 m).

2.3 Solve Problem 2.1 by integrating the Euler-Bernoulli equation twice, and satisfy appropriate boundary conditions for the evaluation of the constants of integration. Assume $\Delta(x) = \Delta\sin(\pi x/2L_o)$.

2.4 Solve Problem 2.1 by using a simplified nonlinear equivalent system and pseudolinear analysis to solve the simplified nonlinear system.

2.5 A uniform cantilever beam loaded by two concentrated loads P_1 and P_2 is shown in Figure 2.13. The cross-sectional dimensions are shown in the same figure. For the indicated loading conditions, determine the horizontal displacement Δ_C, vertical displacement δ_C, and rotation θ_C at the free end C of the beam by using a simplified nonlinear equivalent system loaded with an equivalent load P_e at the free end. The modulus of elasticity E = 30,000 ksi.
Answer: $\Delta_C = 28.70$ in., $\delta_C = 64.47$ in., $\theta_C = 59.23°$.

2.6 The uniform cantilever beam in Figure 2.1b is loaded with a concentrated load P at the free end B and a distributed load $w_oL = 3P$ so that $PL^2/EI = 1.642$. Verify the results shown in Table 2.3 by using pseudolinear analysis. The length L = 1000 in. (25.4 m) and EI = 180,000 kip-in^2 (516,541 Nm^2).

2.7 Solve Problem 2.6 by using a simplified nonlinear equivalent system loaded with an equivalent concentrated load P_e at the free end.

2.8 A uniform cantilever beam is loaded as shown in Figure 2.2. The stiffness EI = 180,000 kip-in.^{2} (516,541 Nm^2), L = 1000 in. (25.4 m), $P_1 = 0.5$ kips (2224 N), $P_2 = 1.0$ kip (4448 N), and assume $k_1 = k_2 = 1$ in Equation 2.22. By using a simplified equivalent nonlinear system loaded with an equivalent concentrated load P_e at the free end, determine the horizontal displacement Δ at the free end A of the member. Compare the results with the results obtained in Section 2.3.

2.9 The doubly tapered cantilever beam in Figure 1.3 is loaded with a uniformly distributed load $w_o = 10.0$ lb/in. (1751.268 N/m) as shown. The stiffness $EI_B = 180{,}000$ kip-in.^{2} (516,541 Nm^2), length L = 1000 in.

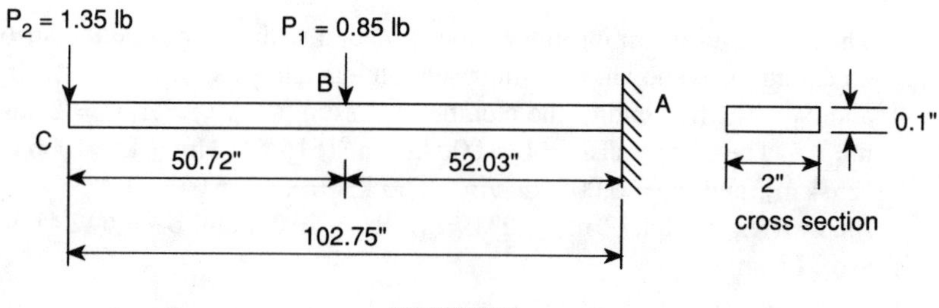

FIGURE 2.13.

(25.4 m), and taper n = 1.5. Determine the horizontal displacement Δ_B, rotation θ_B, and vertical displacement δ_B at the free end B of the beam by using pseudolinear analysis and $\Delta(x) = \Delta$.
Answer: Δ_B = 557.08 in. (14.1498 m), θ_B = 84.08°, and δ_B = 799.36 in. (20.3037 m).

2.10 Solve Problem 2.9 for w_o = 2.0 lb/in. (350.2536 N/m) and $\Delta(x) = \Delta$
Answer: Δ_B = 121.20 in. (3.0785 m), θ_B = 37.99°, and δ_B = 435.41 in. (11.0594 m).

2.11 Solve Problem 2.9 by assuming that $\Delta(x) = \Delta(x/L_o)$ and compare results.

2.12 Solve Problem 2.9 by using a simplified equivalent nonlinear system of constant stiffness EI_B that is loaded with a concentrated equivalent load P_e at the free end. Compare the results.

2.13 For the tapered cantilever beam in Problem 2.9, determine the horizontal and vertical displacements and the rotation at x = L/2.

2.14 Solve the Problem in Figure 2.3a by using a simplified equivalent nonlinear system of constant stiffness EI_B that is loaded with an equivalent concentrated load P_e at the free end. Compare the results with the results obtained in Section 2.5. The required information regarding this problem is given in the same section.

2.15 Solve Problem 2.14 (Figure 2.3a) by integrating the Euler-Bernoulli equation twice and determining the two constants of integration by applying appropriate boundary conditions. Carry out the required integrations by using Simpson's rule. Compare the results with the ones obtained in Section 2.5 by considering the displacements and rotation of the free end of the member. Assume $\Delta(x) = \Delta$.

2.16 Solve the problem in Figure 2.4a by integrating the Euler-Bernoulli equation twice and determining the two constants of integration by using appropriate boundary conditions. Carry out the required integrations by using Simpson's rule. Compare the results with the results obtained in Section 2.6.

2.17 The cantilever beam in Figure 2.6a is loaded with a trapezoidal distributed loading as shown. By using pseudolinear analysis, determine Δ_B, θ_B, and δ_B at the free end of the member by assuming $\Delta(x) = \Delta$, m = 2, and n = 1.5. The stiffness EI_B = 180,000 kip-in.^{2} (516,541 Nm^2), L = 1000 in. (25.4 m), and w_1 = 0.005 kips/in. (87.6 kN/m).
Answer: Δ_B = 284.72 in. (7.2319 m), θ_B = 56.3°, and δ_B = 642.41 in. (16.3172 m).

2.18 Solve Problem 2.17 by assuming $\Delta(x) = \Delta \sin(\pi x/2L_o)$, and compare the results.

2.19 Solve Problem 2.17 by integrating the Euler-Bernoulli equation twice and determining the two constants of integration by applying appropriate boundary conditions. Use Simpson's rule to carry out the required integrations. Compare the results.

2.20 The uniform simply supported beam in Figure 2.18 is loaded with a uniformly distributed load w as shown. Determine the rotation θ_B of the end support B, the horizontal displacement Δ_A of support A, and the maximum vertical displacement δ_D by using pseudolinear analysis with $\Delta(x) = \Delta$. Length L = 1000 in. (25.4 m), EI = 75×10^3 kip-in.^{2} (215,224 Nm^2), and w = 10 lb/in. (1751.268 N/m).
Answer: θ_B = 76.63°, Δ_A = 440.53 in. (11.1895 m), and δ_D = 367.65 in. (9.3383 m).

2.21 Solve Problem 2.20 by assuming $\Delta(x) = \Delta(x/L_o)^{1/2}$ and compare the results.

2.22 A tapered simply supported beam is loaded with a uniformly distributed load w as shown in Figure 2.9. Determine the horizontal displacement Δ_B of the end support B, the rotation θ_A of the end support A, and the maximum vertical displacement δ_B by using pseudolinear analysis and $\Delta(x) = \Delta$. The stiffness EI_A = 75x10^3 kip-in.^{2} (215,224 Nm^2), L = 1000 in. (25.4 m), w = 5.0 lb/in. (875.634 N/m), and taper n = 1.5.
Answer: Δ_B = 194.21 in. (4.9329 m), θ_A = 57.73°, and δ_D = 262.04 in. (6.6558 m).

2.23 Solve Problem 2.22 by assuming $\Delta(x) = \Delta(x/L_o)$ and compare the results.

2.24 Solve Problem 2.22 by using a simplified nonlinear equivalent system of constant stiffness EI_A loaded with a concentrated load P_e. Compare the results.

2.25 The tapered simply supported beam in Figure 2.11 is loaded with a trapezoidal distributed loading as shown. Determine the value and

location of the maximum vertical displacement δ_D by using pseudolinear analysis with $\Delta(x) = \Delta$. $EI_B = 75{,}000$ kip-in.^{2} (215,224 Nm^2), L = 1000 in. (25.4 m), $w_1 = 0.0075$ kips/in. (1313.451 N/m), and m = n = 2.
Answer: $x_D = 431.55$ in. (10.9614 m), $\delta_D = 293.60$ in. (7.4574 m).

2.26 Solve Problem 2.25 by integrating the Euler-Bernoulli equation twice and determining the two constants of integration by appropriate boundary conditions. Compare the results.

2.27 A tapered cantilever beam is loaded as shown in Figure 2.14. By applying pseudolinear analysis, determine the vertical displacements δ_B and δ_C at points B and C, respectively. Note the deformation configuration shown in Figure 2.14. The stiffness $EI_C = 180{,}000$ kip-in.^{2} (516,541 Nm^2).
Answer: $\delta_B = 238.36$ in. (6.0543 m), $\delta_C = 746.41$ in. (18.9588 m).

2.28 A stepped cantilever beam is loaded by a uniformly distributed load w_o = 5.0 lb/in. (875.634 N/m), as shown in Figure 2.15. Determine the horizontal displacement Δ, rotation θ, and vertical displacement δ at the free end of the beam by using pseudolinear analysis. Assume that L = 1000 in. (25.4 m), $L_1 = L_2 = 500$ in. (12.7 m), EI_1 at the free end is 180,000 kip-in.^{2} (516,541 Nm^2), and n = 2.
Answer: $\Delta = 155.25$ in. (3.9434 m), $\theta = 49.84°$, and $\delta = 460.877$ in. (11.7063 m).

2.29 Solve Problem 2.28 by using a simplified equivalent nonlinear system of constant stiffness EI loaded with an equivalent concentrated load P_e at the free end. Compare the results.

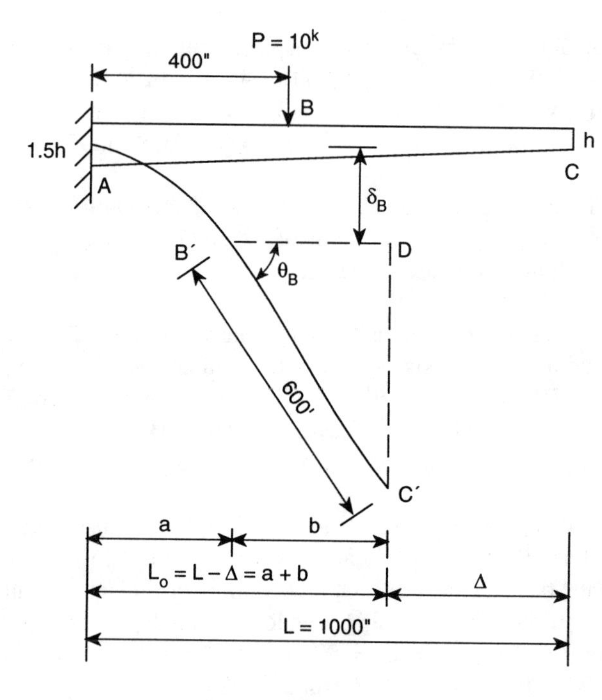

FIGURE 2.14.

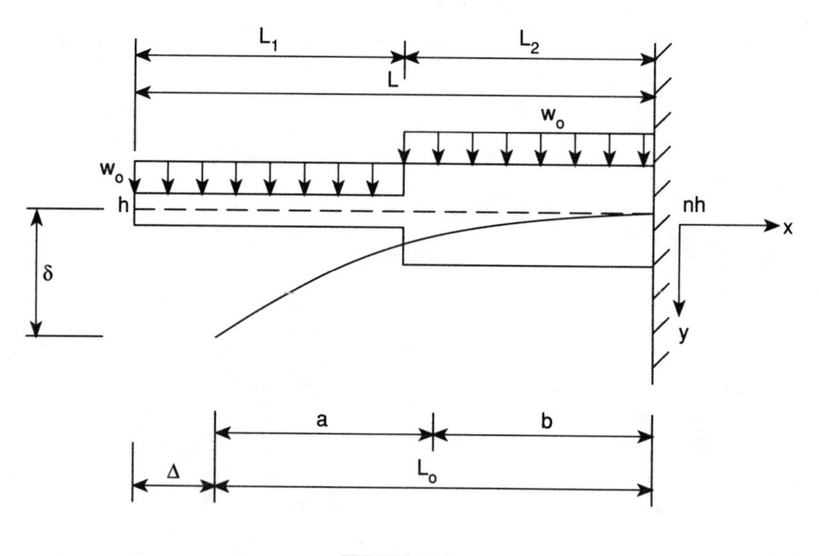

FIGURE 2.15.

Chapter 3

STATICALLY INDETERMINATE FLEXIBLE BARS OF UNIFORM AND VARIABLE THICKNESS

3.1 INTRODUCTION

In the preceding two chapters, the static response of flexible statically determinate beams of either uniform or variable stiffness was investigated in detail. Various cases of stiffness variations and loading conditions along the length of the flexible member were examined. The initial complex mathematical model was simplified by using (1) pseudolinear equivalent systems of constant stiffness solved by linear analysis or (2) simplified nonlinear equivalent systems of constant stiffness that can be solved by either applying pseudolinear analysis or utilizing existing nonlinear methodologies suitable for such types of problems. In both situations, however, the main objective was to simplify the initially complicated problem in order to obtain a reasonably accurate solution with a large reduction in actual effort. The analysis in the preceding two chapters clearly demonstrated the benefits of such an approach. Such benefits become increasingly larger as the initial problem becomes progressively more complex.

In the preceding two chapters it was also demonstrated that the expressions for the bending moment M_x and moment of inertia I_x are, in general, nonlinear functions of the large deformation, and consequently, the Euler-Bernoulli equation becomes an integral equation that is difficult to solve. Therefore, certain simplifications were introduced in order to reduce the complexity of the problem. When the flexible member is statically indeterminate, the additional constraints introduced in the analysis should be taken into consideration. Such constraints usually appear in the form of redundants that satisfy specific boundary conditions with regard to the global deformation of the flexible member. The purpose in this chapter is to extend the theories developed in the preceding two chapters so that they can be used for the solution of statically indeterminate flexible members. The beam can have any arbitrary stiffness variation along its length, and it can be subjected to various types of combined loadings.

3.2 SOLUTION METHODOLOGY

The solution of statically indeterminate flexible members will be illustrated here by considering the uniform single span beam in Figure 3.1, which is loaded with a concentrated load P at distance L_1 from the fixed support of the member. The right support of the member is a roller, which permits this end of the beam to move horizontally. Therefore, there are no axial restraints applied to the member during bending. The large deformation configuration of the beam is also

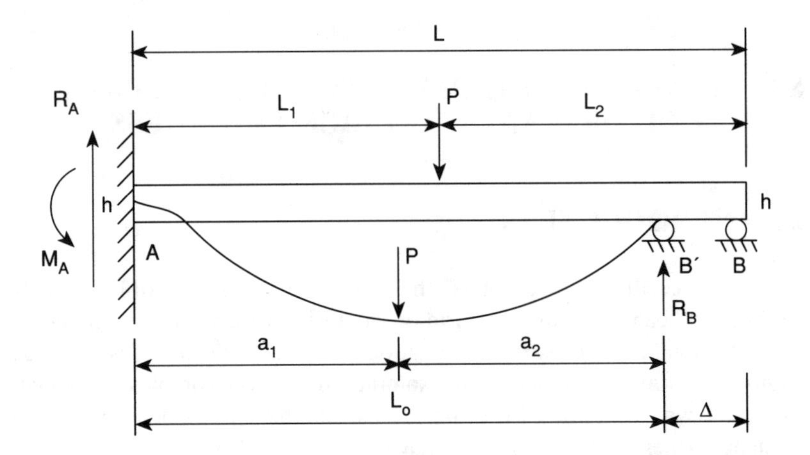

FIGURE3.1. Statically indeterminate beam of uniform stiffness with a concentrated load P.

shown in the same figure. This is a very realistic case of a statically indeterminate beam that is subjected to large deformations. If both ends of the member were fixed, then severe axial restraints would be introduced, and the member would probably fail from large axial stresses long before the deformations become large enough to necessitate large deflection theory.

The modulus of elasticity E and the moment of inertia I of the member are both constant. The governing nonlinear differential equations for the large static deformation of the member may be obtained by using the Euler-Bernoulli equation as it was done in the preceding chapters. However, since the beam is statically indeterminate, the redundant reactions would have to be determined first using appropriate boundary conditions. In this problem, the vertical reaction R_B of the roller support will be taken as the redundant reaction. By utilizing statics, we may write the following equations:

$$R_A + R_B = P \tag{3.1}$$

$$M_A = Pa_1 - R_B L_o \tag{3.2}$$

We also have

$$I = I_B = \frac{bh^3}{12} \tag{3.3}$$

where b is the constant width of the member.

The bending moment M_x at any $0 \le x \le L_o$ may be written as follows:

$$M_x = -(Pa_1 - R_B L_o) + (P - R_B)x \quad 0 \le x \le a_1 \tag{3.4}$$

$$M_x = R_B(L_o - x) \qquad a_1 \le x \le (a_1 + a_2) \tag{3.5}$$

where

$$L_o = (a_1 + a_2)$$

By substituting Equations 3.4 and 3.5 into the Euler-Bernoulli equation we obtain

$$\frac{y_1''}{\left[1+(y_1')^2\right]^{3/2}} = \frac{1}{EI_B}\left[P(a_1 - x) - R_B(L_o - x)\right] \qquad 0 \le x \le a_1 \tag{3.6}$$

$$\frac{y_2''}{\left[1+(y_2')^2\right]^{3/2}} = -\frac{1}{EI_B}\left[R_B(L_o - x)\right] \qquad a_1 \le x \le (a_1 + a_2) \tag{3.7}$$

By integrating Equations 3.6 and 3.7 once we obtain

$$\frac{y_1'}{\left[1+(y_1')^2\right]^{1/2}} = \frac{1}{2EI_B}\left[P(2a_1x - x^2) - R_B(2L_ox - x^2)\right] + C_1 \qquad 0 \le x \le a_1 \tag{3.8}$$

$$\frac{y_2'}{\left[1+(y_2')^2\right]^{1/2}} = -\frac{1}{2EI_B}\left[R_B(2L_ox - x^2)\right] + C_2 \qquad a_1 \le x \le (a_1 + a_2) \tag{3.9}$$

where C_1 and C_2 are the constants of integration. They can be determined from the boundary conditions

$$y_1'(0) = 0 \tag{3.10}$$

$$y_1'(a_1) = y_2'(a_1) \tag{3.11}$$

Therefore, by utilizing the above two boundary conditions we find $C_1 = 0$ and $C_2 = Pa_1/2EI_B$. On this basis, Equations 3.8 and 3.9 yield

$$\frac{y_1'}{\left[1+(y_1')^2\right]^{1/2}} = G_1(x) \qquad 0 \le x \le a_1 \tag{3.12}$$

$$\frac{y_2'}{\left[1+(y_2')^2\right]^{1/2}} = G_2(x) \qquad a_1 \le x \le (a_1 + a_2) \tag{3.13}$$

where

$$G_1(x) = \frac{1}{2EI_B}\left[P(2a_1x - x^2) - R_B(2L_ox - x^2)\right] \tag{3.14}$$

$$G_2(x) = -\frac{1}{2EI_B}\left[R_B(2L_ox - x^2) - Pa_1^2\right] \tag{3.15}$$

By using Equation 1.66 in conjunction with Equations 3.12 and 3.13 we may write the following equations for L, L_1, and L_2:

$$L = \int_0^{a_1} \frac{1}{\left\{1-\left[G_1(x)\right]^2\right\}^{1/2}}dx + \int_{a_1}^{(a_1+a_2)} \frac{1}{\left\{1-\left[G_2(x)\right]^2\right\}^{1/2}}dx \tag{3.16}$$

$$L_1 = \int_0^{a_1} \frac{1}{\left\{1-\left[G_1(X)\right]^2\right\}^{1/2}}dx \tag{3.17}$$

$$L_2 = \int_{a_1}^{(a_1+a_2)} \frac{1}{\left\{1-\left[G_2(x)\right]^2\right\}^{1/2}}dx \tag{3.18}$$

By examining Equations 3.14 and 3.15, we observe that y_1' and y_2' can be determined if the horizontal displacement Δ of support B, and the reaction R_B at the same support, are known. Note that $L_o = (L - \Delta)$.

One way to determine Δ and R_B is to use a trial and error precedure. For example, we may start by assuming a value for the redundant reaction R_B. Then, since $L_o = (L - \Delta) = (a_1 + a_2)$, a trial and error procedure may be initiated in order to determine a_1 and a_2 by satisfying Equations 3.16, 3.17, and 3.18. When this is accomplished, we check if the assumed redundant reaction R_B satisfies the boundary condition of zero displacement at support B. If this condition is not satisfied, then increase or decrease R_B until it is satisfied. The procedure converges fairly fast to the correct value of R_B. It should also be noted at this point that $L_1 = a_1 + \Delta_1$, and $L_2 = a_2 + \Delta_2$, where Δ_1 and Δ_2 are the horizontal displacements of L_1 and L_2, respectively, during the large deformation configuration. Therefore, when a_1 and a_2 are determined using the above trial and error procedure, it may be implied that Δ_1 and Δ_2 are also determined.

With L_o, a_1, and R_B known, the rotations y_1' and y_2' can be determined from Equations 3.12 and 3.13, respectively, and the moment M_x at any $0 \leq x \leq L_o$ may be determined from Equations 3.4 and 3.5. Pseudolinear analysis may now be initiated as discussed in the preceding two chapters, because the moment diagram M_e' of the pseudolinear equivalent system of constant stiffness EI_B may be determined using Equations 1.33 and 1.34. The approximation of the shape of the M_e' diagram with a few straight line segments leads to an equivalent pseudolinear system of constant stiffness EI_B that is loaded by a few concentrated loads. The pseudolinear system may be solved by linear analysis, as it was done in the preceding two chapters.

The deflection y at any $0 \leq x \leq L_o$ could also be obtained by integrating Equations 3.12 and 3.13. The two constants of integration may be determined by satisfying the displacement conditions $y_1(a_1) = y_2(a_1)$ and $y_1(o) = 0$. This procedure would be more tedious compared to pseudolinear analysis.

The required integrations may be carried out by Simpson's rule. This has proven to be reasonably accurate for practical engineering applications.

In the remaining parts of this chapter, the above methodology is applied to statically indeterminate, single span beam problems of uniform and variable stiffness. Various loading conditions are examined, and simplified methods of analysis are introduced in order to reduce the problem complexity.

3.3 UNIFORM SINGLE SPAN BEAMS SUBJECTED TO DISTRIBUTED LOADINGS

Consider the uniform single span beam in Figure 3.2 that is subjected to a uniformly distributed load w_o over its entire length. The beam is fixed at the end A and it is supported by a roller at the end B. The large deformation configuration of the member is shown in the same figure. The modulus of elasticity E and moment of inertia I are assumed to be constant. Since the member is statically indeterminate, the reaction R_B at the end B is taken as the redundant. By satisfying static equilibrium in the vertical direction we write the equation

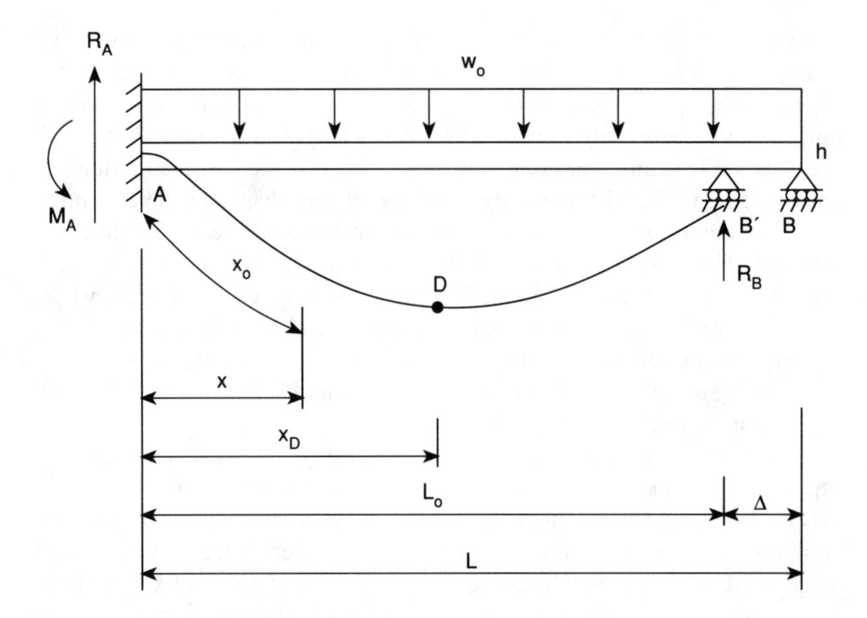

FIGURE 3.2. Uniform statically indeterminate beam subjected to a uniformly distributed load w_o.

$$R_A + R_B = Lw_o \tag{3.19}$$

The bending moment at any $0 \le x \le L_o$ is given by the expression

$$M_x = \frac{w_o}{2}\left[2Lx - LL_o - xx_o\right] + R_B\left[L_o - x\right] \tag{3.20}$$

By utilizing Equation 3.20, the Euler-Bernoulli equation yields

$$\frac{y''}{\left[1+(y')^2\right]^{3/2}} = \frac{w_o}{2EI}\left[LL_o - 2Lx + xx_o\right] - \frac{R_B}{EI}\left[L_o - x\right] \tag{3.21}$$

where

$$x_o(x) = \int_0^x \left\{1 + \left[y'(\xi)\right]^2\right\}^{1/2} d\xi \tag{3.22}$$

and ξ is a dummy variable that depends on x. By substituting Equation 3.22 into Equation 3.21 we obtain

$$\frac{y''}{\left[1+(y')^2\right]^{3/2}} = \frac{w_o}{2EI}\left\{LL_o - 2Lx + x\left[\int_0^x \left\{1+[y'(\xi)]^2\right\}^{1/2} d\xi\right]\right\} - \frac{R_B}{EI}\left[L_o - x\right] \quad (3.23)$$

Equation 3.23 is a nonlinear integral differential equation that is very difficult to solve. Its solution, however, may be simplified again in the same way as in the preceding two chapters. That is, we replace the integral that represents $x_o(x)$ by a function that contains the horizontal displacement $\Delta(x)$. On this basis we may write

$$x_o(x) = x + \Delta(x) \quad (3.24)$$

which is the same as Equation 1.42 in Section 1.7. Various expressions for $\Delta(x)$ are suggested in the same section (see, e.g., Equations 1.49 to 1.52).

A popular expression for $x_o(x)$ that was used extensively in the two preceding chapters and provided very accurate results, is Equation 1.49, i.e.,

$$x_o(x) = x + \Delta \quad (3.25)$$

where Δ is the horizontal movement of the roller support of the member. Thus, with Equation 3.25 in mind, the Euler-Bernoulli equation yields

$$\frac{y''}{\left[1+(y')^2\right]^{3/2}} = \frac{w_o}{2EI}\left[LL_o - 2Lx + x(x+\Delta)\right] - \frac{R_B}{EI}\left[L_o - x\right] \quad (3.26)$$

which is a much easier differential equation to solve.

By integrating Equation 3.26 once and determining the constant of integration by satisfying the boundary condition of zero rotation at x = 0, that is at the fixed end, we obtain

$$y'(x) = \frac{G(x)}{\left\{1-[G(x)]^2\right\}^{1/2}} \quad (3.27)$$

where

$$G(x) = \frac{w_o}{12EI}\left[6LL_o x - 6Lx^2 + 3\Delta x^2 + 2x^3\right] - \frac{R_B}{2EI}\left[2L_o x - x^2\right] \quad (3.28)$$

By using Equation 1.66 we can also write the equation

$$L = \int_0^{(L-\Delta)} \frac{1}{\left\{1-[G(x)]^2\right\}^{1/2}} dx \tag{3.29}$$

The vertical displacement y(x) of the member at any $0 \le x \le L_o$ may be obtained from the equation

$$y(x) = \int_0^x \frac{G(x)}{\left\{1-[G(x)]^2\right\}^{1/2}} dx \tag{3.30}$$

Equations 3.27, 3.29, and 3.30 contain the unknown horizontal displacement Δ and the unknown redundant reaction R_B of the support B of the member. These two quantities may be determined using Equations 3.29 and 3.30 and applying a trial and error procedure as discussed in the preceding section.

Let it be assumed that the flexible beam in Figure 3.2 has a stiffness EI = 75,000 kip-in.^{2} (215,224 Nm^2) and length L = 1000 in. (25.4 m). For this specific beam problem, the methodology discussed above and in the preceding section was used to determine the horizontal displacement Δ of support B, the redundant reaction R_B and rotation θ_B at the same support, the maximum displacement δ_D, and the location x_D of the maximum displacement for various values of the uniformly distributed load w_o. The results are shown in Table 3.1. In the same table, the values of the same quantities were obtained by assuming that $\Delta(x) = \Delta$, $\Delta(x) = \Delta x/L_o$, $\Delta(x) = \Delta[x/L_o]^{1/2}$, and $\Delta(x) = \Delta\sin(\pi x/2L_o)$, in order to compare results. These results are in close agreement, which indicates that the deformation of the flexible member, for all practical purposes, is not very sensitive to the variation assumed for the function $\Delta(x)$.

3.4 VARIABLE THICKNESS SINGLE SPAN BEAMS SUBJECTED TO TRAPEZOIDAL LOADINGS

The case of flexible variable thickness single span beams with trapezoidal loading will be investigated here by considering the doubly tapered beam in Figure 3.3. The beam is fixed at support A and is supported by a roller at the end B. The large deformation configuration of the member is shown in the same figure.

By following the methodologies developed in the preceding two chapters, the variable moment of inertia I_x at any $0 \le x \le L_o$, may be expressed as

$$I_x = I_B\left[1+(n-1)\frac{x}{L_o}\right]^3 \tag{3.31}$$

TABLE 3.1
Variation of Δ, θ_B, δ_D, x_D, and R_B with Uniformly Distributed Load w_o for Various Cases of Horizontal Displacement Function Δ(x)

$\Delta(x) = \Delta$

w_o (lb/in.)	Δ	θ_B (°)	δ_D (in.)	x_D (in.)	R_B (kips)
2.0	42.80	28.56	128.806	558.05	0.7592
5.0	150.55	53.12	232.764	558.043	1.9584
10.0	274.59	70.09	303.079	440.396	4.059

$\Delta(x) = \Delta x/L_o$

w_o (lb/in.)	Δ	θ_B (°)	δ_D (in.)	x_D (in.)	R_B (kips)
2.0	46.35	29.75	133.39	555.98	0.7485
5.0	184.99	58.59	255.212	486.724	1.8594
10.0	357.682	78.18	337.469	395.347	3.6892

$\Delta(x) = \Delta[x/L_o\}^{1/2}$

w_o (lb/in.)	Δ	θ_B (°)	δ_D (in.)	x_D (in.)	R_B (kips)
2.0	44.65	29.20	130.896	556.30	0.7528
5.0	167.21	55.70	244.128	494.594	1.90107
10.0	315.808	74.07	321.00	415.989	3.8520

$\Delta(x) = \Delta\sin(\pi x/2L_o)$

w_o (lb/in.)	Δ	θ_B (°)	δ_D (in.)	x_D (in.)	R_B (kips)
2.0	44.14	28.87	130.260	555.259	0.7523
5.0	161.65	53.90	241.152	491.608	1.89727
10.0	302.10	70.50	317.195	411.27	3.8410

Note: 1 in. = 0.0254 m, 1 lb/in. = 175.1268 N/m, 1 kip = 4448.222 N.

where

$$I_B = \frac{bh^3}{12} \tag{3.32}$$

is the moment of inertia at the end B of the member, b is the constant width, and n is its taper parameter.

The bending moment M_x at any $0 \le x \le L_o$ may be written as

$$M_x = -\frac{w_1}{6L}\left[3Lxx_o + (m-1)xx_o^2\right] + R_B x \tag{3.33}$$

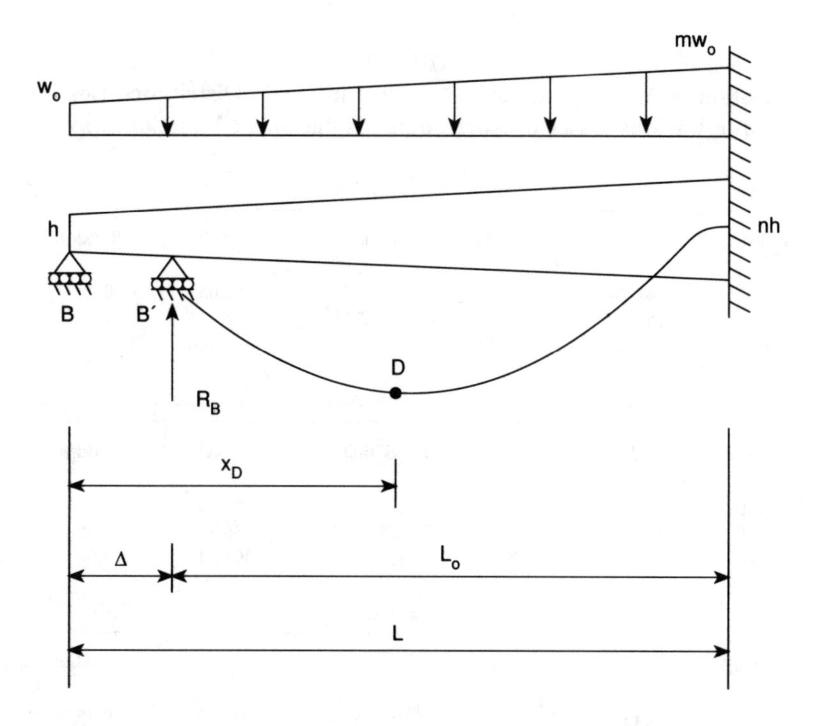

FIGURE 3.3. Statically indeterminate tapered single span beam with a trapezoidal loading.

where m is the trapezoidal load parameter. By employing the Euler-Bernoulli equation we find

$$\frac{y''}{\left[1+(y')^2\right]^{3/2}} = \frac{w_1 L_o^2}{6EI_B L}\left[\frac{3Lxx_o + (m-1)xx_o^2}{\{L_o + (n-1)x\}^3}\right] - \frac{R_B L_o^3}{EI_B}\left[\frac{x}{\{L_o + (n-1)x\}^3}\right] \tag{3.34}$$

If $x_o = x + \Delta(x)$ is substituted in Equation 3.34 we find

$$\frac{y''}{\left[1+(y')^2\right]^{3/2}} = \frac{w_1 L_o^2}{6EI_B L}\left\{\frac{3Lx[x+\Delta(x)] + (m-1)x[x+\Delta(x)]^2}{\{L_o + (n-1)x\}^3}\right\} - \frac{R_B L_o^3}{EI_B}\left[\frac{x}{\{L_o + (n-1)x\}^3}\right] \tag{3.35}$$

Equation 3.35 is written in terms of the horizontal displacement function $\Delta(x)$, which assumably takes any of the forms given by Equations 1.49 to 1.52. We assume here that $\Delta(x) = \text{constant} = \Delta$, and we integrate Equation 3.35 once. The constant of integration may be determined by satisfying the boundary condition of zero rotation at the fixed end. This procedure yields

$$\frac{y'}{\left[1+(y')^2\right]^{1/2}} = G(x) \tag{3.36}$$

where $G(x)$, in nondimensional form, is given by the expression

$$
\begin{aligned}
&G(\xi) = \\
&\frac{w_1 L_o^2}{12(n-1)^4 EI_B L}\Big[\Big((m-1)L_o^2\{-(n-1)^3\xi^3 - 3(n-1)^2\xi^2[(n-1)\xi+1] \\
&6(n-1)\xi[(n-1)\xi+1]^2 - 6[(n-1)\xi+1]^2 \ln[(n-1)L_o\xi + L_o]\} \\
&L_o[3L + 2(m-1)\Delta]\{-3(n-1)^3\xi^3 - 2(n-1)\xi^3 \\
&+2(n-1)[(n-1)\xi+1]^2 \ln[(n-1)\xi L_o + L_o]\} \\
&[3L\Delta + (m-1)\Delta^2]\{-2(n-1)^3\xi - (n-1)^2\}\Big) \\
&\frac{1}{[(n-1)\xi+1]^2} - \Big((m-1)L_o^2 \\
&\{-(n-1)^3 - 3n(n-1)^2 + 6n^2(n-1) - 6n^2 \ln[nL_o]\} \\
&+L_o[3L + 2(m-1)\Delta]\{-3(n-1)^3 - 2(n-1)^2 + 2n^2(n-1)\ln[nL_o]\} \\
&+[3L\Delta + (m-1)\Delta^2]\{-2(n-1)^3 - (n-1)^2\}\Big)\frac{1}{n^2}\Big] \\
&-\frac{R_B L_o^2}{2(n-1)^2 EI_B}\left[\frac{2(n-1)\xi + 1}{[(n-1)\xi+1]^2} - \frac{2(n-1)+1}{n^2}\right]
\end{aligned}
\tag{3.37}
$$

In Equation 3.37 the variable $\xi = x/L_o$. If the statically indeterminate beam is of uniform stiffness, that is $n = 1$, Equation 3.37 yields

$$G(\xi)=\frac{w_1L_o^2}{72EI_BL}\Big[\big((3L_o^2(m-1)\xi^4+4L_o[3L+2(m-1)\Delta]\xi^3$$

$$6\big[3L\Delta+2(m-1)\Delta^2\big]\xi^2\big)-\big(3L_o^2(m-1)$$

$$4L_o[3L+2(m-1)\Delta]+6\big[3L\Delta+(m-1)\Delta^2\big]\big)\Big] \tag{3.38}$$

$$-\frac{R_BL_o^2}{2EI_B}\big[\xi^2-1\big]$$

We also have the equation

$$L=(L-\Delta)\int_0^1\frac{1}{\left\{1-[G(\xi)]^2\right\}^{1/2}}d\xi \tag{3.39}$$

In the above equations, the horizontal displacement Δ and the redundant reaction R_B at the end B are not known. They can be determined, however, by trial and error procedure as discussed in the preceding sections. That is, we may assume a value for the redundant reaction R_B, and then use Equation 3.39 to determine Δ by a trial and error. When this is done, we check to find out if the displacement condition at the end B of the beam is satisfied. If not, we repeat the procedure with new values for R_B until the displacement condition is satisfied. With known Δ and R_B, the values of y' at any $0 \le x \le L_o$ may be determined from Equation 3.36. Equation 3.36 may be also written as

$$y'(x)=\frac{G(x)}{\left\{1-[G(x)]^2\right\}^{1/2}} \tag{3.40}$$

The large deflection $y(x)$ at any $0 \le x \le L_o$ may be determined by (1) integrating Equation 3.40 once and determining the constant of integration by applying the boundary condition of zero displacement at the fixed end A or (2) by applying pseudolinear analysis as discussed in Section 3.2 and in the preceding two chapters. Pseudolinear analysis would be the easiest to use.

Numerical results are obtained here by assuming that EI_B = 75,000 kip-in^2 (215,224 Nm^2), L = 1000 in. (25.4 m), taper parameter n = 2, and load parameter m = 2. In Table 3.2 the values of Δ, θ_B, x_D, and y_{max} are obtained for various values of the trapezoidal load w_o. Note that x_D gives the location of the maximum displacement y_{max}. In Table 3.3 the values of the same quantities are obtained for various values of the taper parameter n. For these results, it was assumed that m = 2 and w_o = 0.01 kip/in. (175,126.8 N/m).

TABLE 3.2
Values of Δ, θ_A, θ_B, x_D, and y_{max} for Various Values of the Loading w_o and m = n = 2

w_o (kips/in.)	Δ (in.)	θ_A (°)	θ_B (°)	x_D (in.)	y_{max} (in.)
0.0025	13.50	0.0	19.41	632.35	69.75
0.0050	47.08	0.0	36.08	616.54	128.80
0.0075	88.30	0.0	49.00	597.62	173.92
0.0100	129.28	0.0	58.68	577.98	207.96
0.0125	166.61	0.0	65.87	559.79	234.01

Note: 1 in. = 0.0254 m, 1 kip/in. = 175,126.8 N/m.

TABLE 3.3
Values of Δ, θ_A, θ_B, x_D, and y_{max} for Various Values of the Taper Parameter n, and with m = 2 and w_o = 0.01 kips/in.

n	Δ (in.)	θ_A (°)	θ_B (°)	x_D (in.)	y_{max} (in.)
1.0	352.31	0	76.96	393.67	338.14
1.5	219.28	0	69.72	504.34	270.56
2.0	129.28	0	58.68	577.98	207.96
3.0	41.50	0	38.42	657.05	116.52

Note: 1 in. = 0.0254 m, 1 kip/in. = 175,126.8 N/m.

3.5 SIMPLIFIED NONLINEAR EQUIVALENT SYSTEMS FOR STATICALLY INDETERMINATE FLEXIBLE BEAM

In Sections 1.6 and 1.9, the derivation of simplified nonlinear equivalent systems of constant stiffness for statically determinate beam problems was discussed. Such nonlinear equivalent systems can greatly simplify the solution of beam problems with complicated stiffness variations along the length of the member and subjected to complicated loading conditions. The utilization of simplified nonlinear equivalent systems for statically determinate beam problems was amply illustrated in the two preceding chapters.

For statically determinate problems, simplified equivalent nonlinear systems may be easily obtained, because in Equation 1.39 the bending moment M_x at any $0 \leq x \leq L$ of the original system can be easily derived using statics. When the member is statically indeterminate, the redundant reactions should be determined before we are in a position to determine the moment diagram M_e from Equation 1.39. However, we could overcome this difficulty by utilizing principles of structural mechanics that are commonly used to solve statically indeterminate problems. The procedure will be illustrated here with an example (see also Fertis[6,7] and Fertis and Keene[32]).

Consider the doubly tapered single span beam in Figure 3.4a with a trapezoidal loading w_0. This is the same problem as the one shown in Figure 3.3. Since the beam is statically indeterminate, the reaction R_B at the end B of the member is taken as the redundant. Since we are concerned with the derivation of a simplified nonlinear equivalent system of constant stiffness EI_B, the redundant reaction R_B can be determined by following the well known theories of linear structural mechanics found in books dealing with this subject (see for example Fertis[6,7] and Fertis and Keene[32]). This is permissible because the nonlinear differential equations given by Equations 1.37 and 1.38 are equivalent. Therefore, the member in Figure 3.4a may be replaced by two cantilever beams as shown in Figures 3.4b and 3.4c. The cantilever beam in Figure 3.4b is loaded with the trapezoidal loading, and the one in Figure 3.4c is loaded with the redundant reaction R_B at the free end B.

Since the problems in Figures 3.4b and 3.4c are statically determinate, the moment M_x at any $0 \le x \le L$ may be determined using statics. Therefore, Equation 1.39 may be applied for each case in order to determine the bending moment M_e of the equivalent system of constant stiffness EI_B. Since we have two problems, we find the corresponding M_e for each problem. Note that the M_e for the problem in Figure 3.4c will be in terms of the redundant reaction R_B. For each case, the approximation of the shape of M_e with a few straight line segments, as was done in the preceding chapters, leads to an equivalent system of constant stiffness EI_B that is loaded with a few concentrated loads.

By using the equivalent system and applying linear analysis, say the moment area method, the vertical deflection δ'_B at the end B of the problem in Figure 3.4b may be determined. This is convenient because the M_e/EI_B diagram is known. In the same manner, the vertical deflection δ''_B at the end B of the problem in Figure 3.4c may be determined using the equivalent system of constant stiffness EI_B for this problem. The redundant reaction R_B can be determined by applying the boundary condition

$$\delta'_B + \delta''_B = 0 \tag{3.41}$$

Here we have one equation with R_B as the unknown, and therefore R_B can be determined.

In order to get numerical results, we assume that EI_B = 75,000 kip-in^2 (215,322 Nm^2), L = 1000 in. (25.4 m), w_0 = 0.01 kips/in. (1751.268 N/m), and m = n = 2. By considering the problem in Figure 3.4b, the bending moment M_x at any $0 \le x \le L$ may be determined using statics. If preferred, an equation for M_x may be written by using a free body diagram and applying statics. The values of M_x at intervals of 100 in. (2.54 m) are given in Table 3.4. In the same

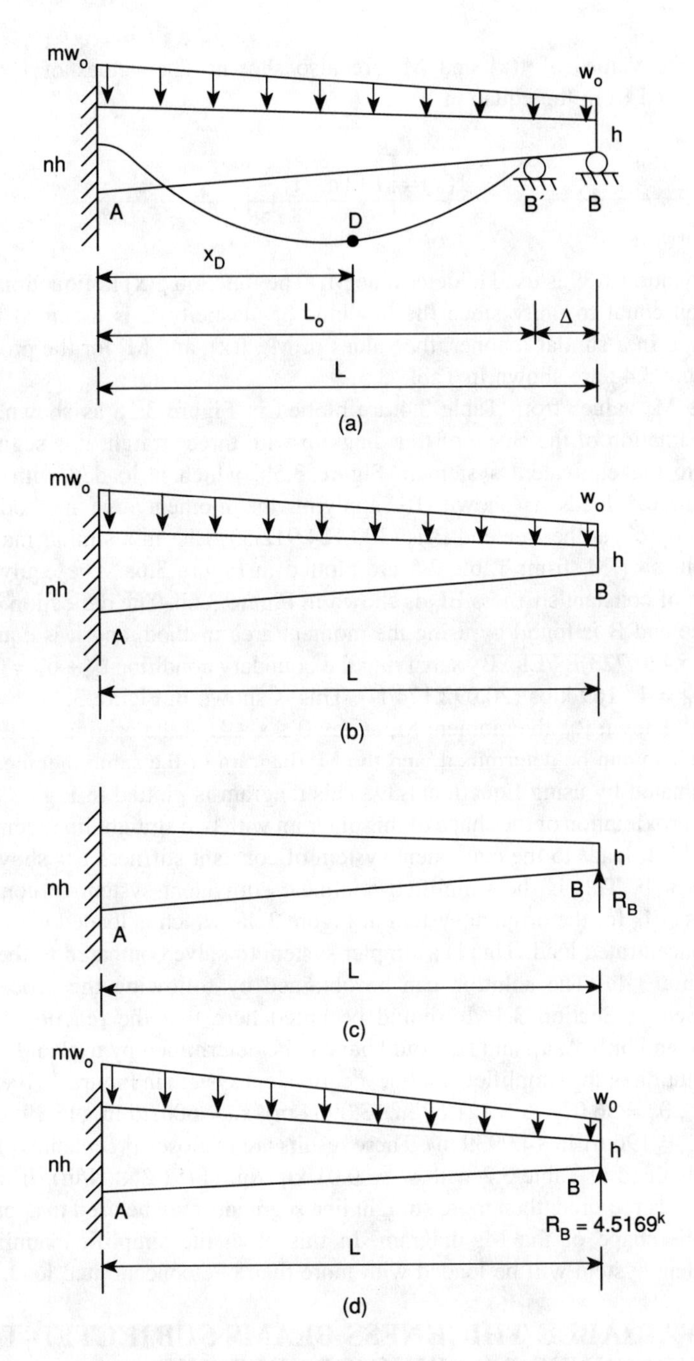

FIGURE 3.4. (a) Initial statically indeterminate member of variable stiffness; (b) cantilever beam loaded with a trapezoidal loading; (c) cantilever beam loaded with the redundant reaction R_B; (d) initial variable stiffness beam with known R_B.

table, the values of f(x) and M_e are also shown. The values of f(x) are determined from the equation

$$f(x) = \left[1 + (n-1)\frac{x}{L}\right]^3 \tag{3.42}$$

and Equation 1.39 is used to determine M_e. The function g(x) in Equation 1.39 is taken equal to unity since the modulus of elasticity E is assumed to be constant. In a similar manner, the values of M_x, f(x), and M_e for the problem in Figure 3.4c are shown in Table 3.5.

The M_e values from Table 3.4 are plotted in Figure 3.5a as shown. The approximation of the shape of this diagram with three straight line segments leads to the equivalent system in Figure 3.5b, which is loaded with three concentrated loads as shown. By applying the moment area method, the deflection δ'_B at the free end B is $(300{,}124{,}912.53)/EI_B$. In a similar manner, the values of M_e from Table 3.5 are plotted in Figure 3.6a. The equivalent system of constant stiffness EI_B is shown in Figure 3.6b. The deflection δ''_B at the free end B is found by using the moment area method, and it is equal to $(66{,}445{,}420.723)R_B/EI_B$. By satisfying the boundary condition $\delta'_B + \delta''_B = 0$, we find $R_B = 4.5169$ kips (20,092.174 N). This is shown in Figure 3.4d.

With known R_B, the moment M_x at any $0 \le x \le L$ of the original system in Figure 3.4a can be determined, and the M_e diagram of the same member can be evaluated by using Equation 1.39. This diagram is plotted in Figure 3.7a. The approximation of the shape of this diagram with two straight line segments EF and FB, leads to the equivalent system of constant stiffness EI_B shown in Figure 3.7b. This is the simplified nonlinear equivalent system of constant stiffness EI_B for the original system in Figure 3.4a, which is loaded with only one concentrated load. This is a simpler system to solve compared to the one in Figure 3.4a. The solution can be obtained by following the procedure discussed in Section 3.1. It should be noted here that the reaction R_B is known, and only Δ, a_1, and L_0 would have to be determined by trial and error. The solution of the simplified nonlinear equivalent system in Figure 3.7b yields $\theta_A = 0°$, $\theta_B = 56.07°$, $\Delta = 121.67$ in. (3.0904 m), $x_D = 606.25$ in. (15.3988 m), and $y_{max} = 196.61$ in. (4.9939 m). These results are in close agreement with the ones obtained in Table 3.2 with $w_o = 0.01$ kips/in. (1751.268 N/m). If better accuracy is required, then more straight line segments may be used to approximate the shape of the M_e diagram. In this case, the simplified nonlinear equivalent system will be loaded with more than one concentrated load.

3.6 VARIABLE THICKNESS BEAMS SUBJECTED TO COMBINED LOADING CONDITIONS

In order to stress the importance and convenience of simplified nonlinear equivalent systems of constant stiffness, the statically indeterminate beam in

TABLE 3.4
Values of M_x, f(x), and M_e for the Problem in Figure 3.4b

x (in.)	M_x (kip-in.)	f(x)	$M_e = M_x/f(x)$ (kip-in.)
0	0.0	0.0	0.0
100	–66.667	1.331	–50.09
200	–266.667	1.728	–154.32
300	–600.000	2.197	–273.10
400	–1066.667	2.744	–388.73
500	–1666.667	3.375	–493.83
600	–2400.000	4.096	–585.94
700	–3266.667	4.913	–664.90
800	–4266.667	5.832	–731.60
900	–5400.000	6.859	–787.29
1000	–6666.667	8.000	–833.33

Note: 1 in. = 0.0254 m, 1 kip-in. = 112.9848 Nm.

TABLE 3.5
Values of M_x, f(x), and M_e for the Problem in Figure 3.4c

x (in.)	M_x (kip-in.)	f(x)	$M_e = M_x/f(x)$ (kip-in.)
0	0	0.0	0.0
100	100 R_B	1.331	75.13 R_B
200	200 R_B	1.728	115.74 R_B
300	300 R_B	2.197	136.55 R_B
400	400 R_B	2.744	145.77 R_B
500	500 R_B	3.375	148.15 R_B
600	600 R_B	4.096	146.48 R_B
700	700 R_B	4.913	142.48 R_B
800	800 R_B	5.832	137.17 R_B
900	900 R_B	6.859	131.21 R_B
1000	1000 R_B	8.000	125.00 R_B

Note: 1 in. = 0.0254 m, 1 kip-in. = 112.9848 Nm.

Figure 3.8a is considered. The beam is doubly tapered, and it is loaded with a partial, uniformly distributed load w and a concentrated load P as shown. A direct solution of this problem by applying the Euler-Bernoulli nonlinear differential equation would be extremely difficult because this equation would have to be applied three times. The utilization of pseudolinear equivalent systems of constant stiffness would be less cumbersome, but the required trial and error procedure may become rather tedious.

For such types of problems, a practical solution would be the utilization of equivalent nonlinear systems of constant stiffness as discussed in the preceding

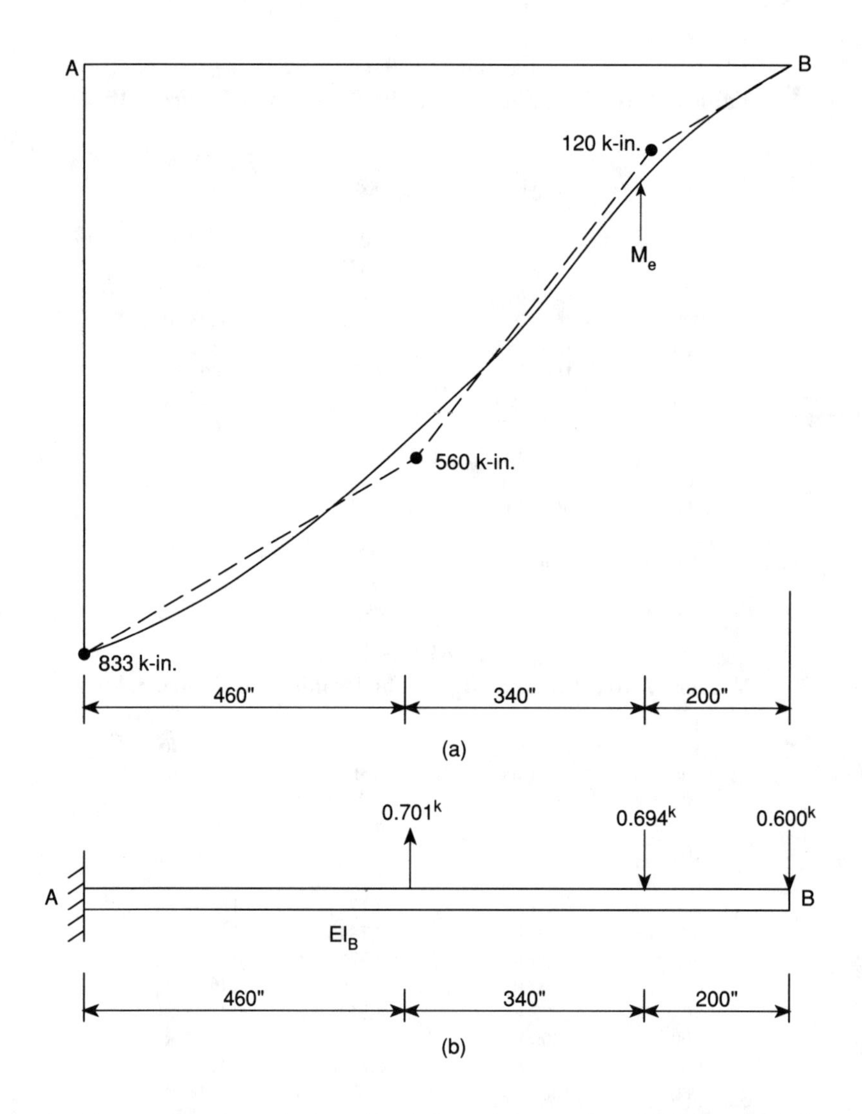

FIGURE 3.5. (a) M_e diagram for the problem in Figure 3.4b approximated with three straight line segments; (b) equivalent system of constant stiffness EI_B for the problem in Figure 3.4b (1 in. = 0.0254 m, 1 kip-in. = 112.9848 Nm, 1 kip = 4448.222 N).

section. It was shown that reasonably accurate results can be obtained when the nonlinear equivalent system is loaded with only one concentrated load. On this basis, instead of solving the original complicated problem in Figure 3.8a, we can now solve a uniform stiffness member that is loaded with a concentrated load P_e at some point along its span length. Sections 3.2 and 3.5 discuss the solution of such types of problems.

Let it be assumed that the beam in Figure 3.8a has a length L = 1000 in. (25.4 m) and a stiffness EI_B, at the end B, equal to 15,000 kip-in.^{2} (43,045 Nm^2). The

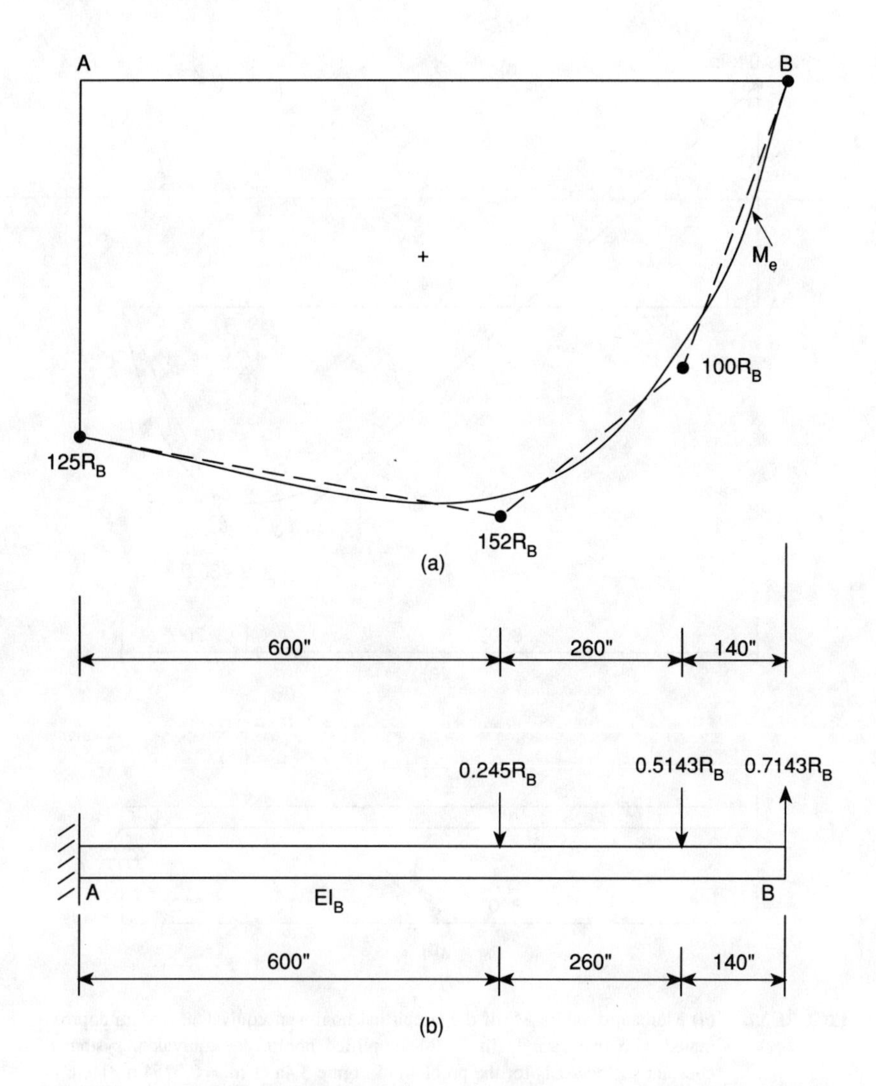

FIGURE 3.6. (a) M_e diagram for the problem in Figure 3.4c approximated with three straight line segments; (b) equivalent system of constant stiffness EI_B for the problem in Figure 3.4c (1 in. = 0.0254 m, 1 kip = 4448.222 N).

distributed load w = 0.01 kips/in. (1751.268 N/m), P = 2 kips (8.8 kN), L_1 = 500 in. (12.7 m), L_2 = 100 in. (2.54 m), and taper n = 2. Since the member in Figure 3.8a is statically indeterminate, the redundant reaction R_B at the end B of the member must be determined first. When this is accomplished, the simplified nonlinear equivalent system of constant stiffness can be easily established.

The redundant reaction R_B can be determined in a manner similar to the one used in the preceding section, by employing the two cantilever beams loaded as shown in Figures 3.8b and 3.8c. By applying the method of the equivalent

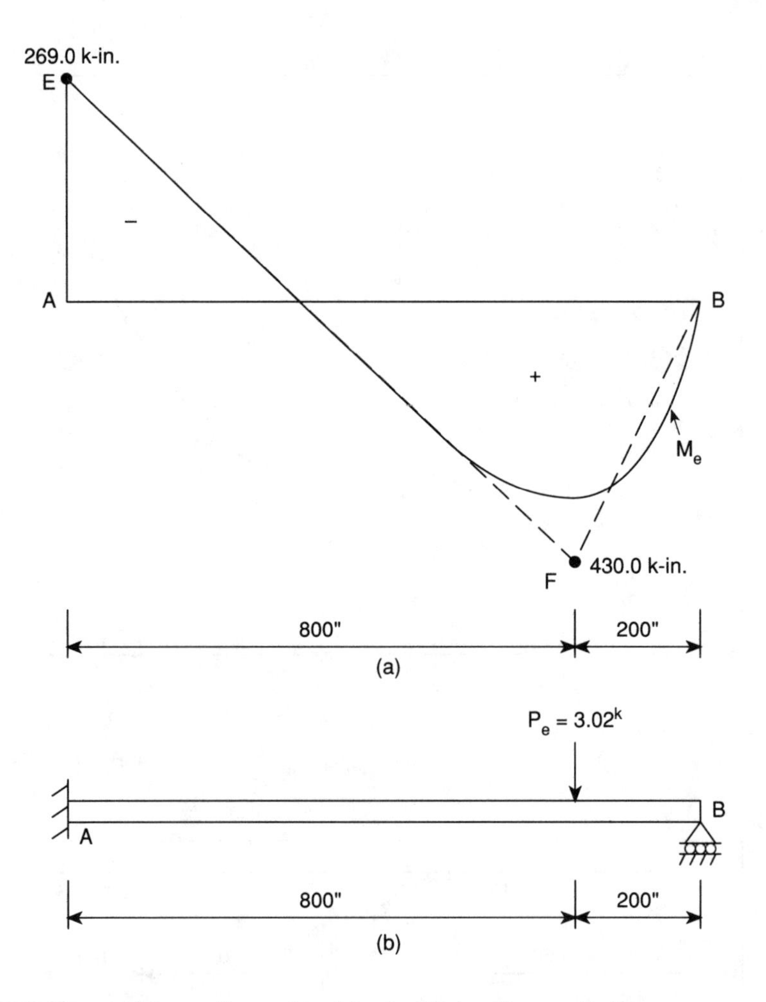

FIGURE 3.7. (a) Moment diagram M_e of the simplified nonlinear equivalent system approximated with two straight lines; (b) simplified nonlinear equivalent system of constant stiffness EI_B for the problem in Figure 3.4a (1 in. = 0.0254 m, 1 kip-in. = 112.9848 Nm).

systems as in the preceding section, we determine the vertical displacements δ'_B and δ''_B at the free end B of the beams in Figures 3.8b and 3.8c, respectively. Note that δ''_B would be in terms of the unknown redundant reaction R_B. By satisfying the boundary condition of zero vertical displacement at the end B, that is $\delta'_B + \delta''_B = 0$, the redundant reaction may be determined. This procedure yields $R_B = 1.0073$ kips (4480 N).

With known R_B, the moment M_x at any $0 \le x \le L$ of the original member in Figure 3.8a may be determined. The simplified nonlinear equivalent system of constant stiffness EI_B for the member in Figure 3.8a now can be obtained easily since M_x is known. The bending moment diagram of this system may be determined by using Equation 1.39 with $g(x) = 1$, since the modulus of

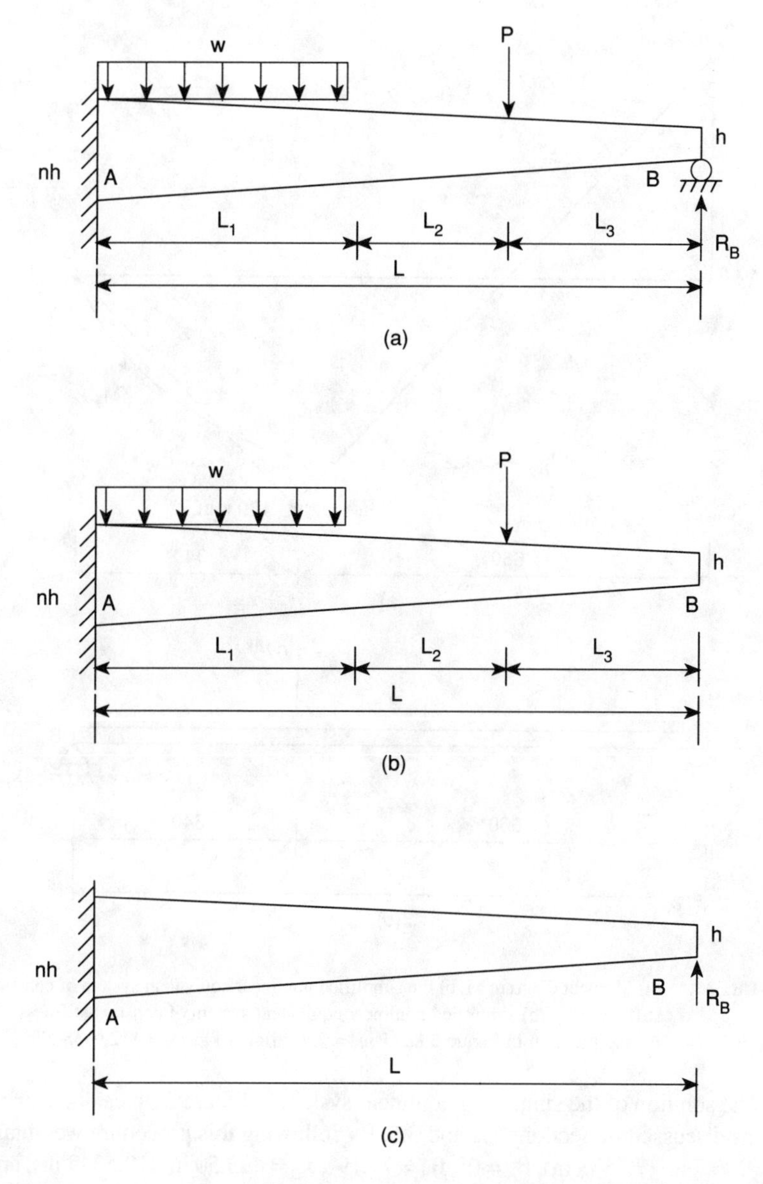

FIGURE 3.8. (a) Initial variable stiffness statically indeterminate member subjected to a combined loading; (b) cantilever beam subjected to the combined loading; (c) cantilever beam loaded with the redundant reaction R_B.

elasticity E is assumed to be constant. The M_e diagram is shown plotted in Figure 3.9a. The approximation of its shape with two straight line segments as shown leads to the simplified nonlinear equivalent system of constant stiffness EI_B shown in Figure 3.9b. This is a much simpler problem to solve compared to the one in Figure 3.8a. Note that all supporting reactions and moments of the problem in Figure 3.9b are known.

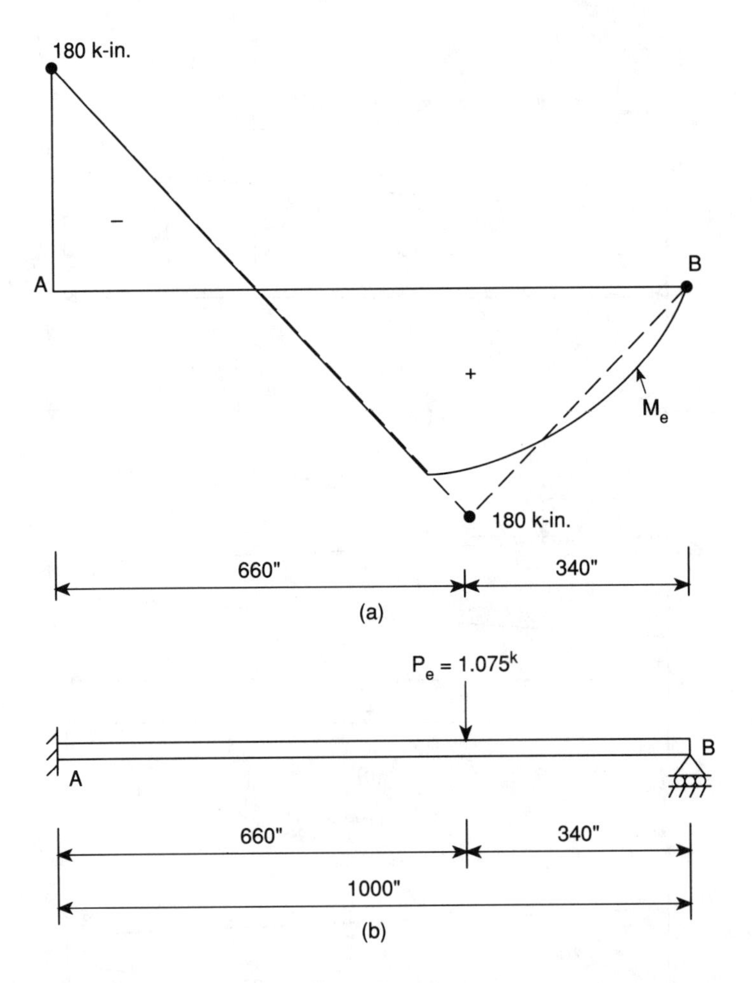

FIGURE 3.9. (a) Moment diagram M_e of the simplified nonlinear equivalent system of constant stiffness EI_B; (b) simplified nonlinear equivalent system of constant stiffness EI_B for the problem in Figure 3.8a (1 in. = 0.0254 m, 1 kip-in. = 112.9848 Nm).

The solution of the simplified nonlinear system in Figure 3.9b can be carried out as discussed in Sections 3.2 and 3.5. By following this procedure we obtain $\Delta = 290.7$ in. (7.3838 m), $\theta_A = 0°$, $\theta_B = 77.19°$, $x_D = 493.50$ in. (12.5349 m), and $y_{max} = 303.36$ in. (7.7053 m), where x_D denotes the location of the maximum vertical displacement y_{max}. For practical purposes, these results should be within acceptable error as it was demonstrated in the preceding section. The shape of the M_e diagram with two straight lines is usually sufficient for practical applications, because the error that is introduced in the approximation of M_e becomes much smaller for the rotation and deflection, since mathematically you have to integrate for rotation and deflection. Integration processes usually reduce the error, while differentiation processes increase it.

As a second example regarding the derivation of simplified nonlinear equivalent systems of constant stiffness, consider the statically indeterminate flexible

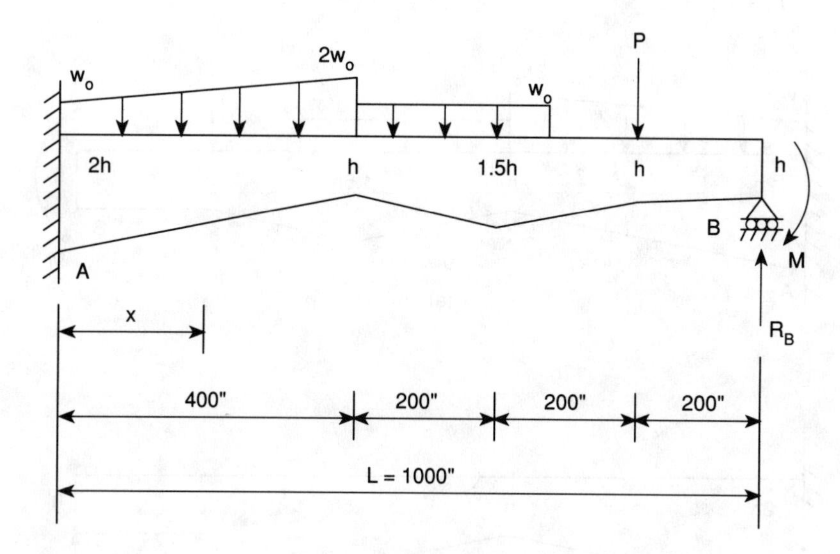

FIGURE 3.10. Statically indeterminate single span beam subjected to combined loading.

beam loaded as shown in Figure 3.10. The stiffness EI_B at the end B is 75,000 kip-in.^{2} (215,224 Nm^2), L = 1000 in. (25.4 m), P = 3.0 kips (13,344.666 N), w_o = 0.01 kips/in. (1751.268 N/m), and M = 1500 kip-in. (169,477.2 Nm). The derivation of the simplified nonlinear system may be initiated by considering the reaction R_B at the end support B as the redundant. On this basis, the two statically determinate beams in Figures 3.11a and 3.11b would have to be solved using the method of the equivalent systems and linear analysis as discussed in the preceding example of this section. Table 3.6 gives the values of M_x, f(x), and $M_e = M_x/f(x)$ for the problem in Figure 3.11a. The diagram is shown plotted in Figure 3.12a, and the approximation of its shape with six straight line segments leads to the equivalent system of constant stiffness EI_B shown in Figure 3.12b. In a similar manner, the M_x, f(x), and M_e for the problem in Figure 3.11b are shown in Table 3.7, and the equivalent system of constant stiffness EI_B is derived as shown in Figure 3.13. Note that the concentrated loads in Figure 3.13b are in terms of the redundant reaction R_B. This reaction can be determined by satisfying the boundary condition

$$\delta_B = \delta'_B + \delta''_B = 0 \tag{3.43}$$

where δ_B is the vertical displacement at the end B of the problem in Figure 3.10, δ'_B is the displacement at B of the cantilever beam in Figure 3.11a, and δ''_B is the vertical displacement at B of the cantilever beam in Figure 3.11b. The moment area method, that is linear analysis, may be used to determine δ'_B and δ''_B, as discussed earlier.

By calculating δ'_B and δ''_B as suggested, and substituting into Equation 3.43, we can determine R_B. The value of R_B = 4.7089 kips (20,946.23 N). With known R_B, the values of M_x, f(x), and $M_e = M_x/f(x)$ for the initial problem in

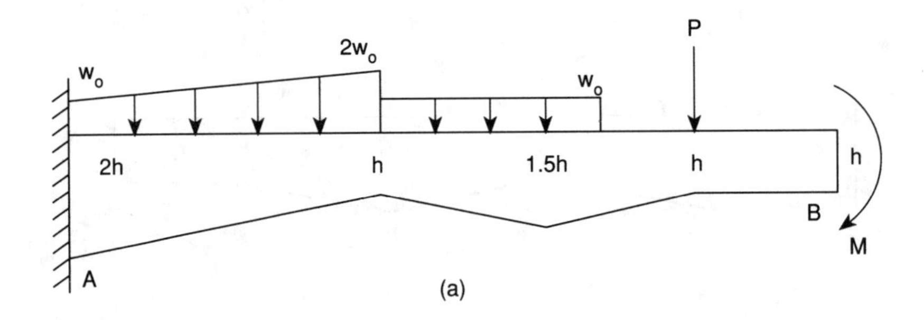

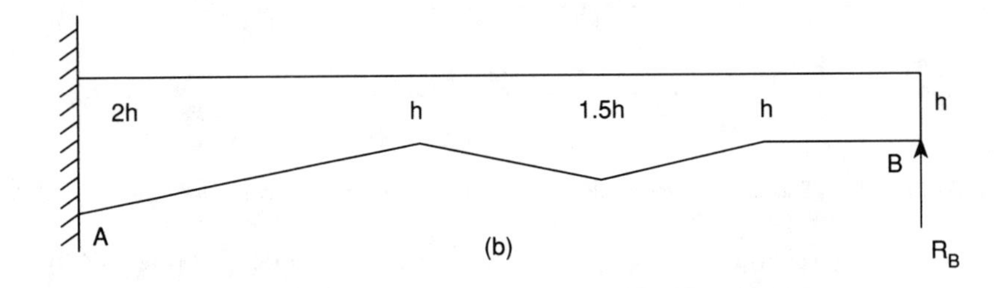

FIGURE 3.11. (a) Cantilever beam subjected to the combined loading of the beam in Figure 3.10; (b) cantilever beam loaded with the redundant reaction R_B at the free end B.

TABLE 3.6
Values of M_x, f(x), and M_e for the Problem in Figure 3.11a

x (in.)	M_x (kip-in.)	f(x)	$M_e = M_x/f(x)$ (kip-in.)
0	6745	8.0	843.0
100	5620	5.3594	1049.0
200	4619	3.3750	1369.0
300	3768	1.9531	1929.0
400	3092	1.0000	3092.0
500	2562	1.9531	1312.0
600	2132	3.3750	632.0
700	1800	1.9531	922.0
800	1500	1.0000	1500.0
900	1500	1.0000	1500.0
1000	1500	1.0000	1500.0

Note: 1 in. = 0.0254 m, 1 kip-in. = 112.9848 Nm.

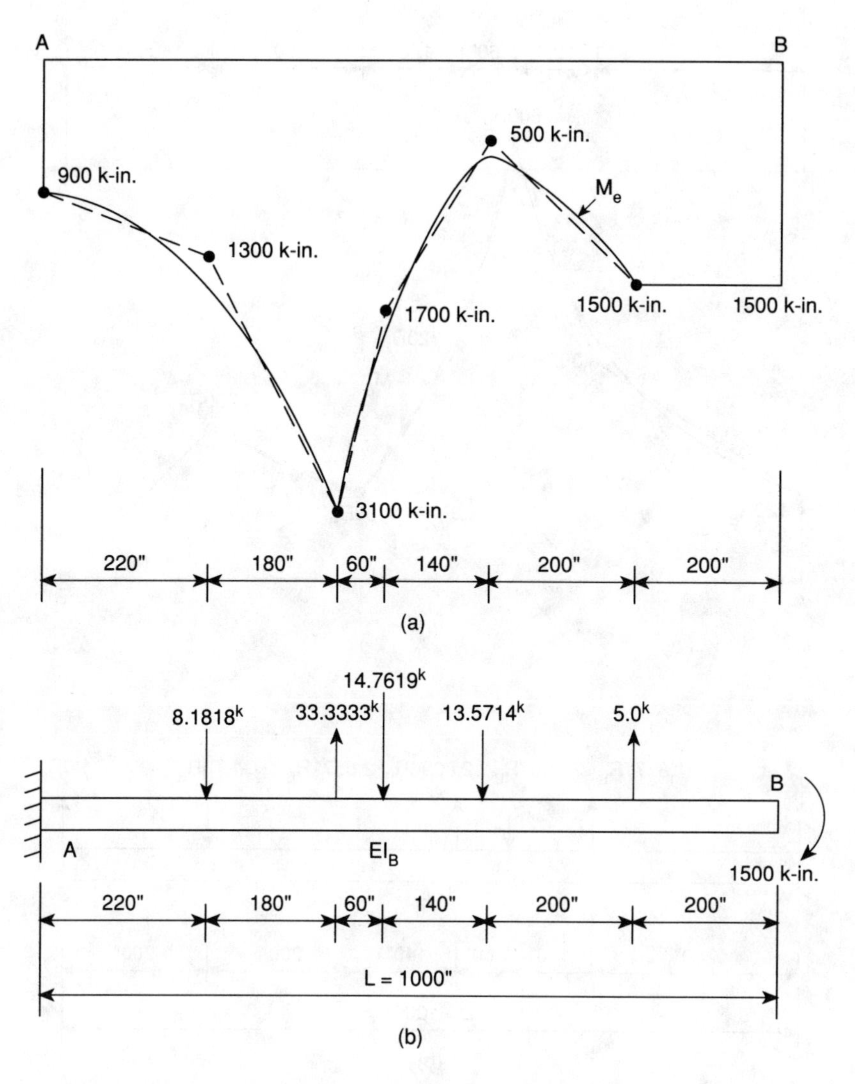

FIGURE 3.12. (a) M_e diagram for the cantilever beam in Figure 3.11a; (b) equivalent system of constant stiffness EI_B for the cantilever beam in Figure 3.11a (1 in. = 0.0254 m, 1 kip = 4448.222 N, 1 k-in. = 112.9848 Nm).

Figure 3.10 may be determined, and they are shown in Table 3.8. The M_e diagram is shown plotted in Figure 3.14a. The approximation of its shape with two straight line segments as shown leads to the simplified nonlinear equivalent system of constant stiffness EI_B shown in Figure 3.14b. The simplified nonlinear system is loaded with only one concentrated load P_e = 4.5998 kips (20,460.93 N), and its nonlinear solution would be a great deal simpler than the solution of the original system in Figure 3.10.

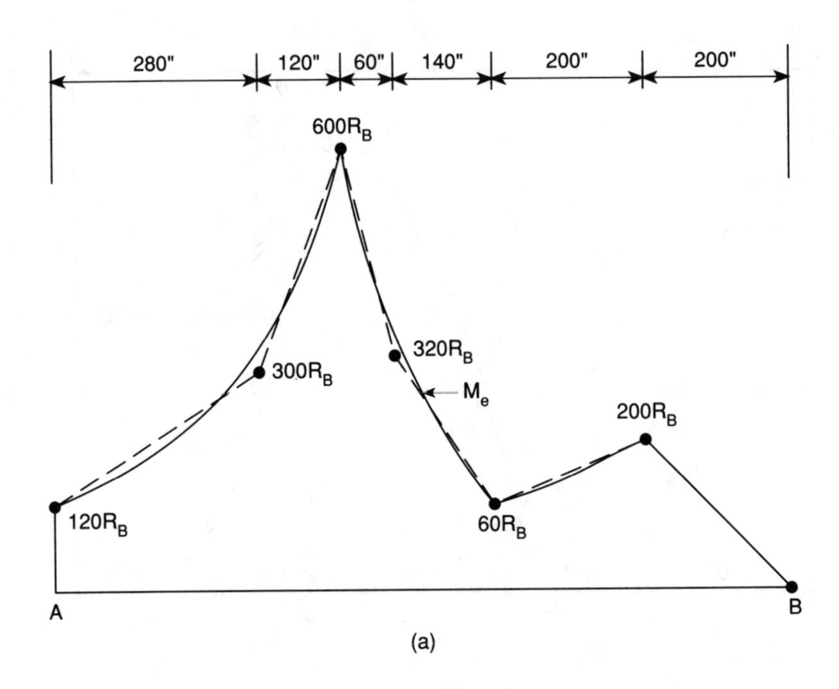

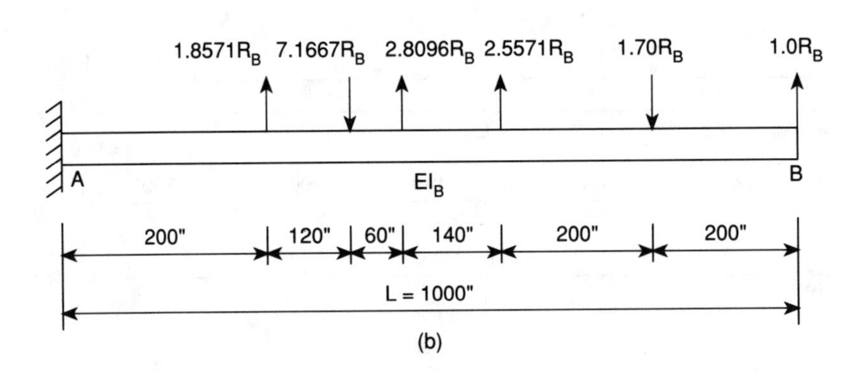

FIGURE 3.13. (a) M_e diagram for the cantilever beam in Figure 3.11b; (b) equivalent system of constant stiffness EI_B for the cantilever beam in Figure 3.11b (1 in. = 0.0254 m, 1 kip= 4448.222 N, 1 kip-in. = 112.9848 Nm).

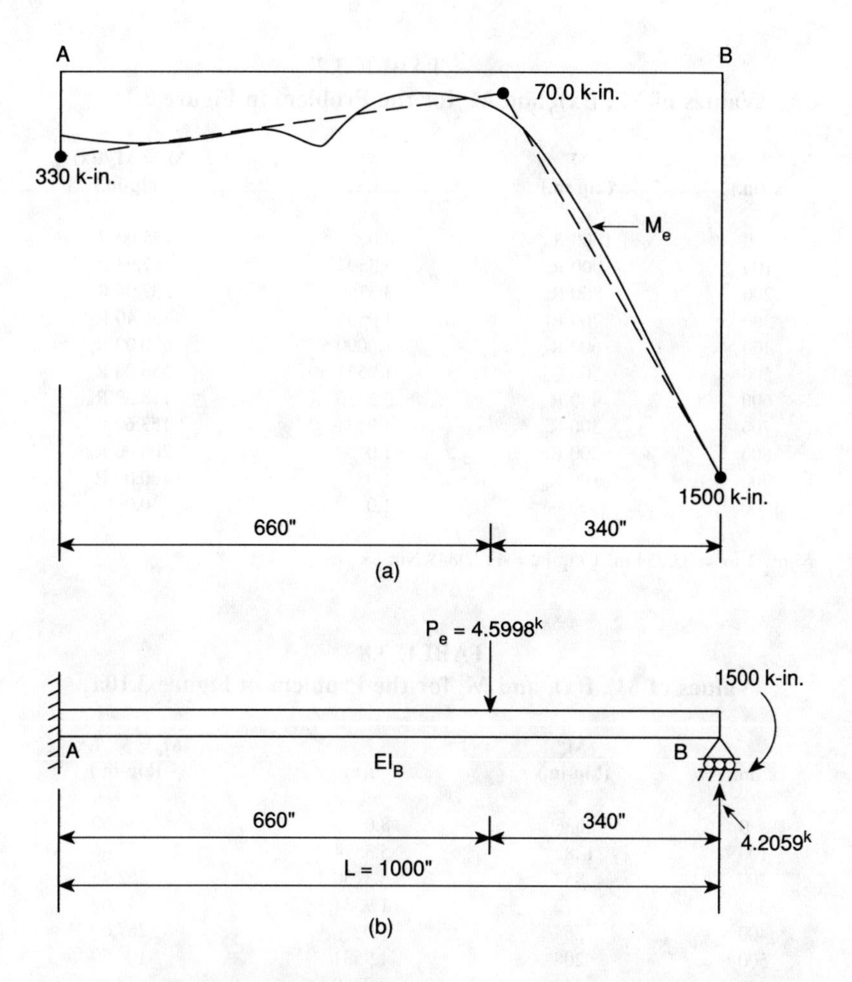

FIGURE 3.14. (a) M_e diagram of the original system in Figure 3.10 approximated with two straight line segments; (b) simplified nonlinear equivalent system of constant stiffness EI_B for the problem in Figure 3.10 (1 in. = 0.0254 m, 1 kip = 4448.222 N, 1 kip-in. = 112.9848 Nm).

TABLE 3.7
Values of M_x, f(x), and M_e for the Problem in Figure 3.11b

x (in.)	M_x (kip-in.)	f(x)	$M_e = M_x/f(x)$ (kip-in.)
0	1000 R_B	8.0	125.00 R_B
100	900 R_B	5.3594	167.93 R_B
200	800 R_B	3.3750	237.04 R_B
300	700 R_B	1.9531	358.40 R_B
400	600 R_B	1.0000	600.00 R_B
500	500 R_B	1.9531	256.00 R_B
600	400 R_B	3.3750	118.52 R_B
700	300 R_B	1.9531	153.60 R_B
800	200 R_B	1.0	200.00 R_B
900	100 R_B	1.0	100.00 R_B
1000	0	1.0	0.0

Note: 1 in. = 0.0254 m, 1 kip-in. = 112.9848 Nm.

TABLE 3.8
Values of M_x, f(x), and M_e for the Problem in Figure 3.10a

x (in.)	M_x (kip-in.)	f(x)	$M_e = M_x/f(x)$ (kip-in.)
0	2036	8.0	254.50
100	1382	5.3594	257.86
200	852	3.3750	252.44
300	472	1.9531	241.67
400	267	1.0	267.00
500	208	1.9531	106.50
600	248	3.3750	73.48
700	387	1.9531	198.15
800	558	1.0	558.00
900	1029	1.0	1029.00
1000	1500	1.0	1500.00

Note: 1 in. = 0.0254 m, 1 kip-in. = 112.9848 Nm.

PROBLEMS

3.1 The statically indeterminate flexible beam in Figure 3.1 is loaded with a concentrated load P = 3.5 kips (15,568.777 N) at a distance L_1 = 600 in. (15.24 m). The length L of the member is 1000 in. (25.4 m), and the stiffness EI = 75,000 kip-in.^{2} (215,224 Nm^2). Determine the rotation θ_B at the end B of the member, its maximum vertical displacement y_{max}, and the location x_D of the maximum vertical displacement. Use pseudolinear analysis.

3.2 Solve Problem 3.1 by integrating the Euler-Bernoulli equation and satisfying appropriate boundary conditions to determine the constants of integration and the redundant reaction R_B at support B. Compare the results.

3.3 Solve Problem 3.1 by assuming that P = 1.5 kips (6672.33 N) by (1) using nonlinear analysis and (2) using linear analysis. Compare the results.

3.4 The statically indeterminate beam in Figure 3.2 is loaded with a uniformly distributed load w_o = 5 lb/in. (875.634 N/m). The length L = 1000 in. (25.4 m), and EI = 75,000 kip-in.^{2} (215,224 Nm^2). Determine Δ, θ_B, δ_D, x_D, and R_B, using pseudolinear analysis. Assume $\Delta(x)$ = constant = Δ.
Answer: Δ = 150.55 in., θ_B = 53.12°, δ_D = 232.764 in., and R_B = 1.9584 kips.

3.5 Solve Problem 3.4 by assuming $\Delta(x) = \Delta x/L_o$ and $\Delta\sin(\pi x/2L_o)$, and compare the results.

3.6 The statically indeterminate beam in Figure 3.3 is loaded with a trapezoidal loading of w_o = 0.0125 kips/in. (2,189.085 N/m). Determine Δ, θ_A, θ_B, x_D, and maximum vertical displacement y_{max} using pseudolinear analysis. The length L = 1000 in. (25.4 m), EI_B = 75,000 kip-in.^{2} (215,224 Nm^2), and m = n = 2. Assume $\Delta(x)$ = constant = Δ.
Answer: Δ = 166.61 in., θ_A = 0°, θ_B = 65.87°, x_D = 559.79 in., and y_{max} = 234.01 in.

3.7 Solve Problem 3.6 with n = 3 and w_o = 0.01 kip/in. (1751.268 N/m).
Answer: Δ = 41.50 in., θ_B = 38.42°, x_D = 657.05 in., and y_{max} = 116.52 in.

3.8 Solve Problem 3.6 by using a simplified nonlinear system of constant stiffness EI_B that is loaded with only one concentrated load. Compare the results.

3.9 Solve the simplified nonlinear system in Figure 3.14b by applying pseudolinear analysis.

Chapter 4

VIBRATION THEORY OF FLEXIBLE BARS

4.1 INTRODUCTION

The objectives in this chapter is to derive the nonlinear differential equations of motion that can be used for the vibration analysis of flexible bars of uniform and/or variable cross section along the length of the member. In addition, solution methodologies have been developed that greatly simplify the solution of complicated flexible beam problems.

Free vibrations, in general, are taking place from the static equilibrium position of the member. For a flexible member, the static equilibrium position is associated with large static amplitudes, and the differential equation of motion that expresses the free vibration of the member becomes nonlinear. If the vibrational amplitudes from the static equilibrium position are small, then the free frequencies of vibration are independent of the amplitude of vibration, but they do depend upon the static amplitude that defines the static equilibrium position. If, however, the frequency amplitudes are also large, then the free frequencies of vibration will be dependent on both static and vibrational amplitudes.

The methodologies developed in this chapter deal primarily with small oscillation vibration superimposed on large static displacements that define the static equilibrium position of the flexible member. This is not unreasonable, because a large family of vibration problems are falling into this category. However, the nonlinear differential equation of motion that can be used for large amplitude vibrations is derived in this chapter.

The methodologies developed in the preceding three chapters of the book, provide a great deal of information that can be used to simplify the solution for the vibration of flexible beam problems. In the analysis, it is assumed that the moment curvature relationship obeys the Euler-Bernoulli law, the material is linearly elastic, shear deformation and rotatory inertia are neglected, and damping is not taken into consideration. The moment of inertia of the member may vary in any arbitrary manner along its length, and weights, other than the weight of the member, that are securely attached to the member and participate in its vibrational motion may be included. These are common assumptions for small vibration analysis, but the effect of rotatory and longitudinal inertia may be significant for large amplitude vibration.

4.2 THE NONLINEAR DIFFERENTIAL EQUATION OF MOTION

We recall here the Euler-Bernoulli Law that is given by the following nonlinear differential equation:

$$\frac{y''}{\left[1+(y')^2\right]^{3/2}} = -\frac{M_x}{E_x I_x} \tag{4.1}$$

By multiplying both sides of Equation 4.1 by $E_x I_x$ we get

$$E_x I_x \frac{y''}{\left[1+(y')^2\right]^{3/2}} = -M_x \tag{4.2}$$

Differentiating both sides of Equation 4.2 with respect to x, we get

$$\frac{d}{dx}\left\{E_x I_x \frac{y''}{\left[1+(y')^2\right]^{3/2}}\right\} = -\frac{dM_x}{dx} \tag{4.3}$$

or

$$\frac{d}{dx}\left\{E_x I_x \frac{y''}{\left[1+(y')^2\right]^{3/2}}\right\} = -V(x) \tag{4.4}$$

The shear force V(x) at any cross section must be the same whether it is defined in the reference (undeformed) configuration or the deformed configuration, i.e., $V(x_o) = V(x[x_o])$. Thus, Equation 4.4 may be written as

$$\frac{d}{dx}\left\{E_x I_x \frac{y''}{\left[1+(y')^2\right]^{3/2}}\right\} = -\frac{V(x_o)}{\cos\theta} \tag{4.5}$$

By differentiating Equation 4.5 with respect to x we obtain the expression

$$\frac{d^2}{dx^2}\left\{E_x I_x \frac{y''}{\left[1+(y')^2\right]^{3/2}}\right\} = -\frac{d}{dx}\left\{\frac{V(x_o)}{\cos\theta}\right\} \tag{4.6}$$

Since the expression for the transverse weight is well defined in the undeformed configuration, we can rewrite Equation 4.6 as follows:

$$\frac{d^2}{dx^2}\left\{E_x I_x \frac{y''}{\left[1+(y')^2\right]^{3/2}}\right\} = -\frac{1}{\cos\theta}\frac{d}{dx_o}\left\{\frac{V(x_o)}{\cos\theta}\right\} \tag{4.7}$$

For example, for distributed weight, the expression of the shear force $V(x_o)$ is

$$V(x_o) = -w(x_o)\cos\theta \tag{4.8}$$

Performing the indicated differentiation in Equation 4.7 we find

$$\frac{d^2}{dx^2}\left\{E_x I_x \frac{y''}{\left[1+(y')^2\right]^{3/2}}\right\} = -\frac{1}{\cos\theta} w(x_o) \tag{4.9}$$

where

$$x_o(x) = \int_0^x \sqrt{1+(y')^2(\xi)}d\xi \tag{4.10}$$

and

$$\cos\theta = \frac{1}{\sqrt{1+(y')^2}} \tag{4.11}$$

For uniformly distributed loading we have $w(x_o) = w_o$, and Equation 4.9 yields

$$\frac{d^2}{dx^2}\left\{E_x I_x \frac{y''}{\left[1+(y')^2\right]^{3/2}}\right\} = -\frac{1}{\cos\theta} w_o \tag{4.12}$$

or

$$\frac{d^2}{dx^2}\left\{E_x I_x \frac{y''}{\left[1+(y')^2\right]^{3/2}}\right\} = -\left[1+(y')^2\right]^{1/2} w_o \tag{4.13}$$

For transverse free vibration, the weight w_o is replaced by the inertia force w_{in}, i.e.,

$$w_{in} = m\frac{d^2y}{dt^2} \tag{4.14}$$

where m is the uniform mass density and the d^2y/dt^2 is the relative accelaration.

Therefore, the transverse free vibration of a flexible member with a uniformly distributed weight or mass is given by the equation

$$\frac{d^2}{dx^2}\left\{E_x I_x \frac{y''}{\left[1+(y')^2\right]^{3/2}}\right\}+\sqrt{1+(y')^2}\, m \frac{d^2 y}{dt^2}=0 \tag{4.15}$$

In general, for arbitrarily distributed weight, the mass m is replaced with an appropriate expression $m(x_o)$. Note that m is a function of x_o. Thus, the nonlinear differential equation of motion becomes

$$\frac{d^2}{dx^2}\left\{E_x I_x \frac{y''}{\left[1+(y')^2\right]^{3/2}}\right\}+\sqrt{1+(y')^2}\, m(x_o) \frac{d^2 y}{dt^2}=0 \tag{4.16}$$

Equation 4.16 cannot be solved by separating the variables. In addition, the natural frequencies of vibration are amplitude dependent. The natural modes of vibration could be assumed, then the natural frequencies of vibration and the corresponding mode shapes could be obtained by the perturbation method. This procedure is usually used to obtain the fundamental natural frequency. For higher frequencies, this method becomes extremely difficult.

The difficulties associated with the solution of Equation 4.16 may be reduced by taking into consideration that the vibration of the flexible bar is taking place from the static equilibrium position, defined as $y_s(x)$. The vibrational amplitudes will be assumed to be small, (see also Fertis and Afonta[21,22]). It is further considered that

1. The slope of the dynamic amplitude is negligibly small.
2. The large static equilibrium configuration is obtained by static analysis.
3. The total weight acting on the deformed segment is approximated by a function of the horizontal displacement. This is done by replacing the arc length position in the deformed segment by the equation

$$x_o(x) = x + \Delta(x) \tag{4.17}$$

4. The law of variation of the height h(x) of the cross section in the deformed configuration is an integral equation which depends on the deformation. This integral will be replaced with a function that contains the horizontal displacement function $\Delta(x)$.
5. The time dependent is harmonic, i.e., $\ddot{y}_d = -\omega^2 y_d$, where y_d is the dynamic displacement from the static equilibrium position.

4.3 DIFFERENTIAL EQUATION OF MOTION FOR SMALL AMPLITUDE VIBRATIONS

Consider the flexible cantilever beam in Figure 4.1a, where w_o is the uniform weight of the beam. This weight may include other weights attached to the member which participate in the vibrational motion of the beam. An

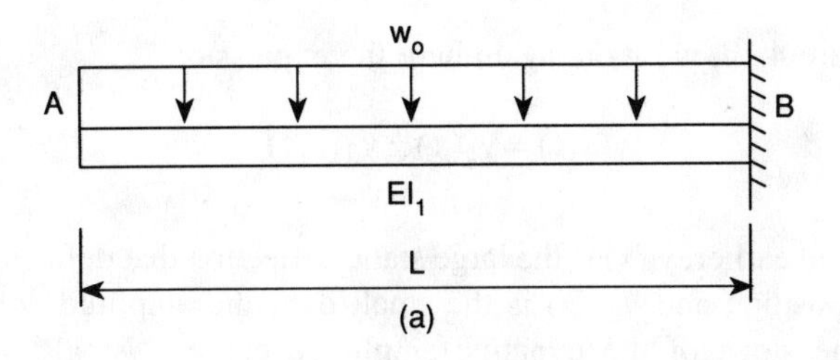

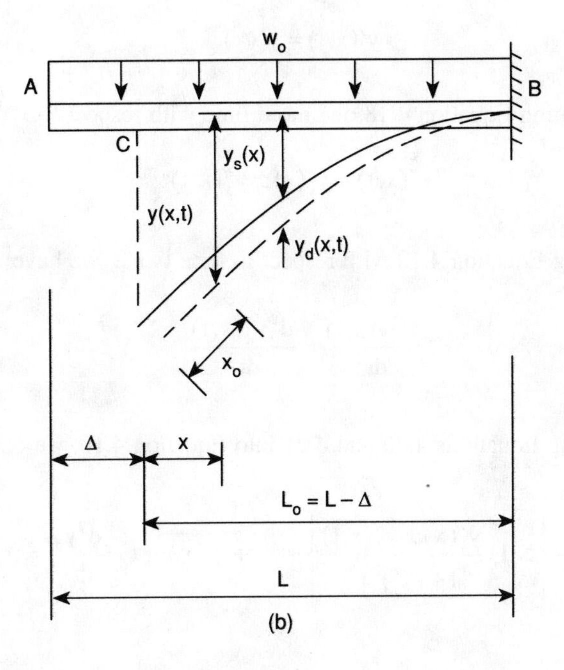

FIGURE 4.1. (a) Original undeformed uniform flexible cantilever beam; (b) deformed configuration of the uniform flexible cantilever beam.

example of such condition would be the girder of a highway bridge and the deck of the bridge. If the deck is attached to the girders by shear conectors, then the whole weight of the deck participates in the vibrational motion of the girders and must be taken into consideration in the vibration analysis of the girders, (see also Fertis[6,7]).

The deformation configuration of the member is shown in Figure 4.1b. In this figure, $y_s(x)$ is the large static deformation that defines the static equilibrium position. The dynamic amplitude, which represents the small vibration of the member from the static equilibrium position, is denoted by $y_d(x,t)$. From the undeformed straight configuration of the member, the amplitude is defined by $y(x,t) = [y_s(x)+y_d(x,t)]$. The rest of the notation is similar to the one used in the preceding three chapters, and it is self explanatory.

From Figure 4.1b, we write again here the expression

$$y(x,t) = y_s(x) \pm y_d(x,t) \tag{4.18}$$

where, as stated earlier, $y_s(x)$ is the large static deflection that defines the static equilibrium position and $y_d(x,t)$ is the small dynamic amplitude of the free vibration. The slope of the dynamic amplitude curve is small when it is compared to the large static slope, i.e., $y_s'(x) >> y_d'(x,t)$. Thus, from Equation 4.18, by differentiation, we obtain

$$y'(x,t) = y_s'(x) \tag{4.19}$$

By differentiating Equation 4.18 one more time with respect to x we find

$$y''(x,t) = y_s''(x) \pm y_d''(x,t) \tag{4.20}$$

Differentiating Equation 4.18 with respect to time twice, we have

$$\frac{d^2y(x,t)}{dt^2} = \frac{d^2y_d(x,t)}{dt^2} \tag{4.21}$$

By substituting Equations 4.20 and 4.21 into Equation 4.16, we obtain

$$\frac{d^2}{dx^2}\left\{E_xI_x \frac{y_s''(x) \pm y_d''(x,t)}{\left[1+(y_s')^2\right]^{3/2}}\right\} + \sqrt{1+(y_s')^2}\,m(x_o)\frac{d^2y_d}{dt^2} = 0 \tag{4.22}$$

or

$$\begin{gathered}\frac{d^2}{dx^2}\left\{E_xI_x \frac{y_s''(x)}{\left[1+(y_s')^2\right]^{3/2}}\right\} \pm \frac{d^2}{dx^2}\left\{E_xI_x \frac{y_d''(x)}{\left[1+(y_s')^2\right]^{3/2}}\right\} \\ + \sqrt{1+(y_s')^2}\,m(x_o)\frac{d^2y_d}{dt^2} = 0\end{gathered} \tag{4.23}$$

From the large static equilibrium configuration we have

$$\frac{d^2}{dx^2}\left\{E_xI_x \frac{y_s''(x)}{\left[1+(y_s')^2\right]^{3/2}}\right\} = -\frac{w(x_o)}{\cos\theta} \tag{4.24}$$

and from the dynamic equilibrium configuration we have

$$\frac{d^2}{dx^2}\left\{E_x I_x \frac{y_s''(x)}{\left[1+(y_s')^2\right]^{3/2}}\right\} = -\sqrt{1+(y_s')^2}\,m(x_o)\frac{d^2 y_d}{dt^2} \tag{4.25}$$

Since the time function is harmonic we may write

$$\frac{d^2 y_d(x,t)}{dt^2} = -\omega^2 y_d(x,t) \tag{4.26}$$

where ω is the free vibration in radians per second. By substituting Equation 4.26 into Equation 4.25 we find

$$\frac{d^2}{dx^2}\left\{E_x I_x \frac{y_d''(x)}{\left[1+(y_s')^2\right]^{3/2}}\right\} - \left\{\sqrt{1+(y_s')^2}\,m(x_o)\right\}\omega^2 y_d(x,t) = 0 \tag{4.27}$$

Equation 4.27 may be also written in terms of an equivalent variable stiffness $I_e(x)$ and an equivalent variable mass density $m_e(x)$ as follows

$$\frac{d^2}{dx^2}\left\{E_x I_e(x) y_d''(x)\right\} - m_e(x)\omega^2 y_d(x) = 0 \tag{4.28}$$

where

$$I_e(x) = \frac{I_x}{\left[1+(y_s')^2\right]^{3/2}} \tag{4.29}$$

$$m_e(x) = \sqrt{1+(y_s')^2}\,m(x_o) \tag{4.30}$$

and I_x is the moment of inertia of the original member at any $0 \le x \le L_o$. The quantities $I_e(x)$ and $m_e(x)$ take into account the change of mass and moment of inertia due to large static deformation. Therefore, Equation 4.28 may be thought of as representing a straight beam of length L_o, which vibrates with the same frequencies of vibration as the initial member from its static equilibrium position $y_s(x)$. The variation of its equivalent moment of inertia $I_e(x)$ and equivalent mass $m_e(x)$ are given by Equations 4.29 and 4.30, respectively. Therefore, in summary, the following two differential equations define completely the transverse vibration due to bending of a flexible member undergoing large static deformation $y_s(x)$ with small amplitudes of vibration:

$$\frac{d^2}{dx^2}\left\{\frac{E_x I_x y_s''(x)}{\left[1+(y_s')^2\right]^{3/2}}\right\} = -w_o(x_o)\sqrt{1+(y_s')^2} \tag{4.31}$$

$$\frac{d^2}{dx^2}\left\{E_x I_e(x) y_d''(x)\right\} - m_e(x)\omega^2 y_d(x) = 0 \tag{4.32}$$

Equations 4.31 and 4.32 must be solved simultaneously for the frequencies of vibration. Equation 4.31 is the large static equilibrium equation which defines the large static configuration. Equation 4.31 will be difficult to solve because it involves a fourth order nonlinear differential equation. However, Equation 4.31 is equivalent to the Euler-Bernoulli's equation given below:

$$\frac{y_s''}{\left[1+(y_s')^2\right]^{3/2}} = -\frac{M_x}{E_x I_x} \tag{4.33}$$

where M_x is the bending moment of the member and $E_x I_x$ is its bending stiffness.

Instead of solving Equation 4.31, we are now in a position to solve Equation 4.33, which utilizes the suggestions made in statements 3 and 4 at the end of the preceding section. In general, M_x and I_x are integral equations which depend on the large static deformation of the member. Therefore, both M_x and I_x may be simplified by replacing the integtal equation with a function of the unknown horizontal displacement $\Delta(x)$, as discussed in the first chapter.

When $y_s(x)$ and $y'_s(x)$ are known, then Equation 4.32 may be solved for the computation of the natural frequencies of vibration and the corresponding mode shapes. Equation 4.32 may be thought of as representing the transversed bending vibration of a pseudovariable stiffness member of equivalent mass density m_e and equivalent moment of inertia I_e. The quantities $I_e(x)$ and $m_e(x)$ define the geometry and mass of an equivalent straight member. The depth $h_e(x)$ of the equivalent straight member is given by the equation

$$h_e(x) = \frac{h(x)}{\left[1+(y_s')^2\right]^{1/2}} \tag{4.34}$$

where h(x) represents the variation in depth of the original member. The depth of the original member may have any arbitrary variation along its length.

The solution of the differential equation of motion given by Equation 4.32 may be obtained using known methods of analysis for free vibration of beams.

The application of the finite difference method, or utilization of Galerkin's finite element method, should yield reasonable results. The same methods could be used to solve Equation 4.27, which is the nonlinear differential equation of motion representing directly the original problem. The solution of Equation 4.28, however, would be more convenient. Galerkin's finite element method (GFEM) with equivalent uniform stiffness and mass will be discussed here.

4.4 GFEM

GFEM may be used to solve Equation 4.27 or Equation 4.28 for the computation of the natural frequencies and the corresponding mode shapes of flexible members. In order to apply Galerkin's method, we need to calculate the interpolation functions. In the consistent method, the interpolation function is obtained by solving the equation

$$\frac{d^2}{dx^2}\left\{E_x I_x \frac{y_d''(x)}{\left[1+(y')^2\right]^{3/2}}\right\} = 0 \tag{4.35}$$

Since y_d cannot be solved explicitly, it complicates the solution of Galerkin's method. In order to avoid this problem, we develop an equivalent uniform stiffness and mass approach so that uniform shape functions can be used.

In this method, an equivalent uniform stiffness and an equivalent uniform mass are defined for each element. Based on the differential equation for the kth element, the equivalent uniform stiffness and mass are defined by using Galerkin's method with uniform shape function. It should be noted here that the solution of Equation 4.27 represents the solution of the original member as a curved beam because of its large static deformation. The solution of Equation 4.28 represents the solution of an equivalent straight beam at length L_o with an equivalent stiffness $I_e(x)$ and an equivalent mass $m_e(x)$.

The main reason in using the equivalent stiffness and mass approach in Galerkin's method is because it can replace the complex variable stiffness and equivalent masses with uniform stiffness and uniform mass density. The methodology is as follows:

The beam is divided into M elements. There are a total of M + 1 node points. Each element has two degrees of freedom per node, vertical translation, and rotation. Now the differential equation for the kth element can be expressed as

$$E\tilde{J}_k y_d'''' - \beta_k \omega^2 y_d = 0 \qquad k = 1, 2, 3, \ldots, M \tag{4.36}$$

where $\tilde{J}_k$ is the equivalent uniform stiffness of the kth element and β_k is the equivalent uniform mass of the same element.

The expressions for $\tilde{J}_k$ and β_k are as follows:

$$\tilde{J}_k = 1/2I_1\left\{\frac{f(x_i)}{\left\{1+\left[y_s'(x_i)\right]^2\right\}^{3/2}} + \frac{f(x_j)}{\left\{1+\left[y_s'(x_j)\right]^2\right\}^{3/2}}\right\} \tag{4.37}$$

$$\beta_k = 1/2m(x_o)\left\{\sqrt{1+\left[y_s'(x_i)\right]^2} + \sqrt{1+\left[y_s'(x_j)\right]^2}\right\} \tag{4.38}$$

If $\beta_k = m(x_o)$, then the effect of change due to large static curvature is ignored.

Galerkin's method is used in order to obtain an approximate solution to a differential equation. It is accomplished by requiring that the error between the approximate solution and the true solution is orthogonal to the function used in the approximation.

If we start with a differential equation

$$E\tilde{J}_k y_d'''' - \beta_k \omega^2 y_d = 0 \tag{4.39}$$

and approximate the solution by

$$y_d(x) = \sum +_i U_i \tag{4.40}$$

then the equation becomes

$$E\tilde{J}_k\left[+_i U_i\right]'''' \quad \beta_k \omega^2\left[+_i U_i\right] = \varepsilon \tag{4.41}$$

where ε is the residual or error because the solution is only approximate. The purpose here is to make the error as small as possible. One way to achieve this is by requiring that

$$\int_R +_i \varepsilon dR = 0 \tag{4.42}$$

for each of the basis function $\mathbf{L}_i$. The integral states that the basis function must be orthogonal to the error.

By substituting ε into Equation 4.42 we find

$$\int_0^{L_o} +_i\left[E\tilde{J}_k y_d'''' - \beta_k \omega^2 y_d\right]dx = 0 \tag{4.43}$$

The interpolation function for y_d is defined over a single element and therefore Equation 4.43 must be written in terms of a summation, i.e.,

$$\sum_{k=1}^{M} \int_0^{L_0} +_i \left[E\tilde{J}_k y_d'''' - \beta_k \omega^2 y_d \right] dx = 0 \tag{4.44}$$

The integral must be reduced into one containing first and second derivatives before we can define the stiffness and mass matrices. Integrating by parts, element by element, we have

$$\begin{aligned} &E\tilde{J}\left\{ \left[+_i^k \frac{d}{dx}[y_d''] \right]_0^{L_k} - \left[\frac{d}{dx} +_i^k [y_d''] \right]_0^{L_k} \right. \\ &\left. + \int_0^{L_k} \frac{d^2}{dx^2} +_i^k [y_d''] dx \right\} - \omega^2 \beta_k \int_0^{L_k} \frac{d^2}{dx^2} +_i^k [y_d] dx = 0 \end{aligned} \tag{4.45}$$

When the summation over the elements is completed, the first two terms in Equation 4.45 will drop out, and we obtain

$$E\tilde{J}_k \int_0^{L_k} \frac{d^2}{dx^2} [+_k]\{y_d''\} dx - \omega^2 \beta_k \int_0^{L_k} [+_k]\{y_d\} dx = 0 \tag{4.46}$$

If $y_d(x) = [\mathbf{L}]\{U\}$, then Equation 4.46 becomes

$$E\tilde{J}_k \int_0^{L_k} [+_k'']^t [+_k'']\{U\} dx - \omega^2 \beta_k \int_0^{L_k} [+_k]^t [+_k]\{U\} dx = 0 \tag{4.47}$$

From Equation 4.47 the stiffness and mass matrices are defined as follows:

$$[K]\{U\} - \omega^2 [M]\{U\} = 0 \tag{4.48}$$

where

$$\left[K^k\right] = E\tilde{J}_k \int_0^{L_k} [+_k'']^t [+_k''] dx \tag{4.49}$$

$$\left[M^k\right] = \beta_k \int_0^{L_k} [+_k]^t [+_k] dx = 0 \tag{4.50}$$

The derivation of the element stiffness and mass matrices may be carried out as follows.

Consider an element in bending vibration where $\tilde{J}_k$ and β_k are of equivalent uniform bending stiffness and equivalent uniformly distributed mass density, respectively, for the kth element. The interpolation or shape function is obtained by solving the equation

$$E\tilde{J}_k \frac{d^4 y_d(x)}{dx^4} = 0 \tag{4.51}$$

By integrating Equation 4.51 four times we obtain

$$y_d(x) = C_1 x^3 + C_2 x^2 + C_3 x + C_4 \tag{4.52}$$

The constants C_1, C_2, C_3, and C_4 are determined from the geometric boundary conditions at each node. There are two degrees of freedom per node, vertical translation, and rotation. These boundary conditions are

$$y_d(0) = y_1 \tag{4.53}$$

$$\frac{dy_d(0)}{dx} = \Theta_1 \tag{4.54}$$

$$y_d(L_{k.}) = y_2 \tag{4.55}$$

$$\frac{dy_d(L_k)}{dx} = \Theta_2 \tag{4.56}$$

By applying the above boundary conditions to Equation 4.52), we obtain the following expression for $y_d(x)$:

$$y_d(x) = +_1(x)y_1 + +_2(x)L_k\Theta_1 + +_3(x)y_2 + +_4(x)L_k\Theta_2 \tag{4.57}$$

where

$$+_1(x) = 1 - 3\left[\frac{x}{L_k}\right]^2 + 2\left[\frac{x}{L_k}\right]^3 \tag{4.58}$$

$$+_2(x) = \frac{x}{L_k} - 2\left[\frac{x}{L_k}\right]^2 + \left[\frac{x}{L_k}\right]^3 \tag{4.59}$$

$$+_3(x) = 3\left[\frac{x}{L_k}\right]^2 - 2\left[\frac{x}{L_k}\right]^3 \tag{4.60}$$

$$+_4(x) = -\left[\frac{x}{L_k}\right]^2 + \left[\frac{x}{L_k}\right]^3 \tag{4.61}$$

which are known as Hermite cubics.

The stiffness and mass matrices are derived by using Equations 4.49 and 4.50. Another important factor in this procedure is that both the stiffness and mass matrices for each element have the same form. They are as follows:

$$\left[K^k\right] = \frac{E\tilde{J}_k}{L_k^3}\begin{bmatrix} 12 & 6 & -12 & 6 \\ & 4 & -6 & 2 \\ & & 12 & -6 \\ \text{symm} & & & 4 \end{bmatrix} \tag{4.62}$$

$$\left[M^k\right] = \frac{\beta_k}{420L_k}\begin{bmatrix} 156 & 22 & 54 & -13 \\ & 4 & 13 & -3 \\ & & 156 & -22 \\ \text{symm} & & & 4 \end{bmatrix} \tag{4.63}$$

Where $E\tilde{J}_k$ and β_k are the equivalent uniform bending stiffness and equivalent uniformly distributed mass density, respectively, for the kth element.

After the individual stiffness and mass matrices are formed, they are assembled. This is accomplished by adding the contributions from all the elements, i.e.,

$$\left[K_s\right] = \sum_{k-1}^{M}[K] \tag{4.64}$$

$$\left[M_s\right] = \sum_{k-1}^{M}[M] \tag{4.65}$$

The equation of motion becomes

$$[K_s]\{U\} - \omega^2[M_s]\{U\} = 0 \quad (4.66)$$

In order to compute the natural frequencies of vibration and the corresponding mode shapes, we must solve the eigenvalue problem

$$[K]\{U\} = \omega^2[M]\{U\} \quad (4.67)$$

where [K] is the stiffness matrix, [M] is the mass matrix, and {U} represents the nodal coordinates. By multiplying both sides of Equation 4.67 by $[K]^{-1}$, we obtain

$$\{U\} = \omega^2[K]^{-1}[M]\{U\} \quad (4.68)$$

Equation 4.68 may be written in a more standard form as

$$[A - \Lambda I]\{U\} = 0 \quad (4.69)$$

where

$$\Lambda = \frac{1}{\omega^2} \quad (4.70)$$

A canned eigensolver may be used to determine the eigenvalues and corresponding eigenvectors, and thus Equation 4.69 may be solved to obtain the natural frequencies of flexible beams.

4.5 VIBRATION OF UNIFORM FLEXIBLE CANTILEVER BEAMS BY USING AN EQUIVALENT STRAIGHT BEAM AND GFEM

Consider the uniform flexible cantilever beam in Figure 4.1a, where w_o is its distributed weight, and possibly other additional attached weights participating in the vibrational motion of the member. Its deformed configuration is shown in Figure 4.1b. Let it be assumed that the length L = 1000 in. (25.4 m), $EI_1 = EI = 180 \times 10^3$ kip-in.^{2} (516,541 Nm^2), and w_o = 1.5 lb/in. (262.6902 N/m).

The large static deflection analysis will be carried out first in order to establish the static equilibrium position $y_s(x)$ of the member. Since the member is uniform, we have I_x = Constant = I. The moment M_x at any $0 \leq x \leq L_o$ is given by the equation

$$M_x = -\frac{w_o x}{2} x_o(x) \quad (4.71)$$

with

$$x_o(x) = \int_0^x \left\{1 + \left[y_s'(\xi)\right]^2\right\}^{1/2} d\xi \tag{4.72}$$

and ξ is a dummy variable. The coordinate $x_o(x)$, as stated earlier, is a function of the horizontal deformation $\Delta(x)$ and it can be written as

$$x_o(x) = x + \Delta(x) \tag{4.73}$$

By assuming $\Delta(x)$ = constant = Δ, as in the preceding chapters, Equation 4.73 yields

$$x_o(x) = x + \Delta \tag{4.74}$$

therefore, the moment equation given by Equation 4.71 yields

$$M_x = -\frac{w_o x}{2}(x + \Delta) \tag{4.75}$$

Other expressions for $\Delta(x)$, such as $\Delta(x/L_o)$, $\Delta[x/L_o]^{1/2}$, and $\Delta\sin(\pi x/2L_o)$ may be used. The accuracy of the method, however, is not very sensitive to the expression used for $\Delta(x)$.

By subsituting Equation 4.75 into the Euler-Bernoulli equation, that is Equation 4.33), we obtain

$$\frac{y_s''}{\left[1 + (y_s')^2\right]^{3/2}} = \frac{w_o x}{2EI}(x + \Delta) \tag{4.76}$$

By integrating Equation 4.76 once and satisfying the boundary condition of zero rotation at $x = L_o$, we obtain

$$\frac{y_s'}{\left[1 + (y_s')^2\right]^{1/2}} = \frac{w_o}{12EI}\left[2x^3 + 3\Delta x^2 - 2(L - \Delta)^3 - 3\Delta(L - \Delta)^2\right] \tag{4.77}$$

Solution of Equation 4.77 for $y_s'(x)$ yields the expression

$$y_s'(x) = \frac{Q(x)}{\left\{1 - [Q(x)]^2\right\}^{1/2}} \tag{4.78}$$

where

$$Q(x) = \frac{w_0}{12EI}\left[2x^3 + 3\Delta x^2 - 2(L-\Delta)^3 - 3\Delta(L-\Delta)^2\right] \tag{4.79}$$

It should be noted that Equation 4.78 is in terms of the unknown horizontal displacement Δ of the free end of the member. A trial and error procedure, as done in preceding chapters, may be used here to obtain Δ by using the equation

$$L = \int_0^{L_0} \left\{1 + \left[y_s'(x)\right]^2\right\}^{1/2} dx \tag{4.80}$$

By making the change of variables $x = \xi L_0$, $dx = L_0 d\xi$, Equation 4.80 yields

$$L = L_0 \int_0^1 \left\{1 + \left[y_s'(\xi)\right]^2\right\}^{1/2} d\xi \tag{4.81}$$

Equation 4.81 is easier to handle. The trial and error procedure yields Δ = 277.25 in. (7.0422 m), and $L_0 = (L - \Delta) = 1000 - 277.25 = 722.75$ in. (18.3579 m). At the free end, the static rotation θ_s may be obtained from Equation 4.78. This yields $\theta_s = \tan^{-1} y_s' = 55.69°$.

The static equilibrium position y_s may be completely established since y'_s is known from Equation 4.78. This is done in one of two ways: (1) by integrating Equation 4.78 once and satisfying the boundary condition of zero deflection at the fixed end or (2) by using pseudolinear equivalent systems as discussed in the preceding chapters. At the free end of the beam the vertical deflection δ_A is found to be 638.0 in. (16.2052 m).

Now that the static equilibrium position y_s is completely established, the differential equation of motion given by Equation 4.28 may be used to determine the natural frequencies of vibration of the member and the corresponding mode shapes. Equation 4.28 represents a pseudovariable stiffness equivalent straight beam of length L_0, as shown in Figure 4.2a. Its equivalent depth $h_e(x)$, moment of inertia $I_e(x)$, and mass $m_e(x)$ are given by the expressions

$$h_e(x) = h_1\left\{1 - [Q(x)]^2\right\}^{1/2} \tag{4.82}$$

$$I_e(x) = I_1\left\{1 - [Q(x)]^2\right\}^{3/2} \tag{4.83}$$

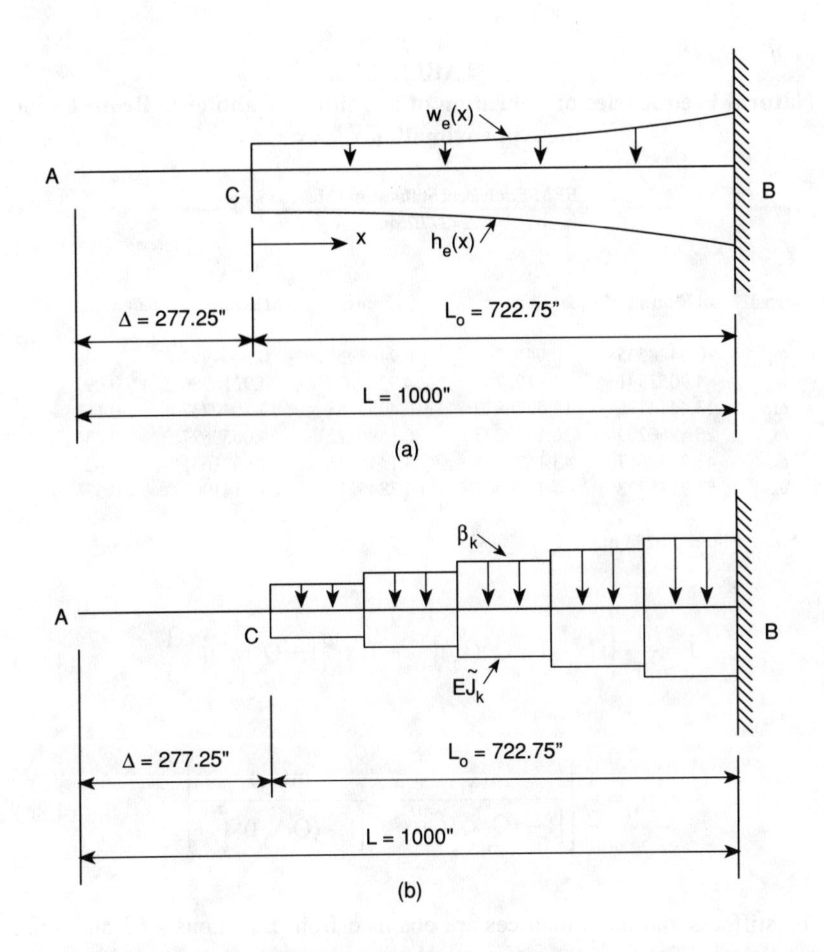

FIGURE 4.2. (a) Equivalent pseudovariable straight cantilever beam of length L_o, undergoing small amplitude vibrations; (b) equivalent piecewise uniform straight cantilever beam with equivalent piecewise uniform mass.

$$m_e(x) = \frac{m(x)}{\left\{1 - [Q(x)]^2\right\}^{1/2}} \tag{4.84}$$

where Q(x) is given by Equation 4.79.

The pseudovariable stiffness member in Figure 4.2a may be solved by using the GFEM in conjunction with equivalent uniform stiffness and equivalent uniform mass, as shown in Figure 4.2b.

In order to apply GFEM with equivalent uniform stiffness $E\tilde{J}$ and equivalent uniform mass β, the beam is subdivided into M elements. The equivalent uniform stiffness and equivalent uniform mass density are defined by Equations 4.37 and 4.38, respectively, which in this case take the form

TABLE 4.1
Natural Frequencies of Vibration of a Uniform Cantilever Beam Using the Approximation $x_o = x + \Delta$

FEM/Equivalent Stiffness and Mass
Δ=277.25 in.

ω (rps)	10-Element	20-Element	40-Element	FDM 81 elements	Relative difference (%)
ω_1	0.9428535	0.9453264	0.9468956	0.9464998	0.040
ω_2	4.9057541	4.9192247	4.9261675	4.9216986	0.092
ω_3	13.4414196	13.4236879	13.4253788	13.4040737	0.159
ω_4	26.4260209	26.1823273	26.1596222	26.0850677	0.286
ω_5	43.7238617	43.1523285	43.1515045	42.9671631	0.429
ω_6	63.9175772	64.4496307	64.3845215	64.0141907	0.578

Note: 1 in. = 0.0254 m.

$$\tilde{J} = \frac{1}{2I_1}\left\{f(x_i)\left[1 - Q^2(x_i)\right]^{3/2} + f(x_j)\left[1 + Q^2(x_j)\right]^{3/2}\right\} \tag{4.85}$$

$$\beta_k = \frac{1}{2}\left\{\frac{m(x_i)}{\left[1 - (Q(x_i))^2\right]^{1/2}} + \frac{m(x_j)}{\left[1 - (Q(x_j))^2\right]^{1/2}}\right\} \tag{4.86}$$

The stiffness and mass matrices are obtained from Equations 4.62 and 4.63, respectively. The element stiffness and mass matrices are assembled by using Equations 4.64 and 4.65, respectively, and using the boundary conditions of zero vertical displacement and zero rotation at the fixed end. The eigenvalue problem is then solved in order to determine the natural frequencies of vibration of the member and the corresponding mode shapes. The results are shown in Table 4.1 by considering M = 10, 20, and 40 elements. The schematic representation of the first three modes of vibration are shown in Figure 4.3b. Note in this figure that the vibration is taking place with respect to the static equilibrium position y_s.

The pseudovariable system in Figure 4.2a was also solved by using the finite difference method (FDM) with 81 elements. The results are shown in Tables 4.1 and 4.2. Both methods provided good agreement for the results obtained. In Table 4.2 we should note that the case of $\Delta(x) = \Delta$ represents an upper bound for the results, while all other cases of $\Delta(x)$ accurately represent the correct variation of $\Delta(x)$. The results, however, indicate that the large deformation of the flexible bar is not very sensitive to the variation of $\Delta(x)$.

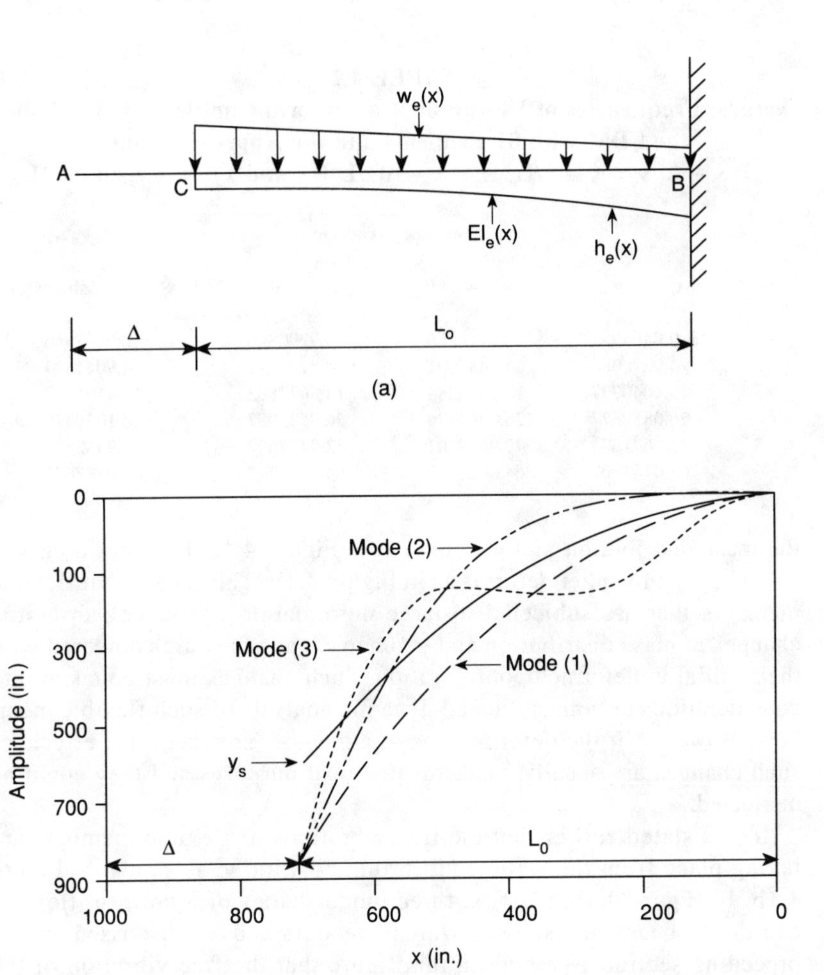

FIGURE 4.3. (a) Pseudovariable straight cantilever beam of length L_0, undergoing small amplitude vibrations; (b) first three mode shapes of the beam with $\Delta(x) = \Delta$, and w_o = 1.5 lb/in. (2.6269 N/m).

Note in Table 4.1 that the column designated as FDM shows the results obtained using the FDM with 81 elements.

4.6 EFFECT OF MASS POSITION CHANGE DURING LARGE DEFORMATION

When the deformation of a member is large, see for example Figure 4.1b, the initially straight member, in its deformed configuration, becomes a curved member. Therefore, the initially straight configuration of

TABLE 4.2
Natural Frequencies of Vibration of a Uniform Cantilever Beam Using the FDM with 81 Elements and the Approximations $x_o = x + \Delta$, $x_o = x + \Delta/L_o$, $x_o = x + \Delta[x/L_o]^{1/2}$, and $x_o = x + \Delta\sin \pi x/2L_o$

	Frequency (rad/sec)			
Mode	$x_o = x + \Delta$	$x_o = x + \Delta x/L_o$	$x_o = x + \Delta[x/L_o]^{1/2}$	$x_o = x + \Delta\sin(\pi x/2L_o)$
1	0.9464998	0.9178566	0.9290826	0.9118916
2	4.9216986	4.8487091	4.8703041	4.9044514
3	13.4040737	13.3545284	13.3677883	13.4105587
4	26.0850677	26.0493774	26.0579987	26.1030426
5	42.9671631	42.9428101	42.9477692	42.9932709
6	64.0141907	64.0025330	64.0033722	64.0487671

the mass distribution and geometry, see Figure 4.1a, becomes a curved one when the member deforms as in Figure 4.1b. This means that flexible members that are subjected to large deformations, produce substantial changes in mass distribution and geometry when they are compared with their initial undeformed configuration. Such changes must be taken into consideration for both static and dynamic analysis of such flexible members. However, if the deformations of a member are small, the effects of such changes are usually small for practical purposes and they could be neglected.

It was stated earlier that the free vibrations of flexible members are taking place from the static equilibrium position y_s as shown in Figure 4.1b. In Figure 4.3b, the first three mode shapes of a uniform flexible cantilever beam are shown, which are obtained as discussed in the preceding section. We note in this figure that the free vibration of the member is taking place with respect to the static equilibrium position y_s as shown. At the position y_s, the geometric distribution of mass $m(x_o)$ of the member is substantially changed when it is compared to its initial straight configuration.

In order to clarify this point further, we again write Equation 4.27, i.e.,

$$\frac{d^2}{dx^2}\left\{E_x I_x \frac{y_d''(x,t)}{\left[1+(y_s')^2\right]^{3/2}}\right\} - \left(\sqrt{1+(y_s')^2}\right) m(x_o)\omega^2 y_d(x,t) = 0 \tag{4.87}$$

In the second term of Equation 4.87 the quantity $[1 + (y_s')^2]^{1/2}$ takes into consideration the geometric configuration of the mass $m(x_o)$ at the large static equilibrium position y_s. If the static deformation y_s is rather small, or moderately large, the quantity $[1 + (y_s')^2]^{1/2}$ could be assumed equal to unity, and Equation 4.87 yields

$$\frac{d^2}{dx^2}\left\{E_xI_x\frac{y_d''(x,t)}{\left[1+(y_s')^2\right]^{3/2}}\right\}-m(x_o)\omega^2y_d(x,t)=0 \tag{4.88}$$

Equations 4.87 and 4.88 will be used here in order to demonstrate the error that is introduced in the vibration analysis of flexible members when the changes of the mass geometry during large deformation are not taken into consideration.

Consider the tapered cantilever beam in Figure 4.4a, where the weight w_o = 5 lb/in. (875.59 N/m), represents the weight of the beam and other possible weights attached to the member and participating in its vibrational motion. The Length L = 1000 in. (25.4 m), stiffness EI_A = 180,000 kip-in^2 (516,541 Nm^2), and taper n = 1.5. This problem will be solved for the frequencies of vibration and the corresponding mode shapes by using Equations 4.87 and 4.88 and the results will be compared. The static equilibrium position y_s will be established using Equation 4.33 which is the Euler-Bernoulli equation, as discussed in the preceding section.

The variation of the moment of inertia I_x at any $0 \le x \le (L-\Delta)$ is given by the expression

$$I_x=\frac{bh^3}{12}\left[1+(n-1)\frac{x}{L-\Delta}\right]^3=I_Af(x) \tag{4.89}$$

where

$$I_A=\frac{bh^3}{12} \tag{4.90}$$

$$f(x)=\left[1+(n-1)\frac{x}{L-\Delta}\right]^3 \tag{4.91}$$

and b is the constant width of the member. Note that I_A is the moment of inertia at the free end of the bar and f(x) represents its variation along its length.

The moment M_x of weight w_o at any $0 \le x \le L_o$, where $L_o = (L-\Delta)$, is

$$M_x=w_ox_o\frac{x}{2} \tag{4.92}$$

where

$$x_o=x+\Delta(x) \tag{4.93}$$

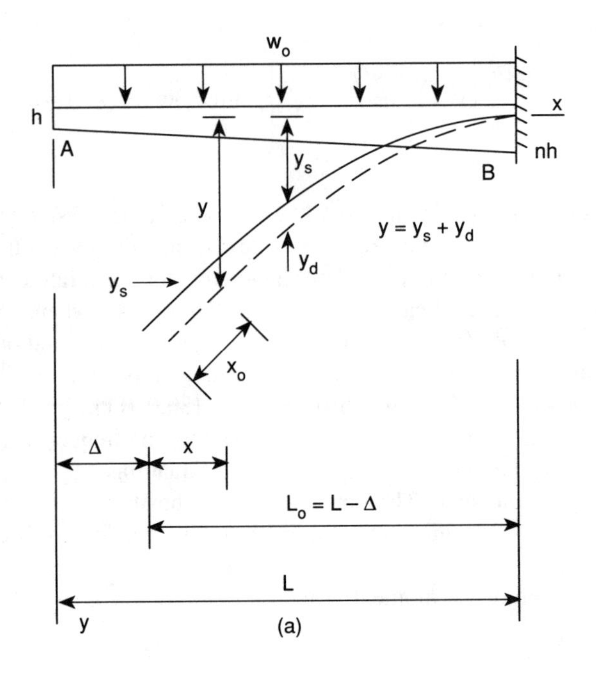

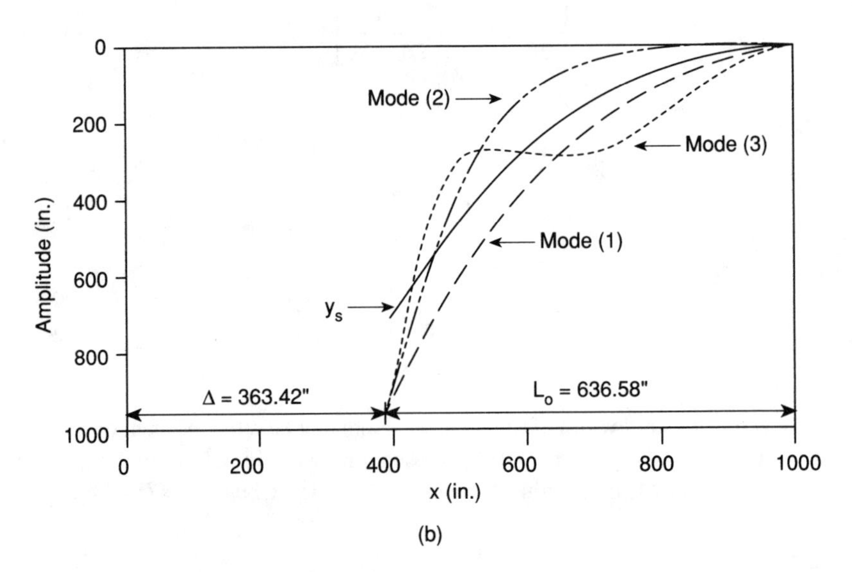

FIGURE 4.4. (a) Tapered cantilever member; (b) first three mode shapes of the member (1 in. = 0.0254 m).

and $\Delta(x)$ is a function representing the horizontal displacement of the bar at any x. By substituting Equation 4.93 into Equation 4.92 we have

$$M_x = w_o[x + \Delta(x)]\frac{x}{2} \tag{4.94}$$

It was demomstrated in preceding sections that a reasonable solution to the problem may be obtained by assuming that $\Delta(x)$ = constant = Δ, which is the horizontal displacement Δ of the free end of the member.

By using Equations 4.89, 4.90, 4.91, and 4.94, and assuming $\Delta(x) = \Delta$, Equation 4.33 yields

$$\frac{y_s''}{\left[1+(y_s')^2\right]^{3/2}} = -\frac{w_o(L-\Delta)^3}{2EI_A}\cdot\frac{(x^2+\Delta x)}{[(n-1)x+(L-\Delta)]^3} \tag{4.95}$$

In the above equation, the factor

$$M_e = \frac{M_x}{f(x)} = \frac{w_o(L-\Delta)^3}{2}\cdot\frac{(x^2+\Delta x)}{[(n-1)x+(L-\Delta)]^3} \tag{4.96}$$

represents the moment at any $0 \le x \le L$ of the equivalent nonlinear system of constant stiffness EI_A, and the factor

$$M_e' = \frac{z_e M_x}{f(x)} = \frac{z_e w_o(L-\Delta)^3}{2}\cdot\frac{(x^2+\Delta x)}{[(n-1)x+(L-\Delta)]^3} \tag{4.97}$$

where

$$z_e = \left[1+(y_s')^2\right]^{3/2} \tag{4.98}$$

The term M_e' represents the moment at any $0 \le x \le L_o$ of an equivalent linear system of constant stiffness EI_A.

By integrating Equation 4.95 once and determining the constant of integration by satisfying the boundary condition of zero rotation at $x = (L-\Delta)$, we find

$$y_s'(x) = \frac{Q(x)}{\left\{1-[Q(x)]^2\right\}^{1/2}} \tag{4.99}$$

where

$$Q(x) = -\frac{w_o(L-\Delta)^3}{2EI_A}\left\{\frac{-3(n-1)^2x^2 - 2(n-1)x[L-\Delta+\Delta(n-1)]}{2(n-1)^3[(n-1)x+(L-\Delta)]^2}\right.$$

$$+\frac{-(L-\Delta)(n-1)\Delta + 2[(n-1)x+(L-\Delta)]^2 \ln[(n-1)x-(L-\Delta)]}{2(n-1)^3[(n-1)x+(L-\Delta)]^2}$$

$$-\frac{-(L-\Delta)^2[3(n-1)^2+2(n-1)] - \Delta(L-\Delta)[2(n-1)^2-(n-1)]}{2(n-1)^3n^2(L-\Delta)^2} \tag{4.100}$$

$$\left.+\frac{2n^2(L-\Delta)^2\ln[n(L-\Delta)]}{2(n-1)^3n^2(L-\Delta)^2}\right\}$$

The horizontal displacement Δ in Equation 4.100 may be determined from the equation

$$L = \int_0^{(L-\Delta)} \left[1+(y_s')^2\right]^{1/2} dx \tag{4.101}$$

by trial and error. That is, assume a value of Δ in Equation 4.101 and integrate to determine the length L. The procedure may be repeated for various values of Δ until the correct length L is obtained. With known Δ, the values of y_s' at any point $0 \le x \le Lo$ may be computed from Equation 4.99.

With known y_s' the static equilibrium position y_s may be determined by (1) integrating Equation 4.99 once and satisfying the boundary condition of zero deflection at $x = (L-\Delta)$, or (2) by following pseudolinear analysis as discussed in preceding sections. The second procedure that utilizes Equation 4.97) is the most convenient to use. On this basis, the static equilibrium position y_s may be established.

With known y_s', Equations 4.87 and 4.88 may be used to determine the frequencies of vibration of the flexible member and its corresponding mode shapes. Any known method of analysis that is appropriate for such types of problems may be used for this purpose. For the results obtained here, the FEM and the FDM, as well, are used and the results are compared. The FEM was carried out by using 10 elements, and 81 elements were used to carry out the FDM. The results for the first six natural frequencies of vibration are shown in Table 4.3. This table shows that both FEM and FDM yield similar results for the solution of Equation 4.87. The relative difference is rather small. The fifth column of Table 4.3 shows the results obtained by solving Equation 4.88, where the change in mass geometry during deformation is not taken into consideration. The FDM with 81 elements was used for this purpose. The last column in this table shows the large error involved when Equation 4.88 is used. Therefore, it may be concluded that the change in mass geometry during the

TABLE 4.3
Values of the First Six Natural Frequencies of Vibration of a Tapered Flexible Cantilever Beam by Using Equations 4.87 and 4.88

1 ω_n (rps)	2 FDM Equation 4.87 81 elements	3 FEM Equation 4.87 10 elements	4 % Difference	5 FDM Equation 4.88 81 elements	6 % Error
ω_1	0.8954	0.8957	0.0335	1.3707	53.08
ω_2	3.8808	3.8896	0.2268	5.3082	36.78
ω_3	9.8536	9.8990	0.4607	13.0459	32.40
ω_4	18.7886	18.9436	0.8250	24.6507	31.20
ω_5	30.6729	31.1738	1.6330	40.1203	30.80
ω_6	45.4691	46.7509	2.8191	59.4082	30.66

large static deformation of a member, must be taken into consideration when vibration analysis is considered.

The first three mode shapes of the member that are obtained from the solution of Equation 4.87 are shown in Figure 4.4b. In this figure, the curve y_s represents the static equilibrium position of the member.

4.7 GALERKIN'S CONSISTENT FEM

The application of the Galerkin's consistent FEM for the solution of Equation 4.28 is briefly discussed here. We approximate the solution $y_d(x)$ in Equation 4.28 by

$$y_d(x) = \sum N_i U_i \tag{4.102}$$

which yields

$$\left\{EI_e(x)\left[N_i U_i\right]''\right\}'' - m_e(x)\omega^2\left[N_i U_i\right] = \varepsilon \tag{4.103}$$

In order to make the error ε as small as possible, we require

$$\int_R N_i \varepsilon dR = 0 \tag{4.104}$$

for each of the basis function N_i. The basis function must be orthogonal to the error.

By substituting ε from Equation 4.103 into Equation 4.104, we obtain

$$\int_0^{L_o} N_i \left\{ \left[EI_e(x) y_d'' \right]'' - m_e(x)\omega^2 y_d \right\} dx = 0 \tag{4.105}$$

Since the interpolation function for y_d is defined over a single element, Equation 4.105 should be written as a summation:

$$\sum_{k=1}^{M} \int_0^{L_o} N_i \left\{ \left[EI_e(x) y_d'' \right]'' - m_e(x)\omega^2 y_d \right\} dx = 0 \tag{4.106}$$

By integrating by parts and element by element, and completing the summation over the elements in the resulting expression, we obtain

$$\int_0^{L_k} \frac{d^2}{dx^2} [N_k] \{ EI_e(x) y_d'' \} dx - \omega^2 \int_0^{L_k} [N_k] \{ m_e(x) y_d \} dx = 0 \tag{4.107}$$

With $y_d(x) = [N]\{U\}$, Equation 4.107 yields

$$[K]\{U\} - \omega^2 [M]\{U\} = 0 \tag{4.108}$$

where [K] and [M] are the stiffness and mass matrices, respectively, defined as

$$\left[K^k\right] = \int_0^{L_k} EI_e(x) \left[N_k''\right]^t \left[N_k''\right] dx \tag{4.109}$$

$$\left[M^k\right] = \int_0^{L_k} m_e(x) \left[N_k\right]^t \left[N_k\right] dx \tag{4.110}$$

In order to derive the element stiffness and mass, we rewrite Equation 4.103 as

$$EI_1 \left[\frac{y_d''}{\Pi(x)} \right]'' - m_e(x) z_e^{1/3} \omega^2 y_d = 0 \tag{4.111}$$

where y'_s is the slope of the large static deflection and

$$II(x) = \frac{z_e}{f(x)} \tag{4.112}$$

$$z_e = \left\{1 + \left[y_s'(x)\right]^2\right\}^{3/2} \tag{4.113}$$

For uniform cross sections we have f(x) = 1.

The interpolation or shape functions which are required for the implementation of Galerkin's method may be obtained by solving explicitly for y_d the equation

$$EI_1\left[\frac{y_d''}{II(x)}\right] = 0 \tag{4.114}$$

Since the depth function h(x) may be complicated, the integration of Equation 4.114 explicitly for y_d may become impossible, and a curve-fitting technique such as the least squares may have to be utilized. We rewrite Equation 4.114 as

$$EI_1\left\{\frac{y_d''}{\left[II_a(x) + E_r(x)\right]}\right\}'' = 0 \tag{4.115}$$

where $II_a(x)$ is a polynomial and $E_r(x)$ is the error fuction to be minimized. In general, $II_a(x)$ may be assumed as

$$II_a(x) = a_o + a_1x + a_2x^2 + a_3x^3 + a_4x^4 \tag{4.116}$$

By considering many elements the error function may be minimized, that is $E_r(x) = 0$, and Equation 4.115 yields

$$EI_1\left[\frac{y_d''}{II_a(x)}\right] = 0 \tag{4.117}$$

The second, third, and fourth integrations of Equation 4.117 yields the following expressions

$$y_d'' = \left[C_1 + C_2x\right]II_a(x) \tag{4.118}$$

$$y'_d = C_1 f_1(x) + C_2 f_2(x) + C_3 \tag{4.119}$$

$$y_d = C_1 g_1(x) + C_2 g_2(x) + C_3 x + C_4 \tag{4.120}$$

where

$$f_1(x) = \frac{x}{60}(60a_o + 30a_1 x + 20a_2 x^2 + 15a_3 x^3 + 12a_4 x^4) \tag{4.121}$$

$$f_2(x) = \frac{x^2}{60}(30a_o + 20a_1 x + 15a_2 x^2 + 12a_3 x^3 + 10a_1 x^4) \tag{4.122}$$

$$g_1(x) = \frac{x^2}{60}(30a_o + 10a_1 x + 5a_2 x^2 + 3a_3 x^3 + 2a_4 x^4) \tag{4.123}$$

$$g_2(x) = \frac{x^3}{420}(70a_o + 35a_1 x + 21a_2 x^2 + 14a_3 x^3 + 10a_4 x^4) \tag{4.124}$$

The constants C_1, C_2, C_3, and C_4 may be determined for each beam element by using the boundary conditions of rotation and vertical translation at each node, i.e.,

$$y_d\left(L_i^k\right) = y_1 \qquad y_d\left(L_j^k\right) = y_2$$

$$y'_d\left(L_i^k\right) = \theta_1 \qquad y'_d\left(L_j^k\right) = \theta_2$$

Application of the above boundary conditions yields the matrix equation

$$\{y\} = [A]\{C\} \tag{4.125}$$

where

$$[A] = \begin{bmatrix} g_1^k(L_i^k) & g_2^k(L_i^k) & L_i^k & 1.0 \\ L_o f_1^k(L_i^k) & L_o f_2^k(L_i^k) & L_o & 0 \\ g_1^k(L_j^k) & g_2^k(L_j^k) & L_j^k & 1.0 \\ L_o f_1^k(L_j^k) & L_o f_1^k(L_j^k) & L_o & 0 \end{bmatrix} \tag{4.126}$$

$$\{y\} = \begin{bmatrix} y_1 \\ \theta_1 L_o \\ y_2 \\ \theta_2 L_o \end{bmatrix} \qquad \{C\} = \begin{bmatrix} C_1 \\ C_2 \\ C_3 \\ C_4 \end{bmatrix} \tag{4.127}$$

In the above two equations, L_i^k and L_j^k are the positions of the ith and jth nodes, respectively, of the kth element, and g_1^k, g_2^k, f_1^k, and f_2^k, are the g_1, g_2, f_1, and f_2, of the kth element.

By solving the matrix equation $\{C\} = [A]^{-1}\{y\}$ or $\{C\} = [B]\{y\}$, the constants C_1, C_2, C_3, and C_4 may be determined, and Equation 4.120 yields

$$y_d(x) = N_1(x)y_1 + N_2(x)\theta_2 L_o + N_3(x)y_2 + N_4(x)\theta_4 L_o \tag{4.128}$$

where

$$\begin{aligned} N_1(x) = {} & B(1,1)g_1^k(x + L_k) + B(2,1)g_2^k(x + L_k) \\ & + B(3,1)\,(x + L_k) + B(4,1) \end{aligned} \tag{4.129}$$

$$\begin{aligned} N_2(x) = {} & B(1,2)g_1^k(x + L_k) + B(2,2)g_2^k(x + L_k) \\ & + B(3,2)\,(x + L_k) + B(4,2) \end{aligned} \tag{4.130}$$

$$\begin{aligned} N_3(x) = {} & B(1,3)g_1^k(x + L_k) + B(2,3)g_2^k(x + L_k) \\ & + B(3,3)\,(x + L_k) + B(4,3) \end{aligned} \tag{4.131}$$

$$N_4(x) = B(1,4)g_1^k(x + L_k) + B(2,4)g_2^k(x + L_k) + B(3,4)(x + L_k) + B(4,4) \tag{4.132}$$

Equations 4.109 and 4.110 may be used to determine the stiffness and mass matrices, respectively. Simpson's rule may be used to carry out the required integrations. By adding the contributions of all elements, these matrices are assembled as

$$[K_s] = \sum_{k=1}^{M} K_k \tag{4.133}$$

$$[M_s] = \sum_{k=1}^{M} M_k \tag{4.134}$$

where the symbol M in the summation indicates the number of elements.

The natural frequencies of vibration and the corresponding mode shapes may be obtained by solving the following eigenvalue problem:

$$[M]\{\ddot{y}_d\} + [K]\{y_d\} = 0 \tag{4.135}$$

$$-\omega^2[M]\{y_d\} + [K]\{y_d\} = 0 \tag{4.136}$$

$$-\omega^2[K]^{-1}[M]\{y_d\} + \{y_d\} = 0 \tag{4.137}$$

where [K] and [M] are the stiffness and mass matrix, respectively, of the structure, ω are the free frequencies of vibration, and y_d represents the mode shapes. These frequency equations may be also represented by the convenient form given by Equation 4.69.

4.8 VIBRATION OF UNIFORM FLEXIBLE CANTILEVER BEAMS BY USING GALERKIN'S CONSISTENT FEM

Galerkin's consistent FEM will be applied here by considering the uniform flexible cantilever beam shown in Figure 4.1a. The same problem was solved in Section 4.5 by using Galerkin's method with equivalent uniform mass and stiffness, which permits the utilization of uniform shape functions.

The stiffness EI = 180,000 k-in.^{2} (516,541 Nm^2), the length L = 1000 in. (25.4 m), and weight w_o = 1.5 lb/in. (262.6902 N/m). The static equilibrium position y_s in Figure 4.1b may be established by using the Euler-Bernoulli equation, which is given by Equation 4.33. The solution of this equation, by assuming that $\Delta(x)$ = constant = Δ, was carried out in Section 4.5 yielding

$$y_s'(x) = \frac{Q(x)}{\left\{1-[Q(x)]^2\right\}^{1/2}} \tag{4.137a}$$

where

$$Q(x) = \frac{w_o}{12EI}\left[2x^3 + 3\Delta x^2 - 2(L-\Delta)^3 - 3\Delta(L-\Delta)^2\right] \tag{4.138}$$

The trial and error solution of Equation 4.137a was carried out by using Equation 4.81), and the results are Δ = 277.25 in. (7.0422 m), and $L_o = (L - \Delta)$ = 722.75 in. (18.3579 m). At the free end A of the member the static rotation $\theta_{sA} = \tan^{-1}y'_{sA}$ = 55.69° and the static vertical deflection δ_{sA} = 638.00 in. (16.2052 m). From Equation 4.137a, the static equilibrium position y_s may be obtained by either integrating Equation 4.137a once and satisfying the boundary condition of zero deflection at the fixed end, or by using pseudolinear analysis as discussed in preceding chapters.

With y_s determined as explained above, the natural frequencies of vibration and their corresponding mode shapes may be obtained by solving Equation 4.28. The equivalent depth $h_e(x)$, moment of inertia $I_e(x)$ and mass density $m_e(x)$, are given by Equations 4.34, 4.29, and 4.30, respectively. By using Equation 4.137a, we write

$$h_e(x) = h\left\{1-[Q(x)]^2\right\}^{1/2} \tag{4.139}$$

$$I_e(x) = I\left\{1-[Q(x)]^2\right\}^{3/2} \tag{4.140}$$

$$m_e(x) = \frac{m(x)}{\left\{1-[Q(x)]^2\right\}^{1/2}} \tag{4.141}$$

where Q(x) is given by Equation 4.138.

By using 10 elements, the approximation of $\Pi(x)$ by $\Pi_a(x)$ yields the following piecewise polynomials for each element:

$$\Pi_1(x) = 5.587401 - 0.3344214e-5x - 0.2486714e-4x^2$$
$$-0.7009043e-7x^3 + 0.2188282e-9x^4 \tag{4.142}$$
$$x \in [0, 100]$$

$$\Pi_2(x) = 5.569538 + 0.5562873e-3x - 0.3109594e-4x^2$$
$$-0.4223982e-7x^3 + 0.1822909e-9x^4 \tag{4.143}$$
$$x \in [100, 200]$$

$$\Pi_3(x) = 6.491382 - 0.1487796e-1x + 0.6361365e-4x^2$$
$$-0.2921511e-6x^3 + 0.4172325e-9x^4 \tag{4.144}$$
$$x \in [200, 250]$$

$$\Pi_4(x) = 4.498280 + 0.1837430e-1x - 0.1457312e-3x^2$$
$$+0.2967163e-6x^3 - 0.2066294e-9x^4 \tag{4.145}$$
$$x \in [250, 300]$$

$$\Pi_5(x) = -0.580204e2 + 0.7969625x - 0.3777663e-2x^2$$
$$+0.7817675e-5x^3 + 0.6039937e-8x^4 \tag{4.146}$$
$$x \in [300, 350]$$

$$\Pi_6(x) = 0.7355303e1 - 0.1478313e-1x - 0.8016825e-6x^2$$
$$+0.1366492e-7x^3 + 0.1986822e-11x^4 \tag{4.147}$$
$$x \in [350, 400]$$

$$
\begin{aligned}
II_7(x) = &-0.9941851e1 - 0.3571422e - 1x + 0.5934725e - 4x^2 \\
&-0.5613764e - 7x^3 + 0.2657374e - 10x^4 \\
&x \in [400, 480]
\end{aligned} \tag{4.148}
$$

$$
\begin{aligned}
II_8(x) = &-0.4163647e2 + 0.3645583x - 0.1105170e - 2x^2 \\
&+0.1449386e - 5x^3 - 0.7033348e - 9x^4 \\
&x \in [480, 560]
\end{aligned} \tag{4.149}
$$

$$
\begin{aligned}
II_9(x) = &0.1116807e2 - 0.4350773e - 1x + 0.7550647e - 4x^2 \\
&-0.6660819e - 7x^3 + 0.2258033e - 10x^4 \\
&x \in [560, 640]
\end{aligned} \tag{4.150}
$$

$$
\begin{aligned}
II_{10}(x) = &-0.6152563e1 + 0.5468258e - 1x - 0.1323083e - 3x^2 \\
&+0.1278867e - 6x^3 - 0.4228480e - 10x^4 \\
&x \in [640, 722.75]
\end{aligned} \tag{4.151}
$$

With known interpolation functions, the stiffness and mass matrices are obtained from Equations 4.109 and 4.110, respectively, by integration. Simpson's rule may be used for this purpose. The element stiffness and mass matrices are then assembled by using Equations 4.133 and 4.134 respectively. The boundary conditions of zero rotation and zero displacement at the fixed end of the member are used here. The solution of the eigenvalue problem for the first six frequencies of vibration, in radians per second (rps), yields the results shown in Table 4.4. This problem is also solved using the FDM, and the results are shown in the same table. The results of both methods are in close agreement. In Table 4.5, the first six free frequencies of vibration of the cantilever beam are obtained using the FDM with 81 elements. Various cases of the horizontal displacement function $\Delta(x)$ are examined and the results are compared. For all practical purposes, the results obtained for the various cases of $\Delta(x)$ are closely identical, which indicates that the free vibration of the member is not very sensitive to the variation of the function $\Delta(x)$.

TABLE 4.4
Natural Frequencies of Vibration for a Uniform Flexible Cantilever Beam by Using the Consistant GFEM and $\Delta(x)$ = Constant = Δ

ω(rps)	GFEM	FDM	Relative difference (%)
1	0.9466	0.9465	0.010
2	4.9261	4.9217	0.090
3	13.4379	13.4041	0.252
4	26.2340	26.0850	0.571
5	43.3370	42.9671	0.861
6	65.5216	64.0142	2.355

TABLE 4.5
Natural Frequencies Of Vibration for a Uniform Flexible Cantilever Beam by Using the FDM with 81 Elements and Various Functions of $\Delta(x)$

w(rps)	$\Delta(x) = \Delta$	$\Delta(x) = \Delta(x/L_o)$	$\Delta(x) = \Delta(x/L_o)^{1/2}$	$\Delta(x) = \Delta\sin(\pi x/2L_o)$
1	0.9465	0.9178	0.9291	0.9119
2	4.9217	4.8487	4.8703	4.9044
3	13.4041	13.3545	13.3678	13.4105
4	28.0850	26.0494	26.0580	26.1030
5	42.9671	42.9428	42.9477	42.9933
6	64.0142	64.0025	64.0034	64.0487

PROBLEMS

4.1 The uniform flexible cantilever beam in Figure 4.1a has a uniform weight distribution w_o = 1.0 lb/in. (175.1268 N/m), stiffness EI = 180,000 kip-in.^{2} (516,541 Nm^2), and length L = 1000 in. (25.4 m). By using GFEM with uniform stiffness and mass and an equivalent straight beam of length L_o, determine the first six frequencies of vibration and the corresponding mode shapes. Use $\Delta(x)$ = constant = Δ.
Answer: ω_1 = 1.0564 rps, ω_2 = 5.9103 rps, ω_3 = 16.3255 rps, ω_4 = 31.8721 rps, ω_5 = 52.5667 rps, and ω_6 = 78.3691 rps.

4.2 Solve Problem 4.1 by using Galerkin's consistent FEM.

4.3 Solve Problem 4.1 by assuming that $\Delta(x) = \Delta(x/L_o)^{1/2}$

4.4 Solve Problem 4.1 with a weight distribution $w_o = 2.5$ lb/in. (437,817 N/m) and $\Delta(x) = \Delta \sin(\pi x / 2L_o)$.
Answer: $\omega_1 = 0.7808$ rps, $\omega_2 = 3.9429$ rps, $\omega_3 = 10.5353$ rps, $w_4 = 20.362$ rps, $\omega_5 = 33.4284$ rps, and $\omega_6 = 49.7077$ rps.

4.5 The tapered flexible cantilever beam in Figure 4.4a has a uniform weight distribution $w_o = 2.0$ lb/in. (350.2536 N/m) as shown. The stiffness EI_A at the free end A of the member is 180,000 kip-in.^{2} (516,541 Nm^2), the length L = 1000 in. (25.4 m), and the taper n = 1.5. By using an equivalent straight beam of length L_o and GFEM with uniform stiffness and mass, determine the first six frequencies of vibration and their corresponding mode shapes. Use $\Delta(x) = \Delta$.
Answer: $\omega_1 = 1.1649$ rps, $\omega_2 = 5.9745$ rps, $\omega_3 = 15.9398$ rps, $\omega_4 = 30.8116$ rps, $\omega_5 = 50.6009$ rps, and $\omega_6 = 75.2685$ rps.

4.6 Solve Problem 4.5 by using Galerkin's consistent FEM.

4.7 Solve Problem 4.5 by assuming that $\Delta(x) = \Delta(x/L_o)$.

4.8 Solve Problem 4.5 with a weight distribution $w_o = 10.0$ lb/in. (1751.268 N/m).
Answer: $\omega_1 = 0.7964$ rps, $\omega_2 = 2.9830$ rps, $\omega_3 = 6.8480$ rps, $\omega_4 = 12.5469$ rps, $\omega_5 = 20.0401$ rps, and $\omega_6 = 29.2377$ rps.

4.9 Solve Problem 4.5 with $w_o = 5.0$ lb/in. (875.634 N/m), $\Delta(x) = \Delta(x/L_o)$, and taper n = 2.
Answer: $\omega_1 = 1.0657$ rps, $\omega_2 = 4.9360$ rps, $\omega_3 = 12.8067$ rps, $\omega_4 = 24.5416$ rps, $\omega_5 = 40.1462$ rps, and $\omega_6 = 59.5884$ rps.

4.10 Solve Problem 4.9 by using Galerkin's consistent FEM.

Chapter 5

VIBRATION ANALYSIS OF FLEXIBLE BARS

5.1 INTRODUCTION

In the preceding chapter, the nonlinear differential equations of motion for the free vibration of flexible bars of uniform and variable stiffness have been derived. The solution methodologies that have been developed for the solution of such problems are dealing primarily with oscillation vibration that is taking place with respect to the large static equilibrium position of the flexible member. This problem was investigated theoretically in detail, and the developed methodologies were applied to a small number of flexible beam cases. In this chapter, additional cases of flexible beam problems will be investigated using the above methodologies.

5.2 VIBRATION OF UNIFORM FLEXIBLE SIMPLY SUPPORTED BEAMS USING GALERKIN'S CONSISTENT FEM

In this section, Galerkin's consistent FEM will be used to determine the natural frequencies of the uniform flexible simply supported beam in Figure 2.8 of Section 2.9. The distributed loading w acting on the member represents the distribution of its weight and other possible weights that are attached to the beam and participating in its vibrational motion. Such an example of weight distribution would be the girder of a bridge and the attached bridge deck. It is further assumed that the beam length L = 1000 in. (25.4 m), stiffness EI = 75,000 kip-in.^{2} (215,224 Nm^2), and w = 5.0 lb/in. (875.634 N/m).

In order to proceed with the vibration analysis, the static equilibrium position y_s must first be determined. The large deformation configuration shown in Figure 2.8 represents in this case the static equilibrium position y_s. This is the large deformation produced by the weight w of the member, and it may be determined by using the Euler-Bernoulli equation, which is written again here for convenience (see also Equation 4.33).

$$\frac{y_s''}{\left[1+\left(y_s'\right)^2\right]^{3/2}} = -\frac{M_x}{E_x I_x} \tag{5.1}$$

By following the procedure discussed in Section 2.9 and assuming that $\Delta(x) = \Delta x/L_o$, we find that

$$y_s' = \frac{G(x)}{\left\{1-[G(x)]^2\right\}^{1/2}} \tag{5.2}$$

where

$$G(x) = \frac{wL}{24EI(L-\Delta)}\left[6(L-\Delta)x^2 - 4x^3 - (L-\Delta)^3\right] \tag{5.3}$$

By using Equation 2.94 and a trial and error procedure as discussed in preceding chapters, we obtain $\Delta = 411.335$ in. (10.4479 m) and $L_o = (L - \Delta) = 588.665$ in. (14.9521 m). We also find that the maximum rotation $\theta_{max} = 74.24°$ and the vertical displacement δ at the center of the member is 357.68 in. (9.0851 m).

With known Δ, the static deflection y_s at any $0 \le x \le L_o$ may be determined using Equation 5.2 and integrating once, or by using pseudolinear analysis as discussed in the preceding chapters. Pseudolinear analysis would be the easiest to use. The next step, since y_s is known, would be the solution of the differential equation of motion, Equation 4.28, for the computation of the natural frequencies of vibration of the member. We write this equation again here:

$$\frac{d^2}{dx^2}\left\{E_x I_e(x) y_d''(x)\right\} - m_e(x)\omega^2 y_d(x) = 0 \tag{5.4}$$

where

$$I_e(x) = \frac{I_x}{\left[1 + \left(y_s'\right)^2\right]^{3/2}} \tag{5.5}$$

$$m_e(x) = \left[1 + \left(y_s'\right)^2\right]^{1/2} m(x_o) \tag{5.6}$$

Equation 5.4, as discussed in Chapter 4, represents a straight beam of length L_o, which vibrates with the same frequencies ω as it does the initial system in Figure 2.8 from its static equilibrium position y_s. Since y_s is known, the equivalent quantities $I_e(x)$ and $m_e(x)$ which represent the variation of the moment of inertia and mass, respectively, of the equivalent straight beam of length L_o, may be obtained from Equations 5.5 and 5.6. The equivalent straight beam of length L_o is shown in Figure 5.1a and Galerkin's consistent finite element method (FEM) will be used to determine its natural frequencies of vibration. The solution procedure is discussed in Section 4.7.

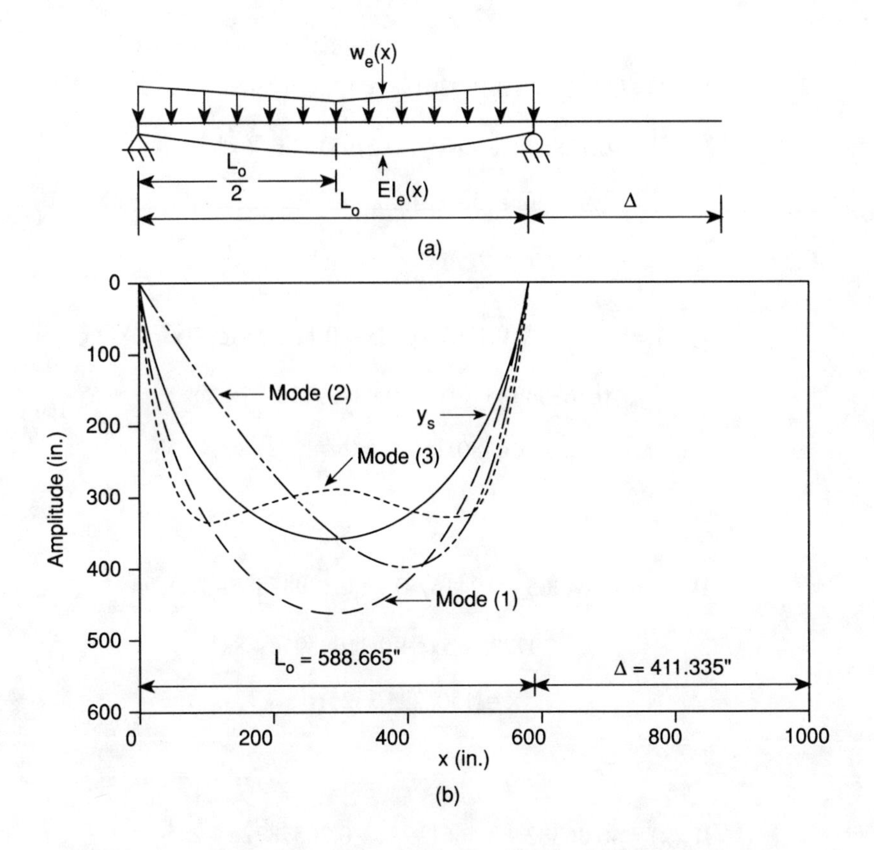

FIGURE 5.1. (a) Equivalent pseudovariable straight simply supported beam of length L_o; (b) the first three mode shapes of the simply supported beam in Figure 2.8.

By using 10 elements, the approximation of II(x) by $II_a(x)$ yields the following piecewise polynomials for the elements:

$$II_1(x) = 50.23307 + 0.10613650e - 1x - 0.3654956e - 1x^2$$
$$+0.4388151e - 3x^3 + 0.2156067e - 5x^4 \tag{5.7}$$
$$x \in [0, 25]$$

$$II_2(x) = 52.92605 - 0.2441617x - 0.3250092e - 1x^2$$
$$+0.6666961e - 3x^3 - 0.40262886e - 5x^4 \tag{5.8}$$
$$x \in [25, 50]$$

$$\Pi_3(x) = 73.80300 - 0.1993019e1x + 0.2312842e-1x^2$$
$$-0.1286520e-6x^3 + 0.2795133e-6x^4 \tag{5.9}$$
$$x \in [50, 80]$$

$$\Pi_4(x) = 57.48781 - 0.1309552e1x + 0.1250581e-1x^2$$
$$-0.5633438e-4x^3 + 0.9872977e-7x^4 \tag{5.10}$$
$$x \in [80, 150]$$

$$\Pi_5(x) = 18.64365 - 0.2416777x + 0.1298788e-2x^2$$
$$-0.3229773e-5x^3 + 0.3108395e-8x^4 \tag{5.11}$$
$$x \in [150, 294.3325]$$

$$\Pi_6(x) = 40.86382 - 0.4661198x + 0.2057872e-2x^2$$
$$-0.4089440e-5x^3 + 0.3108395e-8x^4 \tag{5.12}$$
$$x \in [294.3325, 438.665]$$

$$\Pi_7(x) = 0.3977887e4 - 0.3535552e2x + 0.1181269x^2$$
$$-0.1759058e-3x^3 + 0.9860649e-7x^4 \tag{5.13}$$
$$x \in [438.665, 508.665]$$

$$\Pi_8(x) = 0.1423585e5 - 0.1195622e3x + 0.3770822x^2$$
$$-0.5295068e-3x^3 - 0.2795133e-6x^4 \tag{5.14}$$
$$x \in [508.665, 538.665]$$

TABLE 5.1
Natural Frequencies of Vibration of the Simply Supported Beam in Figure 2.8 Using the Approximation $\Delta(X) = \Delta x/L_o$

ω(rps)	GFEM	FDM	% Difference
ω_1	1.3111	1.3058	0.406
ω_2	3.1781	3.1559	0.703
ω_3	7.0785	7.0075	1.013
ω_4	12.3616	12.1927	1.385
ω_5	19.0585	18.8458	1.128
ω_6	27.7739	26.9334	3.121

$$\begin{aligned} II_9(x) &= 0.3588342e6 + 0.2630682e4x - 0.7226408e1x^2 \\ &+0.8813839e-2x^3 - 0.4026286e-5x^4 \\ &x \in [538.665, 563.665] \end{aligned} \tag{5.15}$$

$$\begin{aligned} II_{10}(x) &= 0.3351390e6 - 0.2167769e4x + 0.5209093e1x^2 \\ &-0.5501580e-2x^3 + 0.2149963e-5x^4 \\ &x \in [563.665, 588.665] \end{aligned} \tag{5.16}$$

By utilizing the above approximations, the stiffness and mass matrices are obtained using Equations 4.109 and 4.110, respectively, by integration. Simpson's rule may be used to carry out the required integrations. The element stiffness and mass matrices are then assembled by using Equations 4.133 and 4.134, respectively. The boundary conditions of zero deflection at x = 0 and x = L_0 are used for this purpose. The solution of the eigenvalue problem for the first six frequencies of vibration, in radians per second, yields the results shown in Table 5.1. In the same table, the results are compared with the results obtained by using the finite difference method (FDM) with 81 elements. Reasonable agreement between these two methods is obtained. The first three mode shapes of the flexible member are shown in Figure 5.1b. In this figure, the curve y_s is the static equilibrium position

5.3 VIBRATION OF TAPERED FLEXIBLE SIMPLY SUPPORTED BEAMS USING GALERKIN'S CONSISTENT FEM

In this section, the natural frequencies of vibration and the corresponding mode shapes for the flexible tapered simply supported beam in Figure 5.2a

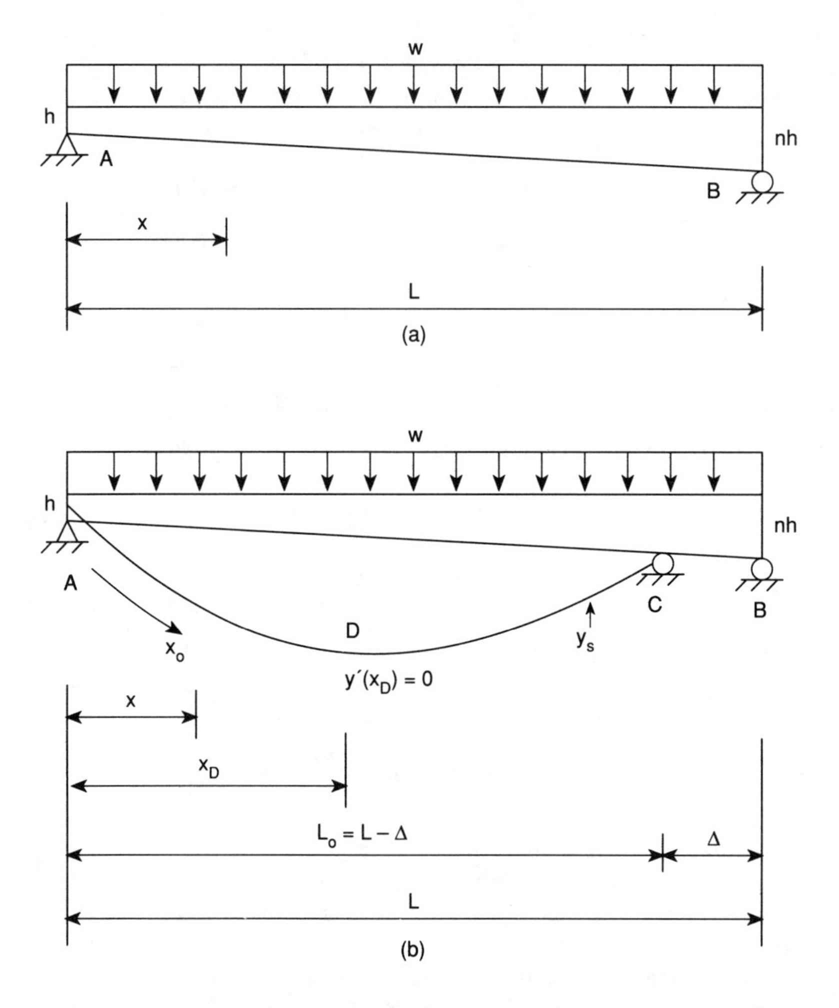

FIGURE 5.2. (a) Tapered flexible simply supported beam; (b) static equilibrium configuration y_s of the tapered simply supported beam.

will be determined using Galerkin's consistent FEM. The uniformly distributed loading w on the beam is assumed to consist of its weight and other possible weights attached to the member. It is further assumed here that the length L = 1000 in. (25.4 m), bending stiffness EI_A at its left support equals 75,000 kip-in.^{2} (215,224 Nm^2), w = 10.0 lb/in. (1751.268 N/m), and taper n = 1.5.

The static equilibrium position y_s of the member is shown in Figure 5.2b. The position y_s may be determined using the Euler-Bernoulli equation as discussed in the preceding section. The stiffness EI_x of the member is given by the equation

$$EI_x = EI_A f(x) \tag{5.17}$$

where

$$f(x) = \left[1 + (n-1)\frac{x}{(L-\Delta)}\right]^3 \tag{5.18}$$

$$I_A = \frac{bh^3}{12} \tag{5.19}$$

and b and E are the constant width and constant modulus of elasticity, respectively. We assume here that the horizontal displacement function $\Delta(x)$ = constant = Δ, where Δ is the horizontal displacement of the right end support B.

The solution for the computation of y_s was carried out in Section 2.10. The bending moment at any $0 \le x \le L_0$ is given by Equation 2.98. In this equation, the coordinate x_0 will be assumed as $x_0 = (x + \Delta)$, as stated above. On this basis the Euler-Bernoulli equation yields the results shown in Equation 2.102. The rotation $y_s'(x)$ at any $0 \le x \le L_0$ is obtained by integrating Equation 2.102 once and satisfying the boundary condition $y'(x_D) = 0$, where x_D defines the point D of maximum vertical deflection. The result is given by Equation 2.103, where H(x) and C are as shown by Equations 2.105 and 2.104, respectively. A trial and error procedure is performed as in the preceding section to determine the unknown displacement Δ, and consequently the length $L_0 = (L - \Delta)$, of the equivalent straight beam in Figure 5.3a. In this figure, the variation of the equivalent moment of inertia $I_e(x)$ and equivalent mass $m_e(x)$ at any $0 \le x \le L_0$ may be determined from Equations 5.5 and 5.6, respectively, since y_s' is known.

The above trial and error procedure yielded Δ = 326.59 in. (8.2954 m), and L_0 = 673.41 in. (17.1046 m). We also found that the maximum vertical static displacement y_{max} = 326.98 in. (8.3053 m), and the maximum rotation θ_{max} = 73.49°. With known y'_s, the computation of the static equilibrium position y_s can be carried out as discussed in the preceding section.

The solution of the equivalent straight beam in Figure 5.3a for the computaion of its natural frequencies of vibration and its corresponding mode shapes will be carried out as in the preceding section by using the differential equation of motion given by Equation 4.28 and applying Galerkin's consistent Finite Element method. The approximation of II(x) by $II_a(x)$ with ten piecewise polynomials yields the following equations:

$$\begin{gathered} II_1(x) = 43.56606 - 0.9044605e-1x - 0.3744401e-1x^2 \\ +0.5495874e-3x^3 + 0.2848886e-5x^4 \\ x \in [0, 20] \end{gathered} \tag{5.20}$$

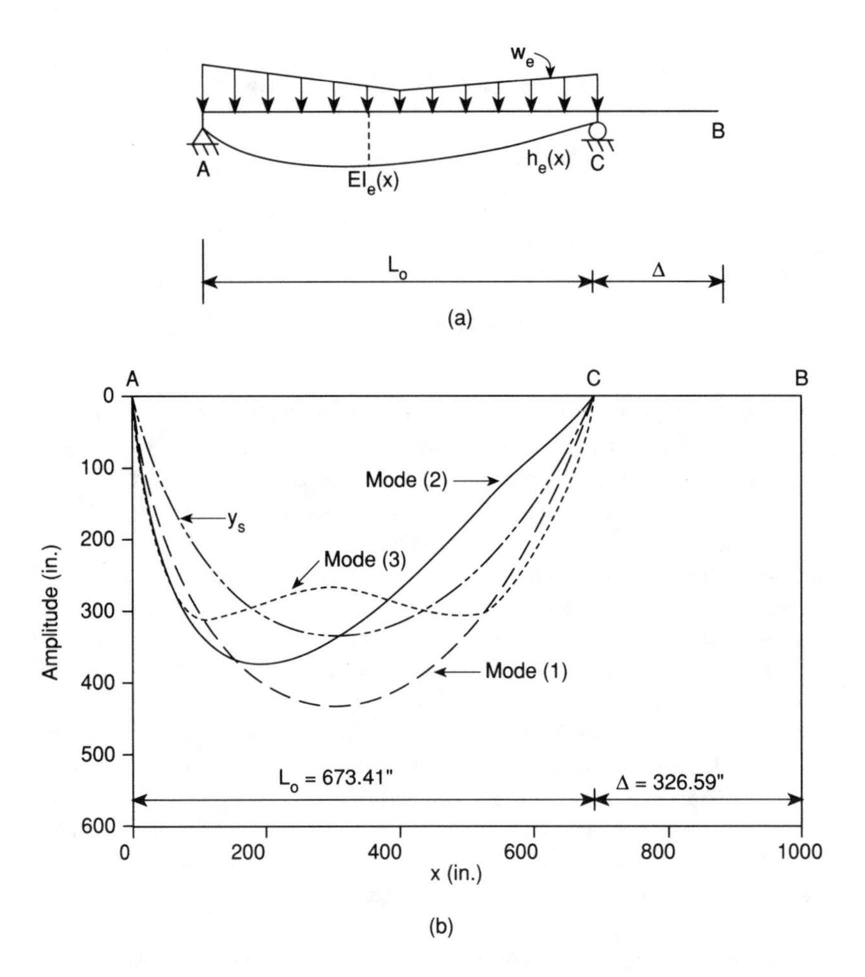

FIGURE 5.3. (a) Pseudovariable straight simply supported beam of length L_o undergoing small amplitude vibrations; (b) the first three mode shapes of the flexible beam vibrating about its static equilibrium position y_s (1 in. = 0.0254 m).

$$\begin{aligned} II_2(x) &= 0.4457658 - 0.1849683x - 0.3898381e1x^2 \\ &\quad +0.9259398e-3x^3 - 0.6627420e-5x^4 \\ &\quad x \in [20, 40] \end{aligned} \tag{5.21}$$

$$\begin{aligned} II_3(x) &= 0.5894749e2 - 0.1656230e1x + 0.1820775e \\ &\quad -1x^2 - 0.7397476x^3 \\ &\quad x \in [40, 60] \end{aligned} \tag{5.22}$$

$$
\begin{aligned}
II_4(x) &= 0.468796e2 - 0.1104543e1x + 0.9775192e \\
&\quad -2x^2 - 0.3082772e - 4x^3 \\
&\quad x \in [60, 100]
\end{aligned} \tag{5.23}
$$

$$
\begin{aligned}
II_5(x) &= 0.3487944e2 - 0.7535314x + 0.6782375e - 2x^2 \\
&\quad -0.2873024e - 4x^3 + 0.4735049e - 7x^4 \\
&\quad x \in [100, 150]
\end{aligned} \tag{5.24}
$$

$$
\begin{aligned}
II_6(x) &= 0.1164433e2 - 0.1380063x + 0.6077050e - 3x^2 \\
&\quad -0.9354993e - 6x^3 \\
&\quad x \in [150, 200]
\end{aligned} \tag{5.25}
$$

$$
\begin{aligned}
II_7(x) &= 0.8461912e1 - 0.9467101e - 1x + 0.4451540e - 3x^2 \\
&\quad -0.9747137e - 6x^3 + 0.8320968e - 9x^4 \\
&\quad x \in [200, 300]
\end{aligned} \tag{5.26}
$$

$$
\begin{aligned}
II_8(x) &= 0.4226931e1 - 0.3480356e - 1x + 0.1263683e - 3x^2 \\
&\quad -0.2173161e - 6x^3 + 0.1550188e - 9x^4 \\
&\quad x \in [300, 450]
\end{aligned} \tag{5.27}
$$

$$
\begin{aligned}
II_9(x) &= -0.3767786e2 + 0.3061981x - 0.912711e - 3x^2 \\
&\quad +0.1187784e - 5x^3 - 0.5563956e - 9x^4 \\
&\quad x \in [450, 580]
\end{aligned} \tag{5.28}
$$

TABLE 5.2
Natural Frequencies of Vibration in rps for a Tapered Flexible Simply Supported Beam by using the Approximation $\Delta(x) = \Delta$

ω (rps)	GFEM	FDM	% Difference
ω_1	1.0930	1.0893	0.339
ω_2	2.9813	2.9632	0.611
ω_3	6.6267	6.5661	0.923
ω_4	11.6286	11.5034	1.088
ω_5	17.8152	17.8279	−0.071
ω_6	25.1959	25.5183	−1.263

$$\begin{gathered} \mathrm{II}_{10}(x) = 0.2957112e3 - 0.1683401e1x + 0.3444572e-2x^2 \\ -0.2927629e-5x^3 + 0.8376366e-9x^4 \\ x \in [580, 673.41] \end{gathered} \tag{5.29}$$

The stiffness and mass matrices are determined from Equations 4.109 and 4.110, respectively, by integration, and they are assembled by using Equations 4.133 and 4.134. The boundary conditions used are zero deflection at $x = 0$ and $x = L_0$. The solution of the eigenvalue problem yields the results shown in Table 5.2. In the same table, the first six frequencies were also calculated using the FDM with 81 elements for comparison purposes. For practical applications the results from both methods are in close agreement.

5.4 VIBRATION OF TAPERED FLEXIBLE SIMPLY SUPPORTED BEAMS USING GALERKIN'S FEM WITH EQUIVALENT UNIFORM STIFFNESS AND MASS

For many practical problems, the application of Galerkin's consistent FEM may become extremely difficult. This difficulty arises, primarily, from the complicated mathematical functions that are associated with the equivalent moment of inertia $I_e(x)$ and equivalent mass $m_e(x)$ in the differential equation of motion given by Equation 4.28. In order to facilitate the application of the FEM, an equivalent uniform stiffness and an equivalent uniform mass methodology may be employed. This procedure was thoroughly discussed in Sections 4.4 and 4.5.

In this section, Galerkin's FEM with equivalent uniform stiffness and mass will be used to determine the free frequincies of vibration and the associated mode shapes for the tapered flexible simply supported beam in Figure 5.2a. The same problem was solved in Section 5.3 by using the Galerkin's consistent FEM. As in the preceding section, we assume here that L = 1000 in. (25.4 m),

EI_A = 75,000 kip-in.^{2} (215,224 Nm^2), w = 10.0 lb/in. (1751.268 N/m), taper n = 1.5, and $\Delta(x)$ = constant = Δ. The static equilibrim position y_s and slope y_s' are established in the preceding section where Δ = 326.59 in. (8.2954 m), and L_o = 673.41 in. (17.1046 m).

The equivalent depth $h_e(x)$, moment of inertia $I_e(x)$, and mass $m_e(x)$ at any $0 \le x \le L_o$ of the equivalent straight beam of length L_o may be determined from Equations 4.34, 4.29, and 4.30, respectively. Since y_s is known, these equations yield

$$h_e(x) = h^3\sqrt{f(x)}\left\{1 - [p(x)]^2\right\}^{1/2} \tag{5.30}$$

$$I_e(x) = I_A f(x)\left\{1 - [p(x)]^2\right\}^{3/2} \tag{5.31}$$

$$m_e(x) = \frac{m}{\left\{1 - [p(x)]^2\right\}^{1/2}} \tag{5.32}$$

where

$$\begin{aligned} p(x) = \frac{2w(L-\Delta)^3}{EI_A}\Bigg\{ & \frac{-0.5x(L-\Delta) - 0.75x^2 + 0.5(L-\Delta)^2}{[0.5x + (L-\Delta)]^2} \\ & + \frac{2[0.5x + (L-\Delta)]^2 \ln[0.5x + (L-\Delta)]}{[0.5x + (L-\Delta)]^2} \\ & \frac{-0.5x(L-\Delta) - 0.75x^2 + 0.5(L-\Delta)^2}{[0.5x + (L-\Delta)]^2} \\ & + \frac{2[0.5x + (L-\Delta)]^2 \ln[0.5x + (L-\Delta)]}{[0.5x + (L-\Delta)]^2}\Bigg\} \end{aligned} \tag{5.33}$$

The equivalent straight beam of length L_o is shown in Figure 5.4a. This is the same beam as the one in Figure 5.3a of the preceding section. This beam is solved here by using Equation 4.28 and applying GFEM with uniform equivalent mass as shown in Figure 5.4b. The procedure is thoroughly explained in Section 4.4 and with an application in Section 4.5.

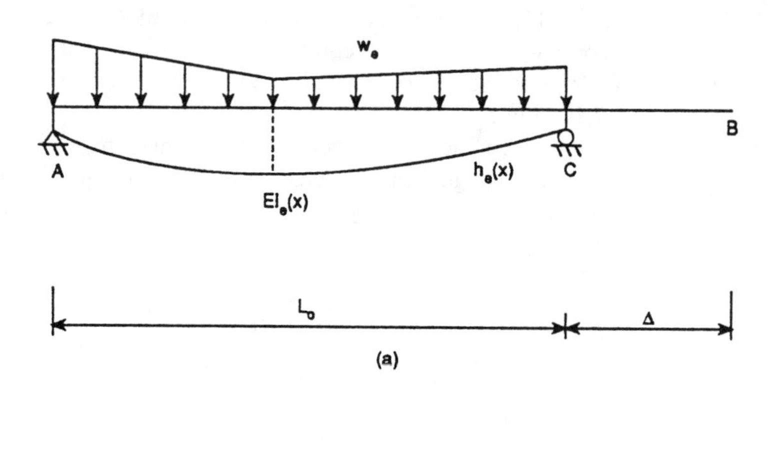

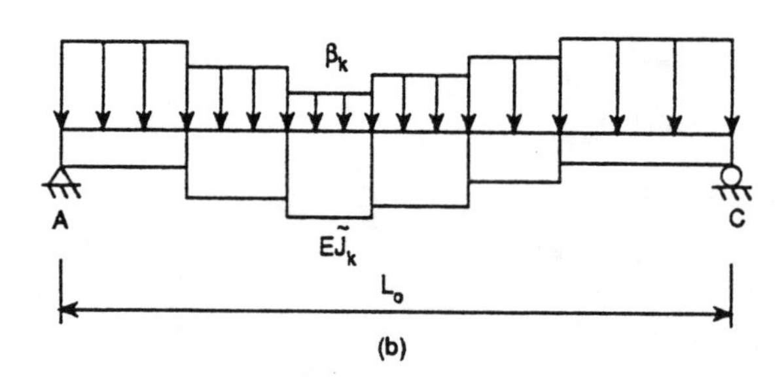

FIGURE 5.4. (a) Equivalent straight simply supported beam of length L_o; (b) equivalent piecewise uniform straight simply supported beam with equivalent piecewise uniform mass.

Table 5.3 shows the values of the first six free frequencies of vibration that have been determined by solving the equivalent straight beam in Figure 5.4a and 5.4b with 12, 22, and 40 elements. The results show that the utilization of 22 elements yields an accurate solution of the problem for practical applications. Similar observations can be made if the results are compared with the results obtained in Table 5.2 where the consistent GFEM was used. In Table 5.3 the results are also compared with the ones obtained by solving the problem in Figure 5.4a with the FDM and 81 elements. The agreement of the results is reasonable for practical purposes. The first three mode shapes of the member are shown in Figure 5.5b, where the curve y_s represents the static equilibrium position.

5.5 VIBRATION OF FLEXIBLE BARS WITH PIECEWISE THICKNESS VARIATION AND PIECEWISE UNIFORM WEIGHT

The vibration analysis of flexible members with complicated stiffness and weight variations along their length becomes convenient if the solution of

TABLE 5.3
Natural Frequencies of Vibration for a Tapered Flexible Simply Supported Beam by Using the Approximation Δ(x) = Δ

	GFEM				
ω (rps)	12 Elements	22 Elements	40 Elements	FDM	% Difference
ω_1	1.0955	1.0940	1.0927	1.0893	0.315
ω_2	3.0713	2.9992	2.9816	2.9632	0.618
ω_3	6.8513	6.6673	6.6194	6.5661	0.810
ω_4	12.3459	11.7284	11.6385	11.5034	1.174
ω_5	19.8877	18.2372	18.1244	17.8279	1.663
ω_6	29.9358	26.1894	26.1040	25.5183	2.295

Equation 4.28, which represents an equivalent straight beam of length L_o, is carried out by using GFEM with equivalent uniform stiffness and mass as discussed in Section 4.4. In order to illustrate the methodology, consider the doubly tapered cantilever beam in Figure 5.6 with thickness variations and weight distributions as shown. Numerical results will be obtained by assuming that the length of the member L = 1000 in. (25.4 m), the stiffness EI_A = 180,000 kip-in.^{2} (516,541 Nm^2), $L_1 = L_2$ = 500 in. (12.70 m), w_1 = 5.0 lb/in. (875.634 N/m), w_2 = 10.0 lb/in. (1,751.268 N/m), $\alpha = 0.75$, and taper n = 2.

The static analysis that is required for the determination of the large static equilibrium position y_s will be carried out using the Euler-Bernoulli equation as in the preceding sections. The variation of the moment of inertia I_x along the length L_o of the member is given by the equations

$$I_1(x) = I_A f_1(x) \qquad 0 \le x \le a \tag{5.34}$$

$$I_2(x) = I_A f_2(x) \qquad a \le x \le (a+b) \tag{5.35}$$

where $I_A = bh^3/12$, b is the constant width of the member and

$$f_1(x) = \left[1 + \frac{5x}{a}\right]^3 \tag{5.36}$$

$$f_2(x) = \left[q + \frac{rx}{b}\right]^3 \tag{5.37}$$

$$s = \left[\frac{(n+1)\alpha}{2} - 1\right] \tag{5.38}$$

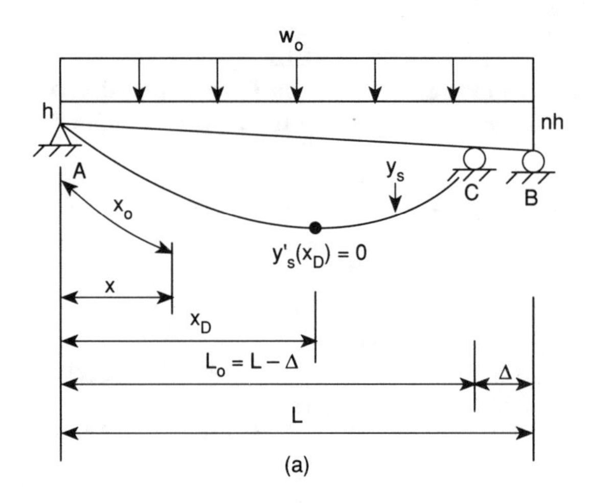

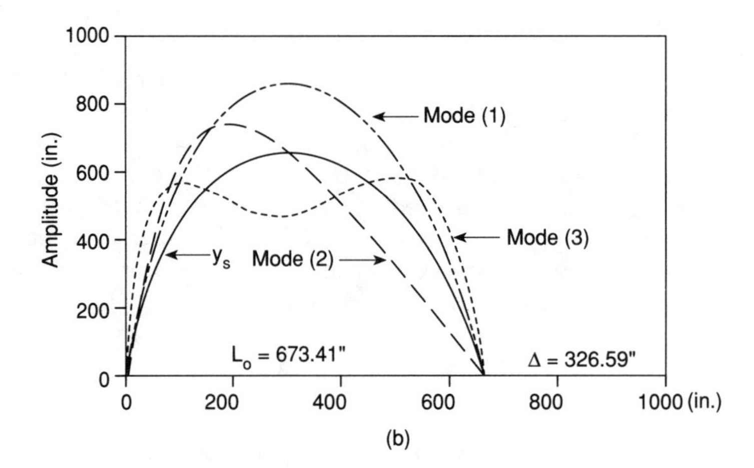

FIGURE 5.5. (a) Original flexible simply supported tapered beam and its deformed static equilibrium position y_s; (b) first three mode shapes of the flexible simply supported beam.

$$q = \left\{ \frac{(n+1)\alpha}{2} - \left[\frac{n(1-\alpha)}{2} + \frac{(n-\alpha)}{2} \right] \left(\frac{a}{b} \right) \right\} \tag{5.39}$$

$$r = \left[\frac{n(1-\alpha) + (n-\alpha)}{2} \right] \tag{5.40}$$

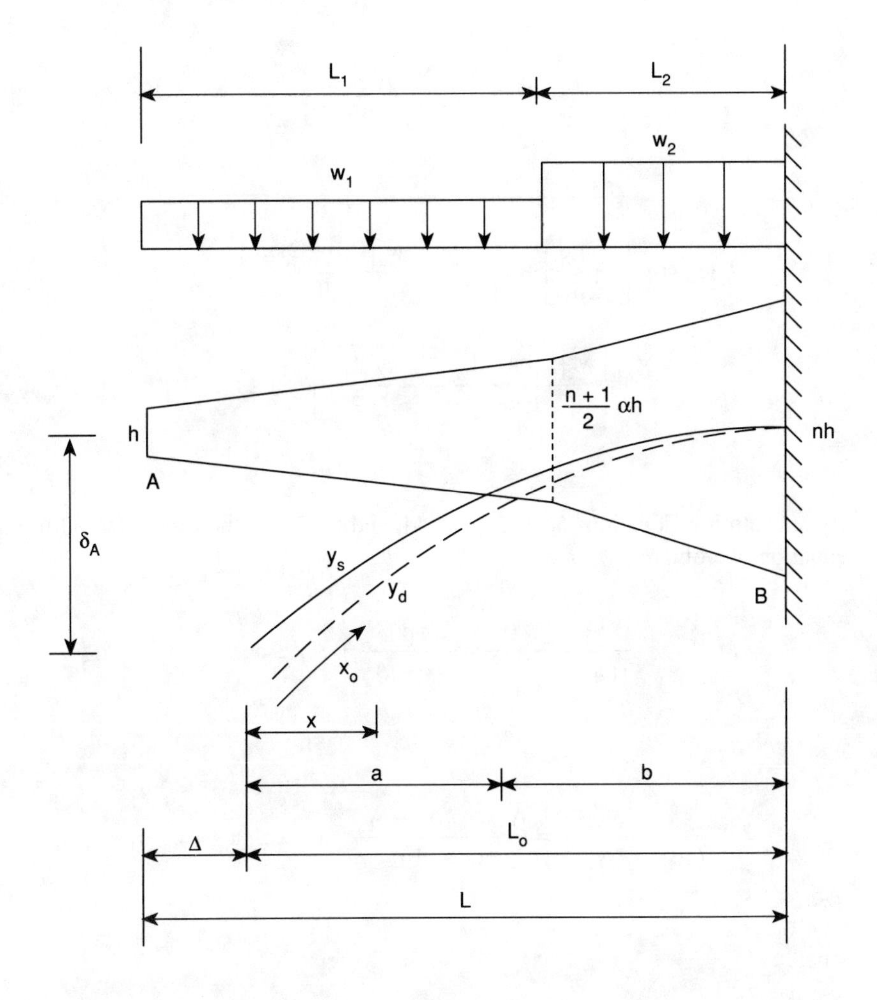

FIGURE 5.6. Flexible cantilever beam with piecewise thickness variation and piecewise uniform weight.

The bending moment M_x along the length of the member may be obtained from the equations

$$M_1(x) = \frac{w_1 x}{2} x_{o1}(x) \qquad 0 \le x_{o1} \le L_1 \tag{5.41}$$

$$M_2(x) = w_1 L_1\left(x - \frac{\alpha}{2}\right) - \frac{w_2(x-a)}{2} x_{o2}(x) \quad L_1 \le x_{o2} \le L \tag{5.42}$$

By assuming that $x_{o1}(x) = L_1x/a$ and $x_{o2}(x) = L_2x/b + (L_1b - L_2a)/b$ and substituting into Equations 5.41 and 5.42, we obtain

$$M_1(x) = \frac{w_1 L_1 x^2}{2a} \quad 0 \le x \le a \tag{5.43}$$

$$M_2(x) = x^2\left[\frac{w_2 L_2}{2b}\right] + x\left\{w_1 L_1 + \frac{w_2[L_1 b - 2L_2 a]}{2b}\right\} + \left\{\frac{w_2[L_1 ba - L_2 a^2]}{2b} + \frac{w_1 L_1 a}{2}\right\} \quad a \le x \le (a+b) \tag{5.44}$$

By substituting Equations 5.41, 5.34, 5.44, and 5.35 into the Euler-Bernoulli equation, we obtain

$$\frac{y_{s1}''}{\left[1 + (y_{s1}')^2\right]^{3/2}} = \frac{A_1 x^2}{(sx + a)^3} \quad 0 \le x \le a \tag{5.45}$$

$$\frac{y_{s2}''}{\left[1 + (y_{s2}')^2\right]^{3/2}} = \frac{A_2 x^2 + A_3 + A_4}{(rx + qb)^3} \quad a \le x \le (a+b) \tag{5.46}$$

where

$$A_1 = -\frac{w_1 a^2 L_1}{2EI_A} \tag{5.47}$$

$$A_2 = -\frac{w_2 L_2 b^2}{2EI_A} \tag{5.48}$$

$$A_3 = \frac{b^2}{2EI_A}\left[2bw_1 L_1 - w_2(L_1 b - 2L_2 a)\right] \tag{5.49}$$

$$A_4 = \frac{b^2}{2EI_A}\left[w_1L_1ab + w_2(L_1ab - L_2a^2)\right] \tag{5.50}$$

By integrating Equations 5.45 and 5.46 once, and satisfying the continuity condition $y'_{s1}(a) = y'_{s2}(a)$ and the boundary condition $y'_{s2}(L_o) = 0$ for the evaluation of the constants of integration, we find

$$\frac{y'_{s1}}{\left[1+(y'_{s1})^2\right]^{1/2}} = G_1(\xi) \tag{5.51}$$

$$\frac{y'_{s2}}{\left[1+(y'_{s2})^2\right]^{1/2}} = G_2(\xi) \tag{5.52}$$

where

$$\begin{aligned} G_1(\xi) = A_1 &\left\{\frac{-3s^2\xi^2 - 2s\xi a_o + 2(s\xi + a_o)^2 \ln(sL_o\xi + a_oL_o)}{2s^3(s\xi + a_o)^2}\right. \\ &\left. - \frac{-3s^2a_o^2 - 2sa_o^2 + 2(sa_o + a_o)^2 \ln(sL_oa_o + a_oL_o)}{2s^3(sa_o + a_o)^2}\right\} \\ &+ \left\{\left[\frac{A_2\left[-3r^2a_o^2 - 2ra_oqb_o + 2(ra_o + qb_o)^2 \ln(rL_oa_o + qb_oL_o)\right]}{2r^3(ra_o + qb_o)^2}\right.\right. \\ &\left. + \frac{A'_3(-2r^2a_o - rqb_o) - A'_4r^2}{2r^3(ra_o + qb_o)^2}\right] \\ &- \left[\frac{A_2\left[-3r^2 - 2rqb_o + 2(r + qb_o)^2 \ln(rL_o + qb_oL_o)\right]}{2r^3(r + qb_o)^2}\right. \\ &\left.\left. + \frac{A'_3(-2r^2 - rqb_o) - A'_4r^2}{2r^3(r + qb_o)^2}\right]\right\} \end{aligned} \tag{5.53}$$

$$G_2(\xi) = \left\{ \left[\frac{A_2\left[-3r^2\xi^2 - 2r\xi qb_o + 2(r\xi + qb_o)^2 \ln(rL_o\xi + qb_oL_o)\right]}{2r^3(r\xi + qb_o)^2} \right. \right.$$

$$\left. + \frac{A_3'(-2r^2\xi - rqb_o) - A_4'r^2}{2r^3(ra_o + qb_o)^2} \right]$$

$$- \left[\frac{A_2\left[-3r^2 - 2rqb_o + 2(r + qb_o)\ln(rL_o + qb_oL_o)\right]}{2r^3(r + qb_o)^2} \right. \tag{5.54}$$

$$\left. \left. + \frac{A_3'(-2r^2 - rqb_o) - A_4'r^2}{2r^3(r + qb_o)^2} \right] \right\}$$

where

$$A_3' = \frac{A_3}{L_o} \tag{5.55}$$

$$A_4' = \frac{A_4}{L_o^2} \tag{5.56}$$

In the above equations the dimensionless parameter $\xi = x/L_o$ is introduced for convenience. Also $a_o = a/L_o$ and $b_o = b/L_o$, and consequently $(a_o + b_o) = 1$. In the above derivations it was assumed that $x_o = [x + \Delta(x)] = (x + \Delta)$, which implies that $\Delta(x) = \text{constant} = \Delta$.

Equations 5.51 and 5.52 are in terms of the unkown parameters a_o, b_o, and L_o. These parameters can be determined by using the expressions

$$L = L_o \int_0^{a_o} \frac{1}{\left\{1 - [G_1(\xi)]^2\right\}^{1/2}} d\xi + \int_{a_o}^{1} \frac{1}{\left\{1 - [G_2(\xi)]^2\right\}^{1/2}} d\xi \tag{5.57}$$

$$L_1 = L_o \int_0^{a_o} \frac{1}{\left\{1 - [G_1(\xi)]^2\right\}^{1/2}} d\xi \tag{5.58}$$

$$L_2 = L_o \int_{a_o}^{1} \frac{1}{\left\{1 - [G_2(\xi)]^2\right\}^{1/2}} d\xi \tag{5.59}$$

and applying a trial and error procedure as in earlier sections. The trial and error procedure yields the results $a_o = 0.3431$, $b_o = 0.6569$, and $\Delta = 380.12$ in. (9.655 m); therefore, $L_o = (L - \Delta) = 619.88$ in. (15.745 m). At the free end, the rotation $\theta_A = 67.307°$, and the vertical displacement $\delta_A = 701.993$ in. (17.8306 m).

With known Δ and L_o, the values of y'_{s1}, and y'_{s2} can be determined from Equations 5.51 and 5.52, respectively. The static equilibrium position y_s can be determined as discussed in the preceding sections. Pseudolinear analysis would be the easiest to use since y'_{s1} and y'_{s2} are known.

Vibration analysis may now be performed by using Equation 4.28, which represents an equivalent straight beam. The equivalent parameters $I_e(x)$ and $m_e(x)$ are given by Equations 4.29 and 4.30, respectively. For the problem in question, these parameters may be written as

$$I_e(x) = \frac{I_x}{\left[1 + (y'_{s1})^2\right]^{3/2}} \qquad 0 \le x \le a \tag{5.60}$$

$$I_e(x) = \frac{I_x}{\left[1 + (y'_{s2})^2\right]^{3/2}} \qquad a \le x \le L_o \tag{5.61}$$

$$m_e(x) = \left[1 + (y'_{s1})^2\right]^{1/2} m(x_o) \qquad 0 \le x_o \le L_1 \tag{5.62}$$

$$m_e(x) = \left[1 + (y'_{s2})^2\right]^{1/2} m(x_o) \qquad L_1 \le x_o \le L \tag{5.63}$$

where y'_{s1} and y'_{s2} are given by Equations 5.51 and 5.52, respectively.

The equivalent pseudovariable straight beam of length L_o that is represented by Equation 4.28 is shown in Figure 5.7a. Its solution for the computation of its natural frequencies of vibration and the corresponding mode shapes is obtained using the GFEM with equivalent uniform stiffness and uniform mass as shown in Figure 5.7b. The solution is obtained by using 10, 20, and 40 elements, in order to compare results. The values of the first six frequencies of vibration are shown in Table 5.4.

5.6 ADDITIONAL DISCUSSION AND REMARKS

The results and discussions in this chapter, and in the preceding one, show that the undamped free frequencies of vibration of uniform and variable stiffness members and their corresponding mode shapes may be determined rather conveniently in three main steps. Since free vibrations are taking place from the static equilibrium position of the flexible member, the first step would be the utilization of the Euler-Bernoulli nonlinear differential equation in order to establish the static equilibrium position y_s. For this purpose, the utilization

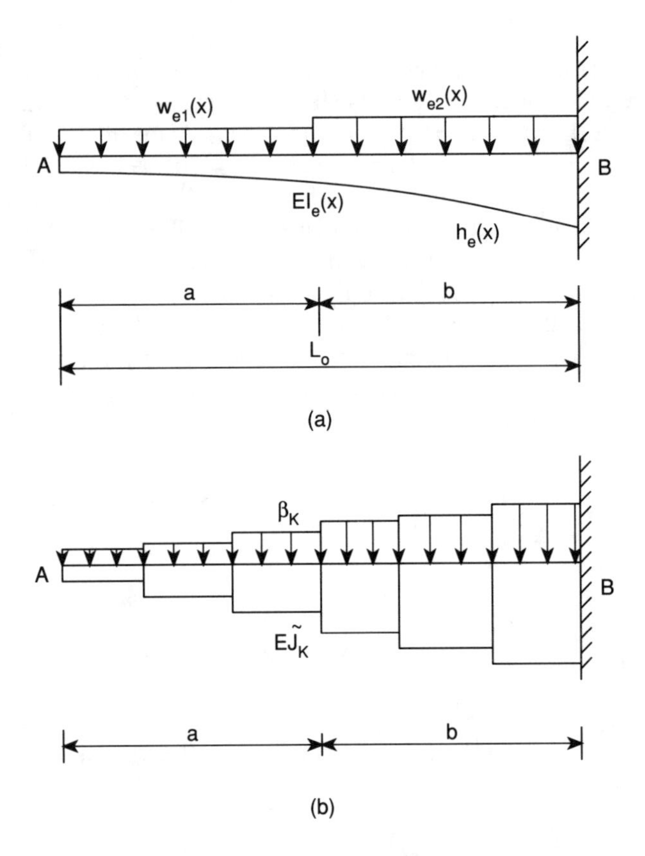

FIGURE 5.7. (a) Equivalent pseudovariable straight cantilever beam of length L_0; (b) equivalent piecewise uniform straight cantilever beam with equivalent piecewise uniform mass.

of appropriate equivalent systems as discussed in the first three chapters will greatly simplify the solution of this problem.

From the static equilibrium position, since the static deflection y_s is large, the member vibrates as a curved beam. Therefore, the purpose of this step is to derive an equivalent straight beam of length $L_0 = (L - \Delta)$, which vibrates tranversly in the same way as the original member does about its static equilibrium position y_s. The moment of inertia $I_e(x)$, mass $m_e(x)$, and depth $h_e(x)$ of the equivalent straight beam of length L_0 are appropriately defined so that the equivalent straight member vibrates with the same frequencies of vibration as the initial flexible member. An equivalent differential equation of motion is derived for this purpose. Although the static deflection y_s is large, the amplitudes of vibration are assumed to be small and damping is neglected. This is not a severe limitation, because many engineering structures and their components are associated with small amplitude vibrations.

The third and final step is to solve the differential equation of motion of the equivalent straight beam. The methods used here for this purpose are Galerkin's

TABLE 5.4
The First Six Frequencies of Vibration for the Flexible Beam in Figure 5.6 by Using the Pseudovariable Straight Beam in Figure 5.7

	GFEM		
ω (rps)	10 Elements	20 Elements	40 Elements
ω_1	1.0811	1.0702	1.0635
ω_2	3.6672	3.7434	3.6182
ω_3	9.3737	9.6181	9.1359
ω_4	17.8072	18.2174	17.4293
ω_5	28.6329	29.6438	28.2691
ω_6	44.7309	45.4646	42.1944

consistent FEM and the GFEM with equivalent uniform stiffness and uniform mass for each element. The FDM is also used in order to compare results. The results illustrate that all threee methods are reliable and in excellent agreement for practical applications. The results also show that the effect of mass position change during the large static deformation y_s must be taken into consideration. The error is very large if this is not included in the analysis (see Section 4.6).

PROBLEMS

5.1 The uniform flexible simply supported beam in Figure 2.8 has a uniform weight distribution w = 2.0 lb/in. (350.2536 N/m), stiffness EI = 75,000 kip-in.^{2} (215,224 Nm^2), and length L = 1000 in. (25.4 m). By using Galerkin's consistent finite element and an equivalent straight beam of length L_o, determine the first six frequencies of vibration and the corresponding mode shapes. Use $\Delta(x)$ = constant = Δ.
Answer: ω_1 = 1.3935 rps, ω_2 = 4.7824 rps, ω_3 = 10.7121 rps, ω_4 = 18.9972 rps, ω_5 = 29.6284 rps, and ω_6 = 42.5886 rps.

5.2 Solve Problem 5.1 by using the horizontal displacement functions $\Delta(x) = \Delta x/L_o$ and $\Delta(x) = \Delta(x/L_o)^{1/2}$ and compare the results.

5.3 Solve Problem 5.1 by using the horizontal displacement functions $\Delta(x) = \Delta(x/L_o)$ and $\Delta(x) = \Delta \sin(\pi x/2L_o)$ and compare the results.

5.4 Solve Problem 5.1 by using a uniform weight distribution w = 5.0 lb/in. (875.634 N/m).
Answer: ω_1 = 1.1064 rps, ω_2 = 3.0854 rps, ω_3 = 6.8738 rps, ω_4 = 12.0995 rps, ω_5 = 29.6284 rps, and ω_6 = 42.5886 rps.

5.5 Solve Problem 5.4 by using $\Delta(x) = \Delta(x/L_o)$ and $\Delta(x) = \Delta(x/L_o)^{1/2}$ and compare the results.

5.6 Solve Problem 5.4 by using $\Delta(x) = \Delta(x/L_o)^{1/2}$ and $\Delta(x) = \Delta\sin(\pi x/2L_o)$ and compare the results.

5.7 Solve Problem 5.1 by using a uniform weight distribution w = 10.0 lb/in. (1751.268 N/m).
Answer: $\omega_1 = 0.9767$ rps, $\omega_2 = 2.2596$ rps, $\omega_3 = 5.0121$ rps, $\omega_4 = 8.6750$ rps, $\omega_5 = 13.3764$ rps, and $\omega_6 = 19.0850$ rps.

5.8 Solve Problem 5.7 by using $\Delta(x) = \Delta(x/L_o)$ and $\Delta(x) = \Delta(x/L_o)^{1/2}$ and compare the results.

5.9 Solve Problem 5.1 by using the GFEM with equivalent uniform stiffness and uniform mass elements to solve the equivalent straight beam of length L_o and compare the results.

5.10 Solve Problem 5.9 by using a weight distribution w = 5.0 lb/in. (875,634 N/m). Compare the results with the ones obtained in Problem 5.4.

5.11 The tapered flexible simply supported beam in Figure 5.2a is assumed to have the attached uniform weight distribution w = 2.0 lb/in. (350.2536 N/m), stiffness EI_A = 75,000 kip-in.^{2} (215,224 Nm^2), length L = 1000 in. (25.4 m), and taper n = 1.5. By using Galerkin's consistent FEM and an equivalent straight beam of length L_o, determine the first six frequencies of vibration and the corresponding mode shapes. Assume $\Delta(x)$ = constant = Δ.
Answer: $\omega_1 = 1.7410$ rps, $\omega_2 = 6.5320$ rps, $\omega_3 = 14.6451$ rps, $\omega_4 = 25.9885$ rps, $\omega_5 = 40.5423$ rps, and $\omega_6 = 58.2834$ rps.

5.12 Solve Problem 5.11 for $\Delta(x) = \Delta$ and $\Delta(x) = \Delta x/L_o$ and compare the results.

5.13 Solve Problem 5.11 by using a weight distribution w = 5.0 lb/in. (875.634 N/m).
Answer: $\omega_1 = 1.2824$ rps, $\omega_2 = 4.1460$ rps, $\omega_3 = 9.2486$ rps, $\omega_4 = 16.3602$ rps, $\omega_5 = 25.4809$ rps, and $\omega_6 = 36.5939$ rps.

5.14 Solve Problem 5.13 by using $\Delta(x) = \Delta$ and $\Delta(x) = \Delta x/L_o$ and compare the results.

5.15 Solve Problem 5.11 by using GFEM with equivalent uniform stiffness and uniform mass elements to solve the equivalent straight beam of length L_o, and compare the results.

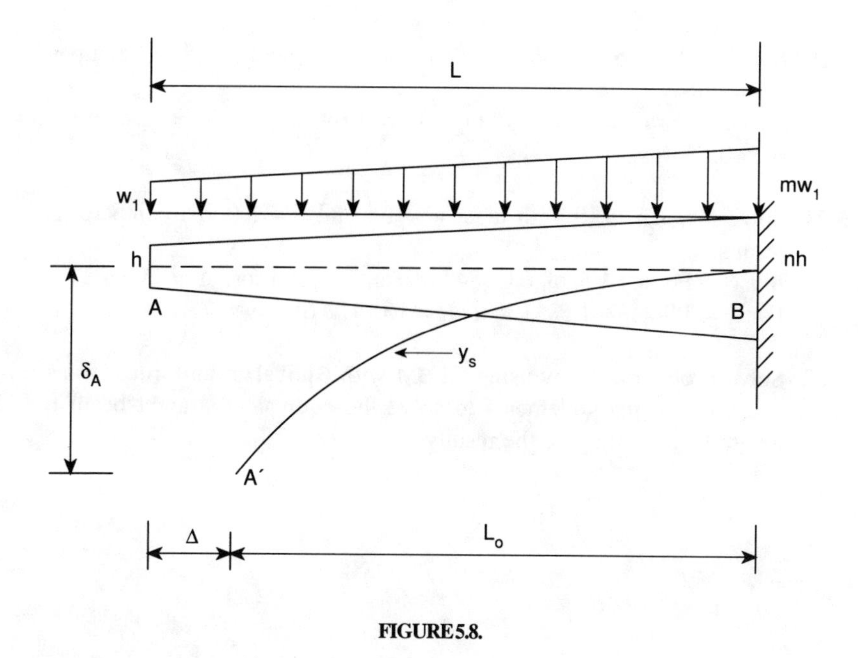

FIGURE 5.8.

5.16 Solve Problem 5.13 by using GFEM with equivalent uniform stiffness and uniform mass elements to solve the equivalent straight beam of length L_o and compare the results.

5.17 Solve Problem 5.15 by using a weight distribution w = 10.0 lb/in. (1,751.268 N/m) and taper n = 2.
Answer: $\omega_1 = 1.2087$ rps, $\omega_2 = 3.6792$ rps, $\omega_3 = 8.1388$ rps, $\omega_4 = 14.3133$ rps, $\omega_5 = 22.2171$ rps, and $\omega_6 = 31.8297$ rps.

5.18 Solve Problem 5.17 by using the horizontal displacement function $\Delta(x) = \Delta x/L_o$ and compare the results.

5.19 The doubly tapered flexible cantilever beam in Figure 5.8 has a trapezoidal weight distribution with weight $w_1 = 5.0$ lb/in. (875.634 N/m), m = 2, n = 2, length L = 1000 in. (25.4 m), and stiffness EI_A at the free end A equal to 180,000 kip-in.^{2} (516,541 Nm^2). By using Galerkin's consistent FEM and an equivalent straight beam of length L_o, determine the first six frequencies of vibration and the corresponding mode shapes. Use $\Delta(x)$ = constant = Δ.
Answer: $\omega_1 = 0.9690$ rps, $\omega_2 = 4.0938$ rps, $\omega_3 = 10.2786$ rps, $\omega_4 = 19.5081$ rps, $\omega_5 = 31.7825$ rps, and $\omega_6 = 47.0738$ rps.

5.20 Solve Problem 5.19 with load variations m = 1 and m = 2 and compare the results.
Answer: For m = 1 we have ω_1 = 1.0813 rps, ω_3 = 12.7356 rps, and ω_6 = 59.0559 rps.

5.21 Solve Problem 5.19 with taper n = 1.5 and n = 3.0 and compare the results.
Answer: For n = 1.5: ω_1 = 0.7844 rps, ω_3 = 7.7384 rps, ω_6 = 35.0562 rps. For n = 3.0: ω_1 = 1.5331 rps, ω_3 = 16.1412 rps, ω_6 = 73.1947 rps.

5.22 Solve Problem 5.19 by using GFEM with equivalent uniform stiffness and uniform mass elements to solve the equivalent straight beam of length L_o and compare the results.

Chapter 6

INELASTIC ANALYSIS OF UNIFORM AND VARIABLE THICKNESS MEMBERS

6.1 INTRODUCTION

Since prismatic and nonprismatic members are commonly used in many engineering structures, such as highway bridges, buildings, space and aircraft structures, as well as for machine elements, it is important to know how these members react to loading conditions that cause the material to be stressed well beyond its elastic limit and all the way to failure. Such analysis is usually complicated, particularly when both the modulus of elasticity E_x and the moment of inertia I_x can vary along the length of the member.

Although mumerical methods such as the FEM and the boundary element method are widely used, it is generally agreed that analytical work based on closed form solutions is definitely needed in order to establish the reliability of results produced by numerical methods. The method of equivalent systems, as developed by Fertis and his collaborators,[5–8,25,32–34] may be effectively used for inelastic analysis, because it provides the required ingredients and methodology for a convenient solution of such complicated problems.

In the first three chapters, the method of equivalent systems was developed and applied for the solution of flexible members subjected to various types of loading conditions and stiffness variations. In Chapters 4 and 5, the free vibration response of such flexible members was investigated. In all cases, however, it was assumed that the material of the member remains linearly elastic, and consequently the modulus of elasticity stays constant.

In this chapter, the material of the member under investigation will be stressed well beyond its elastic limit, practically all the way to failure, thus causing the modulus of elasticity of the member to vary along its length. Since both E and I may vary in this case, the method of the equivalent systems will be adjusted in order to take into consideration the actual variations of both E and I. In this chapter, however, it is assumed that the deformations of the member are small and that the member is absent of any axial effects and axial restraints that cause shifting of the neutral axis at cross sections along the length of the member. Such subjects are discussed in detail in following chapters. Various cases of loading conditions and stiffness variations are examined.

6.2 EQUIVALENT SYSTEMS FOR INELASTIC ANALYSIS

In the first chapter, Section 1.5, the method of equivalent systems was developed in order to provide a convenient approach for the solution of flexible beam problems. The nonlinear Euler-Bernoulli differential equation, Equation

1.3, was used in its entirety for this purpose. It was assumed, however, that the stresses of the material did not exceed its elastic limit, consequently the modulus of elasticity remained constant throughout the length of the flexible member. This implies that the modulus of elasticity function g(x) in Equation 1.24 is equal to unity.

When the material of a member is stressed beyond its elastic limit and its moment of inertia is either uniform or variable, both functions g(x) and f(x) in Equations 1.29 and 1.33 must be known in order to determine the moment diagram M_e or M_e' of the constant stiffness equivalent system. The evaluation of the function g(x) in this case is based on the determination of a reduced modulus E_r by using Timoshenko's method.[35] In order to utilize this method, the stress/strain curve of the material must be known. Such curves can be easily obtained experimentally, and therefore E_r can be determined for any given material.

In this section, and throughout this chapter, the deformations of the member are assumed to be small and the member is free of any axial restraints, or axial loads, that cause shifting of the neutral axis at cross sections along the length of the member. In other words, we assume that the centroidal and neutral axes coincide. This covers a wide range of problems and it deserves to be treated on its own. Inelastic analysis of members with axial loads and axial restraints that are subjected to small and large deformations are considered in subsequent chapters.

The computation of the reduced modulus E_r may be initiated by considering a beam of rectangular cross section as shown in Figure 6.1b, where r is the radius of curvature of the neutral surface produced by the bending moment M. The stress/strain curve representing the mechanical properties of the beam's material is shown in Figure 6.1a. In Figure 6.1b the quatities h_1 and h_2 denote the distances from the neutral axis to the lower and upper surfaces of the member, respectively. If ε_1 and ε_2 in Figure 6.1a represent the elongations of the extreme fibers, we can write the expressions

$$\varepsilon_1 = \frac{h_1}{r} \tag{6.1}$$

$$\varepsilon_2 = \frac{h_2}{r} \tag{6.2}$$

We may also write the expression

$$\varepsilon = \frac{y}{r} \tag{6.3}$$

where ε is the unit elongation of a fiber at a distance y from the neutral axis.

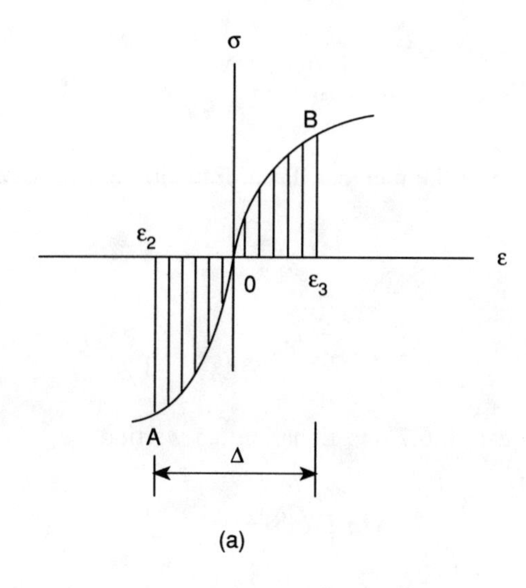

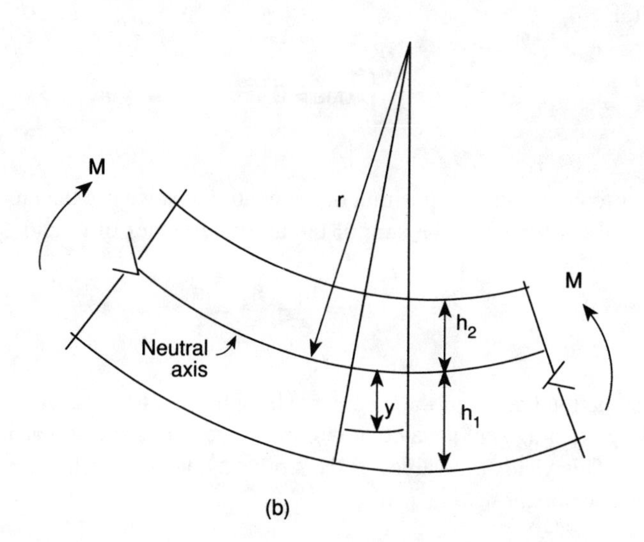

FIGURE 6.1. (a) Stress-strain curve; (b) portion of a member of rectangular cross section.

The position of the neutral axis and the radius of curvature r may be determined from the equations of statics

$$b\int_{-h_2}^{h_1} \sigma dy = 0 \tag{6.4}$$

$$b\int_{-h_2}^{h_1} \sigma y dy = M \tag{6.5}$$

where b is the width of the member. From Equation 6.3 we have

$$y = \varepsilon r \tag{6.6}$$

$$dy = r d\varepsilon \tag{6.7}$$

By substituting Equation 6.7 into Equation 6.5 we find

$$r\int_{\varepsilon_2}^{\varepsilon_1} \sigma d\varepsilon = 0 \tag{6.8}$$

which indicates that the position of the neutral axis, since $r \neq 0$, requires that the integral

$$\int_{\varepsilon_2}^{\varepsilon_1} \sigma d\varepsilon = 0 \tag{6.9}$$

In order to determine the position of the neutral axis we use the curve AOB in Figure 6.1a, where Δ is the sum of the absolute values of ε_1 and ε_2, i.e.,

$$\Delta = \varepsilon_1 - \varepsilon_2 = \frac{h_1}{r} + \frac{h_2}{r} = \frac{h}{r} \tag{6.10}$$

where h is the total depth of the member. The solution of Equation 6.9 may be obtained by defining Δ in a way that will make the two shaded areas in Figure 6.1a equal. This yields the values of the strains ε_1 and ε_2, and from Equations 6.1 and 6.2 we obtain

$$\frac{h_1}{h_2} = \left|\frac{\varepsilon_1}{\varepsilon_2}\right| \tag{6.11}$$

which defines the position of the neutral axis.

Since from Equation 6.3 we note that the strain ε is proportional to the distance y from the neutral axis, we may also conclude that the curve AOB represents the bending stress distribution along the depth h of the member, if h is substituted for Δ.

The radius of curvature r may be determined by substituting Equations 6.6 and 6.7 into Equation 6.5. This yields

$$br^2 \int_{\varepsilon_2}^{\varepsilon_1} \sigma \varepsilon d\varepsilon = M \tag{6.12}$$

or, by using Equation 6.10 and rearranging, we write

$$\frac{bh^3}{12} \cdot \frac{1}{r} \cdot \frac{12}{\Delta^3} \int_{\varepsilon_2}^{\varepsilon_1} \sigma \varepsilon d\varepsilon = M \tag{6.13}$$

The equation that holds for the elastic range is

$$\frac{EI}{r} = M \tag{6.14}$$

If the proportional limit of the material is exceeded, the curvature produced by the bending moment M may be determined from the expression

$$\frac{E_r I}{r} = M \tag{6.15}$$

where E_r is the reduced modulus defined by the expression

$$E_r = \frac{12}{\Delta^3} \int_{\varepsilon_2}^{\varepsilon_1} \sigma \varepsilon d\varepsilon \tag{6.16}$$

The integral in Equation 6.16 represents the moment of the shaded area in Figure 6.1a with respect to the vertical axis through the origin 0. The units of E_r are force per unit area, since the ordinates of the curve in Figure 6.1a represent stresses and the abscissas represent strain, which are the same as the ones for E.

A trial and error procedure may be used to determine E_r for any specific values of h and I at a given cross section of a member. This procedure may be initiated by assuming values of Δ and using the curve in Figure 6.1a to determine for each value of Δ the extreme elongations ε_1 and ε_2. Equation 6.16 may be used to determine E_r. With known E_r, the bending moment M may be determined from Equation 6.15. Note that Equation 6.10 yields $r = h/\Delta$, which defines the value of the curvature r at a cross section of depth h when Δ is known. Since E_r can vary at cross sections along the length of the member, the function g(x) can be determined from the equation

$$g(x) = \frac{E_r}{E} \tag{6.17}$$

where E is the reference elastic modulus.

With known g(x), the inelastic analysis of members of either uniform or variable cross section along their length can be performed using the method of equivalent systems as discussed earlier. This is done in the following sections.

The procedure for the computation of the function g(x) is general, and it can be applied to beams with other types of cross sections.

6.3 INELASTIC ANALYSIS OF PRISMATIC AND NONPRISMATIC CANTILEVER BEAMS

The methodology developed in Section 6.2 will be applied here in order to determine the inelastic response of prismatic and nonprismatic cantilever beams. For this purpose, the tapered cantilevered beam in Figure 6.2b, which is loaded with a uniformly distributed load w as shown is considered. The material of the member is Monel, with a stress/strain curve as shown, in Figure 6.3a. In practice, the yield stress of Monel is equal to 50,000 psi (344,750 kPa) at 70°F, with a modulus of elasticity $E = 26 \times 10^6$ psi (179.27×10^6 kPa). Let it be assumed that the load w = 1600 lb/in. (280,202.88 N/m), the depth h at the free end B is 8 in. (0.2032 m), the width b = 6 in. (0.1524 m), and taper n = 1.5. Under this load, the cantilever beam will be stressed well beyond the elastic limit of its material, and consequently the modulus of elasticity E will vary along its length.

The moment of inertia I_x at any x along the length of the member is given by the expression

$$I_x = \frac{bh^3}{12}\left[\frac{(n-1)(L-x)+L}{L}\right]^3 \tag{6.18}$$

If the taper n = 1, then the cantilever beam is of uniform cross section. By using the moment of inertia $I_B = bh^3/12$ at the free end B of the beam as the reference value, the function f(x) that represents the variation of I_x along the length of the member is

$$f(x) = \left[\frac{(n-1)(L-x)+L}{L}\right]^3 \tag{6.19}$$

The evaluation of the function g(x) that represents the variation of E_x along the length of the member with respect to the elastic value of the modulus E_x, say E_1, is based on the determination of a reduced modulus E_r, using Timoshenko's method[35] as discussed in the preceding section. In order to simplify the procedure, the stress/strain curve of Monel in Figure 6.3a may be

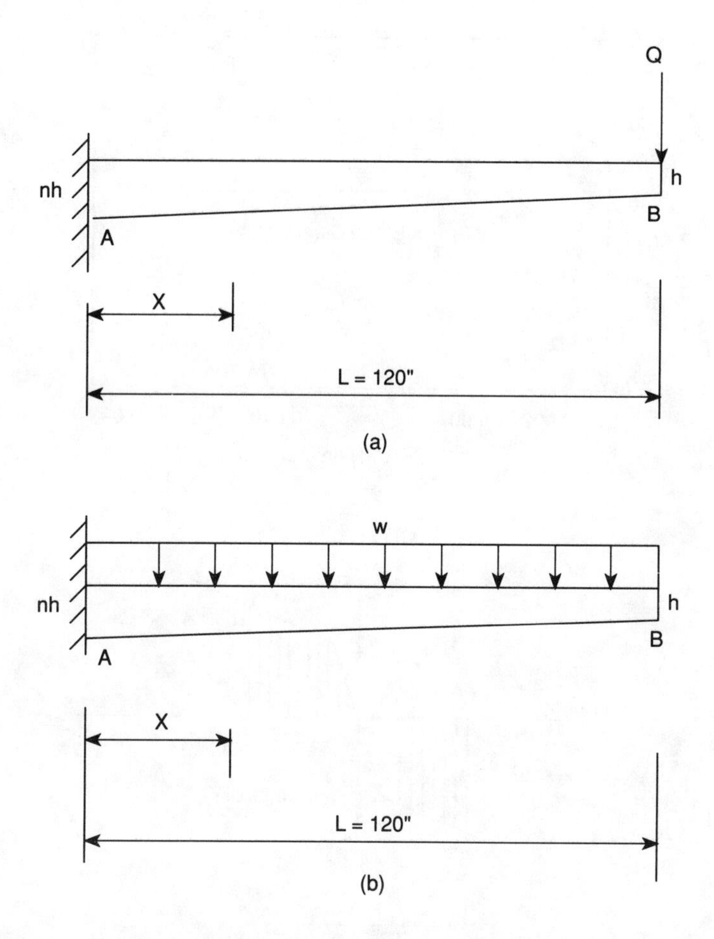

FIGURE 6.2. (a) Tapered cantilever beam loaded with concentrated load Q; (b) tapered cantilever beam loaded with uniformly distributed load w (1 in. = 0.0254 m).

approximated with two, three, or more straight lines, depending upon practical design requirements. A three-line approximation of the stress/strain curve, as shown later, is usually sufficient for practical problems. Tables 6.1 and 6.2 show the corresponding values of the modulus E and bending stress σ associated with the two-, three-, and six-line approximations of the stress/strain curve. The procedure here is illustrated by using first the bilinear approximation of the stress/strain curve, which yields $E_1 = 26 \times 10^6$ psi (179.27×10^6 kPa), $E_2 = 53 \times 10^6$ psi (365.44 kPa), and yield stress $\sigma_y = 50{,}000$ psi (344,750 kPa). Three and six straight line approximations are also used later in this section.

The bilinear approximation of the stress/strain curve of Monel is again shown schematically in Figures 6.3b and 6.3c. The compressive yield strength is assumed to be equal in magnitude to the tensile yield strength, i.e., symmetry. This is not mandatory for this method, but it follows typical engineering practice for steels. The curve AOB in Figure 6.3b represents the bending stress distribution along the depth h of the member, if h is substituted for Δ.

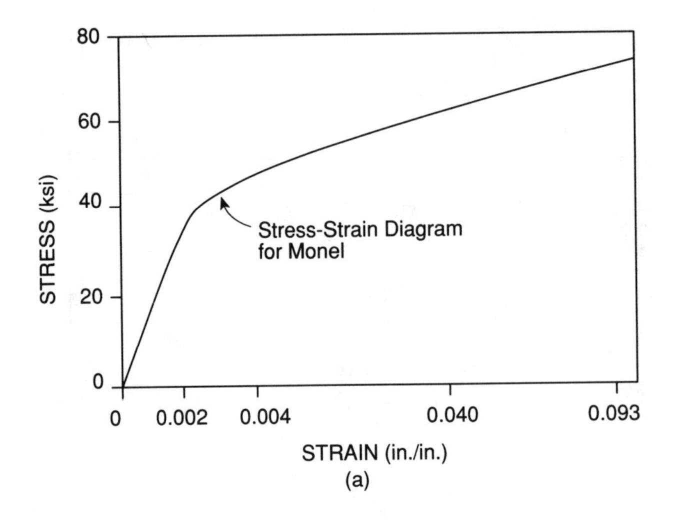

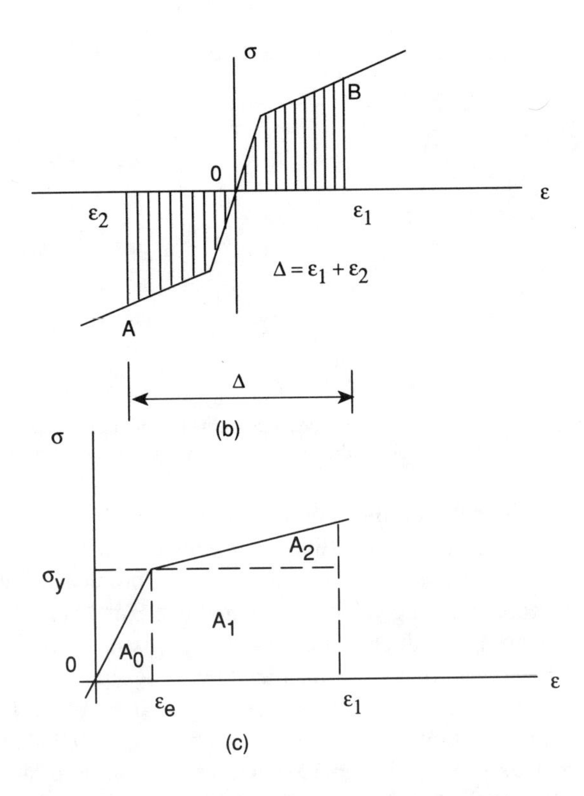

FIGURE 6.3. (a) Stress/strain curve of Monel; (b) bilinear stress/strain curve; (c) areas under the stress/strain curve (1 in. = 0.0254 m, 1 psi = 6.895 kPa).

TABLE 6.1
Values of E for Two-Line, Three-Line, and Six-Line Approximations of the Stress/Strain Curve of Monel

Modulus E (psi)	Two-line approximation	Three-line approximation	Six-line approximation
E_1	26×10^6	22×10^6	30×10^6
E_2	53×10^3	504×10^3	15×10^6
E_3	—	125×10^3	364×10^3
E_4	—	—	400×10^3
E_5	—	—	244×10^3
E_6	—	—	220×10^3

Note: 1 psi = 6.895 kPa.

TABLE 6.2
Values of σ for Two-Line, Three-Line, and Six-Line Approximations of Stress/Strain Curve of Monel

Stress σ (psi)	Two-line approximation	Three-line approximation	Six-line approximation
σ_1	50×10^3	48×10^3	30×10^3
σ_2	—	59×10^3	42×10^3
σ_3	—	—	50×10^3
σ_4	—	—	58×10^3
σ_5	—	—	60.2×10^3

Note: 1 psi = 6.895 kPa.

The trial and error procedure discussed in the preceding section is used here to calculate E_r for any specific value of h and I along the length of the member in Figure 6.2b. That is, at any cross section x from support A in Figure 6.2b, we assume values of Δ and we use the curve AOB in Figure 6.3b or Figure 6.3c to determine for each value of $\Delta = (\varepsilon_1 + \varepsilon_2) = 2\varepsilon$ the extreme strains ε_1 and ε_2. In this case, $\varepsilon_1 = \varepsilon_2 = \varepsilon$. Then, we use Equation 6.16 to determine the corresponding values of E_r and Equation 6.17 to calculate the values of the function g(x). Note that E in Equation 6.17 is the reference value of the modulus E_x. With known E_r, the inelastic or required bending moment M_{req} at the cross section considered can be determined from Equation 6.15. Since the problem is statically determinate, M_{req} would have to be equal to the bending moment M_x that is obtained from elastic analysis. However, when the member is statically indeterminate, the bending moment at a cross section depends on both the external loads and the distribution of the inelastic moment throughout the length of the member. In this case, it may not be reasonable to assume that M_{req} is equal to the elastic moment M_x that is obtained using elastic analysis. This problem is appropriately addressed later in this chapter.

TABLE 6.3
Summary of Reduced Modulus of Elasticity and Required Moment with Respect to Location Along the x-Axis

x (in.)	hx (in.)	I_x (in.^{4})	Δ (in./in.)	Strain ε (in./in.)	E_r (psi)	$M_{req} = M_x$ (in.-lb)
0	12.0	864.00	19.4790×10^{-2}	$\Delta/2$	0.8214×10^6	11.52×10^6
3	11.9	842.58	9.5345×10^{-2}	$\Delta/2$	1.6222×10^6	10.95×10^6
6	11.8	821.52	2.1996×10^{-2}	$\Delta/2$	6.7892×10^6	10.40×10^6
9	11.7	800.81	1.0846×10^{-2}	$\Delta/2$	13.2760×10^6	9.86×10^6
12	11.6	780.45	7.9961×10^{-3}	$\Delta/2$	17.3300×10^6	9.32×10^6
15	11.5	760.44	6.6462×10^{-3}	$\Delta/2$	20.0620×10^6	8.82×10^6
18	11.4	740.77	5.7961×10^{-3}	$\Delta/2$	22.0890×10^6	8.32×10^6
21	11.3	721.45	5.1962×10^{-3}	$\Delta/2$	23.6000×10^6	7.83×10^6
24	11.2	702.46	4.7462×10^{-3}	$\Delta/2$	24.6890×10^6	7.35×10^6
27	11.1	683.82	4.3962×10^{-3}	$\Delta/2$	25.4160×10^6	6.88×10^6
30	11.0	665.50	4.1462×10^{-3}	$\Delta/2$	25.8010×10^6	6.47×10^6
33	10.9	647.51	3.8962×10^{-3}	$\Delta/2$	25.9940×10^6	6.02×10^6
36	10.8	629.86	3.7227×10^{-3}	$\Delta/2$	26.0000×10^6	5.64×10^6

Note: 1 in. = 0.0254 m, 1 psi = 6.895 kPa, 1 lb = 4.448 N.

Table 6.3 summarizes the values of h_x, I_x, Δ, ε, E_r, and $M_{req} = M_x$ at various cross sections x along the length of the member. Note that the stress at locations beyond 36 in. (0.9144 m) is below the yield strength of the material, and consequently $E_r = E$ in these locations. Table 6.4 gives the values of f(x), E_r, g(x), M_x, and $M_e = M_x/f(x)g(x)$ at various locations x of the member. The moment diagram of the equivalent system of constant stiffness $E_B I_B$ is plotted as shown in Figure 6.4a. The approximation of its shape with three straight lines as shown leads to the constant stiffness equivalent system loaded as shown in Figure 6.4b. The deflection at any point x along the length of the member may be determined by using the equivalent system and applying handbook formulas or by using linear methods of elementary mechanics. The results would be almost identical to the results obtained if the original member in Figure 6.2b is solved. This, however, would be an extremely difficult task. By using the equivalent system in Figure 6.4b and applying the moment area method, we find that the vertical deflection y_B at the free end B is 8.76 in. (0.2225 m).

It should be noted here that the juncture points of the straight lines may lie above or below the M_e curve, so that the areas added to the M_e diagram or subtracted from it are approximately balanced. The approximation of the shape of the M_e curve with straight lines results in an equivalent system of constant stiffness that is always loaded with equivalent concentrated loads as shown in Figure 6.4b. Accurate results are obtained, within 2%, if only a few straight lines are used (see, e.g., Fertis[6,7] and Fertis and Keene[32]). If better accuracy is required, more straight lines can be used. The methodology can

TABLE 6.4
Values of f(x), g(x), E_r, M_x, and M_e for Inelastic Analysis

1	2	3	4	5	6
		$E_r \times 10^6$		M_x	$M_e = M_x/f(x)g(x)$
x(in.)	f(x)	(psi)	g(x)	(in.-lb)	(in.-lb)
0	3.38	0.821	0.0316	11.520×10^6	108.040×10^6
6	3.21	6.789	0.2611	10.397×10^6	12.407×10^6
12	3.05	17.330	0.6665	9.323×10^6	4.588×10^6
18	2.89	22.089	0.8496	8.319×10^6	3.384×10^6
24	2.74	24.689	0.9496	7.349×10^6	2.821×10^6
30	2.60	25.801	0.9924	6.472×10^6	2.509×10^6
36	2.46	26.0	1.0	5.645×10^6	2.294×10^6
42	2.33	26.0	1.0	4.867×10^6	2.092×10^6
48	2.20	26.0	1.0	4.147×10^6	1.888×10^6
54	2.07	26.0	1.0	3.485×10^6	1.681×10^6
60	1.95	26.0	1.0	2.880×10^6	1.475×10^6
66	1.84	26.0	1.0	2.333×10^6	1.269×10^6
72	1.73	26.0	1.0	1.843×10^6	1.067×10^6
78	1.62	26.0	1.0	1.411×10^6	0.870×10^6
84	1.52	26.0	1.0	1.037×10^6	0.682×10^6
90	1.42	26.0	1.0	0.720×10^6	0.506×10^6
96	1.33	26.0	1.0	0.461×10^6	0.346×10^6
102	1.24	26.0	1.0	0.259×10^6	0.209×10^6
108	1.16	26.0	1.0	0.115×10^6	0.100×10^6
114	1.08	26.0	1.0	0.029×10^6	0.027×10^6
120	1.00	26.0	1.0	0.0	0.0

Note: 1 in. = 0.0254 m, 1 psi = 6.895 kPa, 1 in.-lb = 0.1130 Nm.

be carried out by hand or by a digital computer. The computer would be helpful in carrying out the required trial and error procedure for the computation of Δ and E_r.

For the bilinear approximation of the stress/strain curve of Monel, the inelastic analysis of the nonprismatic beam in Figure 6.2b was repeated by varying the distributed load w for values of the taper n = 1.5, 1.75, and 2.0, and h = 8 in. (0.2032 m). The results for these three cases of n are shown plotted in Figure 6.5a. The starting point P in these curves represents the transition from elastic to inelastic behavior. Note that as the value of n increases, an additional load w is needed in order to reach this transition stage. We also note that when a certain value of the load w is reached, any further increase in w is associated with large increases in deflection, indicating that the ultimate capacity of the member to resist the load is reached. For example, for n = 1.75, the curve shows that the ultimate load w_u should be about 2000 lb/in. (350,253.6 N/m).

By using the bilinear stress/strain curve approximation of Monel, a similar inelastic analysis was carried out by loading the tapered cantilever beam with a concentrated load Q at the free end B, as shown in Figure 6.2a. The results

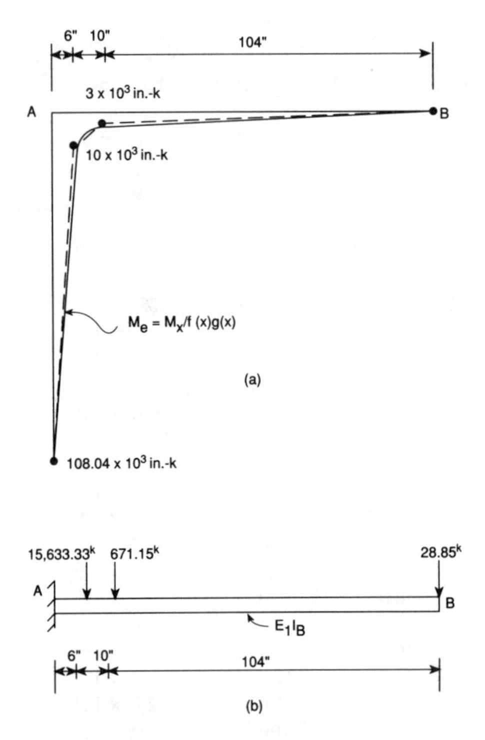

FIGURE 6.4. (a) M_e diagram approximated with three straight lines; (b) equivalent system of constant stiffness E_1I_B (1 in. = 0.0254 m, 1 kip = 4.448 kN, 1 in.-kip = 112.9848 Nm).

are shown in Figure 6.5b, and analogous observations can also be made for this case of loading.

6.4 INELASTIC ANALYSIS USING A THREE-LINE AND A SIX-LINE STRESS/STRAIN CURVE APPROXIMATION

In this section, the inelastic analysis discussed in the preceding sections will be applied to cantilever beams by using a three- and a six-line approximation of the stress/strain curve for Monel and the results will be compared. This will provide a basis which will help us to determine how many straight lines are required in the approximation of the stress/strain curve for a reasonably accurate solution. This is important information for the practicing design engineer who needs to make a decision regarding the liability of his design.

Consider again the tapered cantilever beam in Figure 6.2a that is loaded with concentrated load Q at the free end B. We assume that Q = 125.0 kips (556×10^3N), h = 8.0 in. (0.2032 m), b = 6.0 in. (0.1524 m), and taper n = 1.75. The three-line approximation of the stress/strain curve of Monel is used here to carry out the inelastic analysis. The values of E_1, E_2, and E_3, as well as stresses

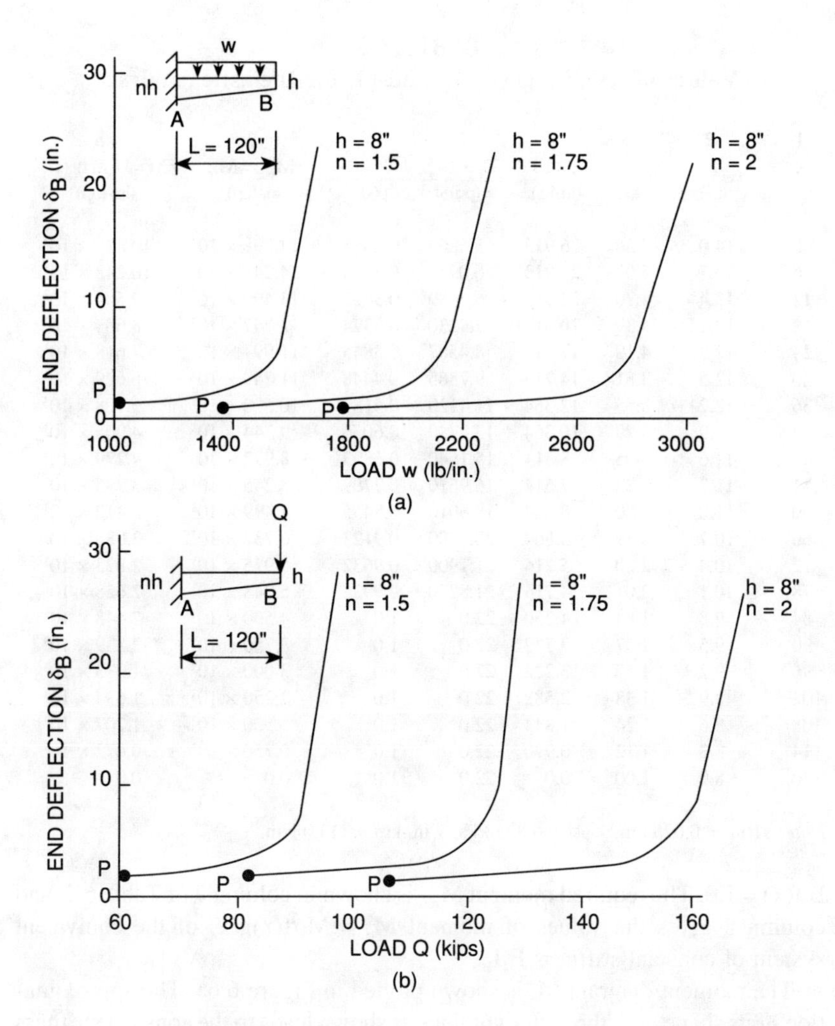

FIGURE 6.5. (a) Load deflection curves for bilinear approximation of the stress/strain curve and uniformly distributed load w, for n = 1.5, 1.75, and 2; (b) load deflection curves for bilinear approximation of the stress/strain curve and concentrated load Q for n = 1.5, 1.75, and 2 (1 in. = 0.0254 m, 1 lb = 4.448 N, 1 lb/in. = 175.1268 N/m).

σ_1 and σ_2, corresponding to this approximation are shown in Tables 6.1 and 6.2. By following the procedure discussed in the preceding section, we obtain the results shown in Table 6.5. The values of the reduced modulus E_r and function g(x) at various sections along the length of the member are shown in Columns 5 and 6, respectively. Note that for values of x ≥ 84.0 in. (2.1336 m) the modulus of elasticity is constant and equal to its reference elastic value E_1 = 22 × 10^6 psi (151.70 × 10^6 kPa), consequently, g(x) = 1.0 for x ≥ 84.0 in. (2.1336 m). The reference value of the moment of inertia I_B is at the free end B, and it is equal to 256.0 in.^{4} (106.555 × 10^{-6} m^4). Therefore, at the free end

TABLE 6.5
Values of f(x), E_r, g(x), M_x, and M_e for Inelastic Analysis

1	2	3	4	5	6	7	8
x (in.)	h_x (in.)	f(x)	$\Delta \times 10^{-3}$ (in./in.)	$E_r \times 10^6$ (psi)	g(x)	$M_{req} = M_x$ (in.-kip)	$M_e = M_x/f(x)g(x)$ (in.-kip)
0	14.0	5.36	26.913	5.6863	0.2585	14.998×10^3	10.827×10^3
6	13.7	5.02	24.913	6.0940	0.2770	14.248×10^3	10.242×10^3
12	13.4	4.70	22.613	6.6489	0.3022	13.499×10^3	9.504×10^3
18	13.1	4.39	20.013	7.4230	0.3374	12.747×10^3	8.604×10^3
24	12.8	4.10	17.313	8.4587	0.3845	11.997×10^3	7.618×10^3
30	12.5	3.81	14.713	9.7865	0.4448	11.249×10^3	6.629×10^3
36	12.2	3.55	12.364	11.4120	0.5187	10.500×10^3	5.708×10^3
42	11.9	3.29	10.364	13.2780	0.6036	9.744×10^3	4.905×10^3
48	11.6	3.05	8.814	15.1640	0.6893	8.992×10^3	4.279×10^3
54	11.3	2.82	7.614	16.9610	0.7709	8.245×10^3	3.795×10^3
60	11.0	2.60	6.664	18.6010	0.8455	7.499×10^3	3.412×10^3
66	10.7	2.39	5.864	20.0700	0.9123	6.737×10^3	3.086×10^3
72	10.4	2.20	5.214	21.1900	0.9632	5.975×10^3	2.823×10^3
78	10.1	2.01	4.714	21.8270	0.9921	5.248×10^3	2.628×10^3
84	9.8	1.84	4.260	22.0	1.0	4.500×10^3	2.448×10^3
90	9.5	1.67	3.777	22.0	1.0	3.750×10^3	2.239×10^3
96	9.2	1.52	3.222	22.0	1.0	3.000×10^3	1.973×10^3
102	8.9	1.38	2.582	22.0	1.0	2.250×10^3	1.634×10^3
108	8.6	1.24	1.844	22.0	1.0	1.500×10^3	1.207×10^3
114	8.3	1.12	0.990	22.0	1.0	0.750×10^3	0.672×10^3
120	8.0	1.00	0.0	22.0	1.0	0.0	0.0

Note: 1 in. = 0.0254 m, 1 psi = 6.895 kPa, 1 in.-kip = 113.0 Nm.

B f(x) = 1.0. The required moment M_{req} is shown in column 7 of Table 6.5, and column 8 gives the values of moment $M_e = M_x/f(x)g(x)$ of the equivalent system of constant stiffness E_1I_B.

The moment diagram M_e is shown plotted in Figure 6.6a. The approximation of its shape with three straight lines as shown leads to the constant stiffness equivalent system shown in Figure 6.6b. The deflections and rotations at any distance x from the fixed end of the initial variable stiffness member in Figure 6.2a may be determined by using the constant stiffness equivalent system in Figure 6.6b and applying known methods of linear elementary mechanics. For example, in this case, by using the moment area method to solve the equivalent system in Figure 6.6b, we find that the vertical deflection δ_B at the free end B is 7.79 in. (0.1979 m). The rotation θ_B at the same end is 9.4426×10^{-2} radians or 5.4130°.

The inelastic analysis of the variable stiffness cantilever beam in Figure 6.2a is repeated here by varying the concentrated load Q and using the beam cases with taper n = 1.25, 1.5, 1.75, and 2.0. The three-line approximation of the stress/strain curve is also used here. The results are shown plotted in Figure 6.7. The starting point P in these curves again represents the transition from elastic

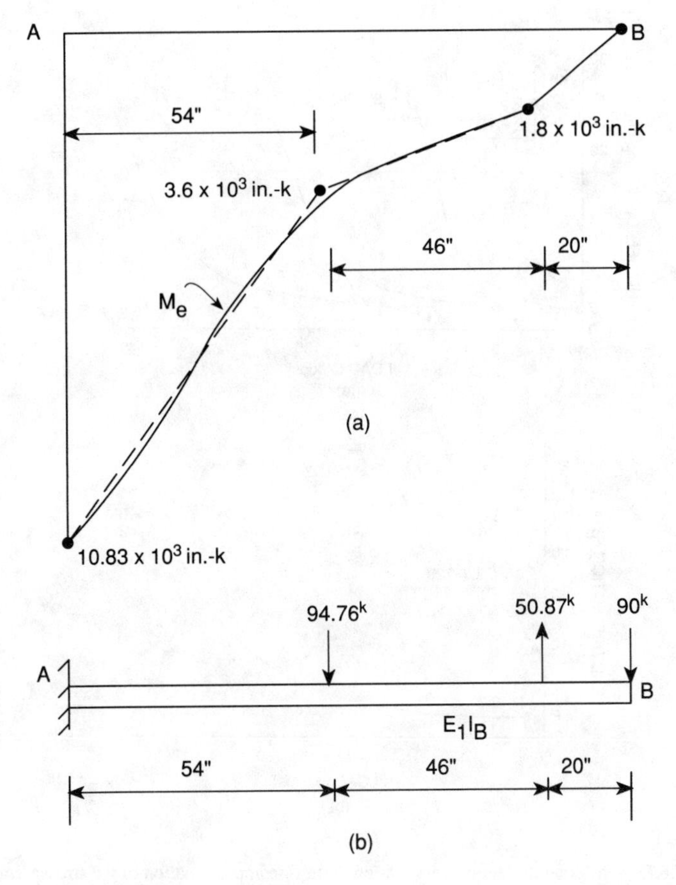

FIGURE 6.6. (a) M_e diagram with its shape approximated with three straight lines; (b) equivalent system of constant stiffness E_1I_B (1 in. = 0.0254 m, 1 kip = 4.448 kN, 1 in.-kip = 112.9848 Nm).

to inelastic behavior. We also note that for each beam and taper n case, there exists a critical value of the load Q which results in large increases of vertical deflection with small increases of Q from its critical value. This information is useful to the design engineer in establishing failure criteria and ultimate and allowable design loads.

The same beam cases are solved by using the six-line approximation of the stress/strain curve of Monel. The results are shown plotted in Figure 6.8, and the observation and remarks made by using the three-line approximation of the stress/strain curve apply as well here. Figure 6.9 shows a comparison of the results obtained using the three-line approximation of the stress/strain curve of Monel and the six-line approximation of the same curve. The three-line approximation is represented by the dashed line, while the solid line is used for the six-line approximation. For all practical purposes, the results for both cases are considered identical. This shows that for Monel a three-line approximation

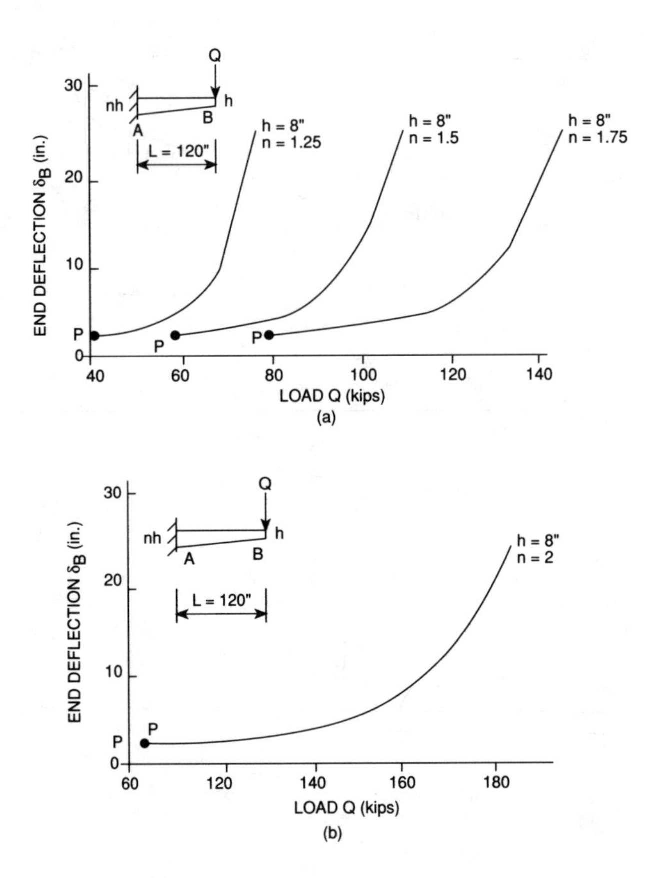

FIGURE 6.7. (a) Load deflection curves for a three-line approximation of the stress/strain curve of Monel and taper n = 1.25, 1.50, and 1.75; (b) load deflection curve for a three-line approximation of the stress-strain curve of Monel and n = 2.0 (1 in. = 0.0254 m, 1 kip = 4.448 kN).

of its stress/strain curve should be sufficiently accurate for practical inelastic analysis.

The bilinear approximation of the stress/strain curve of Monel, as discussed earlier, also provided reasonable results when it was applied to the solution of the cantilever problem in Figure 6.2b. From the results shown in Figure 6.5, we may conclude that the evaluation of the ultimate load Q_u may be obtained with reasonable accuracy for practical applications and the load deflection curve is steeper beyond Q_u. This, however, should not be of considerable disadvantage for the practicing design engineer, because most of his interest is concentrated on a reasonably accurate evaluation of the ultimate load Q_u in order to determine allowable design loads. Therefore, a bilinear approximation of the stress/strain curve could be sufficient for most cases. Such simplifications, however, are important, because when they are combined with the utilization of equivalent systems, they result in savings in computer time and consequently a more economic and safe design.

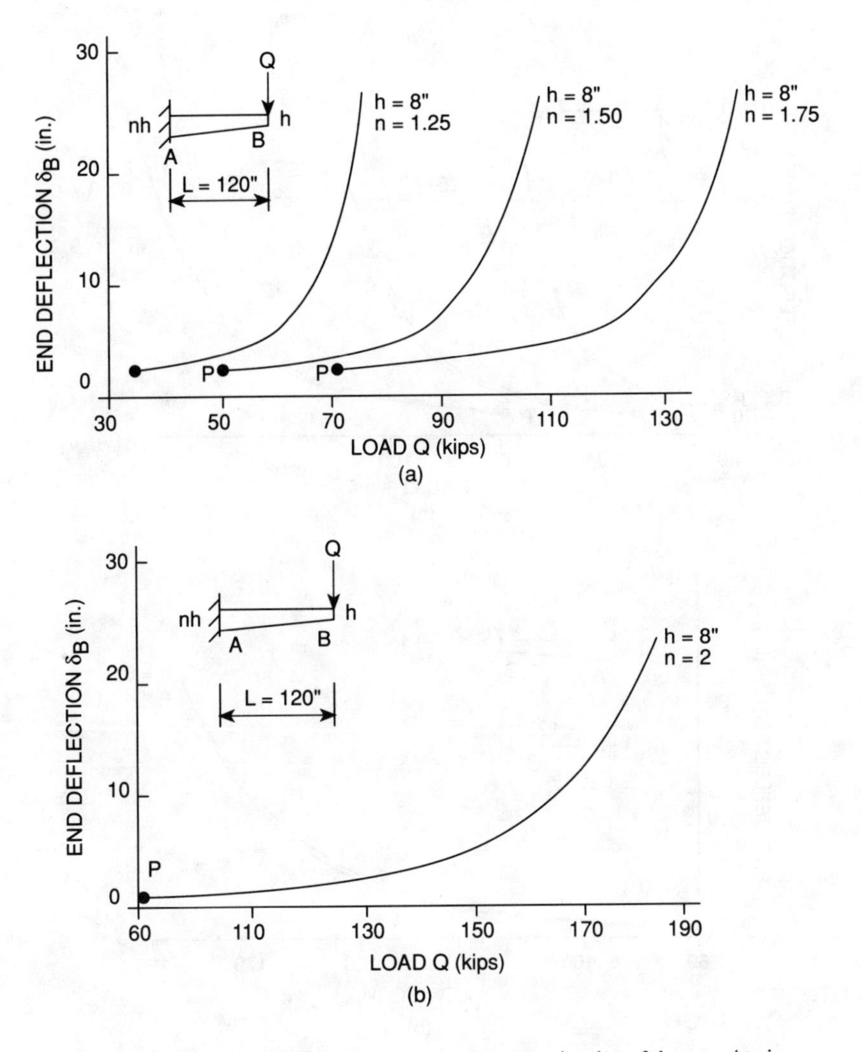

FIGURE 6.8. (a) Load deflection curves for a six-line approximation of the stress/strain curve of Monel and n = 1.25, 1.50, and 1.75; (b) load deflection curve for a six-line approximation of the stress/strain curve of Monel and n = 2.0 (1 in. = 0.0254 m, 1 kip = 4.448 kN).

The six-line approximation of the stress/strain curve of Monel was also used to analyze the cantilever beam in Figure 6.2a in order to study the variation of the modulus function g(x) along the length of the member for various values of the concentrated load Q. Figure 6.10a shows the variation of g(x) when h = 8.0 in. (0.2032 m), b = 6.0 in. (0.1524 m), and n = 1.5 for the indicated values of the concentrated load Q. We observe that as Q increases, the values of g(x) along the length of the member decrease accordingly and more sections of the member enter the inelastic stage. The same observations are made from the results in Figure 6.10b, where the depth of the member at fixed end A is increased by using the taper n = 2.0.

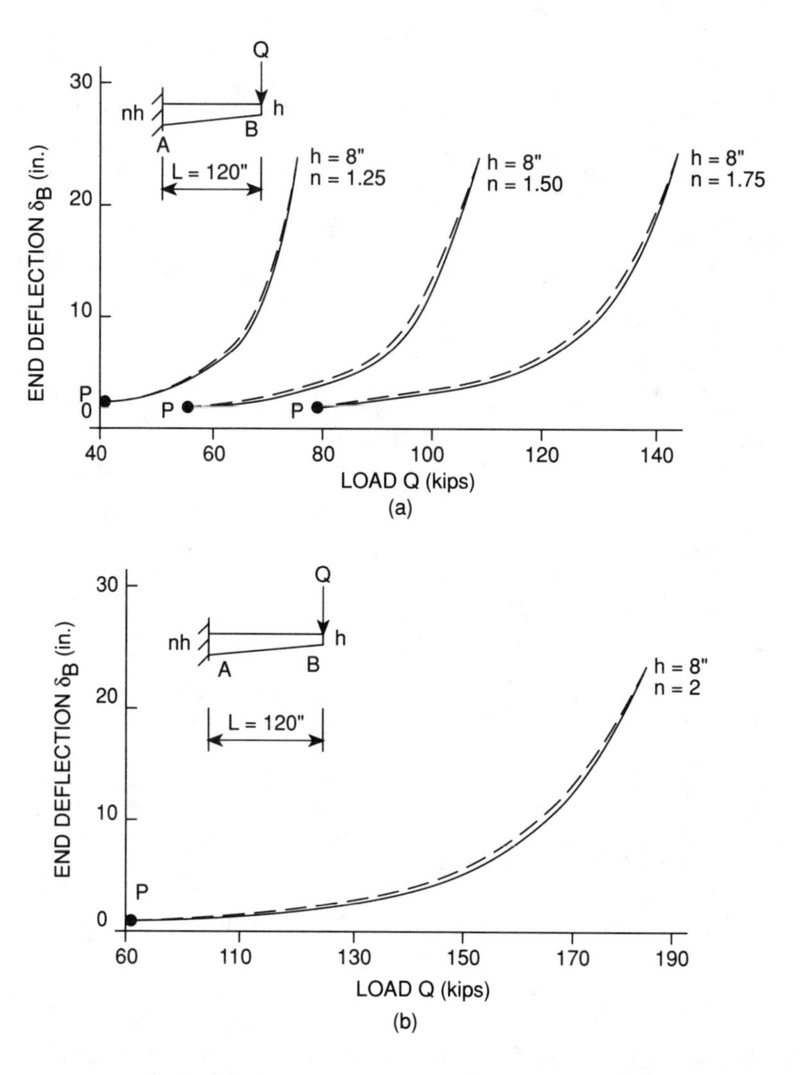

FIGURE 6.9. (a) Comparison of results for three-line and six-line approximations of the stress/strain curve of Monel and n = 1.25, 1.50, and 1.75; (b) comparison of results for three-line and six-line approximations of the stress/strain curve of Monel and n = 2.0 (1 in. = 0.0254 m, 1 kip = 4.448 kN).

6.5 INELASTIC ANALYSIS OF UNIFORM SIMPLY SUPPORTED BEAMS

In this section, the inelastic analysis will be applied for the solution of uniform simply supported beams loaded as shown in Figures 6.11a and 6.11b. The width b = 6.0 in. (0.1524 m), depth h = 8.0 in. (0.2032 m), and length L = 120.0 in. (3.048 m). The material of the member is Monel, and an inelastic analysis is performed here by using the bilinear approximation of the stress/strain curve of Monel.

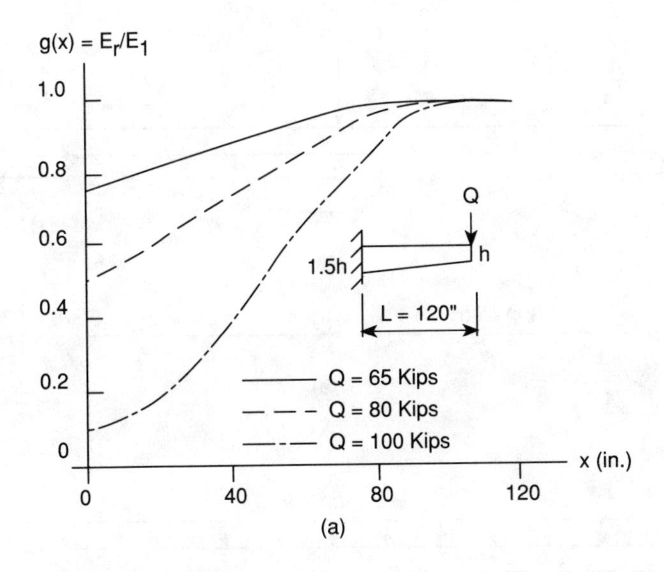

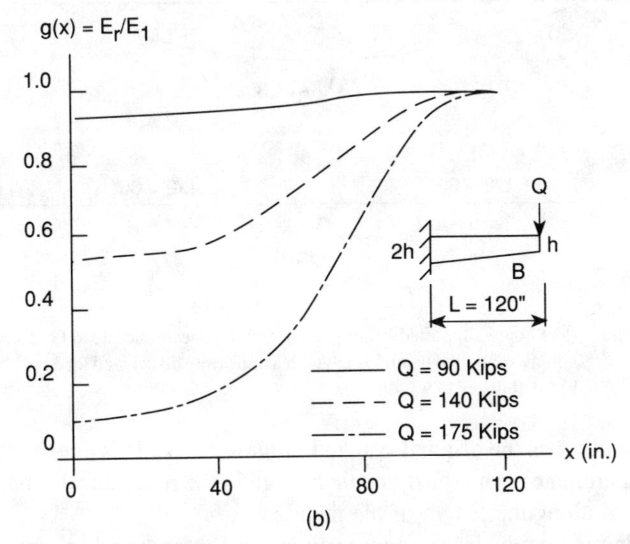

FIGURE 6.10. (a) Curves representing the variation of the function g(x) along the length of the member with increasing Q and n = 1.75; (b) curves representing the variation of the function g(x) with increasing Q and n = 2.0 (1 in. = 0.0254 m, 1 kip = 4.448 kN).

By using the loading case in Figure 6.11a and applying the procedure discussed in the preceding sections, the calculated values of Δ, E_r, g(x), $M_{req} = M_x$, and $M_c = M_x/f(x)g(x)$ at various sections along the length of the member, are shown in Table 6.6. The moment diagram M_e of the equivalent system of constant stiffness E_1I, where $E_1 = 26 \times 10^6$ psi (179.27×10^6 kPa) and I = 256.0 in.^{4} (106.555×10^{-6} m^4), is shown plotted in Figure 6.12a. The approximation of its shape with six straight lines leads to the equivalent system shown in Figure 6.12b. The loading on this equivalent system is dramatically different

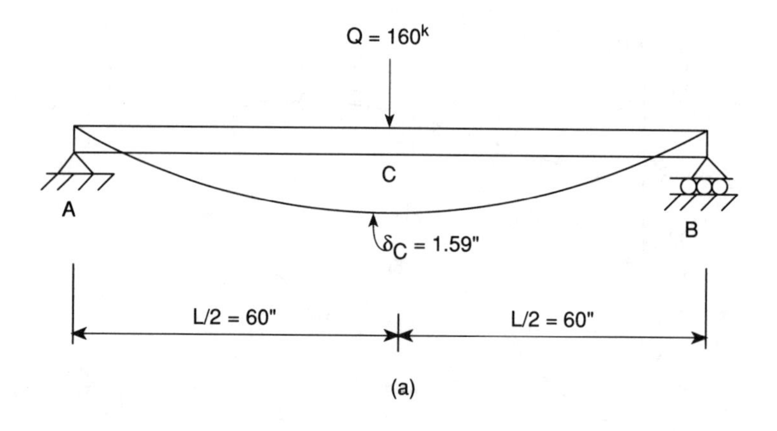

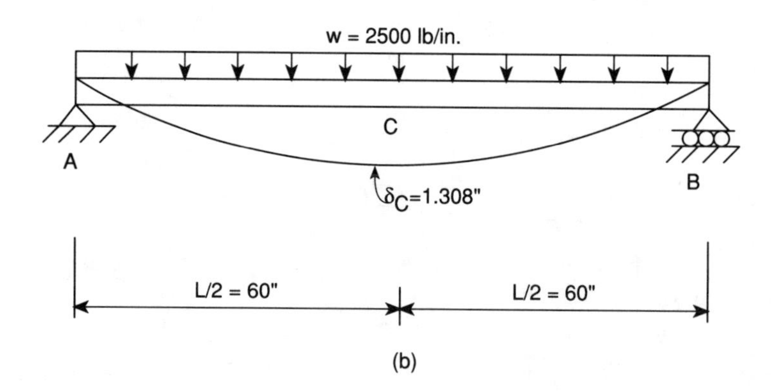

FIGURE 6.11. (a) Simply supported beam loaded with a concentrated load Q at midspan; (b) simply supported beam loaded with a uniformly distributed load w (1 kip = 4.448 kN, 1 lb/in. = 175.1268 N/m).

than the loading on the original system in Figure 6.11a. This drastic difference in loading eminates from the dramatic change in the mechanical property E of the material along the length of the member.

By using the equivalent system in Figure 6.12b and applying linear small deflection theory, in this case the moment area method, the vertical deflection curve of the member is plotted as shown in Figure 6.11a. The maximum vertical deflection δ_C occurs at the center C of the member, and it is equal to 1.59 in. (0.0404 m). The solution of the problem in Figure 6.11a by applying elastic analysis yields $\delta_C = 0.865$ in. (0.022 m), yielding an error of –45.6% when it is compared with the value obtained by inelastic analysis.

For the beam case in Figure 6.11b, the inelastic analysis yields the values of Δ, E_r, g(x), $M_{req} = M_x$, and M_e, shown in Table 6.7. A two-line approximation of the stress/strain curve of Monel is also used here. The moment diagram M_e of the equivalent system of constant stiffness E_1I is shown in Figure 6.13a. The

TABLE 6.6
Values of E_r, g(x), M_{req}, M_e, and f(x) = 1, for the Simply Supported Beam Loaded as Shown in Figure 6-11a

1 x (in.)	2 h_x (in.)	3 I (in.^{4})	4 $\Delta\times10^{-3}$ (in./in.)	5 $E_r\times10^6$ (psi)	6 g(x)	7 $M_{req} = M_x\times10^6$ (in.-lb)	8 $M_e = M_x/f(x)g(x)$ (in.-lb)
0	8	256	0.0	26.0	1.0	0.0	0.0
6	8	256	0.5769	26.0	1.0	0.480	0.480×10^6
12	8	256	1.1538	26.0	1.0	0.960	0.960×10^6
18	8	256	1.7308	26.0	1.0	1.440	1.440×10^6
24	8	256	2.3077	26.0	1.0	1.920	1.920×10^6
30	8	256	2.8846	26.0	1.0	2.400	2.400×10^6
36	8	256	3.4615	26.0	1.0	2.880	2.880×10^6
42	8	256	4.0461	25.906	0.996	3.354	3.366×10^6
48	8	256	4.9461	24.214	0.931	3.833	4.115×10^6
54	8	256	6.9962	19.282	0.742	4.317	5.821×10^6
57	8	256	9.8961	14.397	0.554	4.559	8.234×10^6
60	8	256	53.8960	2.783	0.107	4.800	44.841×10^6
63	8	256	9.8961	14.397	0.554	4.559	8.234×10^6
66	8	256	6.9962	19.282	0.742	4.317	5.821×10^6
72	8	256	4.9461	24.214	0.931	3.833	4.115×10^6
78	8	256	4.0461	25.906	0.996	3.354	3.366×10^6
84	8	256	3.4615	26.0	1.0	2.880	2.880×10^6
90	8	256	2.8846	26.0	1.0	2.400	2.400×10^6
96	8	256	2.3077	26.0	1.0	1.920	1.920×10^6
102	8	256	1.7308	26.0	1.0	1.440	1.440×10^6
108	8	256	1.1538	26.0	1.0	0.960	0.960×10^6
114	8	256	0.5769	26.0	1.0	0.480	0.480×10^6
120	8	256	0.0	26.0	1.0	0.0	0.0

Note: 1 in. = 0.0254 m, 1 psi = 6.895 kPa, 1 in.-lb = 0.11298 Nm.

approximation of its shape with six straight lines as shown leads to the equivalent system of constant stiffness E_1I shown in Figure 6.13b. By using this system and applying the moment area method, the vertical deflection curve is plotted as shown in Figure 6.11b. The maximum value $\delta_C = 1.308$ in. (0.0332 m) occurs at the midspan point C. If elastic analysis is used to solve the problem in Figure 6.11b, we find $\delta_C = 1.009$ in. (0.1524 m), which is 22.86% smaller compared to the value obtained by inelastic analysis.

6.6 INELASTIC ANALYSIS OF VARIABLE THICKNESS SIMPLY SUPPORTED BEAMS

In this section, simply supported beams of variable thickness and of rectangular cross section, which are loaded with a uniformly distributed load w as shown in Figure 6.14a, are considered,. We consider the case where w = 2500 lb/in. (437,817 N/m), depth $h_{A,}$ at support A, is 12.0 in. (0.3048 m), depth h_B

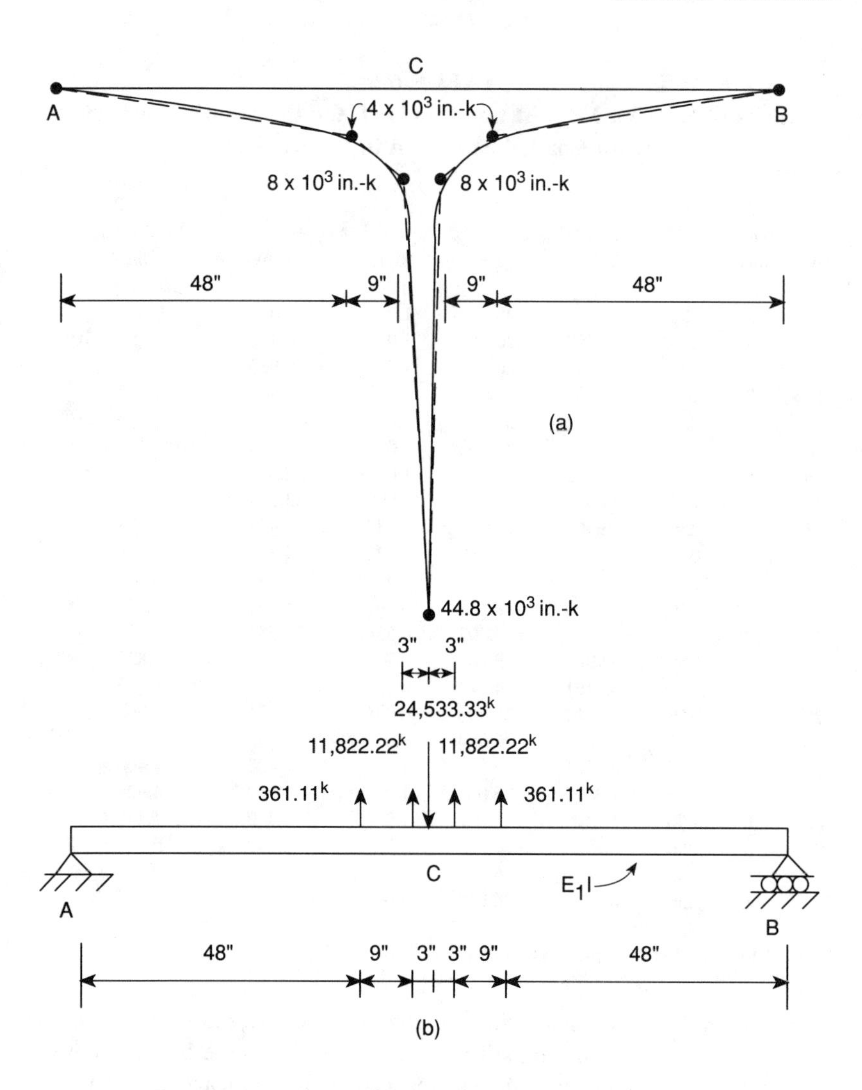

FIGURE 6.12. (a) M_e diagram with its shape approximated with six straight lines; (b) equivalent system of constant stiffness E_1I (1 in. = 0.0254 m, 1 kip = 4.448kN, 1 in.-kip = 113.0 Nm).

at support B is 8.0 in. (0.2032 m), width b = 6.0 in. (0.1524 m), and length L = 120.0 in. (3.048 m). The material of the member is mild steel with a stress/strain curve as shown in Figure 6.14b. The approximation of its shape with four and six straight line segments is shown in the same figure. The corresponding values of the modulus E and stress σ for these two approximation types are shown in Table 6.8.

The inelastic analysis discussed in the preceding sections is applied here to solve the problem in Figure 6.14a. In this case, since the depth is variable, the moment of inertia function f(x) will vary along the length of the member. The

TABLE 6.7
Values of f(x), E_r, g(x), M_{req}, and M_e for the Simply Supported Beam Loaded as Shown in Figure 6-11b

1 x (in.)	2 h_x (in.)	3 f(x)	4 $\Delta \times 10^{-3}$ (in./in.)	5 $E_r \times 10^6$ (psi)	6 g(x)	7 $M_{re} = M_x \times 10^6$ (in.-lb)	8 $M_e = M_x/f(x)g(x)$ (in.-lb)
0	8	1	0.0	26.0	1.0	0.0	0.0
6	8	1	1.0276	26.0	1.0	0.855	0.855×10^6
12	8	1	1.9421	26.0	1.0	1.620	1.620×10^6
18	8	1	2.7584	26.0	1.0	2.295	2.295×10^6
24	8	1	3.4615	26.0	1.0	2.880	2.880×10^6
30	8	1	4.0962	25.858	0.9945	3.389	3.408×10^6
36	8	1	4.7962	24.571	0.9451	3.771	3.990×10^6
42	8	1	5.7961	22.082	0.8493	4.096	4.822×10^6
48	8	1	6.9962	19.282	0.7416	4.317	5.821×10^6
54	8	1	8.2961	16.787	0.6457	4.457	6.902×10^6
60	8	1	8.8961	15.813	0.6082	4.502	7.402×10^6
66	8	1	8.2961	16.787	0.6457	4.457	6.902×10^6
72	8	1	6.9962	19.282	0.7416	4.317	5.821×10^6
78	8	1	5.7961	22.082	0.8493	4.096	4.822×10^6
84	8	1	4.7962	24.574	0.9451	3.771	3.990×10^6
90	8	1	4.0962	25.858	0.9945	3.389	3.408×10^6
96	8	1	3.4615	26.0	1.0	2.880	2.880×10^6
102	8	1	2.7584	26.0	1.0	2.295	2.295×10^6
108	8	1	1.9471	26.0	1.0	1.620	1.620×10^6
114	8	1	1.0276	26.0	1.0	0.855	0.855×10^6
120	8	1	0.0	26.0	1.0	0.0	0.0

Note: 1 in. = 0.0254 m, 1 psi = 6.895 kPa, 1 in.-lb = 0.11298 Nm.

moment of inertia $I_B = 256.0$ in.^{4} (106.555×10^{-6} m^4) is taken as the reference value. By performing the inelastic analysis, the values of f(x), Δ, E_r, g(x), $M_{req} = M_x$, and $M_e = M_x/f(x)g(x)$ are shown in Table 6.9. The values of the moment M_e of the equivalent system of constant stiffness E_1I_B, where $E_1 = 94.366 \times 10^5$ psi (650.6306×10^5 kPa) and $I_B = 256.0$ in.^{4} (106.555×10^{-6} m^4), are shown in column 8 of the same table.

With known M_e, the equivalent system that is loaded with concentrated loads and has constant stiffness E_1I_B can be determined as in the preceding sections (see, i.e., Figure 6.13). By using the constant stiffness equivalent system and applying the moment area method, or other methods of linear mechanics if it is preferable, we find the results shown in Table 6.10. Column 4 in this table shows that the maximum vertical deflection is 1.7977 in. (0.0457 m), and it occurs at 66 in. (1.6764 m) from the left support. The maximum rotation θ, column 3 in the same table, occurs at the right support and it is equal to 0.052571 rads, or 3.0136°.

The inelastic analysis for the nonprismatic member in Figure 6.14a is repeated again by varying the magnitude of the uniformly distributed load w.

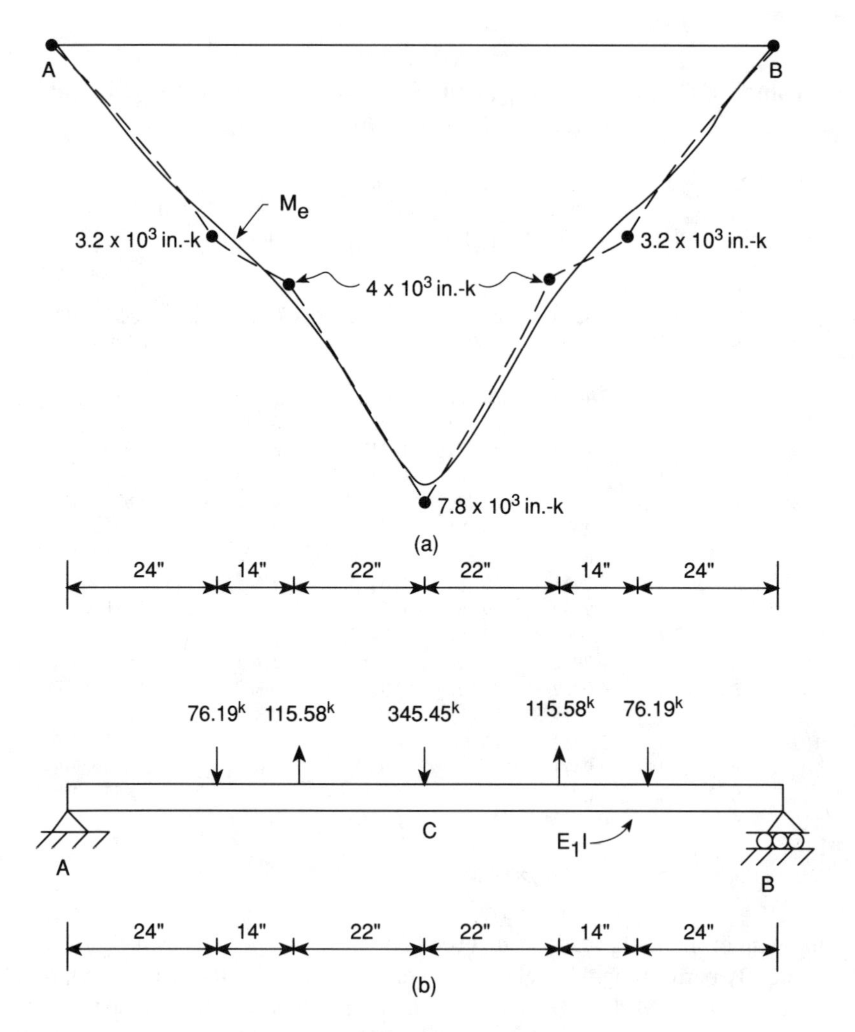

FIGURE 6.13. (a) M_e diagram with its shape approximated with six straight lines; (b) equivalent system of constant stiffness E_1I (1 in. = 0.0254 m, 1 kip = 4.448 kN, 1 in.-kip = 112.98 Nm).

The curves 1, 2, and 3 in Figure 6.15 illustrate the variation of the vertical deflection at the quarter length points 1, 2, and 3, respectively, with increasing w in Figure 6.14a. Sharp increases in deflection are observed for values of w > 3000 lb/in. (525,380.4 N/m), which indicates that the ultimate capacity of the member to resist stress and deformation is reached. The dashed lines in Figure 6.15 represent the results obtained by using the six-line approximation of the stress/strain curve, while the solid lines illustrate the results obtained by using the four-line approximation of the stress/strain curve. For practical purposes, these results are in close agreement.

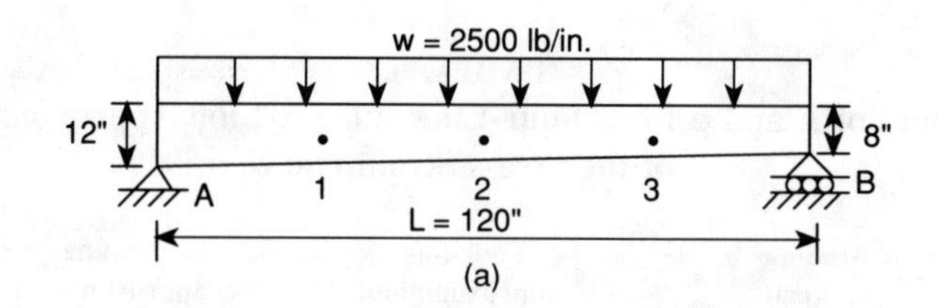

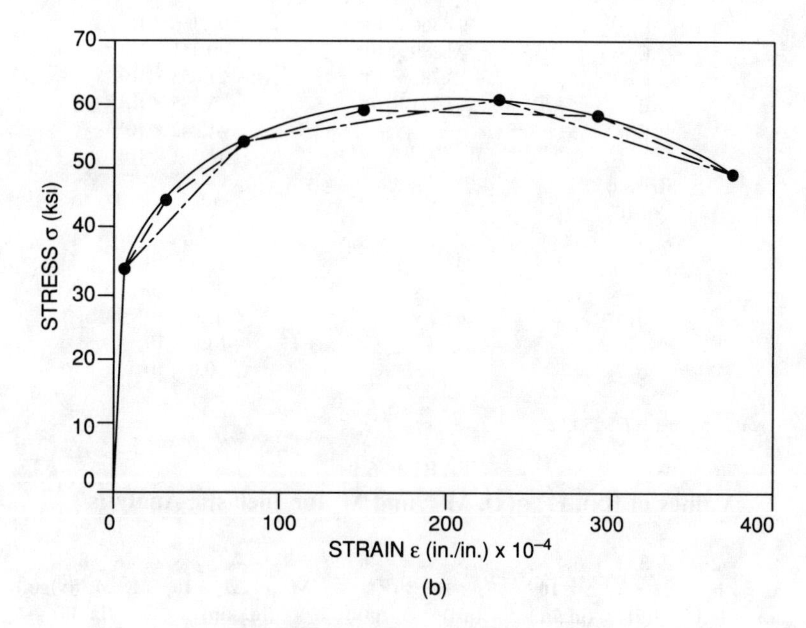

FIGURE 6.14. (a) Tapered simply supported beam loaded with a uniformly distributed load w; (b) stress/strain diagram of the material of the member with its shape approximated with four and six straight lines (1 in. = 0.0254 m, 1 lb/in. = 175.1268 N/m, 1 psi = 6895 kPa).

6.7 INELASTIC ANALYSIS OF UNIFORM AND VARIABLE THICKNESS STATICALLY INDETERMINATE BEAMS

Inelastic analysis will be applied here to statically indeterminate members that are fixed on one end and simply supported at the other end. The member can be of uniform or variable depth along its length. In fact it can have any arbitrary variation in thickness along its length without complicating the procedure. This is an appropriate statement, because in all cases examined earlier the $M_e = M_x/f(x)g(x)$ diagram was obtained by considering cross sections along the length of the member and calculating M_x, $f(x)$, and $g(x)$. The calculated M_e values were then plotted, and the shape of M_e was approximated by straight

TABLE 6.8
Values of E and σ for a Four-Line and Six-Line Approximation of the Stress/Strain Curve

Modulus E (psi)	Four-line approximation	Six-line approximation
E_1	94.366×10^5	94.366×10^5
E_2	28.289×10^4	40.351×10^4
E_3	3.785×10^4	21.053×10^4
E_4	-7.971×10^4	3.785×10^4
E_5		-1.942×10^4
		-11.561×10^4
Stress σ (psi)		
σ_1	33.5×10^3	33.5×10^3
σ_2	55.0×10^3	45.0×10^3
σ_3	61.0×10^3	55.0×10^3
σ_4		61.0×10^3
σ_5		60.0×10^3

Note: 1 psi = 6.895 kPa.

TABLE 6.9
Values of f(x), E_r, g(x), M_{req} and M_e for Inelastic Analysis

1 x (in.)	2 h_x (in.)	3 f(x)	4 $\Delta \times 10^{-3}$ (in./in.)	5 $E_r \times 10^6$ (psi)	6 g(x)	7 $M_{req} = M_x \times 10^6$ (in.-kip)	8 $M_e = M_x/f(x)g(x)$ (in.-lb)
0	12.00	3.38	0.0	9.4366	1.0	0.0	0.0
6	11.80	3.21	1.3014	9.4366	1.0	–0.8550	-0.2664×10^6
12	11.60	3.05	2.5516	9.4366	1.0	–1.6200	-0.5314×10^6
18	11.40	2.89	3.7427	9.4366	1.0	–2.2950	-0.7931×10^6
24	11.20	2.74	4.8660	9.4366	1.0	–2.8800	-1.0496×10^6
30	11.00	2.60	5.9116	9.4366	1.0	–3.3750	-1.2983×10^6
36	10.80	2.46	6.8685	9.4366	1.0	–3.7800	-1.5364×10^6
42	10.60	2.33	7.8000	9.3293	0.9886	–4.0950	-1.7776×10^6
48	10.40	2.20	8.9999	8.8677	0.9397	–4.3200	-2.0906×10^6
54	10.20	2.07	10.5000	8.1524	0.8639	–4.4550	-2.4868×10^6
60	10.00	1.95	12.2500	7.3499	0.7789	–4.5000	-2.9593×10^6
66	9.80	1.84	13.8000	6.7239	0.7125	–4.4550	-2.4018×10^6
72	9.60	1.73	14.4500	6.4865	0.6874	–4.3200	-3.6362×10^6
78	9.40	1.62	13.7500	6.7428	0.7145	–4.0950	-3.5337×10^6
84	9.20	1.52	11.9000	7.5030	0.7951	–3.7800	-3.1247×10^6
90	9.00	1.42	9.8500	8.4660	0.8971	–3.3750	-2.6439×10^6
96	8.80	1.33	8.0500	9.2529	0.9805	–2.8800	-2.2099×10^6
102	8.60	1.24	6.5766	9.4366	1.0	–2.2950	-1.8474×10^6
108	8.40	1.16	4.8660	9.4366	1.0	–1.620	-1.3994×10^6
114	8.20	1.08	2.6950	9.4366	1.0	–0.8550	0.7940×10^6
120	8.00	1.00	0.0	9.4366	1.0	0.0	0.0

Note: 1 in. = 0.0254 m, 1 in.-lb = 0.1130 Nm, 1 psi = 6.895 kPa.

TABLE 6.10
Values of Rotation, Deflection and M_e, at Cross Sections along the Length of the Member

1 x (in.)	2 $M_e \times 10^6$ (in.-lb)	3 Rotation (rad)	4 Deflection (in.)
0	0.0	0.040458	0.0
6	–0.2664	0.040127	0.2420
12	–0.5314	0.039136	0.4800
18	–0.7931	0.037491	0.7102
24	–1.0496	0.035201	0.9285
30	–1.2983	0.032284	1.1312
36	–1.5364	0.028762	1.3145
42	–1.7776	0.024658	1.4750
48	–2.0906	0.019856	1.6088
54	–2.4868	0.014174	1.7113
60	–2.9593	0.007421	1.7765
66	–3.4018	–0.000483	1.7977
72	–3.6362	–0.009276	1.7687
78	–3.5337	–0.018244	1.6860
84	–3.1247	–0.026539	1.5513
90	–2.6439	–0.033698	1.3701
96	–2.2099	–0.039712	1.1495
102	–1.8474	–0.044748	0.8958
108	–1.3994	–0.048802	0.6147
114	–0.7940	–0.051553	0.3131
120	0.0	–0.052571	0.0

Note: 1 in. = 0.0254 m, 1 in.-kip = 113.0 Nm.

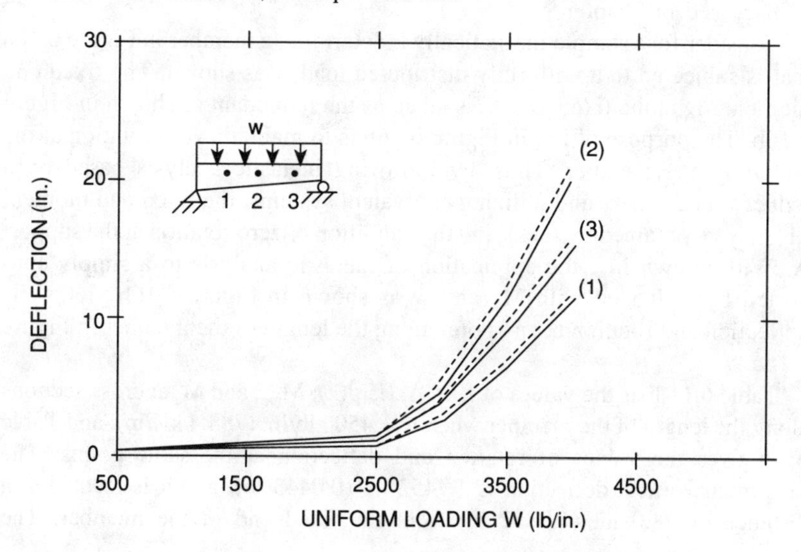

FIGURE 6.15. Load deflection curves for quarter points 1, 2, and 3 in Figure 6.14a (1 in. = 0.0254 m, 1 lb/in. = 175.1268 N/m).

lines. This procedure did not require any mathematical expression for M_e, and consequently the f(x) and g(x) can have any arbitrary variation. This is the main advantage in using the method of equivalent systems for inelastic analysis. The method does not get more complicated for arbitrary variations in beam depth, and it does not require a mathematical expression for the solution of the inelastic problem.

For statically determinate members, the moment at any cross section of the member depends only on the external loads, consequently, it is independent of the inelastic moment developing at cross sections along the length of the beam. Therefore, when the inelastic analysis of the member is based on the computation of a reduced modulus E_r, it would be correct to assume that the required moment M_{req} would have to be equal to the moment M_x that can be obtained from elastic analysis. When the member is statically indeterminate, the bending moment at a cross section depends on both the external load and the distribution of the inelastic moment throughout the length of the member. Therefore, in this case, it may not be reasonable to assume that the required moment M_{req} is equal to the elastic moment M_x obtained by using elastic analysis.

A reasonable approach for statically indeterminate beams would be the utilization of inelastic analysis based on a reduced modulus E_r for the computation of the redundant reactive moments or forces. This procedure should be reasonable as long as the deflection of the member in the inelastic range is not large enough to justify the utilization of large deflection theory. The method of equivalent systems, however, is applicable for large deflection theory (see Chapters 1, 2, and 3) and consequently inelastic analysis as presented here may be extended to include large deformations. This problem is discussed in Chapter 9.

Consider for example the statically indeterminate member in Figure 6.16a that is subjected to a uniformly distributed load w as shown. The fixed end moment M_A at the fixed end A, is taken as the redundant as shown in Figure 6.16b. The purpose of M_A in Figure 6.16b is to maintain zero rotation at the end A of the member. Thus, by following inelastic analysis based on a reduced modulus E_r and utilizing equivalent systems, the fixed end moment M_A may be obtained by satisfying the condition of zero rotation at the support A. With known M_A, the application of inelastic analysis to a simply supported beam loaded with M_A and w as shown in Figure 6.16b yields the deflection and rotation at any point along the length of the member in Figure 6.16a.

Table 6.11 lists the values of f(x), Δ, E_r, g(x), M_{req}, and M_e at cross sections along the length of the member when w = 4500 lb/in. (788.4 kN/m), and Table 6.12 gives the values of rotation and deflection at the same points. The maximum vertical deflection is 1.7452 in. (0.0443 m), and it is located at a distance of 78.0 in. (1.9812 m) from the fixed end of the member. The

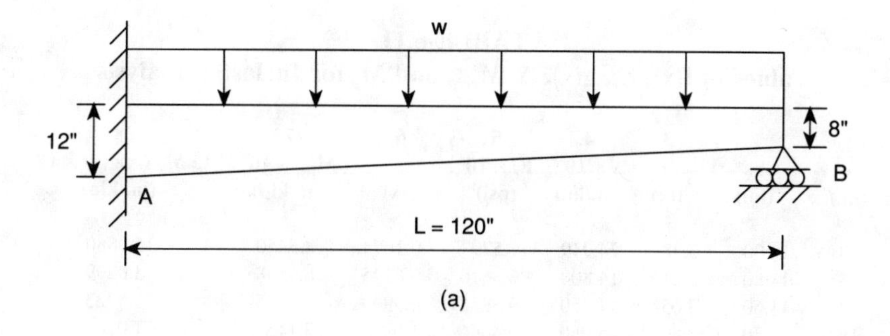

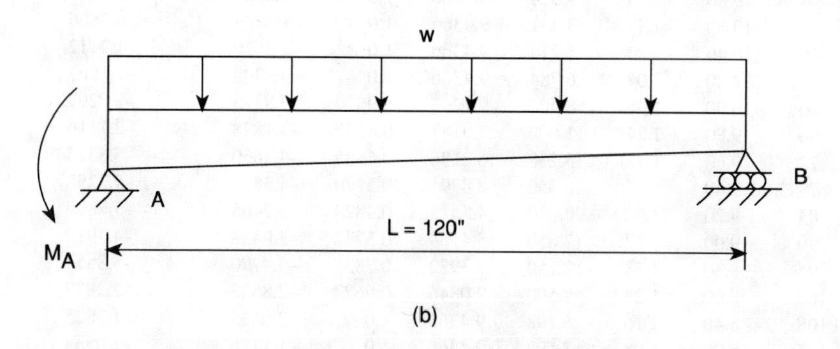

FIGURE 6.16. (a) Tapered statically indeterminate member loaded with a uniformly distributed load w; (b) tapered simply supported member loaded with the redundant moment M_A and the uniformly distributed load w (1 in. = 0.0254 m).

maximum rotation is located at support B and it is equal to 0.059828 rads or 3.4296°.

Figure 6.17a illustrates the variation of the fixed end moment M_A of the member with increasing loading w. This curve shows that up to a loading w of about 3000 lb (13,344.666 N) the variation of M_A is linear, and it becomes nonlinear for values of w > 3000 lb (13,344.666 N). Figure 6.17b shows the variation of the vertical deflection at the quarter points 1, 2, and 3 with increasing loading w. Again these curves become very steep for loadings w > 4500 lb/in. (788.40 kN/m), which indicates that the ultimate capacity of the member to resist stress and deformation is reached.

It should be noted, however, that the above inelastic analysis is restricted to statically indeterminate members, which do not develop axial loads or axial restraints during deformation. Such loading conditions will cause shifting of the neutral axis at cross sections along the length of the member that should be taken into consideration. This problem is treated in the following chapter.

TABLE 6.11
Values of f(x), E_r, g(x), Δ, M_{req}, and M_e for Inelastic Analysis

1 x (in.)	2 h_x (in.)	3 f(x)	4 $\Delta \times 10^{-3}$ (in./in.)	5 $E_r \times 10^6$ (psi)	6 g(x)	7 $M_{req} \times 10^3$ (in.-kip)	8 $M_e = M_x/f(x)g(x) \times 10^3$ (in.-kip)
0	12.00	3.38	74.949	1.5797	0.1674	8.5250	15.0880
6	11.80	3.21	14.800	6.3646	0.6745	6.5598	3.0299
12	11.60	3.05	7.550	9.3888	0.9949	4.7565	1.5723
18	11.40	2.89	5.080	9.4366	1.0	3.1153	1.0766
24	11.20	2.74	2.764	9.4366	1.0	1.6360	0.5962
30	11.00	2.60	0.558	9.4366	1.0	0.3188	0.1226
36	10.80	2.46	1.520	9.4366	1.0	–0.8365	–0.3400
42	10.60	2.33	3.451	9.4366	1.0	–1.8298	–0.7866
48	10.40	2.20	5.214	9.4366	1.0	–2.6610	–1.2112
54	10.20	2.07	6.784	9.4366	1.0	–3.3303	–1.6067
60	10.00	1.95	8.400	9.1247	0.9670	–3.8375	–2.0292
66	9.80	1.84	11.000	7.9147	0.8387	–4.1828	–2.7116
72	9.60	1.73	15.250	6.2136	0.6585	–4.3660	–3.8375
78	9.40	1.62	20.350	4.8791	0.5170	–4.3873	–5.2298
84	9.20	1.52	22.050	4.5513	0.4823	–4.2465	–5.7899
90	9.00	1.42	17.850	5.4564	0.5782	–3.9438	–4.7912
96	8.80	1.33	12.150	7.3932	0.7835	–3.4790	–3.3554
102	8.60	1.24	8.500	9.0846	0.9627	–2.8523	–2.3877
108	8.40	1.16	6.198	9.4366	1.0	–2.0635	–1.7825
114	8.20	1.08	3.507	9.4366	1.0	–1.1128	–1.0333
120	8.00	1.00	0.0	9.4366	1.0	0.0	0.0

Note: 1 in. = 0.0254 m, 1 in.-lb = 0.1130 Nm, 1 psi = 6.895 kPa.

TABLE 6.12
Values of Rotation, Deflection, and M_e at Cross Section along the Length of the Member

1 x (in.)	2 $M_e \times 10)^6$ (in.-lb)	3 Rotation (rad)	4 Deflection (in.)
0	15.0880	0.0	0.0
6	3.0299	0.020280	0.0723
12	1.5723	0.025586	0.2113
18	1.0766	0.028868	0.3751
24	0.5962	0.030945	0.5550
30	0.1226	0.031836	0.7438
36	–0.3400	0.031564	0.9344
42	–0.7866	0.030162	1.1200
48	–1.2112	0.027677	1.2939
54	–1.6067	0.024172	1.4498
60	–2.0292	0.019687	1.5818
66	–2.7116	0.013850	1.6830

TABLE 6.12 (continued)

1 x (in.)	2 $M_e \times 10)^6$ (in.-lb)	3 Rotation (rad)	4 Deflection (in.)
72	–3.8375	0.005793	1.7430
78	–5.2298	–0.005489	1.7452
84	–5.7899	–0.019412	1.6710
90	–4.7912	–0.032740	1.5136
96	–3.3354	–0.042775	1.2857
102	–2.3877	–0.049794	1.0071
108	–1.7825	–0.054960	0.6923
114	–1.0333	–0.058497	0.3513
120	0.0	–0.059828	0.0

Note: 1 in. = 0.0254 m, 1 in.-k = 113.0 Nm.

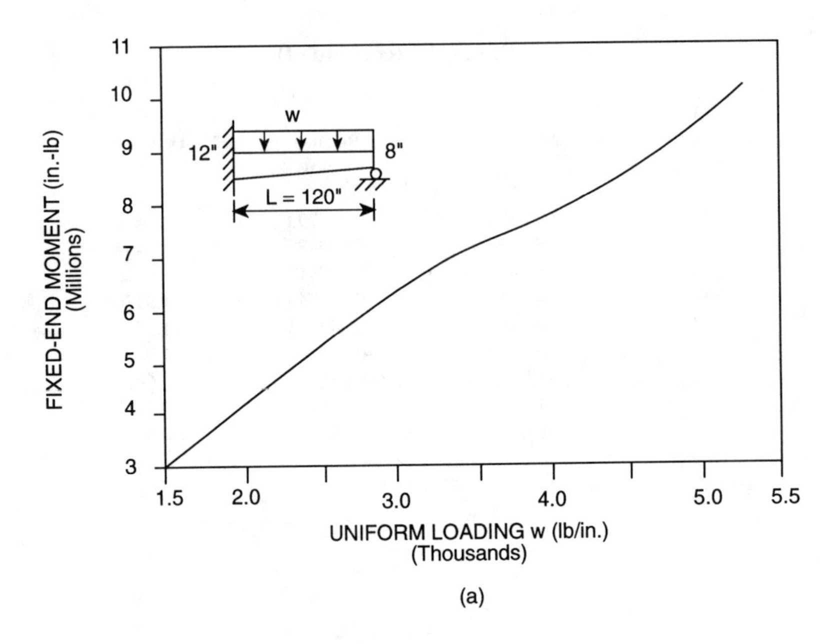

(a)

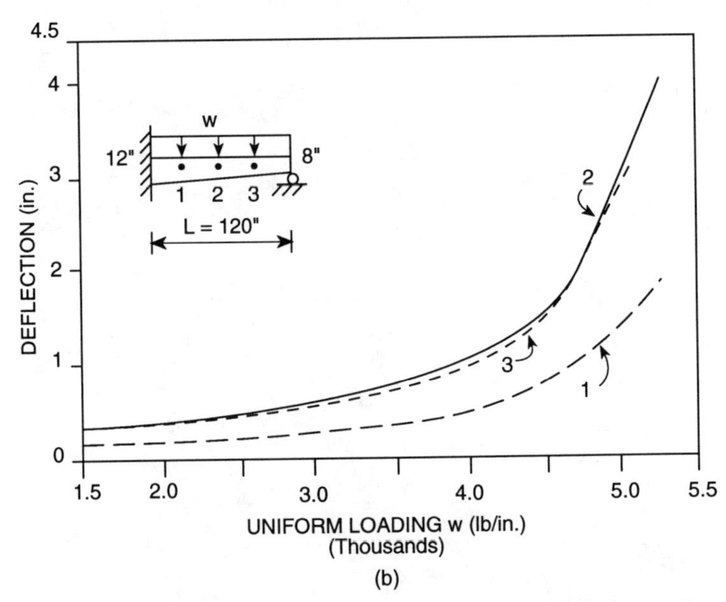

(b)

FIGURE 6.17. (a) Variation of the fixed end moment M_A with increasing loading w. (b) variation of the vertical deflection at quarter points 1, 2, and 3 with increased loading w (1 in. = 0.0254 m, 1 lb/in. = 175.1268 N/m, 1 in.-lb = 0.1130 Nm).

PROBLEMS

6.1 The variable thickness cantilever beam in Figure 6.18 is loaded by a concentrated load of 60,000 lb (266.893 kN) at the free end B. The beam has a rectangular cross section and the thickness varies in a way so that the moment of inertia variation is linear along its length. The material of the member is Monel, $E = 26 \times 10^6$ psi (179.27×10^6 kPa), h = 7.94 in. (0.202 m), width b = 6 in. (0.1524 m), $I_B = 250$ in.^{4} (104.057×10^{-6} m^4), and $I_A = 2I_B$. By applying inelastic analysis and using a bilinear approximation of the stress/strain curve of Monel, determine the values of Δ, E_r, radius of curvature r, g(x), and M_e at intervals of 6.0 in. (0.1524 m) along the length of the member. Also determine an equivalent system of constant stiffness EI_B, loaded with concentrated loads, and the vertical deflection δ_B at the free end B.
Answer: $\delta_B = 3.456$ in. (0.0878 m).

6.2 Solve Problem 6.1 by using a three-line and a six-line approximation of the stress/strain curve of Monel and compare the results.

6.3 Solve Problem 6.1 by assuming that the cantilever beam in Figure 6.18 is loaded with a uniformly distributed load w = 1388.9 lb/in. (243,233.6 N/m).
Answer: $\delta_B = 20.06$ in. (0.5095 m).

6.4 For the cantilever beam loaded as in Problem 6.3, determine the elastic deflection δ_B at the free end B and compare with the value obtained by using inelastic analysis.
Answer: Elastic $\delta_B = 2.68$ in. (0.0681 m).

6.5 For Problem 6.1, plot the load deflection curve by increasing the concentrated load at the free end B to the ultimate load capacity of the member.

6.6 For Problem 6.3, determine the ultimate load capacity w_u of the cantilever beam. Use a two-line and a three-line approximation of the stress/strain curve of Monel and compare the results.

6.7 For Problem 6.3, plot the modulus function g(x) by using a sufficient number of cross sections along the length of the member.

6.8 The uniform simply supported beam in Figure 6.11a is loaded with a concentrated load Q = 180.0 kips (200,679.96 N) at midspan. For the given load Q, reproduce Table 6.6 and compare the results. Also determine the deflection at midspan by using the equivalent system of constant stiffness E_1I. Use a three-line approximation of the stress/strain curve of Monel.

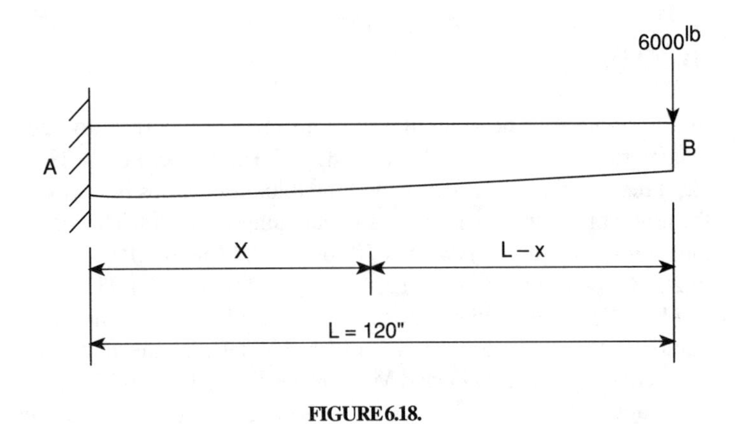

FIGURE 6.18.

6.9 Repeat Problem 6.8 by using the uniform simply supported beam in Figure 6.11b loaded with a uniformly distributed load w = 3000.0 lb/in. (525,380.4 N/m).

6.10 By using the results given in Table 6.9, plot the curves for the reduced modulus E_r, modulus function g(x), and the radius of curvature r.

6.11 By using the results in Table 6.9, plot the curves for M_{req} and M_e in the same figure. Also, by approximating the shape of M_e with straight lines, determine an equivalent system of constant stiffness E_1I_B.

6.12 By using the equivalent system in Problem 6.11, determine the rotations of the member at the end supports and the maximum vertical deflection. Compare the results with the values obtained in Table 6.10.

6.13 Solve the beam problem in Figure 6.14a using the six-line approximation of the stress/strain curve of the mild steel in Figure 6.14b. Compare the results with the results shown in Table 6.10.

6.14 For the statically indeterminate beam in Figure 6.16a, determine the fixed end moment M_A when the uniformly distributed load w = 4000.0 lb/in. (700,507.2 N/m). Compare the results with the values obtained in Figure 6.17a.

6.15 By using the value of the fixed end moment M_A in Problem 6.14, perform an inelastic analysis for the tapered beam in Figure 6.16a to determine the modulus function g(x) and moment M_e of the equivalent system of constant stiffness E_1I_B, where I_B is the moment of inertia at the end B of the member. Use the four-line approximation of the stress/strain curve of the mild steel in Figure 6.14b.

6.16 By using the moment diagram M_e of Problem 6.15, determine an equivalent system of constant stiffness E_1I_B by approximating its shape with straight line segments.

6.17 Use the equivalent system obtained in Problem 6.16 and determine the rotation at the roller support and the vertical deflections at the quarter points. Compare the results with the results obtained in Figure 6.17b.

Chapter 7

INELASTIC ANALYSIS OF UNIFORM AND VARIABLE THICKNESS MEMBERS WITH AXIAL RESTRAINTS

7.1 INTRODUCTION

In the preceding chapter, inelastic analysis was performed for members of uniform and variable thickness with loading conditions that consider coaxiality between neutral and centroidal axes. As stated earlier, such types of problems often confront the practicing engineer, and they deserve special treatment. There are, however, situations where the assumption of coaxiality between neutral and centroidal axes is no longer valid, and the shifting of the neutral axis at cross sections along the length of the member must be taken into consideration. Members that are subjected to axial compressive or tensile loads in addition to transverse loadings, or statically indeterminate problems involving axial restraints, are examples of such types of problems.

The inelastic analysis that is developed in this chapter considers such loading conditions, and the computation of the reduced modulus E_r, and consequently the derivation of the equivalent system of constant stiffness E_1I_1, takes into consideration the shifting of the neutral axis at cross sections along the length of the member. Figure 7.1 illustrates a loading condition for a member of rectangular cross section, where coaxiality between neutral and centroidal axes does not exist. Statically determinate and statically indeterminate prismatic and nonprismatic members with axial restraints and transversed loadings are discussed in this chapter. Both elastic and inelastic ranges are considered for comparison purposes. It should be emphasized, however, that axial restraints become more important in the inelastic range, where deformations are larger and variations in the axial restraints and neutral axis position along the length of the member are very distinct.

7.2 EQUIVALENT SYSTEMS FOR INELASTIC ANALYSIS OF MEMBERS WITH AXIAL RESTRAINTS

When a member is subjected to external loads that produce axial restraints, or to external loads acting in both transversed and axial directions, the strains ε_1 and ε_2 in Figures 6.1a or Figure 7.1a will not be equal, and the neutral axis will not coincide with the centroidal axis of the rectangular cross section. Under these conditions, the computation of the reduced modulus E_r, and consequently the derivation of the constant stiffness equivalent system, should take into consideration the shifting of the neutral axis at sections along the length of the member.

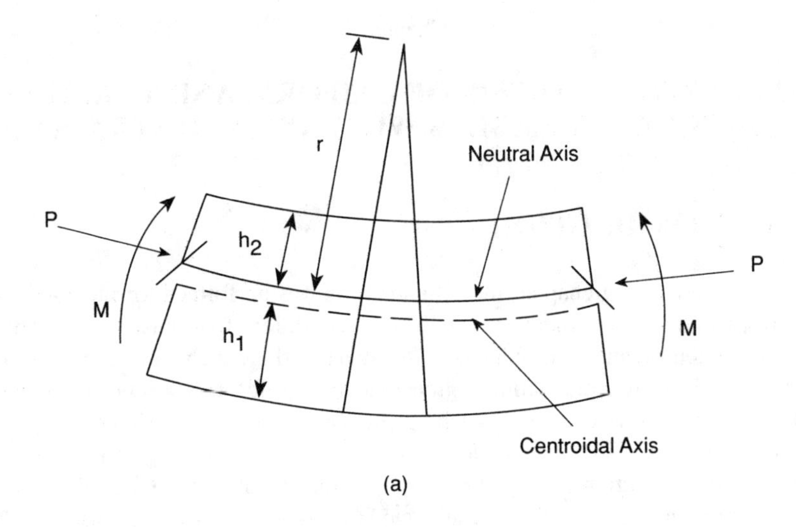

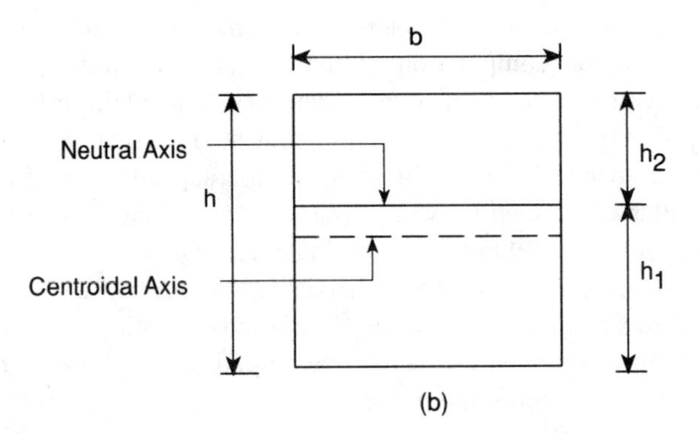

FIGURE 7.1. (a) Free body diagram of a portion of a member where neutral and centroidal axes do not coincide; (b) a rectangular cross section of the member.

Let it be assumed that, in addition to the transversed loading, the member is also subjected to an axial tensile or comprehensive force P. In Figure 7.1, h_1 and h_2 are the distances from the neutral axis to the lower and upper surfaces of the member, respectively. Thus, $\varepsilon_1 = h_1/r$, $\varepsilon_2 = h_2/r$, and $r = h/\Delta$, where r is the radius of curvature. At any distance y from the centroidal axis, the total longitudinal strain ε may be expressed as

$$\varepsilon = \varepsilon_0 + \frac{y}{r} \tag{7.1}$$

where

$$\varepsilon_0 = \frac{P}{bhE} \tag{7.2}$$

b is the width of the cross section, y/r is the strain due to bending, and E is the modulus of elasticity that can be either elastic or inelastic. Thus, from Equation 7.1

$$y = r(\varepsilon - \varepsilon_0) \tag{7.3}$$

$$dy = r d\varepsilon \tag{7.4}$$

and

$$P = b\int_{-h_2}^{h_1} \sigma dy = br\int_{\varepsilon_2}^{\varepsilon_1} \sigma d\varepsilon = \frac{bh}{\Delta}\int_{\varepsilon_2}^{\varepsilon_1} \sigma d\varepsilon \tag{7.5}$$

The integral in Equation 7.5 represents the area under the stress/strain curve.

The expression for the internal bending moment may be written as

$$M = b\int_{-h_2}^{h_1} \sigma y dy = br^2 \int_{\varepsilon_2}^{\varepsilon_1} \sigma(\varepsilon - \varepsilon_0) d\varepsilon \tag{7.6}$$

or

$$M + br^2\varepsilon_0 \int_{\varepsilon_2}^{\varepsilon_1} \sigma d\varepsilon = br^2 \int_{\varepsilon_2}^{\varepsilon_1} \sigma\varepsilon d\varepsilon \tag{7.7}$$

In Equation 7.7 M is the bending moment with respect to the centroidal axis, and

$$br^2 \int_{\varepsilon_2}^{\varepsilon_1} \sigma\varepsilon_0 d\varepsilon \tag{7.7a}$$

represents the additional moment due to the axial load that causes the shift of the neutral axis. Therefore, Equation 7.7 gives the bending moment at any cross section of the member with respect to the neutral axis. Due to the presence of the axial force P, the neutral axis will not coincide with the centroidal axis.

Equation 7.7 may be rewritten as

$$M + \varepsilon_0 rP = br^2 \int_{\varepsilon_2}^{\varepsilon_1} \sigma\varepsilon d\varepsilon \tag{7.8}$$

$$M + \varepsilon_0 \frac{h}{\Delta} P = br^2 \int_{\varepsilon_2}^{\varepsilon_1} \sigma\varepsilon d\varepsilon \tag{7.9}$$

$$M + \varepsilon_0 \frac{h}{\Delta} P = \frac{I}{r} \frac{12}{\Delta^3} \int_{\varepsilon_2}^{\varepsilon_1} \sigma\varepsilon d\varepsilon \tag{7.10}$$

or

$$M + \varepsilon_0 \frac{h}{\Delta} P = \frac{I}{r} E_e \tag{7.11}$$

where

$$E_e = \frac{12}{\Delta^3} \int_{\varepsilon_2}^{\varepsilon_1} \sigma\varepsilon d\varepsilon \tag{7.12}$$

$$I = \frac{bh^3}{12} \tag{7.13}$$

In Equation 7.11 the quantity E_e may be thought of as an equivalent modulus, which corrects for the effect of the shifting of the neutral axis and allows us to maintain coaxiality in the analysis between neutral and centroidal axes when the equivalent system of constant stiffness $E_1 I_1$ is used. In order to determine E_e, however, the position of the neutral axis at sections along the length of the member must be known.

The position of the neutral axis may be determined by trial and error. For example, we can start the procedure by assuming a shape for the deflection curve v of the member. This could be accomplished by using the deflection curve obtained from the inelastic analysis, with the axial force being equal to zero as discussed in Chapter 6. This is usually an excellent first trial deflection, and it makes the procedure converge rather rapidly. The next reasonable step would be to assume values of ε_1 and ε_2 and adjust Δ in Figure 7.2a so that Equation 7.5 is satisfied. The values of ε_1, ε_2, and Δ should be proportioned in a way that will make the sum of the terms on the left side of Equation 7.10 equal to the value

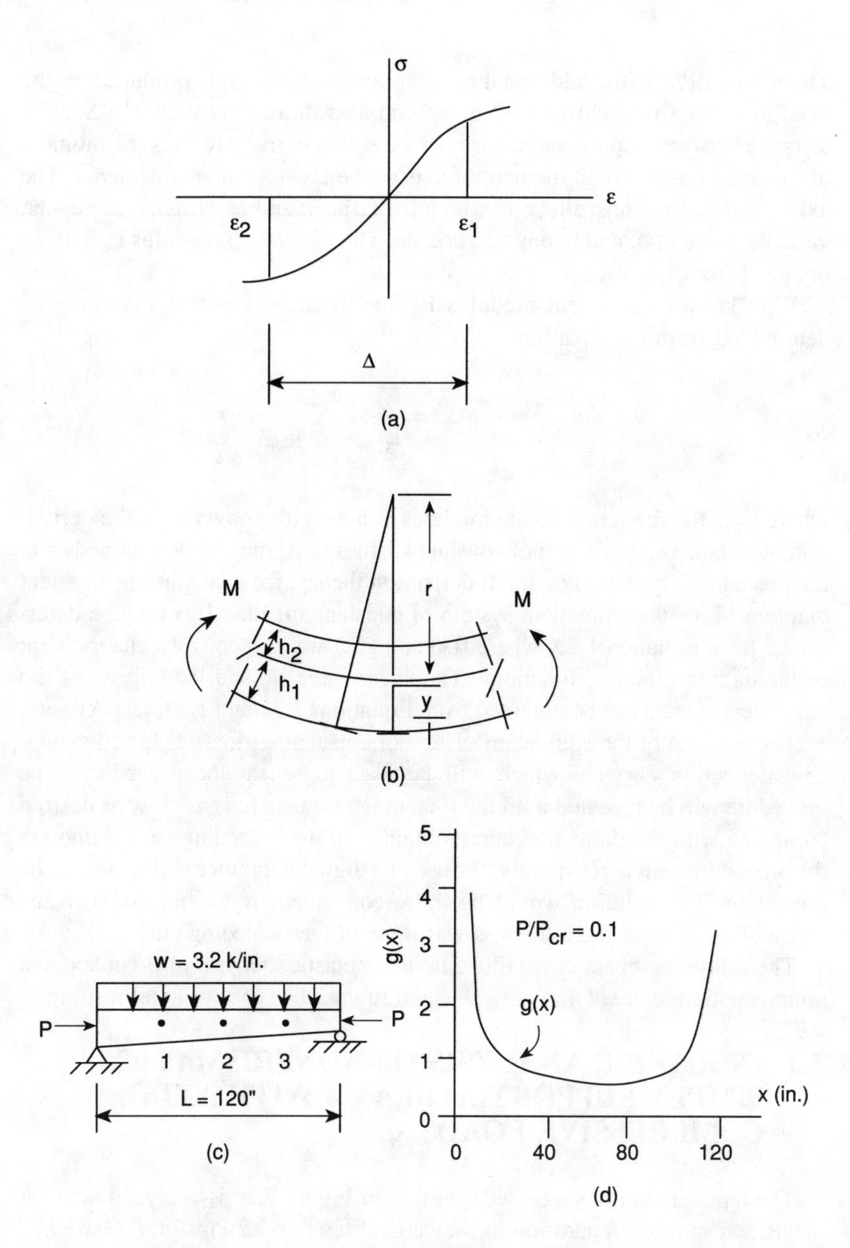

FIGURE 7.2. (a) Stress/strain curve; (b) free body diagram showing r, h_1, h_2, and y; (c) variable thickness member loaded with a uniformly distributed load w = 3.2 kip/in.; (d) variation of modulus function g(x) along the length of the member (1 in. = 0.0254 m, 1 kip/in. = 175.1268 kN/m).

obtained from the right side of the equation. We may assume as many values of ε_1, ε_2, and Δ as it is necessary to satisfy Equations 7.5 and 7.10.

Note that M in Equation 7.10 or Equation 7.11 is the bending moment produced by the external load by including the moment Pv of the axial load.

The term $\varepsilon_o hP/\Delta$ is the additional bending moment, which is produced by the axial load due to the shifting of the neutral axis. In reality, since $\varepsilon_o h/\Delta$ is the distance between neutral and centroidal axes, this term represents the moment of the axial load P about the neutral axis of the cross section considered. The axial load at sections along the length of the member will be somewhat variable, since ε, Δ, and h may be variable. The equivalent modulus E_e may be obtained from Equation 7.12.

With known equivalent modulus E_e, the modulus function g(x) may be determined from the equation

$$g(x) = \frac{E_e}{E_1} \tag{7.14}$$

where E_1 is the reference elastic modulus value. With known g(x), the derivation of an equivalent system of constant stiffness E_1I_1 may be determined as in the preceding chapters. For small deflection theory, for example, the moment diagram M_e of the equivalent system of constant stiffness E_1I_1 may be determined from Equation 1.29, where f(x) and g(x) are, respectively, the moment of inertia and modulus functions. The shear force V_e and loading w_e of the equivalent system can be obtained from Equations 1.31 and 1.32, respectively.

The solution of the equivalent system of constant stiffness E_1I_1 will yield a new deflection curve v which will be used to repeat the procedure. The procedure will be repeated until the final deflection curve is as close as desired compared with the deflection curve obtained in the preceding trial. Although the procedure converges rapidly, the use of a digital computer will facilitatc thc procedure. The solution would be more convenient if we approximate the shape of M_e with straight lines as was done in the preceding chapter.

The following beam cases illustrate the inelastic analysis of prismatic and nonprismatic members that are subjected to various types of axial restraints.

7.3 INELASTIC ANALYSIS OF NONPRISMATIC SIMPLY SUPPORTED BEAMS WITH AXIAL COMPRESSIVE LOADS

The tapered simply supported member in Figure 7.2c is analyzed here for inelastic response. In addition to the vertical load of 3.2 kips/in. (560.63 kN/m), the member is also subjected to an axial force $P = 0.1P_{cr}$, where P_{cr} is the Euler critical load for the simply supported beam of constant depth h = 8.0 in. (0.2032 m). The critical load P_{cr} was found to be equal to 3860.111944 kips (17,177.4982 kN). The material of the member is Monel, and the inelastic analysis is carried out by using the three-line approximation of its stress/strain curve. The required information is given in Tables 6.1 and 6.2. The constant width b of the member is 6 in. (0.1524 m), and the depth h is 8.0 in. (0.2032 m) at the right support and 12.0 in. (0.3048 m) at the left support.

TABLE 7.1
Values of f(x), ε_1, ε_2, Δ, E_e, g(x), M_{req}, and M_e for a Simply Supported Variable Stiffness Member Subjected to a Uniform Vertical Load w and an Axial Compressive Load P

x (in.)	h (in.)	f (x)	$\varepsilon_1 = (h_1/r) \times 10^{-3}$ (in./in.)	$\varepsilon_2 = (h_2/r) \times 10^{-3}$ (in./in.)	$\Delta \times 10^{-3}$ (in./in.)	$E_e \times 10^6$	$g(x) = E_e/E_1$	$M_{req} = M \times 10^6$ (in.-lb)	$M_e \times 10^6$ (in.-lb)
0	12.0	3.3750	—	—	—	0.0	—	0.0	—
6	11.8	3.2090	0.1244	–0.6201	0.7445	51.3655	2.3348	2.6622	0.3550
12	11.6	3.0486	0.4793	–0.9835	1.4628	29.9509	1.3614	2.9477	0.7100
18	11.4	2.8936	0.8186	–1.3316	2.1502	25.8673	1.1758	3.6142	1.0620
24	11.2	2.7440	1.1395	–1.6617	2.8012	24.4036	1.1093	4.2875	1.4080
30	11.0	2.5996	1.4391	–1.9708	3.4098	23.7147	1.0779	4.8922	1.7450
36	10.8	2.4604	1.7191	–2.2623	3.9814	23.2642	1.0575	5.4018	2.0760
42	10.6	2.3262	1.9883	–2.5728	4.5610	22.6453	1.0293	5.8026	2.4230
48	10.4	2.1970	2.2533	–2.9169	5.1702	21.7772	0.9899	6.0890	2.7990
54	10.2	2.0727	2.5402	–3.3020	5.8422	20.5892	0.9359	6.2573	3.2250
60	10.0	1.9531	2.8400	–3.7077	6.5477	19.2637	0.8756	6.3067	3.6870
66	9.8	1.8383	3.1036	–4.0705	7.1742	18.1063	0.8230	6.2377	4.1220
72	9.6	1.7280	3.2079	–4.2299	7.4378	17.6604	0.8027	6.0528	4.3630
78	9.4	1.6222	3.0913	–4.1020	7.1933	18.1024	0.8228	5.7529	4.3090
84	9.2	1.5209	2.7789	–3.7145	6.4934	19.4291	0.8831	5.3391	3.9750
90	9.0	1.4238	2.3959	–3.2279	5.6238	21.1195	0.9600	4.8102	3.5190
96	8.8	1.3310	2.0106	–2.7391	4.7497	22.6623	1.0301	4.1678	3.0390
102	8.6	1.2423	1.5779	–2.2595	3.8374	24.1155	1.0962	3.4221	2.5130
108	8.4	1.1576	1.0649	–1.7611	2.8260	26.1163	1.1871	2.6039	1.8940
114	8.2	1.0769	0.4243	–1.1376	1.5619	35.8741	1.6306	1.8837	1.0720
120	8.0	1.0	—	—	—	—	—	0.0	0.0

Note: 1 in. = 0.0254 m, 1 in.-lb = 0.113 Nm.

TABLE 7.2
Rotations and Vertical Deflections of a Variable Stiffness Member Subject to a Vertical Load w and an Axial Compressive Force P

x (in.)	Rotation $\times 10^{-3}$ fourth trial (rad)	Deflection fourth trial (in.)	Final deflection (in.)
0	22.4258	0.0	0.0
6	22.2370	0.1341	0.1341
12	21.6697	0.2659	0.2658
18	20.7258	0.3930	0.3930
24	19.4101	0.5134	0.5134
30	17.7306	0.6249	0.6248
36	15.6953	0.7251	0.7251
42	13.2988	0.8121	0.8121
48	10.5172	0.8836	0.8835
54	7.3084	0.9371	0.9370
60	3.6266	0.9699	0.9698
66	–0.5336	0.9791	0.9791
72	–5.0538	0.9624	0.9623
78	–9.6737	0.9182	0.9181
84	–14.0864	0.8469	0.8469
90	–18.0787	0.7504	0.7504
96	–21.5720	0.6315	0.6314
102	–24.5293	0.4932	0.4931
108	–26.8768	0.3389	0.3389
114	–28.4567	0.1729	0.1729
120	–29.0459	0.0	0.0

Note: 1 in. = 0.0254 m, 1 in.-lb = 0.113 Nm.

The procedure may be initiated by assuming a shape for the deflection curve v of the member, as discussed in the preceding section. A trial shape could be the one produced by the inelastic analysis of the member with P = 0. For illustration purposes let it be assumed that the values of v at intervals of 6.0 in. (0.1524 m) along the length of the member are as shown in the third column of Table 7.2. These are the results obtained after four trials. The rotations at corresponding points are also shown in the same figure.

The next step is to assume values of ε_1 and ε_2, and consequantly Δ, until Equation 7.5 is satisfied. Consider for example the beam cross section at x = 60 in. (1.524 m) from the left support and assume that $\varepsilon_1 = 2.8400 \times 10^{-3}$, $\varepsilon_2 = 3.7077 \times 10^{-3}$, and $\Delta = \varepsilon_1 + \varepsilon_2 = 6.5477 \times 10^{-3}$. By using the three-line approximation of the stress/strain curve of Monel, we obtain

$$\int_{\varepsilon_2}^{\varepsilon_1} \sigma d\varepsilon = -42.1272(10)^{-3}$$

and

$$P = \frac{bh}{\Delta}\int_{\varepsilon_2}^{\varepsilon_1} \sigma d\varepsilon = \frac{(6)(10)}{6.5477(10)^{-3}}\left[-42.1272(10)^{-3}\right]$$

$$= -386.0336 \text{ kips}$$

This is very close to the value of P = 386.0112 kips (1717.0635 kN), which is actually applied to the member.

By taking the first moment of the area of the stress/strain curve about the σ axis, we find

$$\int_{\varepsilon_2}^{\varepsilon_1} \sigma\varepsilon d\varepsilon = 449.5028(10)^{-6}$$

Thus,

$$E_e = \frac{12}{\Delta}\int_{\varepsilon_2}^{\varepsilon_1} \sigma\varepsilon d\varepsilon = 19.2644(10)^{-6} \text{ psi}$$

From Equation 7.11 we find

$$\frac{I}{r}E_e = \frac{I\Delta E_e}{h} = 6.3068(10)^6 \text{ in.}-\text{lb}$$

The two terms on the left side of Equation 7.11 yield

$$M + \varepsilon_o \frac{h}{\Delta}P = 6134.3537 + 172.40 = 6.3068(10)^6 \text{ in.-lb}$$

which is identical to the value obtained from the right side of Equation 7.11. The moment M is obtained by using statics, and it includes the moment of the distributed load w and the moment Pv of the axial force.

The above procedure may be repeated for various sections along the length of the member. At intervals of 6.0 in. (0.1524 m) along the length of the member, the values of f(x), g(x), ε_1, ε_2, Δ, E_e, M_{req}, and M_e are shown in Table 7.1. The values of the moment M_e of the equivalent system of constant E_1I_1, where $E_1 = 22 \times 10^6$ psi (151.58×10^6 kPa) and $I_1 = 256$ in.^{4} (106.555 $\times 10^{-6}$m^4), are shown in the last column of the table. The moment diagram M_e approximated with four straight line segments is plotted as shown in Figure 7.3a. The equivalent system of constant stiffness E_1I_1 that is loaded with concentrated loads is shown in Figure 7.3b.

Since M_e incorporates the effects of the axial load P, the equivalent loads on the equivalent system also include the effect of P. Therefore, the constant

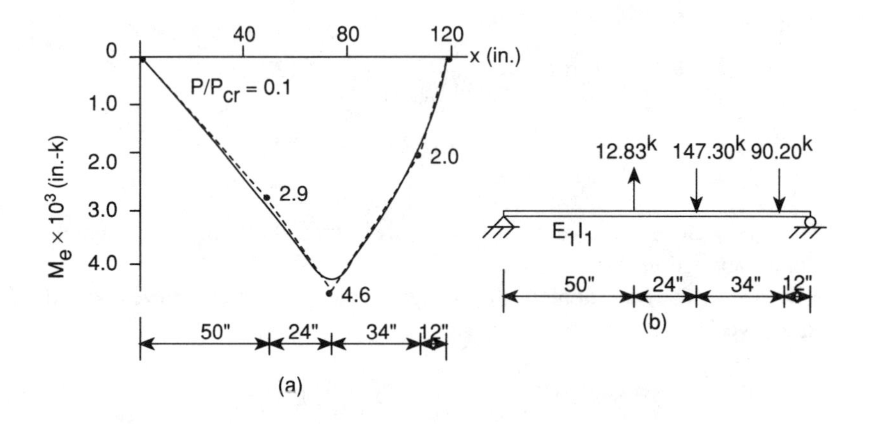

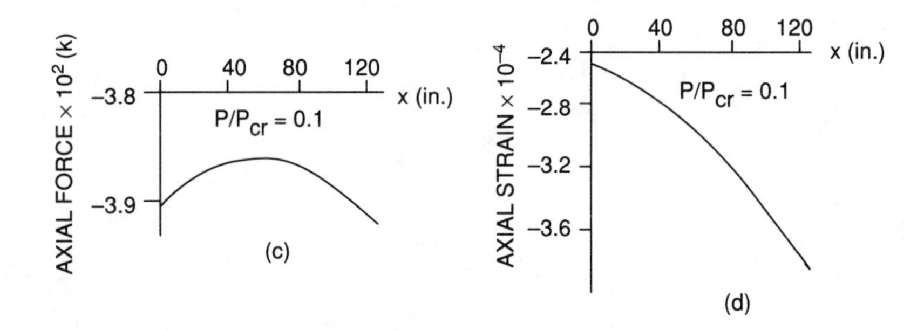

FIGURE 7.3. (a) Moment diagram M_e of the equivalent system of constant stiffness E_1I_1; (b) equivalent system of constant stiffness E_1I_1; (c) axial force variation for $P/P_{cr} = 0.1$; (d) axial strain variation for $P/P_{cr} = 0.1$ (1 in. = 0.0254 m, 1 kip= 4448 N, 1 in.-kip= 113.0 Nm).

stiffness equivalent system in Figure 7.3b may be used to determine the deflections and rotations of the initial variable stiffness member in Figure 7.2c by applying linear analysis. For example, if the moment area method is used, the deflection δ_C at a point C of distance x = 72.0 in. (1.8288 m), would be equal to 0.9623 in. (0.02444 m), which is almost identical to the one obtained in the third column of Table 7.2. The computer results of this final trial are shown in the last column of Table 7.2. The maximum deflection occurs at x = 66 in. (1.6767 m) and it is equal to 0.9791 in. (0.0249 m).

The variation of the axial force along the length of the member and the strain produced by this force are shown in Figures 7.3c and 7.3d, respectively. The presence of the axial force causes the neutral axis to move away from the centroidal axis. With reference to the bottom of the beam, the variation $h_1(x)$ of the position of the neutral axis at cross sections along the length of the

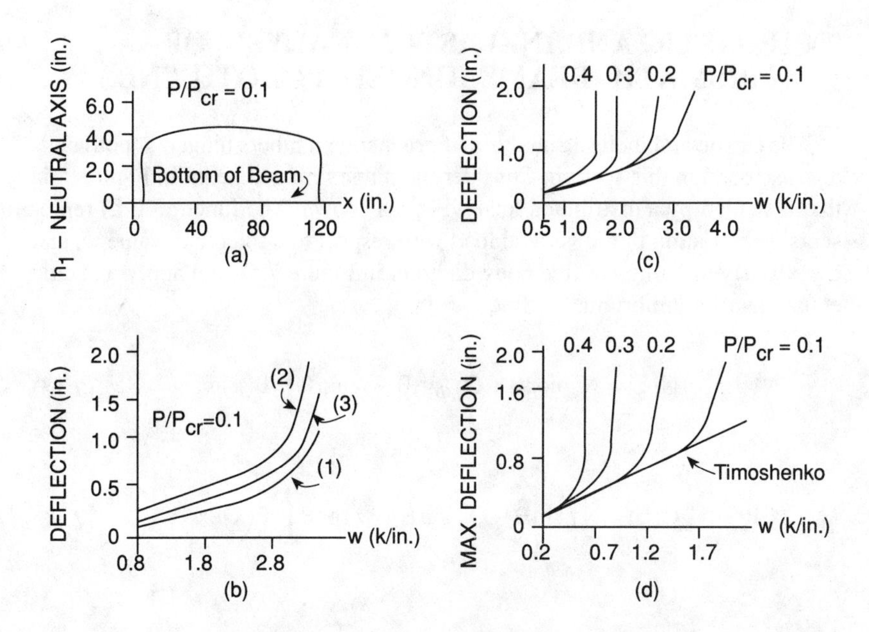

FIGURE7.4. (a) Variation of neutral axis location for P/P_{cr} = 0.1; (b) variation of vertical deflection at points 1, 2, and 3 (Figure 7.2c) with increasing load w for P/P_{cr} = 0.1; (c) variation of vertical deflection at midspan with incresing w for P/P_{cr} = 0.1, 0.2, 0.3, and 0.4; (d) variation of maximum vertical deflection with increasing w for P/P_{cr} = 0.1, 0.2, 0.3, and 0.4 (1 in. = 0.0254 m, 1 kip/in. = 175,126.8 N/m).

member is shown in Figure 7.4a. The variation of the function g(x) is shown in Figure 7.2d. Note that the neutral axis position approaches infinity at cross sections close to the end supports. The variation of the vertical deflection at points 1, 2, and 3, with increasing load w (Figure 7.2c), is shown in Figure 7.4b. Note that the curves become more nonlinear with increasing w, thus establishing a critical value of w where small increases beyond this value of w result in a very large deformation.

Figure 7.4c shows the vertical deflection at midspan with increasing values of the distributed load w, for P/P_{cr} = 0.1, 0.2, 0.3, and 0.4. Note that these curves become steeper with increasing values of the ratios P/P_{cr}, and for each case there is a value of the load w that becomes critical. Any small deviation from this critical value of w results in large increases in the vertical deflection, which indicates that the ultimate capacity of the member to resist load and deformation is reached. Figure 7.4d shows the analogous curves for a prismatic beam of depth h = 8 in. (0.2032 m), width b = 6 in. (0.1524 m), and length L = 120 in. (3.048 m). The curve for P/P_{cr} = 0.1 is also compared with the results obtained by Gere and Timoshenko,[31] where linear analysis was used. The deviation from Timoshenko's straight line becomes more remarkable as the member gets stressed further into the inelastic range.

7.4 ELASTIC AND INELASTIC ANALYSIS OF PRISMATIC BEAMS HINGED AT BOTH ENDS

The elastic and inelastic analysis of prismatic members hinged at both ends are discussed in this section. Consider the hinged/hinged beam in Figure 7.5a that is loaded by a distributed load $wF(\xi)$ as shown. The function $F(\xi)$ represents the variation of the vertical load with respect to a reference value w, and $\xi = x/L$. By utilizing the free body diagram in Figure 7.5b and applying static equations of equilibrium, we find

$$N(\xi) = N_A \cos\theta + Q_A \sin\theta - w\sin\theta \int_0^x F(x)dx \tag{7.15}$$

$$Q(\xi) = N_A \sin\theta - Q_A \sin\theta + w\cos\theta \int_0^x F(x)dx \tag{7.16}$$

$$M(\xi) = -Q_A x + N_A v + w\bar{x} \int_0^x F(x)dx \tag{7.17}$$

The notation used in the above equations is as shown in Figures 7.5a and 7.5b. Since the member is hinged at both ends, a horizontal reaction should be developed at these ends during deformation of the member. Consequently, a normal force N at cross sections along the length of the member should also be present. Thus, the normal strain $\varepsilon(\xi)$ at cross sections along the length of the member may be determined from the equation

$$\varepsilon(\xi) = \frac{N(\xi)}{AE} \tag{7.18}$$

where A is the cross-sectional area of the member, E is the modulus of elasticity, and $N(\xi)$ is the axial force.

By following the analysis given by Eringen,[36] we may write

$$(1+\varepsilon) = \frac{1+u_{,\xi}}{\cos\theta} \tag{7.19}$$

$$u_{,\xi} = (1+\varepsilon)\cos\theta - 1 \tag{7.20}$$

where u is the axial displacement. Indices after a comma represent differentiation.

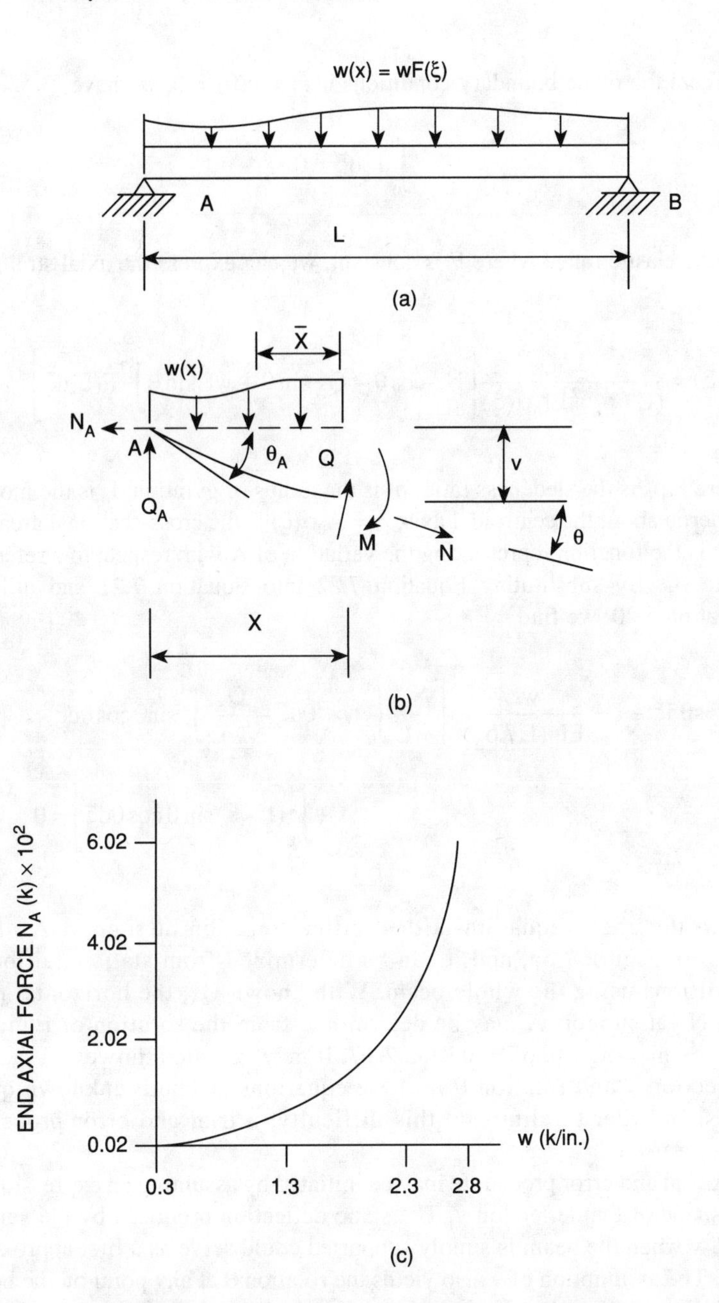

FIGURE 7.5. (a) Hinged/hinged beam loaded with a distributed vertical load w(x); (b) free body diagram of a portion of the member; (c) variation of the end horizontal reaction N_A with increasing load w(x) (1 kip= 4448 N, 1 kip/in. = 175,126.8 N/m).

From the plane boundary conditions u(1) = u(0) = 0, we have

$$\int_0^1 u,_\xi d\xi = 0 \tag{7.21}$$

For the elastic range where E is constant, we can express the axial strain $\varepsilon(\xi)$ as

$$\varepsilon(\xi) = \frac{L^2}{(L/\rho_0)^2 EI_0\alpha(\xi)}\left[N_A \cos\theta - Q_A \sin\theta + wL\sin\theta\int_0^\xi F(\xi)d\xi\right] \tag{7.22}$$

where L/ρ_0 is the sledeness ratio, ρ_0 is the radius of gyration, I_0 is the moment of inertia about the centroidal axis, $A = A_0\alpha(\xi)$ is the cross-sectional area, and $\alpha(\xi)$ is the function representing the variation of A with respect to a reference value A_0. By substituting Equation 7.22 into Equation 7.21 and utilizing Equation 7.20, we find

$$\int_0^1 \cos\theta d\xi - 1 + \frac{wL^3}{EI_0(L/\rho_0)^2}\left[\frac{N_A}{wL}\int_0^1 \cos^2\theta d\xi - \frac{Q_A}{wL}\int_0^1 \sin\theta\cos\theta d\xi + \int_0^1 (1-\xi)\sin\theta\cos\theta d\xi\right] = 0 \tag{7.23}$$

Q_A in the above equation is the vertical reaction at support A of the beam in Figure 7.5a, and it can be determined from static equilibrium conditions using the whole beam. With known Q_A, the horizontal reaction N_A at support A may be determined from the solution of Equation 7.23 by making use of Equation 7.17. It may be noted, however, that the deflection v and rotation θ in these equations are both unknown quantities. In order to eliminate this difficulty, a trial and error procedure may be used.

A trial and error procedure may be initiated by assuming an expression for the shape of the deflection v. The static deflection produced by the vertical load w when the beam is simply supported could serve as a first approximation. The assumption of v also yields the rotation θ at any point on the beam. With known θ, the value of N_A may be determined from Equation 7.23, and the value of M may be determined from Equation 7.17. At this point, the procedure explained in Section 7.3 may be applied to determine a new deflection v using equivalent systems. Only the elastic portion of the stress/strain curve should be used here since we are dealing with the elastic range. If the new v is not the same as the v assumed in the preceding case, the procedure may be repeated by using the new v until the v obtained is about the same as the one of the preceding trial.

Equation 7.23, however, cannot be used when the material of the member is stressed beyond its elastic limit. In this case, the strains produced by the axial force must satisfy the equation

$$\int_0^1 \{[1+\varepsilon(\xi)]\cos\theta - 1\}d(\xi) = 0 \tag{7.24}$$

where the strain $\varepsilon(\xi)$ may be obtained from Equation 7.18.

A trial and error procedure may also be initiated here by assuming a shape for the deflection v. The reaction Q_A in Figure 7.5b may be determined from the beam in Figure 7.5a by applying statics. The next step is to assume a value for N_A and determine $N(\xi)$ from Equation 7.15 and $\varepsilon(\xi)$ from Equation 7.18. Then we check if Equation 7.24 is satisfied. If not, we repeat the procedure until Equation 7.24 is satisfied. When this is completed, we apply the procedure discussed in Section 7.3 to obtain a new v. The inelastic part of the stress/strain curve is used here in conjunction with the method of equivalent systems. If the new v is not the same as the preceding one, the procedure is repeated using the new v. The procedure may be repeated until the new v is almost indentical to the one obtained from the preceding trial.

As an illustration, consider the member in Figure 7.5a and assume that the load w is uniform and equal to 2400 lb/in. (420,304.32 N/m). The depth h of the member is 8 in. (0.2032 m), its width b = 6 in. (0.1524 m), and its length L = 120 in. (3.048 m). By applying the above procedure, the final values for ε_1, ε_2, Δ, E_e, g(x), M_{req}, and M_e are shown in Table 7.3. The whole procedure can be easily computerized for convenience. By using the values of M_e from Table 7.3, the moment diagram M_e of the equivalent system of constant stiffness E_1I_1, where $E_1 = 22 \times 10^6$ psi ($151{,}684.65 \times 10^6$ Pa) and $I_1 = 256$ in.^{4} (100.0×10^{-6} m^4) may be plotted. The approximation of its shape with straight line segments will yield the equivalent sytem of constant stiffness E_1I_1. The rotations and deflections at any point along the length of the member may be determined using the equivalent system and applying well-known methods of linear analysis that can be found in books of elementary mechanics.

The results in Table 7.4 are obtained using the moment diagram M_e of the equivalent system of constant stiffness E_1I_1 and applying the moment area method. In Table 7.4 we observe that the maximum vertical displacement occurs at midspan, and it is equal to 1.1857 in. (0.03012 m). At the same point, we note that the rotation is zero, as should be expected. The maximum rotation occurs at the end supports of the member and it is equal to 0.0308672 rads or 1.7686°. The above results are obtained using the three-line approximation of the stress/strain curve of Monel (see Tables 6.1 and 6.2 for the required information regarding the modulus E and bending stress σ).

TABLE 7.3
Values of ε_1, ε_2, Δ, E_e, g(x), M_{req}, and M_e for a Uniform Hinged/Hinged Beam Loaded with a Uniform Vertical Load w

x (in.)	h (in.)	f (x)	$\varepsilon_1 = (h_1/r) \times 10^{-3}$ (in./in.)	$\varepsilon_2 = (h_2/r) \times 10^{-3}$ (in./in.)	$\Delta \times 10^{-3}$ (in./in.)	$E_e \times 10^6$	$g(x) = E_e/E_1$	$M_{req} = M \times 10^6$ (in.-lb)	$M_e \times 10^6$ (in.-lb)
0	8.0	1.0	—	—	—	—	—	0.0	0.0
6	8.0	1.0	0.7863	–0.3080	1.0943	34.7170	1.5780	1.2158	0.7700
12	8.0	1.0	1.2733	–0.7951	2.0684	25.6390	1.1654	1.6970	1.4560
18	8.0	1.0	1.7009	–1.2227	2.9236	23.8764	1.0853	2.2338	2.0580
24	8.0	1.0	2.0698	–1.5915	3.6613	23.2363	1.0562	2.7224	2.5770
30	8.0	1.0	2.4006	–1.9114	4.3120	22.7723	1.0351	3.1422	3.0350
36	8.0	1.0	2.7507	–2.2087	4.9594	21.9649	0.9984	3.4858	3.4910
42	8.0	1.0	3.1142	–2.5026	5.6168	20.8735	0.9488	3.7518	3.9540
48	8.0	1.0	3.4646	–2.7862	6.2509	19.6997	0.8954	3.9405	4.4000
54	8.0	1.0	3.7598	–3.0255	6.7853	18.6638	0.8484	4.0525	4.7760
60	8.0	1.0	3.8603	–3.1070	6.9673	18.3428	0.8338	4.0896	4.9040
66	8.0	1.0	3.7599	–3.0255	6.7854	18.6637	0.8483	4.0525	4.7760
72	8.0	1.0	3.4646	–2.7862	6.2508	19.6999	0.8954	3.9405	4.4000
78	8.0	1.0	3.1143	–2.5026	5.6169	20.8734	0.9488	3.7518	3.9540
84	8.0	1.0	2.7505	–2.2087	4.9592	21.9651	0.9984	3.4857	3.4910
90	8.0	1.0	2.4005	–1.9114	4.3119	22.7724	1.0351	3.1422	3.0350
96	8.0	1.0	2.0699	–1.5916	3.6614	23.2362	1.0562	2.7225	2.5770
102	8.0	1.0	1.7010	–1.2227	2.9237	23.8762	1.0583	2.2339	2.0580
108	8.0	1.0	1.2733	–0.7951	2.0684	25.6390	1.1654	1.6970	1.4560
114	8.0	1.0	0.7862	–0.3080	1.0942	34.7199	1.5782	1.2157	0.7700
120	8.0	1.0	—	—	—	—	—	0.0	0.0

Note: 1 in. = 0.0254 m,1 in.-lb = 0.113 Nm.

TABLE 7.4
Rotations and Vertical Deflections of a Uniform Hinged/Hinged Beam Loaded with a Uniformly Distributed Vertical Load w

x (in.)	$M_e \times 10^6$ (in.-lb)	Rotation $\times$ 10^{-3} (rad)	Deflection v (in.)
0	0.0	30.8672	0.0
6	0.7700	30.4517	0.1843
12	1.4560	29.2660	0.3634
18	2.0580	27.3942	0.5334
24	2.5770	24.9252	0.6904
30	3.0350	21.9359	0.8309
36	3.4910	18.4597	0.9521
42	3.9540	14.4940	1.0510
48	4.4000	10.0440	1.1246
54	4.7760	5.1563	1.1702
60	4.9040	0.0	1.1857
66	4.7760	–5.1563	1.1702
72	4.4000	–10.0440	1.1246
78	3.9540	–14.4940	1.0510
84	3.4910	–18.4597	0.9521
90	3.0350	–21.9359	0.8309
96	2.5770	–24.9252	0.6904
102	2.0580	–27.3942	0.5334
108	1.4560	–29.2660	0.3634
114	0.7700	–30.4517	0.1843
120	0.0	–30.8672	0.0

Note: 1 in. = 0.0254 m, 1 in.-lb = 0.113 Nm.

Figure 7.5c illustrates the variation of the horizontal reaction N_A in Figure 7.5b with increasing uniformly distributed load w. Note that N_A increases very fast for values of w > 2300 lb/in. (402,791.64 N/m). Figure 7.6a illustrates the variation of the maximum vertical deflection at midspan for the cases of a hinged/hinged beam and a hinged/roller beam. The results are about identical and the curve is linear for loads w smaller than about 1200 lb/in. (210,152.2 N/m). For larger values of w, the curves are nonlinear and the difference in results increases as the load w increases beyond the value of 1200 lb/in. (210,152.2 N/m). The member with both ends hinged will experience appreciably smaller vertical deflection at midspan. Again we note that there is a critical value of the load w, where the maximum deflection at midspan increases rapidly if this value of w is exceeded. This load may be thought of as the ultimate load w_u, suggesting that the ultimate capacity of the member to resist stress and deformation has been reached.

The variation of the axial force along the length of the hinged/hinged member is shown in Figure 7.6b. It reaches its maximum value at midspan. The shifting of the neutral axis at cross sections along the length of the member is illustrated in Figure 7.6c. The position of the neutral axis is measured by the coordinate h_2, which indicates the location of the neutral axis from the top of

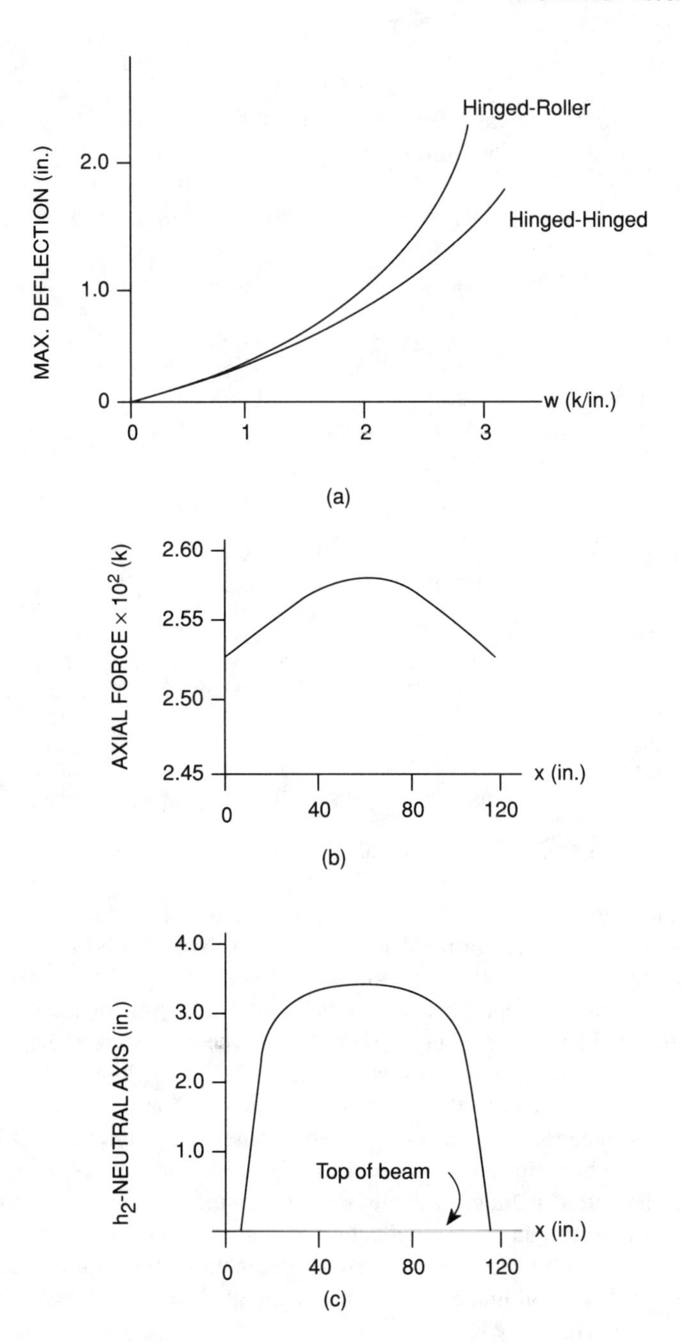

FIGURE 7.6. (a) Variation in maximum vertical deflection at midspan with increasing load w for a hinged/hinged and a hinged/roller member; (b) variation of the axial force with increasing x; (c) variation of the neutral axis position with increasing x (1 in. = 0.0254 m, 1 kip = 4.448 kN, 1 kip/in. = 175,126.8 N/m).

the member as shown in Figure 7.6c. Near the ends of the member, where the bending stresses approach zero values, we note that h_2 approaches infinity.

7.5 ELASTIC AND INELASTIC ANALYSIS OF PRISMATIC BEAMS FIXED AT BOTH ENDS

The inelastic analysis of symmetrical fixed/fixed prismatic members with symmetrical loading is discussed here. The procedure, however, is general and it can be extended to apply for nonprismatic members of various loading conditions or to members with other cross-sectional shapes.

In this section, the elastic and inelastic analysis of fixed/fixed members will be discussed. The analysis is limited to prismatic members with symmetrical loading, but the solution will also apply to nonprismatic members.

Consider a fixed/fixed prismatic member loaded with a uniformly distributed load w as shown in Figure 7.7a. The moments M_A and M_B at the fixed ends A and B, respectively, are taken as the redundants. Due to symmetry, $M_A = M_B$. By considering the free body diagram in Figure 7.7b and applying static equations of equilibrium, we find

$$N(x) = N_A \cos\theta + Q_A \sin\theta - w \sin\theta \int_0^x dx \tag{7.25}$$

$$Q(x) = N_A \sin\theta - Q_A \sin\theta + w \cos\theta \int_0^x dx \tag{7.26}$$

$$M(x) = -Q_A x + N_A v + \frac{wx^2}{2} + M_A \tag{7.27}$$

Equations 7.25, 7.26, and 7.27, are analogous to Equations 7.15, 7.16, and 7.17, respectively, when $x = \xi L$ and $F(x) = 1$. In this case, however, the fixed end moment $M_A = M_B$ is also included and it must be taken into consideration in the trial and error procedure. With M_A and M_B as redundants, the trial and error procedure discussed in Section 7.4 may be used. The procedure may be initiated by assuming a value for the fixed end moment M_A. Since we now have a hinged/hinged beam loaded with the distributed load w, and the redundant moments M_A and M_B, the trial and error procedure discussed in Section 7.4 may be used here to determine the vertical deflection v. In this case, however, we must refer to Equations 7.25 and 7.27 for the computation of N(x) and M(x), respectively. When v is determined, we check if the boundary conditions at the ends A and B ($\theta_A = \theta_B = 0$) are satisfied. If not, the procedure may be repeated with new values of M_A until these boundary conditions are satisfied.

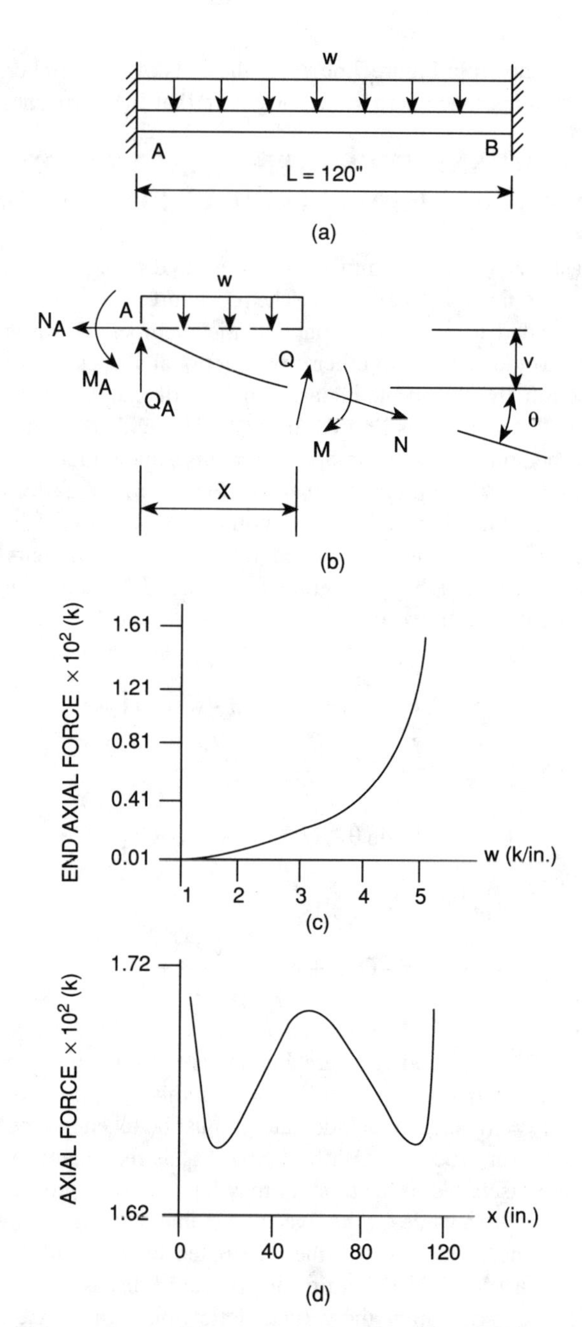

FIGURE 7.7. (a) Fixed/fixed beam loaded with a uniformly distributed load w; (b) free body diagram of a portion of the beam; (c) variation of the end axial reaction with increasing w; (d) variation of the axial force N along the length of the member (1 in. = 0.0254 m, 1 kip/in.= 175,126.8 N/m, 1 kip = 4448 N).

As an illustration regarding the application of the methodology, consider the fixed/fixed beam in Figure 7.7a of length L = 120 in. (3.048 m), which is loaded with a distributed load w = 5500 lb/in. (963.197 kN/m) as shown. By following the trial and error procedure stated above, the values of ε_1, ε_2, Δ, E_e, g(x), M_{req}, and M_e are obtained; they are shown in Table 7.5. This table indicates that the modulus function g(x) has its lowest value at the fixed ends of the member, and it is equal to 0.1236. This, however, should be expected, because the fixed ends of the member are subjected to the highest normal stress distributions.

Since the member is statically indeterminate, we also observe that the required moment M_{req} at cross sections along the length of the member will not, in general, be equal to the bending moment obtained from elastic analysis. For example, from Table 7.5 the required moment M_{req} at the fixed ends of the member is equal to 5.5351×10^6 in.-lb (625.5×10^3 Nm). At the same ends, the bending moment that is calculated by using elastic analysis formulas, that is $wL^2/12$, is 6.0×10^6 in.-lb (677.9×10^3 Nm). This shows that for this cross section the elastic bending moment is 8.4% higher compared to the moment obtained from inelastic analysis.

The last column in Table 7.5 gives the values of the moment $M_e = M_{req}/f(x)g(x)$ of the equivalent system of constant stiffness E_1I, where $E_1 = 22 \times 10^6$ psi ($151{,}684.65 \times 10^6$ Pa), and I = 256.0 in.^{4} (100×10^{-6} m^4). The three-line approximation of the stress/strain curve of Monel was used to obtain the results in Table 7.5. With known M_e, the equivalent system of constant stiffness E_1I may be determined in the usual way. This is done repeatedly in preceding sections and chapters. Rotations and deflections at any point along the length of the member may now be determined using the equivalent system and applying linear elementary analysis. By using the moment area method and the equivalent system, the deflections and rotations at intervals of 6.0 in. (0.1524 m) along the length of the member are determined, and they are shown in Table 7.6. The maximum vertical deflection occurs at midspan and it is equal to 0.9416 in. (0.0239 m). The deflection at midspan is also determined by using the beam in Figure 7.7a and applying elastic analysis. This value is found to be 0.5273 in. (0.0134 m), which is 44.0% smaller compared to the value of 0.9416 in. (0.0239 in.) obtained from inelastic analysis. In other words, the inelastic deflection is 1.7857 times larger than the elastic one.

The variation of the axial forces N_A and N_B ($N_A = N_B$) at the fixed ends A and B, respectively, with increasing vertical load w, is shown in Figure 7.7c. This curve is nonlinear, and it demonstrates very sharp increases in axial force at the higher levels of the loading w. The variation of the normal force N at cross sections along the length of the member is shown in Figure 7.7d, and the variation with increasing w of the end moment M_A or M_B is shown in Figure 7.8a. At the early stages of loading the M_A curve is almost linear, and it becomes increasingly nonlinear as the loading w increases.

The variation of the maximum vertical deflection with increasing vertical loading w is represented by the curve in Figure 7.8b. Again we observe that the

TABLE 7.5
Values of ε_1, ε_2, Δ, E_e, g(x), M_{req}, and M_e for a Uniform Fixed-Fixed Beam Loaded with a Uniform Vertical Load w

x (in.)	h (in.)	f(x)	$\varepsilon_1 = (h_1/r) \times 10^{-3}$ (in./in.)	$\varepsilon_2 = (h_2/r) \times 10^{-3}$ (in./in.)	$\Delta \times 10^{-3}$ (in./in.)	$E_e \times 10^6$ (psi)	$g(x) = E_e/E_1$	$M_{req} \times 10^6$ (in.-lb)	$M_e \times 10^6$ (in.-lb)
0	8.0	1.0	–29.6925	32.8952	–62.5905	2.7636	0.1253	–5.5351	–44.0637
3	8.0	1.0	–6.1989	7.0435	–13.2424	10.8171	0.4917	–4.5838	–9.3226
6	8.0	1.0	–2.5945	2.9632	–5.5576	20.7718	0.9442	–3.6941	–3.9126
12	8.0	1.0	–1.2645	1.5567	–2.8221	22.8132	1.0370	–2.0602	–1.9867
18	8.0	1.0	–0.2300	0.5213	–0.7513	32.0299	1.4559	–0.7701	–0.5290
24	8.0	1.0	0.6650	–0.3737	1.0387	27.3008	1.2409	0.9074	0.7312
30	8.0	1.0	1.4206	–1.1294	2.5500	22.9712	1.0441	1.8745	1.7952
36	8.0	1.0	2.0375	–1.7463	3.7838	22.5012	1.0228	2.7245	2.6638
42	8.0	1.0	2.5859	–2.2630	4.8490	21.8577	0.9935	3.3916	3.4137
48	8.0	1.0	3.2654	–2.8602	6.1257	19.7265	0.8967	3.8668	4.3125
54	8.0	1.0	4.0201	–3.5247	7.5448	17.1910	0.7814	4.1505	5.3115
60	8.0	1.0	4.3877	–3.8489	8.2366	16.1042	0.7320	4.2446	5.7986
66	8.0	1.0	4.0200	–3.5247	7.5447	17.1911	0.7814	4.1505	5.3115
72	8.0	1.0	3.2654	–2.8602	6.1257	19.7265	0.8967	3.8668	4.3125
78	8.0	1.0	2.5859	–2.2631	4.8490	21.8577	0.9935	3.3916	3.4137
84	8.0	1.0	2.0376	–1.7463	3.7838	22.5012	1.0228	2.7245	2.6638
90	8.0	1.0	1.4207	–1.1294	2.5501	22.9711	1.0441	1.8745	1.7953
96	8.0	1.0	0.6650	–0.3738	1.0388	27.2996	1.2409	0.9075	0.7313
102	8.0	1.0	–0.2300	0.5213	–0.7512	32.0327	1.4560	–0.7701	–0.5289
108	8.0	1.0	–1.2654	1.5567	–2.8222	22.8131	1.0370	–2.0602	–1.9868
114	8.0	1.0	–2.5945	2.9631	–5.5576	20.7718	0.9442	–3.6941	–3.9126
117	8.0	1.0	–6.1988	7.0434	–13.2422	10.8172	0.4917	–4.5838	–9.3225
120	8.0	1.0	–29.6960	32.8961	–62.5921	2.7635	0.1256	–5.5351	–44.0648

Note: 1 in. = 0.0254 m, 1 in.-lb = 0.113 Nm.

TABLE 7.6
Rotation and Vertical Deflection of a Uniform Fixed/Fixed Beam Loaded with a Uniform Vertical Load w

x (in.)	$M_e \times 10^6$ (in.-lb)	Rotation $\times$ 10^{-3} (rad)	Deflection v (in.)
0	—	–0.0001	0.0
3	–9.3226	14.2185	0.0213
6	–3.9126	17.7435	0.0693
12	–1.9867	20.8859	0.1852
18	–0.5290	22.2259	0.3145
24	0.7312	22.1182	0.4775
30	1.7952	20.7724	0.5762
36	2.6638	18.3972	0.6937
42	3.4137	15.1600	0.7944
48	4.3125	11.0445	0.8730
54	5.3115	5.9180	0.9239
60	5.7986	0.0	0.9416
66	5.3115	–5.9180	0.9239
72	4.3125	–11.0440	0.8730
78	3.4137	–15.1599	0.7944
84	2.6638	–18.3972	0.6937
90	1.7953	–20.7724	0.5762
96	0.7313	–22.1182	0.4475
102	–0.5289	–22.2261	0.3145
108	–1.9868	–20.8860	0.1852
114	–3.9126	–17.7436	0.0693
117	–9.3225	–14.2186	0.0213
120	—	0.0003	0.0

Note: 1 in. = 0.0254 m, 1 in.-lb = 0.113 Nm.

curve is linear at the early loading stages, and it becomes increasingly nonlinear as the loading w increases. The variation of the modulus function g(x) along the length of the member is represented by the curve in Figure 7.8c, and the shifting of the neutral axis position at cross sections along the length of the member is illustrated by the curves in Figure 7.8d. The dashed line in Figure 7.8d represents the centroidal axis location at cross sections along the length of the member, which remains at a constant distance from the bottom of the beam. The solid line in the same figure represents the shifting of the neutral axis. At cross sections where the normal stresses due to bending are large compared to the normal stresses produced by the axial force the centroidal and neutral axes are closer together, and they are further apart when the opposite occurs.

It should be pointed out that the above methodology is general, and it can be applied to a variety of cross-sectional shapes of the member and to other types of material. However, knowledge of the stress/strain curve of the material, which can be obtained experimentally, is required. Guidelines for the practicing design engineer can be established by the material manufacturer, and/or appropriate material research and development agencies, in order to facilitate design practices and design requirements.

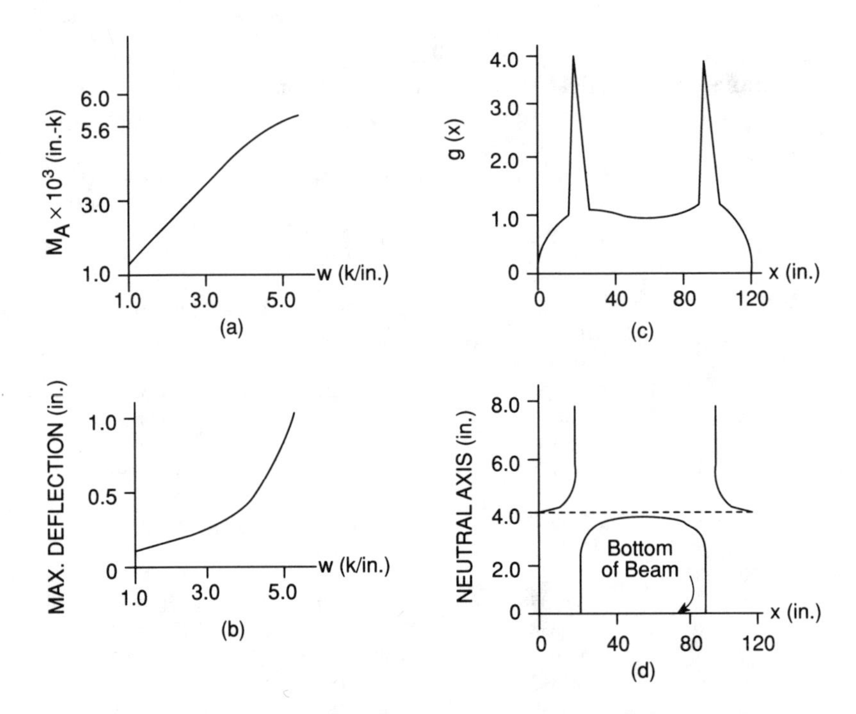

FIGURE 7.8. (a) Variation of the Fixed-end moment with increasing vertical load w; (b) variation of the maximum deflection with increasing w; (c) variation of the modulus function g(x) along the length of the member; (d) shifting of the neutral axis position along the length of the member (1 in. = 0.0254 m, 1 kip/in. = 175,126.8 N/m, 1 in.-kip = 112.9843 Nm).

PROBLEMS

7.1 The uniform simply supported beam in Figure 7.9 is loaded with a uniformly distributed load w = 2.0 kips/in. (350,253.6 N/m) as shown. The length L of the member is 120 in. (3.048 m), width b = 6.0 in. (0.1524 m), and depth h = 8.0 in. (0.2032 m). The member is also subjected to an axial compressive load $P = 0.1\ P_{cr}$, where P_{cr} is the Euler critical load for the beam (P_{cr} = 3860.1119 kips). The material of the member is Monel. By using the three-line approximation of the stress/strain curve of Monel, determine the variation of the modulus function g(x) along the length of the member, and discuss the results.

7.2 By using the results obtained in Problem 7.1, determine an equivalent system of constant stiffness EI, where E is the elastic modulus, and I is the cross-sectional moment of inertia.

7.3 By using the equivalent system obtained in Problem 7.2, determine the vertical deflection at midspan and the rotations at the end supports by applying the moment area method.

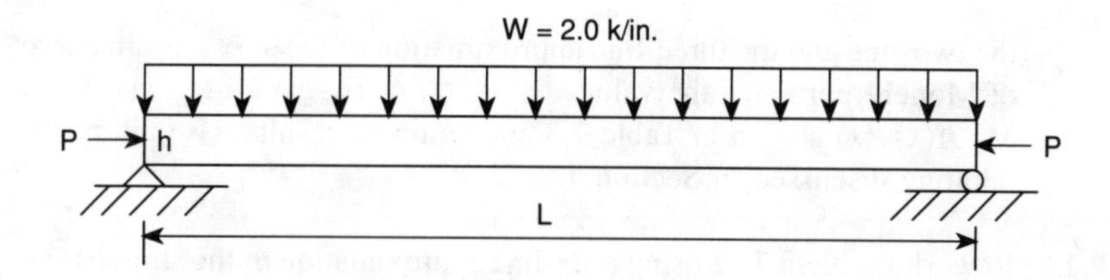

FIGURE 7.9.

7.4 For the beam in Problem 7.1 determine and plot the variation of the midspan vertical deflection with increasing vertical load w when P/P_{cr} = 0.1 and 0.2, and compare the results.

7.5 Repeat Problem 7.4 when P/P_{cr} = 0.1 and 0.3, and compare the results.

7.6 Repeat Problem 7.4 when P/P_{cr} = 0.1 and 0.4, and compare the results.

7.7 The tapered simply supported beam in Figure 7.2c is loaded with a uniformly distributed load w = 2.0 kips/in. (350,253.6 N/m). In addition, the member is subjected to an axial compressive load P = 0.1 P_{cr}, where P_{cr} = 3860.1119 kips (17,177.4982 kN). At the right support the depth h = 8.0 in. (0.2032 m), and at the left support the depth is 1.5h = 12.0 in. (0.3048 m). The width b of the member is 6.0 in. (0.1524 m), and the material of the member is Monel. By using the three-line approximation of the stress/strain curve of Monel, reproduce the quantities shown in Table 7.1 and compare results.

7.8 By using the values of M_e obtained in Problem 7.7, determine an equivalent system of constant stiffness EI, where E is the elastic modulus and I is the moment of inertia of the beam at the right support.

7.9 By using the equivalent system obtained in Problem 7.8 and applying the moment area method, determine the vertical deflection at midspan and the rotations at the end supports. Compare these results with the corresponding values shown in Table 7.2.

7.10 Solve Problem 7.7 by using the bilinear approximation of the stress/strain curve of Monel, and compare the results.

7.11 Solve Problem 7.7 for P/P_{cr} = 0.2 and compare the results.

7.12 The hinged/hinged uniform beam in Figure 7.5a is loaded with a uniformly distributed load w = 2600 lb/in. (455,329.68 N/m). The material of the member is Monel, the length L = 120.0 in. (3.048 m), width b = 6.0 in. (0.1524 m), and depth h = 8.0 in. (0.2032 m). By using the two-line and the three-line approximation of the stress/strain curve of Monel, reproduce the values for ε_1, ε_2, Δ, E_e, g(x), M_{req}, and $M_e = M_{req}/f(x)g(x)$ shown in Table 7.3 and compare results. Use the methodology discussed in Section 7.4.

7.13 Rework Problem 7.12 using a six-line approximation of the stress/strain curve of Monel, and compare the results.

7.14 By using the values of M_e from Problem 7.12, plot the M_e diagram and derive an equivalent system of constant stiffness E_1I_1 that is loaded with equivalent concentrated loads.

7.15 By using the equivalent system obtained in Problem 7.14 and applying the moment area method, determine the rotations at the ends of the hinged/hinged beam and the deflection at midspan. Compare the results with the values shown in Table 7.4.

7.16 Rework Problem 7.15 for a uniformly distributed load w = 1900 lb/in. (332,740.9 N/m). Solve the same problem by using elastic analysis, and compare the results. Compare the deflection at midspan with the value shown in the graph of Figure 7.6a.

7.17 For Problem 7.12, develop a curve representing the variation of the vertical deflection at midspan with increasing uniformly distributed load w. Compare this curve with the one shown in Figure 7.6a.

7.18 By following the procedure discussed in Section 7.5, reproduce the values for ε_1, ε_2, Δ_1, E_e, g(x), M_{req}, and M_e shown in Table 7.5. Also plot the curves that represent the variations of E_e, g(x), M_{req}, and M_e along the length of the member.

7.19 By using the moment diagram M_e from Problem 7.18 and approximating its shape with straight lines, determine an equivalent system of constant stiffness EI that is loaded with equivalent concentrated loads.

7.20 By using the equivalent system derived in Problem 7.19 and applying the moment area method, or a similar method of your choice, determine the vertical deflection at midspan. Compare this value with the value shown in Table 7.6. Also solve the initial problem in Figure 7.7a by using linear elastic analysis, and compare the results.

7.21 Repeat Problem 7.18 by assuming that the vertical uniformly distributed load w = 4000 lb/in. (700,507.2 N/m), and compare the results.

Chapter 8

ELASTIC AND INELASTIC ANALYSIS OF UNIFORM AND VARIABLE THICKNESS PLATES

8.1 INTRODUCTION AND HISTORICAL DEVELOPMENTS

When the thickness along the length and/or width of a plate is variable its closed form solution becomes very complex, even for simple cases. For uniform thickness plates, however, many solutions for elastic analysis are developed in the past, and they are available in the literature. For example, Timoshenko and Weinosky-Krieger[37] developed solutions for the elastic analysis of flat, thin plates with various loadings and boundary conditions by solving the nonhomogeneous partial differential equation

$$\nabla^4 w = \frac{q(x,y)}{D_o} \tag{8.1}$$

where w is the vertical deflection, q(x,y) is the transverse loading, and D_o is the plate rigidity.

For the elastic analysis of plates of variable thickness, only a limited number of closed form solutions are known. The solution of symmetrical circular plates with linear variation in thickness, is given by Olsson[38] in terms of hypergeometric series. This problem is further investigated by Rainville[39] by considering any variation in thickness for the symmetric circular plate. Circular plates of variable thickness have also been investigated by Conway[40,41] who obtained closed form solutions for certain special cases. Olsson[38] investigated symmetrical circular plates with exponentially varying thickness, and Timoshenko[35] catalogued his solutions.

Several graphical methods, such as the parameter method by Favre and Chabloz,[42] have been developed for the solution of circular plates with linearly varying thickness. Olsson[43] has provided a solution for the unsymmetrical bending of circular plates with quadratic thickness variation, and Alwar and Nath[44] have devised a Chebyshev polynomial method to solve the large deflection of circular plates.

The first elastic solutions for a rectangular plate of variable thickness was developed by Olsson;[45] it is only valid for a linearly varying thickness. Reissner[46] improved on this solution, and Olsson[47] extended the solution to include any external loading. Mukhopadhyay[48] and Dey[49] used finite difference techniques to solve the rectangular plate problem. Ohga and Shigematsu[50] used a combination of boundary element transfer matrix methods to solve variable thickness rectangular plates. This method provided a solution for only a special case of variable thickness rectangular plates.

During the past years, Fertis,[7] Kozma and Fertis,[55] Fertis and Mijatov,[51] and Fertis and Lee[52,53,56] developed a convenient and general method to solve variable thickness plates with various boundary conditions and loadings by using equivalent flat plates. This method was also extended later by Fertis and Lee[54,55] to include the inelastic analysis of uniform and variable thickness rectangular and circular plates of various boundary conditions and loadings. Lin[54] also introduced the following differential equation for inelastic analysis of thin plates:

$$D\nabla^4 w = \left(-\frac{\partial^2 M_{I_x}}{\partial x^2} - 2\frac{\partial^2 M_{I_{xy}}}{\partial x \partial y} - \frac{\partial^2 M_{I_y}}{\partial y^2} \right) + q \tag{8.2}$$

where M_{Ix}, M_{Ixy}, and M_{Iy} are defined as the inelastic moments. However, no solution of this equation was provided, even for simple problems.

In this chapter, elastic and inelastic analysis of uniform and variable thickness rectangular and circular plates will be performed. Various cases of loading, thickness variation, and boundary conditions will be investigated. Simplified methods of analysis are also introduced in order to reduce the complexity of the plate problem.

8.2 EQUIVALENT SYSTEMS FOR ELASTIC ANALYSIS OF PLATES OF VARIABLE THICKNESS

8.2.1. RECTANGULAR PLATES OF VARIABLE THICKNESS

Equivalent systems for variable thickness plates in one dimension only, say the y direction of the plate, will be derived in this section. Since the method is general, the boundary conditions of the plate will not be specified here. The method could also be extended to cases where the thickness of the plate can vary in both directions. The derivation of such equivalent systems may be initiated by considering the partial differential equation of thin elastic plates with thickness variation in two dimensions, which is written as follows:

$$\begin{aligned} &D\nabla^4 w + 2\frac{\partial D}{\partial x}\frac{\partial}{\partial x}\nabla^2 w + 2\frac{\partial D}{\partial y}\frac{\partial}{\partial y}\nabla^2 w + \nabla^2 D \nabla^2 w \\ &-(1-\nu)\left[\frac{\partial^2 D}{\partial x^2}\frac{\partial^2 w}{\partial y^2} - 2\frac{\partial^2 D}{\partial x \partial y}\frac{\partial^2 w}{\partial x \partial y} + \frac{\partial^2 D}{\partial y^2}\frac{\partial^2 w}{\partial x^2}\right] = q(x,y) \end{aligned} \tag{8.3}$$

Equation 8.3 is a fourth order partial differential equation in two dimensions with variable coeffecients, which is difficult to solve.

If the plate has variable thickness in one dimension only, say the y direction, Equation 8.3 yields

$$D\nabla^4 w + 2\frac{\partial D}{\partial y}\frac{\partial}{\partial y}\nabla^2 w + \frac{\partial^2 D}{\partial y^2}\left[\frac{\partial^2 w}{\partial y^2} + v\frac{\partial^2 w}{\partial x^2}\right] = q(x,y) \tag{8.4}$$

where

$$\nabla^2 = \frac{\partial^2}{\partial x^2} + \frac{\partial^2}{\partial y^2} \tag{8.5}$$

and

$$\nabla^4 = \frac{\partial^4}{\partial x^4} + 2\frac{\partial^2}{\partial x^2 \partial y^2} + \frac{\partial^4}{\partial y^4} \tag{8.6}$$

The rigidity D of the plate is given by the equation

$$D = \frac{Eh^3}{12(1-v^2)} \tag{8.7}$$

In Equation 8.7 E is the modulus of elasticity and v is the Poisson ratio. In this equation, the thickness h of the plate, as well as the modulus of elasticity E, may vary.

By assuming that the rectangular plate has variable thickness h in the y direction only and constant E, we write

$$h = h_o[1 + \lambda f(y)] \tag{8.8}$$

where h_o is the constant reference thickness, f(y) describes the thickness variation, and λ is a constant that keeps the plate thin enough to fall within the range of thin plate theory. Examples of plate thickness variations are shown in Figure 8.1. By substituting Equation 8.8 into Equation 8.7 we find

$$D = D_o[1 + \lambda f(y)]^3 \tag{8.9}$$

where

$$D_o = \frac{Eh_0^3}{12(1-v^2)} \tag{8.10}$$

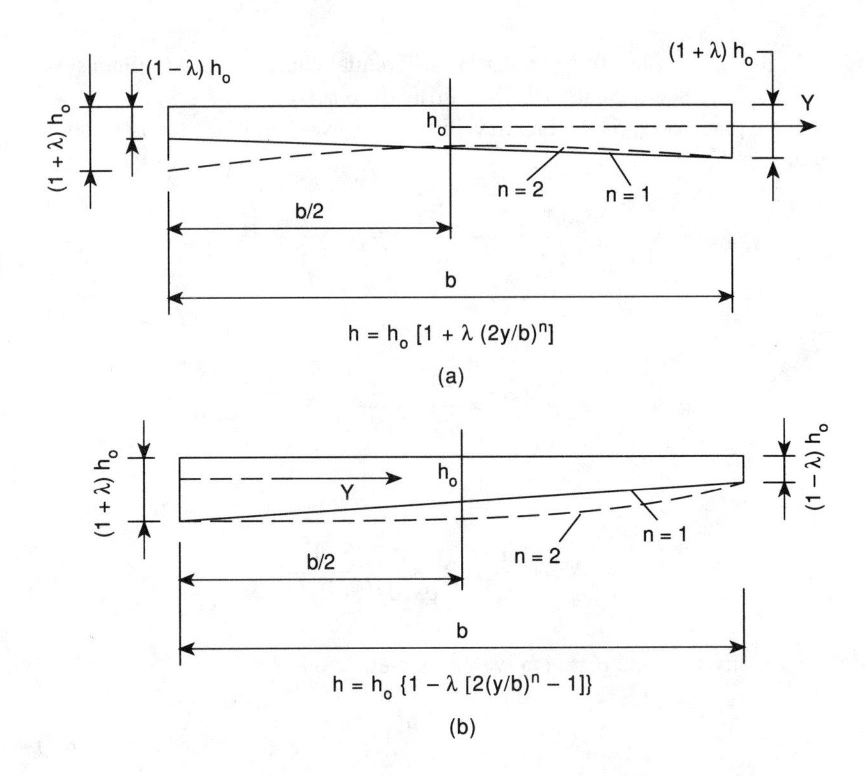

FIGURE 8.1. (a) Variable thickness plate with n = 1 and n = 2; (b) variable thickness plate with n = 1 and n = 2.

By substituting Equation 8.9 into Equation 8.4 and performing the required differentiations, we obtain

$$\nabla^4 w = \frac{q}{D_o[1-\lambda f]^3} - \frac{6\lambda}{[1+\lambda f]}\frac{\partial f}{\partial y}\frac{\partial}{\partial y}\nabla^2 w - \frac{3\lambda}{[1+\lambda f]}\frac{\partial^2 w}{\partial y^2}\left[\frac{\partial^2 w}{\partial y^2} + \nu\frac{\partial^2 w}{\partial x^2}\right]$$
$$-\frac{6\lambda^2}{[1+\lambda f]^2}\left[\frac{\partial f}{\partial y}\right]^2\left[\frac{\partial^2 w}{\partial y^2} + \nu\frac{\partial^2 w}{\partial x^2}\right] \tag{8.11}$$

By using the binomial theorem we may write the following series

$$\frac{1}{[1+\lambda f]^3} = (1 - 3\lambda f + 6\lambda^2 f^2 - 10\lambda^3 f^3 + \ldots) \tag{8.12}$$

$$\frac{1}{[1+\lambda f]^2} = (1 - 2\lambda f + 3\lambda^2 f^2 - 4\lambda^3 f^3 + \ldots) \tag{8.13}$$

$$\frac{1}{[1+\lambda f]} = (1 - \lambda f + \lambda^2 f^2 - \lambda^3 f^3 + \ldots) \tag{8.14}$$

By substituting the above series in Equation 8.11 and rearranging, we obtain

$$\begin{aligned}
\nabla^4 w = {} & \frac{q}{D_o}(1 - 3\lambda f + 6\lambda^2 f^2 - 10\lambda^3 f^3 + \ldots) \\
& -6\lambda \frac{\partial f}{\partial y}\frac{\partial}{\partial y}\nabla^2 w (1 - \lambda f + \lambda^2 f^2 - \lambda^3 f^3 + \ldots) \\
& -3\lambda \frac{\partial^2 f}{\partial y^2}\left(\frac{\partial^2 w}{\partial y^2} + \nu \frac{\partial^2 w}{\partial x^2}\right)(1 - \lambda f + \lambda^2 f^2 - \lambda^3 f^3 + \ldots) \\
& -6\lambda^2 \left(\frac{\partial f}{\partial y}\right)^2 \left(\frac{\partial^2 w}{\partial y^2} + \nu \frac{\partial^2 w}{\partial x^2}\right)(1 - 2\lambda f + 3\lambda^2 f^2 - 4\lambda^3 f^3 + \ldots)
\end{aligned} \tag{8.15a}$$

We can rewrite the above equation so that the right side is arranged in ascending powers of λ, i.e.,

$$\begin{aligned}
\nabla^4 w = {} & \lambda^0 \left(\frac{q}{D_o}\right) \\
& + \lambda \left[-3f\frac{q}{D_o} - 6\frac{\partial f}{\partial y}\frac{\partial}{\partial y}\nabla^2 w - 3\frac{\partial^2 f}{\partial y^2}\left(\frac{\partial^2 w}{\partial y^2} + \nu \frac{\partial^2 w}{\partial x^2}\right)\right] \\
& + \lambda^2 \left\{6f^2\frac{q}{D_o} + 6\frac{\partial f}{\partial y} f \frac{\partial}{\partial y}\nabla^2 w + \left[3\frac{\partial^2 f}{\partial y^2} f - 6\left(\frac{\partial f}{\partial y}\right)^2\right]\left[\frac{\partial^2 w}{\partial y^2} + \nu \frac{\partial^2 w}{\partial x^2}\right]\right\} \\
& + \lambda^3 \left\{-10f^3\frac{q}{D_o} - 6\frac{\partial f}{\partial y} f^2 \frac{\partial}{\partial y}\nabla^2 w - \left[3\frac{\partial^2 f}{\partial y^2} f^2 - 12\left(\frac{\partial f}{\partial y}\right)^2 f\right]\left[\frac{\partial^2 w}{\partial y^2} + \nu \frac{\partial^2 w}{\partial x^2}\right]\right\} \\
& + \lambda^4(\ldots\ldots\ldots\ldots)
\end{aligned} \tag{8.15b}$$

The above differential equation is linear, the principle of superposition applies, and the right side of the equation can be considered a succession of additive loadings, one for each power of λ from 0 to ∞. On this basis, it is reasonable to assume that the solution of the equation can be written as

$$w = \sum_{m=0}^{\infty} \lambda^m w_m \tag{8.16}$$

where w_m represents a series of solutions corresponding to the series of λs appearing on the right side of the equation. Since λ is a constant, we write

$$\nabla^4 w = \nabla^4\left(w_o + \lambda w_1 + \lambda^2 w_2 + \lambda^3 w_3 + \ldots\right) \tag{8.17}$$

$$\frac{\partial}{\partial y}\nabla^2 w = \frac{\partial}{\partial y}\nabla^2\left(w_o + \lambda w_1 + \lambda^2 w_2 + \lambda^3 w_3 + \ldots\right) \tag{8.18}$$

$$\left(\frac{\partial^2 w}{\partial y^2} + \nu\frac{\partial^2 w}{\partial x^2}\right) = \left(\frac{\partial^2}{\partial y^2} + \nu\frac{\partial^2}{\partial x^2}\right)\left(w_o + \lambda w_1 + \lambda^2 w_2 + \lambda^3 w_3 + \ldots\right) \tag{8.19}$$

By substituting Equations 8.17, 8.18, and 8.19 into Equation 8.15, we can write the equation

$$\begin{aligned}
&\nabla^4\left(w_o + \lambda w_1 + \lambda^2 w_2 + \ldots\right) = \frac{q}{D_o}\left[1 - 3\lambda f + 6\lambda^2 f^2 - 10\lambda^3 f^3 + \ldots\right] \\
&-6\lambda\frac{\partial f}{\partial y}\frac{\partial}{\partial y}\nabla^2\left(w_o + \lambda w_1 + \lambda^2 w_2 + \ldots\right)\left(1 - \lambda f + \lambda^2 f^2 - \lambda^3 f^3 + \ldots\right) \\
&-6\lambda^2\left(\frac{\partial f}{\partial y}\right)^2\left(\frac{\partial^2}{\partial y^2} + \nu\frac{\partial^2}{\partial x^2}\right)\left(w_o + \lambda w_1 + \lambda^2 w_2 + \ldots\right)\left(1 - 2\lambda f + 3\lambda^2 f^2 + 4\lambda^3 f^3 + \ldots\right) \\
&-3\lambda\frac{\partial^2 f}{\partial y^2}\left(\frac{\partial^2}{\partial y^2} + \nu\frac{\partial^2}{\partial x^2}\right)\left(w_o + \lambda w_1 + \lambda^2 w_2 + \ldots\right)\left(1 - \lambda f + \lambda^2 f^2 - \lambda^3 f^3 + \ldots\right)
\end{aligned} \tag{8.20}$$

From Equation 8.20, the following series of differential equations may be written:

$$\nabla^4 w_o = \frac{q}{D_o} \tag{8.21a}$$

$$\lambda\nabla^4 w_1 = \lambda\left\{(-3f)\frac{q}{D_o} - 6\frac{\partial f}{\partial y}\frac{\partial}{\partial y}\nabla^2 w_o - 3\frac{\partial^2 f}{\partial y^2}\left(\frac{\partial^2 w_o}{\partial y^2} + \nu\frac{\partial^2 w_o}{\partial x^2}\right)\right\} \tag{8.21b}$$

$$\lambda^2\nabla^4 w_2 = \lambda^2\left\{(6f^2)\frac{q}{D_o} + 6\frac{\partial f}{\partial y}f\frac{\partial}{\partial y}\nabla^2 w_o + 3\frac{\partial^2 f}{\partial y^2}f\left(\frac{\partial^2 w_o}{\partial y^2} + \nu\frac{\partial^2 w_o}{\partial x^2}\right)\right.$$

$$\left. -6\left(\frac{\partial f}{\partial y}\right)^2\left(\frac{\partial^2 w_o}{\partial y^2} + \nu\frac{\partial^2 w_o}{\partial x^2}\right) - 6\frac{\partial f}{\partial y}\frac{\partial}{\partial y}\nabla^2 w_1 - 3\frac{\partial^2 f}{\partial y^2}\left(\frac{\partial^2 w_1}{\partial y^2} + \nu\frac{\partial^2 w_1}{\partial x^2}\right)\right\} \tag{8.21c}$$

$$\lambda^3\nabla^4 w_3 = \lambda^3\left\{-10f^3\frac{q}{D_o} - 6\frac{\partial f}{\partial y}f^2\frac{\partial}{\partial y}\nabla^2 w_o\right.$$

$$-\left[3\frac{\partial^2 f}{\partial y^2}f^2 - 12\left(\frac{\partial f}{\partial y}\right)^2 f\right]\left(\frac{\partial^2 w_o}{\partial y^2} + \nu\frac{\partial^2 w_o}{\partial x^2}\right)$$

$$+6\frac{\partial f}{\partial y}f\frac{\partial}{\partial y}\nabla^2 w_1 + \left[3\frac{\partial^2 f}{\partial y^2}f - 6\left(\frac{\partial f}{\partial y}\right)^2\right]\left(\frac{\partial^2 w_1}{\partial y^2} + \nu\frac{\partial^2 w_1}{\partial x^2}\right) \tag{8.21d}$$

$$\left. -6\frac{\partial f}{\partial y}\frac{\partial}{\partial y}\nabla^2 w_2 - 3\frac{\partial^2 f}{\partial y^2}\left(\frac{\partial^2 w_2}{\partial y^2} + \nu\frac{\partial^2 w_2}{\partial x^2}\right)\right\}$$

$$\lambda^m\nabla^4 w_m = \ldots \tag{8.21e}$$

The first one in the above set of equations represents a flat plate with constant rigidity D_o and loading q that is identical to that of the original variable thickness plate. The remaining equations in this set represent flat plates with differerent loads. These loads can be determined once the deflection from the preceding equation is determined. Thus, Equation 8.21 is a set of equations describing an equivalent system of flat plates that replaces the original variable thickness plate. The solution applies to all boundary conditions and all continuous thickness variations with continuous first and second derivatives. A sufficiently accurate solution, however, can be obtained by using only the first two or three equations from the set of equations given by Equation 8.21, i.e.,

$$w = w_o + \lambda w_1 + \lambda^2 w_2 + \lambda^3 w_3 + \ldots \tag{8.22}$$

where the parameter λ is considered to be smaller than one.

Since the boundary conditions of the plate were not considered in the above derivation of the equivalent flat plates, Equation 8.21 will apply to all possible boundary conditions under which the equation for a flat, thin plate can be solved. Through a suitable choice of the paremeter λ, the function f(y) can be constrained to have the values

$$|f(y)| \le 1 \tag{8.23}$$

along the y axis. Values of λ ranging from 0.1 to 0.4 are examined in detail in the following sections of the chapter. This represents a very sizable variation in the thickness of the plate.

From plate theory, the expressions for the bending moments M_x and M_y may be written as

$$M_x = -D\left[\frac{\partial^2 w}{\partial x^2} + v\frac{\partial^2 w}{\partial y^2}\right] \tag{8.24}$$

$$M_y = -D\left[\frac{\partial^2 w}{\partial y^2} + v\frac{\partial^2 w}{\partial x^2}\right] \tag{8.25}$$

By substituting Equations 8.9 and 8.22 into Equations 8.24 and 8.25 and rearranging, the expressions for the bending moments M_x and M_y in the x and y directions, respectively, may be written as follows:

$$\begin{aligned} M_x = -D_o(1+\lambda f)^3 \Bigg\{ & \left[\frac{\partial^2 w_o}{\partial x^2} + v\frac{\partial^2 w_o}{\partial y^2}\right] + \lambda\left[\frac{\partial^2 w_1}{\partial x^2} + v\frac{\partial^2 w_1}{\partial y^2}\right] \\ & + \lambda^2\left[\frac{\partial^2 w_2}{\partial x^2} + v\frac{\partial^2 w_2}{\partial y^2}\right] + \lambda^3\left[\frac{\partial^2 w_3}{\partial x^2} + v\frac{\partial^2 w_3}{\partial y^2}\right]\Bigg\} \end{aligned} \tag{8.26}$$

$$\begin{aligned} M_y = -D_o(1+\lambda f)^3 \Bigg\{ & \left[\frac{\partial^2 w_o}{\partial y^2} + v\frac{\partial^2 w_o}{\partial x^2}\right] + \lambda\left[\frac{\partial^2 w_1}{\partial y^2} + v\frac{\partial^2 w_1}{\partial x^2}\right] \\ & + \lambda^2\left[\frac{\partial^2 w_2}{\partial y^2} + v\frac{\partial^2 w_2}{\partial x^2}\right] + \lambda^3\left[\frac{\partial^2 w_3}{\partial y^2} + v\frac{\partial^2 w_3}{\partial x^2}\right]\Bigg\} \end{aligned} \tag{8.27}$$

In a similar manner, the bending stresses σ_x and σ_y in the same directions may be written as

$$\sigma_x = -\frac{Eh_o(1+\lambda f)}{2(1-\nu^2)}\left\{\left[\frac{\partial^2 w_o}{\partial x^2}+\nu\frac{\partial^2 w_o}{\partial y^2}\right]+\lambda\left[\frac{\partial^2 w_1}{\partial x^2}+\nu\frac{\partial^2 w_1}{\partial y^2}\right]\right.$$
$$\left.+\lambda^2\left[\frac{\partial^2 w_2}{\partial x^2}+\nu\frac{\partial^2 w_2}{\partial y^2}\right]+\lambda\left[\frac{\partial^2 w_3}{\partial x^2}+\nu\frac{\partial^2 w_3}{\partial y^2}\right]\right\} \tag{8.28}$$

$$\sigma_y = -\frac{Eh_o(1+\lambda f)}{2(1-\nu^2)}\left\{\left[\frac{\partial^2 w_o}{\partial y^2}+\nu\frac{\partial^2 w_o}{\partial x^2}\right]+\lambda\left[\frac{\partial^2 w_1}{\partial y^2}+\nu\frac{\partial^2 w_1}{\partial x^2}\right]\right.$$
$$\left.+\lambda^2\left[\frac{\partial^2 w_2}{\partial y^2}+\nu\frac{\partial^2 w_2}{\partial x^2}\right]+\lambda\left[\frac{\partial^2 w_3}{\partial y^2}+\nu\frac{\partial^2 w_3}{\partial x^2}\right]\right\} \tag{8.29}$$

8.2.2 SYMMETRICAL CIRCULAR PLATES

The differential equation of a symmetrical circular plate with varying thickness in the radial direction R (see, e.g., Figure 8.2a) and symmetrically loaded may be written as

$$M_R + \frac{dM_R}{dR}R - M_T + QR = 0 \tag{8.30}$$

where Q is the shear force per unit length of a circular section of radius R (see Figure 8.2a). The radial moment M_R acting on the side CD of the element ABCD and the tangential moment M_T acting on the side AD of the same element are

$$M_R = -D\left[\frac{d^2w}{dR^2}+\frac{\nu}{R}\frac{dw}{dR}\right] \tag{8.31}$$

$$M_T = -D\left[\frac{1}{R}\frac{dw}{dR}+\nu\frac{d^2w}{dR^2}\right] \tag{8.32}$$

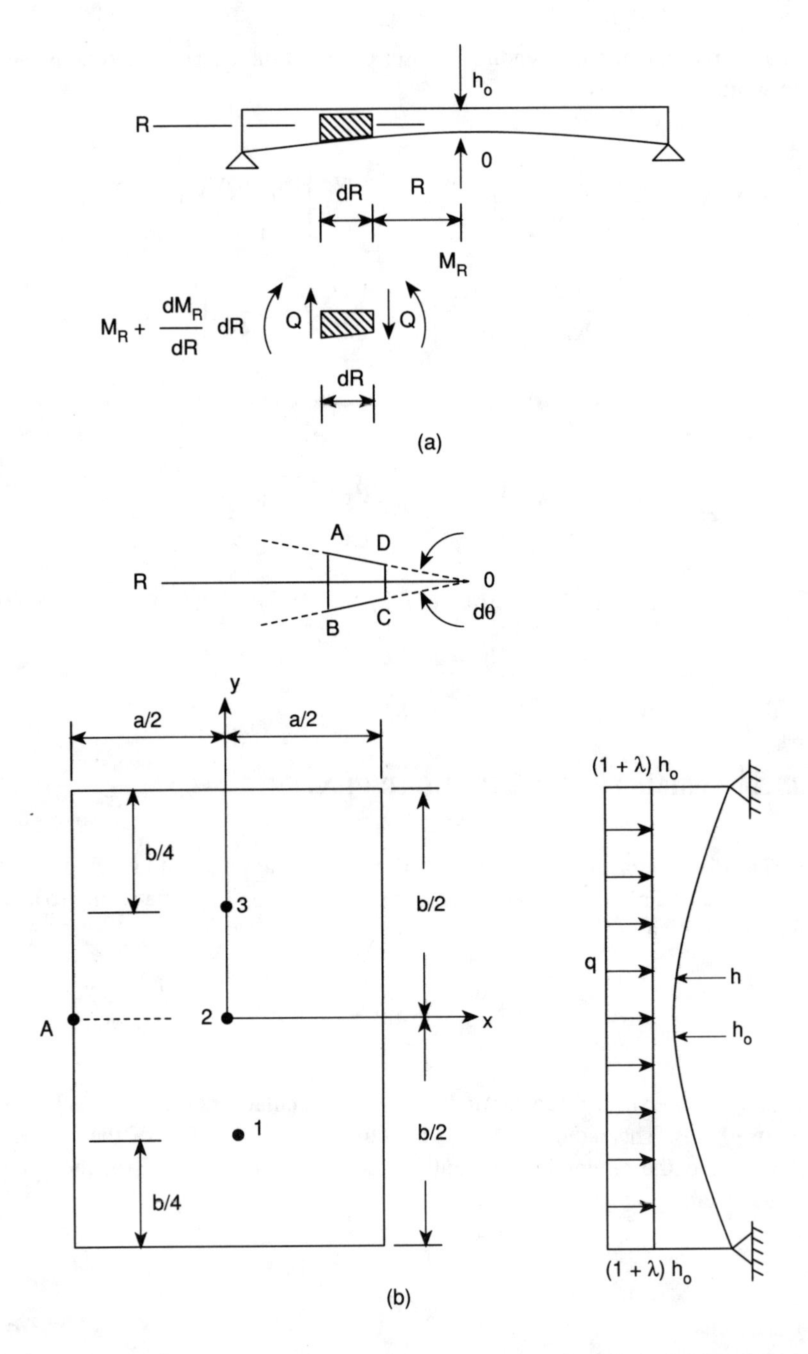

FIGURE 8.2. Cross section of a circular plate of variable thickness in the radial direction and plate element ABCD; (b) simply supported rectangular plate of quadratic thickness variation along the y axis.

The variable rigidity D may be expressed as

$$D = D_o\left[1 + \lambda f(R)\right]^3 \tag{8.33}$$

where $D_o = Eh_o^3/12(1 - \nu^2)$ is a reference value and f(R) is a function representing the plate thickness variation in the radial direction. By using Equations 8.30 through 8.33 and following the procedure used for rectangular plates with varying thickness in one direction only, Equation 8.30 may be replaced by a set of plates of uniform rigidity D_o, which are represented by the following set of differential equations:

$$\frac{d}{dR}\left[\frac{1}{R}\frac{d}{dR}\left(\frac{dw_o}{dR}\right)\right] = \frac{q}{D_o} \tag{8.34a}$$

$$\lambda\left\{\frac{d}{dR}\left[\frac{1}{R}\frac{d}{dR}\left(R\frac{dw_1}{dR}\right)\right]\right\} = \lambda\left\{-\frac{3Q}{D_o}f - 3\lambda\frac{df}{dR}\left(\frac{dw_o^2}{dR^2} + \frac{\nu}{R}\frac{dw_o}{dR}\right)\right\} \tag{8.34b}$$

$$\begin{aligned}\lambda^2\left\{\frac{d}{dR}\left[\frac{1}{R}\frac{d}{dR}\left(R\frac{dw_2}{dR}\right)\right]\right\} = \lambda^2\Bigg\{&-\frac{6Qf^2}{D_o} + 3f\frac{df}{dR}\left(\frac{d^2w_o}{dR^2} + \frac{\nu}{R}\frac{dw_o}{dR}\right) \\ &-3\frac{df}{dR}\left(\frac{d^2w_1}{dR^2} + \frac{\nu}{R}\frac{dw_1}{dR}\right)\Bigg\}\end{aligned} \tag{8.34c}$$

$$\begin{aligned}\lambda^3\left\{\frac{d}{dR}\left[\frac{1}{R}\frac{d}{dR}\left(R\frac{dw_3}{dR}\right)\right]\right\} = \lambda^3\Bigg\{&-\frac{10f^3Q}{D_o} - 3\frac{df}{dR}f^2\left(\frac{d^2w_o}{dR^2} + \frac{\nu}{R}\frac{dw_o}{dR}\right) \\ &+3\frac{df}{dR}f\left(\frac{d^2w_1}{dR^2} + \frac{\nu}{R}\frac{dw_1}{dR}\right) - 3\frac{df}{dR}\left(\frac{d^2w_2}{dR^2} + \frac{\nu}{R}\frac{dw_2}{dR}\right)\Bigg\}\end{aligned} \tag{8.34d}$$

$$\lambda^m\left\{\frac{d}{dR}\left[\frac{1}{R}\frac{d}{dR}\left(R\frac{dw_m}{dR}\right)\right]\right\} = \ldots \tag{8.34e}$$

The set of differential equations given by Equation 8.34 is analogous to Equation 8.21 that was obtained for rectangular plates. An accurate solution is usually obtained using the first two or three equations from the above set.

The bending moments M_T and M_R in the tangential and radial directions, respectively, and the corresponding tangential and radial stresses σ_T and σ_R, respectively, may be determined from the equations

$$M_T = -D_o(1+\lambda f)^3\left\{\frac{1}{R}\frac{dw_o}{dR} + \nu\frac{d^2w_o}{dR^2} + \lambda\left(\frac{1}{R}\frac{dw_1}{dR} + \nu\frac{d^2w_1}{dR^2}\right)\right.$$
$$\left. +\lambda^2\left(\frac{1}{R}\frac{dw_2}{dR} + \nu\frac{d^2w_2}{dR^2}\right) + \lambda^3\left(\frac{1}{R}\frac{dw_3}{dR} + \nu\frac{d^2w_3}{dR^2}\right)\right\} \tag{8.35}$$

$$M_R = -D_o(1+\lambda f)^3\left\{\frac{d^2w_o}{dR^2} + \frac{\nu}{R}\frac{dw_o}{dR} + \lambda\left(\frac{d^2w_1}{dR^2} + \frac{\nu}{R}\frac{dw_1}{dR}\right)\right.$$
$$\left. +\lambda^2\left(\frac{d^2w_2}{dR^2} + \frac{\nu}{R}\frac{dw_2}{dR}\right) + \lambda^3\left(\frac{d^2w_3}{dR^2} + \frac{\nu}{R}\frac{dw_3}{dR}\right)\right\} \tag{8.36}$$

$$\sigma_T = -\frac{Eh_o(1+\lambda f)}{2(1-\nu^2)}\left\{\frac{1}{R}\frac{dw_o}{dR} + \nu\frac{d^2w_o}{dR^2} + \lambda\left(\frac{1}{R}\frac{dw_1}{dR} + \nu\frac{d^2w_1}{dR^2}\right)\right.$$
$$\left. +\lambda^2\left(\frac{1}{R}\frac{dw_2}{dR} + \nu\frac{d^2w_2}{dR^2}\right) + \lambda^3\left(\frac{1}{R}\frac{dw_3}{dR} + \nu\frac{d^2w_3}{dR^2}\right)\right\} \tag{8.37}$$

$$\sigma_R = -\frac{Eh_o(1+\lambda f)}{2(1-\nu^2)}\left\{\frac{d^2w_o}{dR^2} + \frac{\nu}{R}\frac{dw_o}{dR} + \lambda\left(\frac{d^2w_1}{dR^2} + \frac{\nu}{R}\frac{dw_1}{dR}\right)\right.$$
$$\left. +\lambda^2\left(\frac{d^2w_2}{dR^2} + \frac{\nu}{R}\frac{dw_2}{dR}\right) + \lambda^3\left(\frac{d^2w_3}{dR^2} + \frac{\nu}{R}\frac{dw_3}{dR}\right)\right\} \tag{8.38}$$

8.3 EQUIVALENT SYSTEMS FOR RECTANGULAR PLATES WITH QUADRATIC THICKNESS VARIATION

We consider first a thin rectangular plate with quadratic thickness variation h along the y axis as shown in Figure 8.2b. By considering the thickness h_o of the plate along the x axis at y = 0 as the reference value, we write

$$h = h_o\left[1+\lambda\left(\frac{2y}{b}\right)^2\right] \tag{8.39}$$

Therefore,

$$f(y) = \left(\frac{2y}{b}\right)^2 \tag{8.40}$$

$$\frac{\partial f}{\partial y} = \frac{4}{b}\left(\frac{2y}{b}\right) \tag{8.41}$$

$$\frac{\partial^2 f}{\partial y^2} = \frac{8}{b^2} \tag{8.42}$$

By using the above equations and Equation 8.21, the set of differential equations representing the equivalent system for a rectangular plate with quadratic thickness variation, and loaded with a uniformly distributed loading q over its entire area, are as follows:

$$\nabla^4 w_o = \frac{q}{D_o} \tag{8.43a}$$

$$\lambda\nabla^4 w_1 = \lambda\left\{(1-3f)\frac{1}{D_o} - 6\frac{\partial f}{\partial y}\frac{\partial}{\partial y}\nabla^2 w_o - \frac{\partial^2 f}{\partial y^2}\left(\frac{\partial^2 w_o}{\partial y^2} + \nu\frac{\partial^2 w_o}{\partial x^2}\right)\right\} \tag{8.43b}$$

$$\begin{aligned}\lambda^2\nabla^4 w_2 = \lambda^2\Bigg\{ & 6f^2\frac{q}{D_o} + 6\frac{\partial f}{\partial y}f\frac{\partial}{\partial y}\nabla^2 w_o + 3\frac{\partial^2 f}{\partial y^2}f\left(\frac{\partial^2 w_o}{\partial y^2} + \nu\frac{\partial^2 w_o}{\partial x^2}\right) \\ & -6\left[\frac{\partial f}{\partial y}\right]^2\left[\frac{\partial^2 w_o}{\partial y^2} + \nu\frac{\partial^2 w_o}{\partial x^2}\right] - 6\frac{\partial f}{\partial y}\frac{\partial}{\partial y}\nabla^2 w_1 \\ & -3\left(\frac{\partial^2 f}{\partial y^2}\right)\left(\frac{\partial^2 w_1}{\partial y^2} + \nu\frac{\partial^2 w_1}{\partial x^2}\right)\Bigg\}\end{aligned} \tag{8.43c}$$

$$\lambda^3\nabla^4 w_3 = \lambda^3\Bigg\{-10f^3\frac{q}{D_o} - 6\frac{\partial f}{\partial y}f^2\frac{\partial}{\partial y}\nabla^2 w_o$$

$$-\left[3\frac{\partial^2 f}{\partial y^2}f^2 - 12\left(\frac{\partial f}{\partial y}\right)^2 f\right]\left(\frac{\partial^2 w_o}{\partial y^2} + \nu\frac{\partial^2 w_o}{\partial x^2}\right)$$

$$+6\frac{\partial f}{\partial y}f\frac{\partial}{\partial y}\nabla^2 w_1 + \left[3\frac{\partial^2 w_o}{\partial y^2}f - 6\left(\frac{\partial f}{\partial y}\right)^2\right]\left(\frac{\partial^2 w_1}{\partial y^2} + \nu\frac{\partial^2 w_1}{\partial x^2}\right) \tag{8.43d}$$

$$-6\frac{\partial f}{\partial y}\frac{\partial}{\partial y}\nabla^2 w_2 - 3\frac{\partial^2 f}{\partial y^2}\left(\frac{\partial^2 w_2}{\partial y^2} + \nu\frac{\partial^2 w_2}{\partial x^2}\right)\Bigg\}$$

$$\lambda^m\nabla^4 w_m = \ldots\ldots \tag{8.43e}$$

The solution of the first equation, i.e., Equation 8.43a, is well known (see, e.g., Timoshenko[37]). If the origin of the x,y coordinates of the plate is taken at point A in Figure 8.2b, the solution of Equation 8.43a is given by the expression

$$w_o = \frac{qa^4}{D_o}\sum_{n=1,3,\ldots}^{\infty}\left[\frac{4}{n^5\pi^5} + A_{no}\cosh\alpha_n y + B_{no}\alpha_n y\sinh\alpha_n y\right]\sin\alpha_n x \tag{8.44}$$

where

$$A_{no} = \frac{-2(\gamma_n\tanh\gamma_n + 2)}{n^5\pi^5\cosh\gamma_n} \tag{8.45}$$

$$B_{no} = \frac{2}{n^5\pi^5\cosh\gamma_n} \tag{8.46}$$

$$\alpha_n = \frac{n\pi}{a} \qquad \gamma_n = \frac{n\pi b}{2a} \tag{8.47}$$

and

$$\frac{\partial}{\partial y}\nabla^2 w_o = \frac{4qa}{D_o}\sum_{n=1,3,\cdots}^{\infty}\frac{\sinh\alpha_n y}{n^2\pi^2\cosh\gamma_n}\sin\alpha_n x \tag{8.48}$$

$$\left(\frac{\partial^2 w_o}{\partial y^2} + \nu \frac{\partial^2 w_o}{\partial x^2}\right) = \frac{2qa^2}{D_o} \sum_{n=1,3,\ldots}^{\infty} \frac{1}{n^3\pi^3} [-2\nu$$

$$+ \frac{2\nu - (1-\nu)\gamma_n \tanh\gamma_n}{\cosh\gamma_n} \cosh\alpha_n y \tag{8.49}$$

$$\left. + \frac{(1-\nu)}{\cosh\gamma_n} \alpha_n y \sinh\alpha_n y \right] \sin\alpha_n x$$

The results obtained from Equations 8.44 through 8.49 may be used to solve Equation 8.43b and so on. Usually the solution of the first two to three equations from the set of equations given by Equation 8.43 is sufficient for an accurate solution.

A very accurate simplified equivalent system that greatly reduces the mathematical complexity of the problem may be obtained by examining the convergence of the parameter λ (see Fertis and Mijatov[51] and Fertis and Lee[53]) and weighing out the effect of the various terms in Equation 8.21. On this basis, a very accurate simplified equivalent system for a uniformly loaded rectangular plate with quadratic variation in thickness along the y axis is represented by the differential equation

$$\nabla^4 w = (1 - 3\lambda f)\frac{q}{D_o} - \lambda\left\{6\frac{\partial f}{\partial y}\frac{\partial}{\partial y}\nabla^2 w_o + \frac{\partial^2 f}{\partial y^2}\left(\frac{\partial^2 w_o}{\partial y^2} + \nu\frac{\partial^2 w_o}{\partial x^2}\right)\right\} \tag{8.50}$$

This equation combines the first two differential equations from Equation 8.43 into one differential equation, and it represents the differential equation of a flat rectangular plate with new loading that replaces the original variable thickness plate and its load. The values of

$$\frac{\partial}{\partial y}\nabla^2 w_o \tag{8.51a}$$

and

$$\left(\frac{\partial^2 w_o}{\partial y^2} + \nu\frac{\partial^2 w_o}{\partial x^2}\right) \tag{8.51b}$$

may be determined, since w_o can be determined from the equation

$$\nabla^4 w_o = \frac{q}{D_o} \tag{8.52}$$

as shown by Equations 8.44 through 8.49.

Utilization of Equation 8.50 simplifies the solution of the problem and produces reasonable results for a practicing design engineer.

8.4 EQUIVALENT SYSTEMS FOR RECTANGULAR PLATES WITH LINEAR THICKNESS VARIATION

If a uniformly loaded rectangular plate has linearly varying thickness h along the y axis, we have

$$h = h_o\left[1 + \lambda\left(\frac{2y}{b}\right)\right] \tag{8.53}$$

and Equation 8.21 takes the form

$$\nabla^4 w_o = \frac{q}{D_o} \tag{8.54a}$$

$$\lambda\Delta^4 w_1 = \lambda\left[(1-3f)\frac{q}{D_o} - 6\frac{\partial f}{\partial y}\frac{\partial}{\partial y}\nabla^2 w_o\right] \tag{8.54b}$$

$$\lambda^2\nabla^4 w_2 = \lambda^2\left\{6f^2\frac{q}{D_o} + 6\frac{\partial f}{\partial y}f\frac{\partial}{\partial y}\nabla^2 w_o - 6\left(\frac{\partial f}{\partial y}\right)^2\left(\frac{\partial^2 w_o}{\partial y^2} + \nu\frac{\partial^2 w_o}{\partial x^2}\right)\right.$$
$$\left. -6\frac{\partial f}{\partial y}\frac{\partial}{\partial y}\nabla^2 w_1\right\} \tag{8.54c}$$

$$\lambda^3\nabla^4 w_3 = \lambda^3\left\{-10f\frac{q}{D_o} + 6\frac{\partial f}{\partial y}f^2\frac{\partial}{\partial y}\nabla^2 w_o\right.$$
$$+12\left(\frac{\partial f}{\partial y}\right)^2 f\left(\frac{\partial^2 w_o}{\partial y^2} + \nu\frac{\partial^2 w_o}{\partial x^2}\right) + 6\frac{\partial f}{\partial y}f\frac{\partial}{\partial y}\nabla^2 w_1$$
$$\left. -6\left(\frac{\partial f}{\partial y}\right)^2\left(\frac{\partial^2 w_o}{\partial y^2} + \nu\frac{\partial^2 w_o}{\partial x^2}\right) - 6\frac{\partial f}{\partial y}\frac{\partial}{\partial y}\nabla^2 w_2\right\} \tag{8.54d}$$

$$\lambda^m\nabla^4 w_m = \ldots\ldots \tag{8.54e}$$

For this case, a simplified equivalent system of constant thickness h_o, or of constant rigidity D_o, may again be represented by the differential equation

$$\nabla^4 w = (1 - 3\lambda f)\frac{q}{D_o} + 6\lambda\left(\lambda\frac{\partial f}{\partial y}f - \frac{\partial f}{\partial y}\right)\frac{\partial}{\partial y}\nabla^2 w_o \tag{8.55}$$

which combines the first two differential equations of Equation 8.54, plus the second term from the third differential equation. In Equation 8.55 the expression of the derivative

$$\frac{\partial}{\partial y}\nabla^2 w_o \tag{8.56}$$

may be obtained from the solution of the differential equation

$$\nabla^4 w_o = \frac{q}{D_o} \tag{8.57}$$

which is the same as Equation 8.54a. For uniform loading, the solution of Equation 8.57 can be found easily in texts dealing with the theory of thin plates.

Simplified equivalent systems for other cases of plate thickness variations along the y axis may be obtained in a similar manner.

8.5 ELASTIC ANALYSIS OF SIMPLY SUPPORTED RECTANGULAR PLATES WITH QUADRATIC THICKNESS VARIATION

Consider the simply supported rectangular plate in Figure 8.2b, which is loaded with a uniformly distributed load q over the entire area of the plate as shown in the figure. Its quadratic thickness variation h along the y axis is given by Equation 8.39. Therefore, f(y), $\partial f/\partial y$, and $\partial^2 f/\partial y^2$ are given by Equations 8.40, 8.41, and 8.42, respectively. The set of equations that represent equivalent flat plates of rigidity D_o is given by Equation 8.43. The solution here is carried out by using the first four differential equations from the exact solution given by Equation 8.43 and applying Fourier series techniques.

For example, the solution of Equation 8.43a is given by Equation 8.44. With known w_o, we can employ Equations 8.48 and 8.49 and write Equation 8.43b as

$$\nabla^4 w_1 = \frac{48q}{D_o}\sum_{n=1,3,\ldots}^{\infty}\left[C_{o1} + C_{21}y^2 + C_{31}\cosh\alpha_n y + C_{41}y\sinh\alpha_n y\right]\sin\alpha_n x \tag{8.58}$$

where

$$C_{01} = \frac{2\nu a^2}{n^3\pi^3 b^2} \tag{8.59}$$

$$C_{21} = -\frac{1}{n\pi b^2} \tag{8.60}$$

$$C_{31} = -\frac{[2\nu - (1-\nu)\gamma_n \tanh\gamma_n]a^2}{n^3\pi^3 b^2 \cos h\gamma_n} \quad w \tag{8.61}$$

$$C_{41} = -\frac{(5-\nu)}{n^2\pi^3 b^2 \cos h\gamma_n} \tag{8.62}$$

Now we can proceed with the solution of Equation 8.58 to obtain

$$\begin{aligned} w_1 = \frac{48q}{D_o} a^4 \sum_{n=1,3,\cdots}^{\infty} (A_{n1} \cos h\alpha_n y + B_{n1}\alpha_n y \sinh\alpha_n y \\ + K_{01} + K_{21} y^2 + K_{31} y^3 \cos h\alpha_n y + K_{41} y^3 \sinh\alpha_n y) \sin\alpha_n x \end{aligned} \tag{8.63}$$

where

$$\begin{aligned} A_{n1} = \frac{1}{2\cos h\gamma_n} \Bigg[\frac{9(1+\nu) + 2(1+\nu)\gamma_n \tanh\gamma_n}{48 n^5 \pi^5} \\ + \frac{4a^2(2-\nu)}{n^7\pi^7 b^2} \Bigg] - B_{n1}\gamma_n \tanh\gamma_n \end{aligned} \tag{8.64}$$

$$\begin{aligned} B_{n1} = \frac{1}{32\cos h\gamma_n} \Bigg[\frac{(1-\nu)}{n^5\pi^5} \\ - \frac{4a^2[13 - 11\nu + 6(1-\nu)\gamma_n \tanh\gamma_n + 2(1-\nu)\gamma_n^2 \tanh^2\gamma_n]}{n^7\pi^7 b^2} \Bigg] \end{aligned} \tag{8.65}$$

$$K_{01} = \frac{-2(2-\nu)a^2}{n^7\pi^7b^2} \tag{8.66}$$

$$K_{21} = -\frac{1}{n^5\pi^5b^2} \tag{8.67}$$

$$K_{31} = \frac{(5-3\nu)+(1-\nu)\gamma_n \tanh\gamma_n}{8n^5\pi^5b^2 \cos h\gamma_n} \tag{8.68}$$

$$K_{41} = -\frac{(5-\nu)}{24n^4\pi^4ab^2 \cos h\gamma_n} \tag{8.69}$$

With known w_o and w_1, we can procede in a similar manner to determine w_2 from the solution of Equation 8.43c), and so on. When the values of w_o, w_1, w_2, … are determined, Equation 8.22 is used to determine the final values of the vertical displacement w of the rectangular plate.

The problem here is solved using the first four equations from the set in Equation 8.43. The same problem is also solved using the simplified equivalent system given by Equation 8.50 in order to compare the results. Since λ is a constant that keeps the plate thin enough to fall within the range of thin plate theory, the plate problem was solved for $\lambda = 0.1$, 0.2, 0.258, 0.3, and 0.4 for side ratios $b/a = 1$, 1.5, and 2 and Poisson's ratio $v = 0.3$. The results regarding the deflection at the center of the plate, point 2 in Figure 8.2b, are shown in Table 8.1. The results in column 7 of Table 8.1 represent the solution of Equation 8.43d. These values, for all practical purposes, are small compared to the values shown in columns 4, 5, and 6, and they could be neglected. The results obtained using the simplified equivalent system given by Equation 8.50, designated as w_s, are shown in column 8 of Table 8.1, and they are reasonably accurate when they are compared with the exact solution. Tables 8.2 and 8.3 show the results obtained at the center of the plate for the bending moments M_x and M_y and bending stresses σ_x and σ_y, respectively. The symbols M_{x_s}, M_{y_s}, σ_{x_s}, and σ_{y_s} are used to represent the results for the solution of the simplified equivalent system. Note again that the simplified equivalent system yields reasonably accurate results.

In Table 8.4, the vertical deflections w at points 1, 2, and 3 in Figure 8.2b are shown, which are determined by using the equivalent system and the finite difference method with 8×8 mesh. The last column in this table shows the percentage difference, which indicates reasonable agreement in results for practical applications. It should be noted here that the finite difference method

TABLE 8.1
Deflection of Simply Supported Rectangular Plate with Quadratic Thickness Variation and b/a = 1, 1.5, and 2

1 b/a	2 Point	3 λ	4 $\lambda w_o(K)$	5 $\lambda w_1(K)$	6 $\lambda^2 w_2(K)$	7 $\lambda^3 w_3(K)$	8 $w(K)$	9 $w_s(K)$	10 Error (%)
1.0	2	0.1	4.062×10^{-3}	-3.115×10^{-4}	1.516×10^{-5}	-1.624×10^{-7}	3.766×10^{-3}	3.751×10^{-3}	−4.0
1.0	2	0.2	4.062×10^{-3}	-6.231×10^{-4}	6.065×10^{-5}	-1.299×10^{-6}	3.499×10^{-3}	3.439×10^{-3}	−1.71
1.0	2	0.258	4.062×10^{-3}	-8.037×10^{-4}	1.009×10^{-4}	-2.789×10^{-6}	3.357×10^{-3}	3.259×10^{-3}	−2.92
1.0	2	0.3	4.062×10^{-3}	-9.346×10^{-4}	1.365×10^{-4}	-4.385×10^{-6}	3.256×10^{-3}	3.128×10^{-3}	−3.93
1.0	2	0.4	4.062×10^{-3}	-1.246×10^{-3}	2.426×10^{-4}	-1.031×10^{-5}	3.048×10^{-3}	2.816×10^{-3}	−7.61
1.5	2	0.1	7.724×10^{-3}	-5.214×10^{-4}	2.320×10^{-5}	-2.831×10^{-7}	7.226×10^{-3}	7.203×10^{-3}	−0.32
1.5	2	0.2	7.724×10^{-3}	-1.043×10^{-3}	9.279×10^{-5}	-2.264×10^{-6}	6.772×10^{-3}	6.681×10^{-3}	−1.34
1.5	2	0.258	7.724×10^{-3}	-1.345×10^{-3}	1.544×10^{-4}	-4.861×10^{-6}	6.528×10^{-3}	6.379×10^{-3}	−2.28
1.5	2	0.3	7.724×10^{-3}	-1.564×10^{-3}	2.088×10^{-4}	-7.643×10^{-6}	6.361×10^{-3}	6.160×10^{-3}	−3.14
1.5	2	0.4	7.724×10^{-3}	-2.086×10^{-3}	3.712×10^{-4}	-1.812×10^{-5}	5.992×10^{-3}	5.639×10^{-3}	−5.89
2.0	2	0.1	1.013×10^{-2}	-5.565×10^{-4}	2.207×10^{-5}	-3.223×10^{-7}	9.594×10^{-3}	9.572×10^{-3}	−0.23
2.0	2	0.2	1.013×10^{-2}	-1.113×10^{-3}	8.830×10^{-4}	-2.579×10^{-6}	9.101×10^{-3}	9.016×10^{-3}	−0.93
2.0	2	0.258	1.013×10^{-2}	-1.436×10^{-3}	1.469×10^{-4}	-5.535×10^{-6}	8.834×10^{-3}	8.693×10^{-3}	−1.60
2.0	2	0.3	1.013×10^{-2}	-1.670×10^{-3}	1.987×10^{-4}	-8.703×10^{-6}	8.649×10^{-3}	8.459×10^{-3}	−2.20
2.0	2	0.4	1.013×10^{-2}	-2.226×10^{-3}	3.532×10^{-4}	-2.063×10^{-5}	8.235×10^{-3}	7.903×10^{-3}	−4.03

TABLE 8.2
Bending Moments of Simply Supported Rectangular Plate with Quadratic Thickness Variation and b/a = 1, 1.5, and 2

b/a	Point	λ	$M_y(K_1) \times 10^{-2}$	$M_{y_s}(K_1) \times 10^{-2}$	Error (%)	$M_x(K_1) \times 10^{-2}$	$M_{x_s}(K_1) \times 10^{-2}$	Error (%)
1.0	2	0.1	3.682×10^{2}	3.653×10^{2}	–0.78	3.456×10^{2}	3.433×10^{-2}	–0.67
1.0	2	0.2	4.507×10^{2}	4.519×10^{2}	0.27	4.221×10^{2}	4.171×10^{2}	–1.18
1.0	2	0.258	4.420×10^{2}	4.440×10^{2}	0.45	4.074×10^{2}	3.992×10^{2}	–2.01
1.0	2	0.3	4.357×10^{2}	4.384×10^{2}	0.62	3.974×10^{2}	3.863×10^{2}	–2.79
1.0	2	0.4	4.201×10^{2}	4.249×10^{2}	1.14	3.751×10^{2}	3.554×10^{2}	–5.25
1.5	2	0.1	4.926×10^{2}	4.934×10^{2}	0.16	7.658×10^{2}	7.639×10^{2}	–0.25
1.5	2	0.2	4.853×10^{2}	4.883×10^{2}	0.62	7.236×10^{2}	7.163×10^{2}	–1.01
1.5	2	0.258	4.804×10^{2}	4.854×10^{2}	1.04	7.009×10^{2}	6.886×10^{2}	–1.75
1.5	2	0.3	4.765×10^{2}	4.833×10^{2}	1.43	6.852×10^{2}	6.686×10^{2}	–2.42
1.5	2	0.4	4.662×10^{2}	4.787×10^{2}	2.57	6.504×10^{2}	6.209×10^{2}	–4.54
2.0	2	0.1	4.658×10^{2}	4.667×10^{2}	0.19	9.699×10^{2}	9.682×10^{2}	–0.18
2.0	2	0.2	4.662×10^{2}	4.699×10^{2}	0.79	9.264×10^{2}	9.196×10^{2}	0.73
2.0	2	0.258	4.656×10^{2}	4.718×10^{2}	1.33	9.027×10^{2}	8.914×10^{2}	–1.27
2.0	2	0.3	4.648×10^{2}	4.731×10^{2}	1.79	8.862×10^{2}	8.710×10^{2}	–1.72
2.0	2	0.4	4.615×10^{2}	4.763×10^{2}	3.21	8.495×10^{2}	8.224×10^{2}	–3.19

Note: $K_1 = qa^2$.

TABLE 8.3
Bending Stresses of Simply Supported Rectangular Plate with Quadratic Thickness Variation and b/a = 1, 1.5, and 2

b/a	Point	λ	$\sigma_y(K_2) \times 10^{-2}$	$\sigma_{y_s}(K_2) \times 10^{-2}$	Error (%)	$\sigma_x(K_2) \times 10^{-2}$	$\sigma_{x_s}(K_2) \times 10^{-2}$	Error (%)
1.0	2	0.1	2.555×10^{-2}	2.557×10^{-2}	0.08	2.468×10^{-2}	2.462×10^{-2}	–0.24
1.0	2	0.2	2.476×10^{-2}	2.483×10^{-2}	0.28	2.319×10^{-2}	2.292×10^{-2}	–0.49
1.0	2	0.258	2.429×10^{-2}	2.440×10^{-2}	0.45	2.239×10^{-2}	2.194×10^{-2}	–2.00
1.0	2	0.3	2.394×10^{-2}	2.409×10^{-2}	0.63	2.183×10^{-2}	2.122×10^{-2}	–2.79
1.0	2	0.4	2.308×10^{-2}	2.334×10^{-2}	1.13	2.061×10^{-2}	1.953×10^{-2}	–5.24
1.5	2	0.1	2.707×10^{-2}	2.711×10^{-2}	0.15	4.207×10^{-2}	4.197×10^{-2}	–0.24
1.5	2	0.2	2.667×10^{-2}	2.683×10^{-2}	0.60	3.976×10^{-2}	3.935×10^{-2}	–1.03
1.5	2	0.258	2.640×10^{-2}	2.667×10^{-2}	1.02	3.851×10^{-2}	3.784×10^{-2}	–1.74
1.5	2	0.3	2.619×10^{-2}	2.655×10^{-2}	1.41	3.765×10^{-2}	3.674×10^{-2}	–2.42
1.5	2	0.4	2.561×10^{-2}	2.628×10^{-2}	2.62	3.574×10^{-2}	3.412×10^{-2}	–4.53
2.0	2	0.1	2.559×10^{-2}	2.564×10^{-2}	0.20	5.329×10^{-2}	5.320×10^{-2}	–0.17
2.0	2	0.2	2.562×10^{-2}	2.582×10^{-2}	0.78	5.090×10^{-2}	5.053×10^{-2}	–0.73
2.0	2	0.258	2.558×10^{-2}	2.592×10^{-2}	1.33	4.960×10^{-2}	4.900×10^{-2}	–1.21
2.0	2	0.3	2.554×10^{-2}	2.599×10^{-2}	1.76	$4,869 \times 10^{-2}$	4.786×10^{-2}	–1.70
2.0	2	0.4	2.536×10^{-2}	2.617×10^{-2}	3.19	4.668×10^{-2}	4.518×10^{-2}	–3.21

Note: $K_2 = Eqa^2/D_o$.

TABLE 8.4
Comparison of Values of Deflection at Points 1, 2, and 3, for Simply Supported Rectangular Plate with Quadratic Thickness Variation and b/a = 1

		Deflection w		
Point	λ	Equivalent system $w(K) \times 10^{-3}$	Finite differernce mesh 8 × 8 $w(K) \times 10^{-3}$	Error (%)
1	0.1	2.680	2.677	–0.125
2	0.1	3.766	3.762	–0.110
3	0.1	2.680	2.677	–0.125
1	0.2	2.455	2.450	–0.220
2	0.2	3.499	3.492	–0.202
3	0.2	3.455	2.450	–0.220
1	0.3	2.260	2.232	–1.250
2	0.3	3.256	3.206	–1.520
3	0.3	2.260	2.232	–1.250
1	0.4	2.096	2.052	–2.101
2	0.4	3.048	2.979	–2.252
3	0.4	2.096	2.052	–2.101
1	0.5	1.962	1.901	–3.125
2	0.5	2.863	2.770	–3.250
3	0.5	1.962	1.901	–3.125

Note: $K = qa^4/D_o$.

was used to solve the original plate problem in Figure 8.2b and obtain the results in Table 8.4. The finite difference method, however, can also be used to solve the equivalent system. Once the equivalent system is obtained, any known method of analysis for plates can be used for its solution.

8.6 ELASTIC ANALYSIS OF SIMPLY SUPPORTED RECTANGULAR PLATES WITH LINEAR THICKNESS VARIATION

Consider a simply supported rectangular plate that is loaded with a uniformly distributed load q as shown in Figure 8.3a. Its linear thickness variation along the y axis is given by Equation 8.53; therefore,

$$f(y) = \frac{2y}{b} \tag{8.70}$$

The solution is carried out using the first four differential equations from the set of equations given by Equation 8.54. The problem is also solved using the

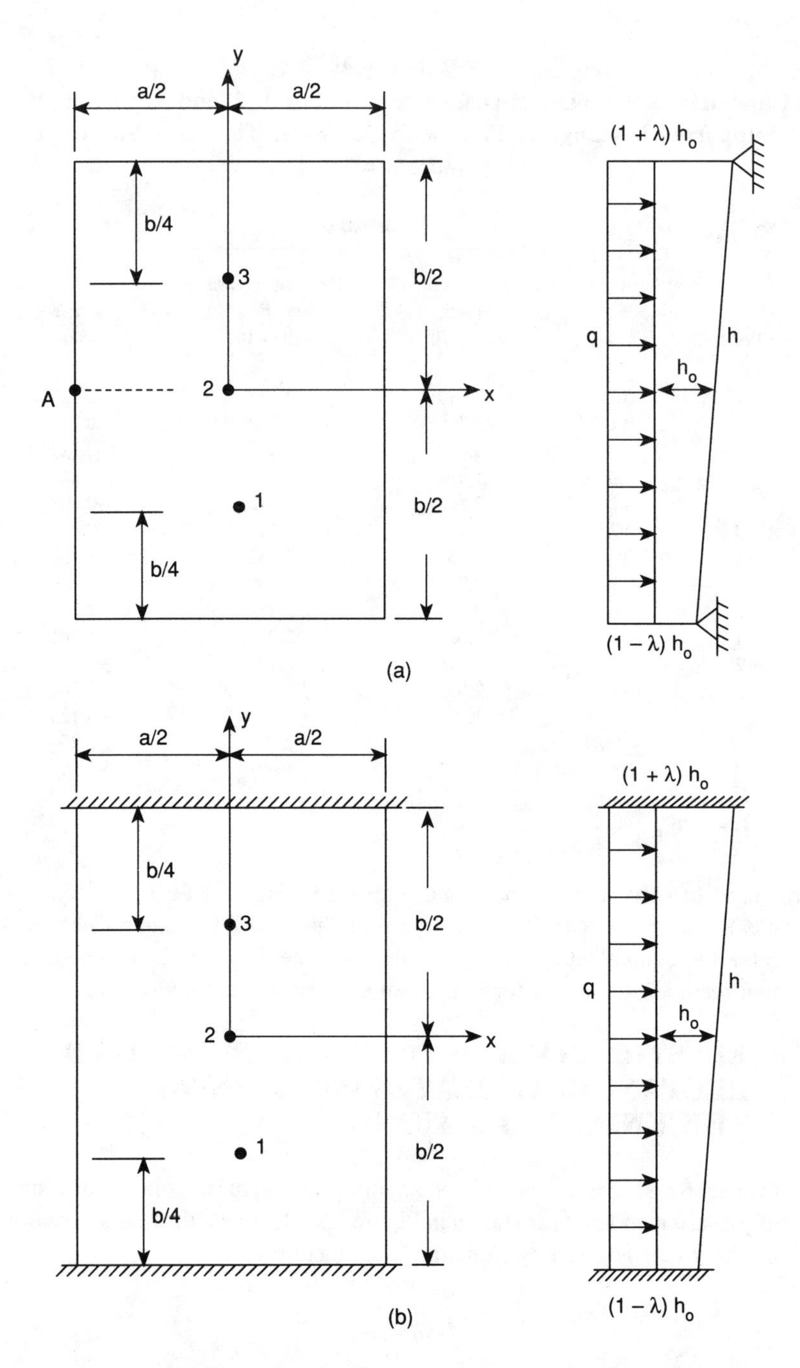

FIGURE 8.3. (a) Simply supported rectangular plate with linear thickness variation along the y axis; (b) rectangular plate with opposite sides a fixed and opposite sides b simply supported.

simplified equivalent system represented by Equation 8.55, and the results are compared. The plate cases examined are for b/a = 1, 1.5, and 2, $\lambda = 0.1, 0.2, 0.258, 0.3$, and 0.4, and Poisson ratio $\nu = 0.3$.

A convenient closed form solution may be obtained using the simplified equivalent system represented by Equation 8.55. By using Equations 8.53 and 8.70, we can write Equation 8.55 as

$$\nabla^4 w = \frac{q}{D_o}\left[1 - \frac{6\lambda y}{b} + \frac{24\lambda^2 y^2}{b^2} - \frac{80\lambda^3 y^3}{b^3}\right] - \frac{12\lambda}{b}\frac{\partial}{\partial y}\nabla^2 w_o \tag{8.71}$$

In the above equation, the deflection w_o may be obtained from the solution of Equation 8.54a, which represents a simply supported rectangular plate of constant rigidity D_o that is loaded with a uniformly distributed load q over the entire area of the plate. This solution can easily be found in the literature.

If the load q is expanded in a Fourier sine series along x, assuming that the origin of the x,y rectangular axes is at point A in Figure 8.3a, we may write Equation 8.54a as

$$\nabla^4 w_o = \frac{4q}{\pi}\sum_{n=1,3,\cdots}^{\infty}\frac{1}{n}\sin\frac{n\pi x}{a} \tag{8.72}$$

The solution of Equation 8.72 is

$$w_o = \sum_{n=1,3,\cdots}^{\infty} Y_n \sin\frac{n\pi x}{a} \tag{8.73}$$

where Y_n is the solution of the ordinary differential equation

$$y_n'''' - 2\alpha_n^2 Y_n'' + \alpha_n^4 Y_n = \frac{4q}{n\pi} \tag{8.74}$$

where

$$\alpha_n = \frac{n\pi}{a} \tag{8.75}$$

From the solution of Equation 8.74, we obtain the solution of Equation 8.72 as

$$w_o = \frac{qa^4}{D_o} \sum_{n=1,3,\cdots}^{\infty} \left[\frac{4}{n^5\pi^5} + A_n \cos h\alpha_n y + B_n \alpha_n y \sinh \alpha_n y \right] \sin \alpha_n x \tag{8.76}$$

By applying the boundary conditions of $w_o = 0$ at $y = \pm$ b/2, we find

$$A_n = \frac{-2(\gamma_n \tanh \gamma_n + 2)}{n^5\pi^5 \cos h\gamma_n} \tag{8.77}$$

$$B_n = \frac{2}{n^5\pi^5 \cos h\gamma_n} \tag{8.78}$$

where

$$\gamma_n = \frac{n\pi b}{2a} \tag{8.79}$$

Therefore,

$$\frac{\partial}{\partial y} \nabla^2 w_o = \frac{4qa}{D_o} \sum_{n=1,3,\cdots}^{\infty} \left[\frac{\sinh \alpha_n y}{n^2\pi^2 \cos h\gamma_n} \right] \sin \alpha_n x \tag{8.80}$$

and

$$\left(\frac{\partial^2 w_o}{\partial y^2} + \nu \frac{\partial^2 w_o}{\partial x^2} \right) = \frac{2qa^2}{D_o \pi^3} \sum_{n=1,3,\cdots}^{\infty} \frac{1}{n^3} \left\{ -2\nu + \frac{[2\nu - (1-\nu)\gamma_n \tanh \gamma_n]}{\cos h\gamma_n} \cosh \alpha_n y + \frac{(1-\nu)}{\cos h\gamma_n} \alpha_n y \sinh \alpha_n y \right\} \sin \alpha_n x \tag{8.81}$$

Equation 8.80 provides the required information for the solution of Equation 8.71, and Equations 8.80 and 8.81 provide the required information to solve the set of differential equations that are included in Equation 8.54.

On this basis, the solution of Equation 8.71 is obtained, and it is given by the equation

$$w = \frac{qa^4}{D_o} \sum_{n=1,3,\cdots}^{\infty} \Big[A_n \cos h\alpha_n y + B_n \alpha_n y \sinh \alpha_n y$$

$$+ C_n \sinh \alpha_n y + D_n \alpha_n y \cos h\alpha_n y$$

$$+ K_o + K_1 y + K_2 y^2 + K_3 y_3^2 \tag{8.82}$$

$$+ K_4 y^2 \sinh \alpha_n y \Big] \sin \alpha_n x$$

where

$$K_o = \frac{4}{n^5 \pi^5} \left[1 + \frac{96\lambda^2 a^2}{n^2 \pi^2 b^2} \right] \tag{8.83}$$

$$K_1 = \frac{-24\lambda}{n^5 \pi^5 b} \left[1 + \frac{120\lambda^2 a^2}{n^2 \pi^2 b^2} \right] \tag{8.84}$$

$$K_2 = \frac{96\lambda^2}{n^5 \pi^5 b^2} \tag{8.85}$$

$$K_3 = -\frac{320\lambda^3}{n^5 \pi^5 b^2} \tag{8.86}$$

$$K_4 = -\frac{6\lambda}{n^4 \pi^4 ab \cos h\gamma_n} \tag{8.87}$$

$$A_n = -\frac{4}{n^5 \pi^5 \cos h\gamma_n} \left[1 + 6\lambda^2 + \frac{96\lambda^2 a^2}{n^2 \pi^2 b^2} \right] - B_n \gamma_n \tanh \gamma_n \tag{8.88}$$

$$B_n = \frac{2}{n^5 \pi^5 \cos h\gamma_n} \left[1 + 6\lambda^2 + \frac{48\lambda^2 a^2}{n^2 \pi^2 b^2} \right] \tag{8.89}$$

$$C_n = \frac{\lambda}{n^5\pi^5 \sinh\gamma_n}\Big[12 + 3\gamma_n \tanh\gamma_n + 40\lambda^2 + \frac{(1440)\lambda^2 a^2}{n^2\pi^2 b^2} - D_n\gamma_n \cot h\gamma_n\Big] \tag{8.90}$$

$$D_n = -\frac{4\lambda}{n^5\pi^5 \sinh\gamma_n}\left[5\lambda^2 - \frac{3a^2}{n^2\pi^2 b^2}\left(\gamma_n \tanh\gamma_n - 20\lambda^2\right)\right] \tag{8.91}$$

The results in Table 8.5 illustrate the results obtained at points 1, 2, and 3 of the plate in Figure 8.3a, for b/a = 1, using the exact equivalent system represented by Equation 8.54 and the simplified equivalent system from the solution of Equation 8.55. The subscript s, such as w_s, is used for the results obtained from the solution of the simplified equivalent system. At the same points, and for the same cases, the results for the bending moments M_x, M_y, M_{x_s}, and M_{y_s} and the bending stresses σ_x, σ_y, σ_{x_s}, and σ_{y_s} are shown in Tables 8.6 and 8.7, respectively. The problem in Figure 8.3a for b/a = 1.0, was also solved by using the finite difference method with 8 × 8 mesh. The results, for various values of λ are shown in Table 8.8. In the same table, the values obtained using the equivalent system represented by Equation 8.54 are also shown. The agreement in results is reasonable for practical applications.

The same problem was solved for b/a = 1.5 and 2.0, and the results are shown in Tables 8.9 through 8.14. In all the cases above, the larger difference in results is associated with points 1 and 3, where the values of deflection, moment, and stress are smaller compared to the values at point 2. It is also important to observe here that if the exact equivalent system is used (Equation 8.54), the solution of the first three differential equations is usually sufficient for an accurate solution of the variable thickness plate problem. It should also be noted here that when $\lambda = 0.4$, the one end of the plate is 2.33 times thicker than the opposite end, and it is three times thicker when $\lambda = 0.5$. These are important observations which allow us to define the largest value of λ, or the limits of λ, for which thin plate theory is still applicable. In practical applications, values of $0 \leq \lambda \leq 0.4$ are usually permissible. For plates with quadratic thickness variation, such as the one in Figure 8.2b, larger values of λ may be used.

8.7 ELASTIC ANALYSIS OF RECTANGULAR PLATES WITH OTHER TYPES OF BOUNDARY CONDITIONS

In this section, the solution of the rectangular plate problem discussed in the preceding section is also carried out by using three additional cases of boundary conditions. In case A, side a and its opposite side are assumed to be fixed, while

TABLE 8.5
Deflections of Simply Supported Rectangular Plate with Linear Thickness and b/a = 1

Point	λ	$w_o(K)$ $\times 10^{-3}$	$\lambda^1 w_1(K)$ $\times 10^{-4}$	$\lambda^2 w_2(K)$ $\times 10^{-5}$	$\lambda^3 w_3(K)$ $\times 10^{-7}$	$w(K)$ $\times 10^{-3}$	$w_s(K)$ $\times 10^{-3}$	Error (%)
1	0.1	2.938	1.87	1.742	−1.23	3.143	3.142	−0.03
2	0.1	4.062	0.0	9.457	0.0	4.072	4.083	0.27
3	0.1	2.938	−1.87	1.742	−1.23	2.769	2.768	−0.04
1	0.2	2.938	3.74	6.969	9.86	3.383	3.379	−0.12
2	0.2	4.062	0.0	3.783	0.0	4.100	4.146	1.12
3	0.2	2.938	−3.74	6.969	−9.86	2.633	2.631	−0.08
1	0.258	2.938	4.83	11.60	21.17	3.539	3.532	−0.20
2	0.258	4.062	0.0	6.295	0.0	4.125	4.201	1.84
3	0.258	2.938	−4.83	11.60	−21.17	2.570	2.567	−0.12
1	0.3	2.938	5.61	15.68	−33.28	3.659	3.649	−0.27
2	0.3	4.062	0.0	8.511	0.0	4.147	4.250	2.48
3	0.3	2.938	−5.61	15.68	−33.28	2.531	2.527	−0.16
1	0.4	2.938	7.48	27.88	78.89	3.973	3.953	−0.50
2	0.4	4.062	0.0	15.13	0.0	4.214	4.340	3.00
3	0.4	2.938	−7.48	27.88	−78.89	2.461	2.457	−0.16

Note: $w = w_0 + \lambda w_1 + \lambda^2 w_2 + \lambda^3 w_3$, $K = Eqa^4/D_o$.

TABLE 8.6
Bending Moments of Simply Supported Rectangular Plate with Linear Thickness and b/a = 1

Point	λ	$M_y(K_1)$ $\times 10^{-2}$	$M_{y_s}(K_1)$ $\times 10^{-2}$	Error (%)	$M_x(K_1)$ $\times 10^{-2}$	$M_{x_s}(K_1)$ $\times 10^{-2}$	Error (%)
1	0.1	4.027	4.005	–0.55	3.391	3.387	–0.12
2	0.1	4.749	4.806	0.36	4.784	4.813	0.61
3	0.1	3.691	3.662	–0.79	3.736	3.729	–0.19
1	0.2	4.087	4.013	-1.81	3.211	3.194	–0.53
2	0.2	4.628	4.859	4.99	4.772	4.884	2.35
3	0.2	3.449	3.313	-3.90	3.924	3.893	–0.79
1	0.258	4.085	3.973	-2.74	3.100	3.075	–0.80
2	0.258	4.522	4.835	6.94	4.760	4.948	3.95
3	0.258	3.298	3.120	-5.37	4.046	3.992	–1.33
1	0.3	4.066	3.925	-3.47	3.016	2.984	–1.06
2	0.3	4.428	4.815	8.74	4.750	5.004	5.34
3	0.3	3.190	2.938	-7.91	4.145	4.067	–1.88
1	0.4	3.962	3.754	–5.25	2.802	2.756	–1.64
2	0.4	4.147	4.573	10.28	4.721	5.171	9.53
3	0.4	2.951	2.631	–10.85	4.422	4.265	-3.55

Note: $K_1 = qa^2$.

TABLE 8.7
Bending Stresses of Simply Supported Rectangular Plate with Linear Thickness and b/a = 1

Point	λ	$\sigma_y(K_2)$ $\times 10^{-2}$	$\sigma_{y_s}(K_2)$ $\times 10^{-2}$	Error (%)	$\sigma_x(K_2)$ $\times 10^{-2}$	$\sigma_{x_s}(K_2)$ $\times 10^{-2}$	Error (%)
1	0.1	2.451	2.438	–0.53	2.065	2.062	–0.15
2	0.1	2.609	2.641	1.23	2.629	2.644	0.57
3	0.1	1.840	1.825	–0.82	1.862	1.859	–0.16
1	0.2	2.772	2.722	–1.80	2.178	2.167	–0.51
2	0.2	2.543	2.670	4.99	2.622	2.684	2.36
3	0.2	1.566	1.505	–3.90	1.782	1.768	–0.79
1	0.258	2.958	2.877	–2.74	2.245	2.227	–0.80
2	0.258	2.484	2.646	6.53	2.616	2.719	3.94
3	0.258	1.422	1.317	–7.38	1.744	1.721	–1.32
1	0.3	3.092	2.985	–3.46	2.293	2.270	–1.00
2	0.3	2.433	2.639	8.49	2.610	2.749	5.33
3	0.3	1.325	1.220	–7.87	1.722	1.690	–1.86
1	0.4	3.401	3.223	–5.23	2.406	2.366	–1.66
2	0.4	2.278	2.513	10.30	2.594	2.841	9.52
3	0.4	1.126	1.004	–10.83	1.687	1.628	–3.50

Note: $K_2 = Eqa^2/D_o$.

TABLE 8.8
Comparison of Values of Deflection at Points 1, 2, and 3 for Simply Supported Plate with Linear Thickness Variation and b/a = 1

		Deflection w		
Point	λ	Equivalent system $w(K) \times 10^{-3}$	Finite difference mesh 8×8 $w(K) \times 10^{-3}$	Error (%)
1	0.1	3.143	3.139	–0.115
2	0.1	4.072	4.068	–0.109
3	0.1	2.769	2.766	–0.125
1	0.2	3.383	3.375	–0.240
2	0.2	4.100	4.091	–0.212
3	0.2	2.633	2.626	–0.260
1	0.3	3.659	3.606	–1.450
2	0.3	4.147	4.084	–1.520
3	0.3	2.531	2.492	–1.550
1	0.4	3.973	3.835	–2.601
2	0.4	4.214	4.098	–2.752
3	0.4	2.461	2.390	–2.901
1	0.5	4.324	4.163	–3.725
2	0.5	4.300	4.134	–3.850
3	0.5	2.423	2.328	–3.925

Note: $K = qa^4/D_o$.

side b and its opposite side are assumed to be simply supported, as shown in Figure 8.3b. Case B assumes that side a and its opposite side are simply supported, while side b and its opposite side are fixed, and case C assumes that all sides of the rectangular plate are fixed. For each case, values of $\lambda = 0.1, 0.2, 0.258, 0.3$, and 0.4 are considered, with b/a = 1.0 and 1.5 and Poisson ratio $\nu = 0.3$.

For case A, the vertical deflections at the quarter points 1, 2, and 3 along the y axis are determined using the exact equivalent system represented by Equation 8.54 and by solving the first four differential equations. The same deflections were also determined using the simplified equivalent system represented by Equation 8.55. The results for b/a = 1.0 are shown in Table 8.15. The subscript s, such as w_s, is used to designate the results obtained using the simplified equivalent system. The bending moments M_x, M_y, M_{x_s}, and M_{y_s} at the same points are shown in Table 8.16, and the bending stresses σ_x, σ_y, σ_{x_s}, and σ_{y_s} are shown in Table 8.17. Reasonable agreement is obtained when the results of the two approaches are compared. The vertical deflections at points 1, 2, and 3 were also determined for b/a = 1, using the finite difference method with 8×8 mesh to solve the original variable thickness plate. The results are shown in Table 8.18, together with the results obtained using the exact equivalent system. Reasonable agreement is obtained when the results from these two

TABLE 8.9
Deflection of Simply Supported Rectangular Plate with Linear Thickness and b/a = 1.5

Point	λ	$w_o(K)$ $\times 10^{-3}$	$\lambda w_1(K)$ $\times 10^{-4}$	$\lambda^2 w_2(K)$ $\times 10^{-5}$	$\lambda^3 w_3(K)$ $\times 10^{-7}$	$w(K)$ $\times 10^{-3}$	$w_s(K)$ $\times 10^{-3}$	Error (%)
1	0.1	5.733	4.85	4.40	4.07	6.262	6.238	−0.38
2	0.1	7.724	0.0	1.78	0.0	7.742	7.745	−0.04
3	0.1	5.733	−4.85	4.40	−4.07	5.292	5.268	−0.45
1	0.2	5.733	9.70	17.59	32.56	6.882	6.781	−1.47
2	0.2	7.724	0.0	3.78	0.0	7.732	7.746	0.18
3	0.2	5.733	−3.74	17.59	−32.56	4.936	4.842	−1.90
1	0.258	5.733	12.51	29.27	69.89	7.284	7.162	−1.67
2	0.258	7.724	0.0	11.87	0.0	7.843	7.866	0.29
3	0.258	5.733	−12.51	29.27	−69.89	4.768	4.671	−2.03
1	0.3	5.733	14.55	39.58	109.9	7.595	7.442	−2.01
2	0.3	7.724	0.0	16.05	0.0	7.885	7.916	0.39
3	0.3	5.733	−14.55	39.58	−109.9	4.663	4.543	−2.58
1	0.4	5.733	19.40	70.36	260.5	8.402	8.152	−2.98
2	0.4	7.724	0.0	28.53	0.0	8.009	8.066	0.71
3	0.4	5.733	−19.40	70.36	−260.5	4.471	4.336	−3.01

Note: $w = w_0 + \lambda w_1 + \lambda^2 w_2 + \lambda^3 w_3,\ K = qa^4/D_o$.

TABLE 8.10
Bending Moments of Simply Supported Rectangular Plate with Linear Thickness and b/a = 1.5

Point	λ	$M_y(K_1)$ $\times 10^{-2}$	$M_{y_s}(K_1)$ $\times 10^{-2}$	Error (%)	$M_x(K_1)$ $\times 10^{-2}$	$M_{x_s}(K_1)$ $\times 10^{-2}$	Error (%)
1	0.1	4.681	4.460	−0.87	5.940	5.912	−0.47
2	0.1	4.932	4.991	1.19	8.116	8.137	0.26
3	0.1	4.174	4.119	−1.32	6.462	6.425	−0.57
1	0.2	4.802	4.663	−2.89	5.662	5.568	−1.67
2	0.2	4.774	5.010	4.94	8.115	8.200	1.05
3	0.2	3.832	3.579	−6.60	6.755	6.584	−2.53
1	0.258	4.825	4.616	−4.33	5.487	5.346	−2.57
2	0.258	4.635	5.027	8.46	8.114	8.256	1.75
3	0.258	3.625	3.309	−8.72	6.954	6.646	−4.43
1	0.3	4.820	4.557	−5.46	5.353	5.176	−3.31
2	0.3	4.512	4.951	9.75	8.114	8.306	2.37
3	0.3	3.477	3.097	−10.94	7.117	6.677	−6.18
1	0.4	4.732	4.343	−8.22	5.006	4.743	−5.25
2	0.4	4.144	4.631	11.76	8.112	8.454	4.22
3	0.4	3.157	2.750	−12.91	7.593	6.706	−11.68

Note: $K_1 = qa^2$.

TABLE 8.11
Bending Stresses of Simply Supported Rectangular Plate with Linear Thickness and b/a = 1.5

Point	λ	$\sigma_y(K_2)$ $\times 10^{-2}$	$\sigma_{y_s}(K_2)$ $\times 10^{-2}$	Error (%)	$\sigma_x(K_2)$ $\times 10^{-2}$	$\sigma_{x_s}(K_2)$ $\times 10^{-2}$	Error (%)
1	0.1	2.850	2.825	–0.88	3.616	3.559	–0.47
2	0.1	2.710	2.742	1.18	4.459	4.471	0.27
3	0.1	2.080	2.053	–1.30	3.220	3.202	–0.56
1	0.2	3.257	3.163	–2.89	3.841	3.777	–1.67
2	0.2	2.623	2.753	4.96	4.459	4.506	1.05
3	0.2	1.740	1.625	–6.61	3.068	2.990	–2.54
1	0.258	3.495	3.343	–4.35	3.974	3.872	–2.57
2	0.258	2.547	2.762	8.46	4.458	4.536	1.75
3	0.258	1.563	1.427	–8.72	2.998	2.865	–4.44
1	0.3	3.665	3.465	–5.46	4.071	3.936	–3.32
2	0.3	2.479	2.720	9.74	4.458	4.564	2.38
3	0.3	1.445	1.287	–10.94	2.957	2.774	6.19
1	0.4	4.063	3.728	–8.25	4.298	4.072	–5.26
2	0.4	2.277	2.545	11.75	4.457	4.645	4.22
3	0.4	1.205	1.050	–12.91	2.897	2.559	–11.67

Note: $K_2 = Eqa^2/D_o$.

TABLE 8.12
Deflections of Simply Supported Rectangular Plate with Linear Thickness and b/a = 2

Point	λ	$w_o(K)$ $\times 10^{-3}$	$\lambda^1 w_1(K)$ $\times 10^{-4}$	$\lambda^2 w_2(K)$ $\times 10^{-5}$	$\lambda^3 w_3(K)$ $\times 10^{-7}$	$w(K)$ $\times 10^{-3}$	$w_s(K)$ $\times 10^{-3}$	Error (%)
1	0.1	7.803	8.03	7.564	7.84	8.683	8.623	–0.69
2	0.1	10.130	0.0	2.497	0.0	10.150	10.140	–0.10
3	0.1	7.803	–8.03	7.564	–7.84	7.075	7.018	–0.81
1	0.2	7.803	16.06	30.250	62.70	9.718	9.575	–1.47
2	0.2	10.130	0.0	9.988	0.0	10.230	10.190	–0.39
3	0.2	7.803	–16.06	30.250	–62.70	6.494	6.395	–1.51
1	0.258	7.803	20.71	50.350	134.6	10.390	10.200	–1.89
2	0.258	10.130	0.0	16.640	0.0	10.290	10.230	–0.58
3	0.258	7.803	–20.71	50.350	–134.6	6.222	6.095	–2.04
1	0.3	7.803	24.08	68.070	–211.6	10.910	10.590	–2.95
2	0.3	10.130	0.0	22.470	0.0	10.350	10.260	–0.87
3	0.3	7.803	–24.08	68.070	–211.6	6.054	5.852	–3.34
1	0.4	7.803	32.11	121.000	501.6	12.270	11.780	–3.99
2	0.4	10.130	0.0	39.950	0.0	10.530	10.370	–1.52
3	0.4	7.803	–32.11	121.000	–501.6	5.752	5.499	–4.39

Note: $w = w_0 + \lambda w_1 + \lambda^2 w_2 + \lambda^3 w_3$, $K = qa^4/D_o$.

TABLE 8.13
Bending Moments of Simply Supported Rectangular Plate with Linear Thickness and b/a = 2

Point	λ	$M_y(K_1)$ ×10^{-2}	$M_{y_s}(K_1)$ ×10^{-2}	Error (%)	$M_x(K_1)$ ×10^{-2}	$M_{x_s}(K_1)$ ×10^{-2}	Error (%)
1	0.1	4.772	4.721	−1.07	7.810	7.754	−0.72
2	0.1	4.583	4.636	1.16	10.170	10.180	0.16
3	0.1	4.274	4.206	−1.59	8.330	8.254	−0.91
1	0.2	4.897	4.725	−3.51	7.523	7.332	−2.54
2	0.2	4.427	4.639	4.79	10.190	10.220	0.30
3	0.2	3.952	3.638	−7.95	8.641	8.292	−4.04
1	0.258	4.923	4.663	−5.28	7.335	7.046	−3.94
2	0.258	4.289	4.642	8.23	10.210	10.260	0.49
3	0.258	3.764	3.399	−9.62	8.869	8.240	−7.09
1	0.3	4.920	4.593	−6.65	7.186	6.823	−5.05
2	0.3	4.167	4.644	11.45	10.230	10.290	0.59
3	0.3	3.634	3.126	−13.97	9.063	8.165	−9.91
1	0.4	4.838	4.354	−10.00	6.784	6.247	−7.92
2	0.4	3.804	4.271	12.27	10.270	10.270	1.17
3	0.4	3.375	2.836	−16.97	9.669	6.851	−12.77

Note: $K_1 = qa^2$.

TABLE 8.14
Bending Stresses of Simply Supported Rectangular Plate with Linear Thickness and b/a = 2

Point	λ	$\sigma_y(K_2)$ $\times 10^{-2}$	$\sigma_{y_s}(K_2)$ $\times 10^{-2}$	Error (%)	$\sigma_x(K_2)$ $\times 10^{-2}$	$\sigma_{x_s}(K_2)$ $\times 10^{-2}$	Error (%)
1	0.1	2.905	2.874	−1.07	4.755	4.721	−0.72
2	0.1	2.518	2.547	1.15	5.590	5.594	0.07
3	0.1	2.130	2.095	−1.64	4.151	4.113	−0.92
1	0.2	3.322	3.205	−3.52	5.103	4.974	−2.53
2	0.2	2.433	2.549	4.77	5.601	5.617	0.29
3	0.2	1.795	1.652	−7.97	3.924	3.765	−4.05
1	0.258	3.566	3.377	−5.30	5.312	5.103	−3.93
2	0.258	2.357	2.550	8.19	5.611	5.637	0.46
3	0.258	1.623	1.467	−9.62	3.823	3.552	−7.09
1	0.3	3.742	3.493	−6.65	5.464	5.189	−5.03
2	0.3	2.290	2.552	11.44	5.619	5.655	0.64
3	0.3	1.510	1.300	−13.91	3.765	3.392	−9.91
1	0.4	4.153	3.738	−9.99	5.824	5.363	−7.92
2	0.4	2.090	2.347	12.30	5.644	5.708	1.13
3	0.4	1.288	1.069	−16.97	3.689	3.218	−12.76

Note: $K_2 = Eqa^2/D_{o}$

TABLE 8.15
Deflections of Rectangular Plate with Opposite Sides a Fixed and Opposite Sides b Simply Supported, Linear Thickness Variation and b/a = 1

Point	λ	$w_o(K)$ $\times 10^{-3}$	$\lambda w_1(K)$ $\times 10^{-4}$	$\lambda^2 w_2(K)$ $\times 10^{-5}$	$\lambda^3 w_3(K)$ $\times 10^{-7}$	$w(K)$ $\times 10^{-3}$	$w_s(K)$ $\times 10^{-3}$	Error (%)
1	0.1	1.117	1.044	1.110	0.445	1.232	1.229	−0.24
2	0.1	1.917	0.0	0.682	0.0	1.924	1.928	0.21
3	0.1	1.117	−1.044	1.110	−0.445	1.023	1.020	−0.29
1	0.2	1.117	2.088	4.442	3.562	1.370	1.356	−1.02
2	0.2	1.917	0.0	2.728	0.0	1.944	1.962	0.93
3	0.2	1.117	−2.088	4.442	−3.562	0.982	0.939	−1.37
1	0.258	1.117	2.694	7.389	7.646	1.461	1.438	−1.57
2	0.258	1.917	0.0	4.540	0.0	1.963	1.992	1.48
3	0.258	1.117	−2.694	7.389	−7.646	0.920	0.910	−1.10
1	0.3	1.117	3.132	9.993	12.02	1.531	1.507	−1.60
2	0.3	1.917	0.0	6.138	0.0	1.979	2.008	1.51
3	0.3	1.117	−3.132	9.993	−12.02	0.902	0.887	−1.70
1	0.4	1.117	4.176	17.77	1.715	1.715	1.680	−2.02
2	0.4	1.917	0.0	10.91	0.0	2.026	2.059	1.65
3	0.4	1.117	−4.176	17.77	−1.715	0.874	0.856	−2.10

Note: $w = w_0 + \lambda w_1 + \lambda^2 w_2 + \lambda^3 w_3$, $K = qa^4/D_o$.

TABLE 8.16
Bending Moments of Rectangular Plate with Opposite Sides a Fixed and Opposite Sides b Simply Supposted, Linear Thickness Variation and b/a = 1

Point	λ	$M_y(K_1)$ $\times 10^{-2}$	$M_{y_s}(K_1)$ $\times 10^{-2}$	Error (%)	$M_x(K_1)$ $\times 10^{-2}$	$M_{x_s}(K_1)$ $\times 10^{-2}$	Error (%)
1	0.1	1.587	1.580	–0.44	1.228	1.225	–0.58
2	0.1	3.297	3.334	1.12	2.436	2.442	0.70
3	0.1	0.893	0.882	–1.28	1.124	1.111	–1.16
1	0.2	1.847	1.830	–1.00	1.260	1.254	–0.51
2	0.2	3.215	3.295	2.51	2.428	2.494	2.72
3	0.2	0.499	0.488	–2.38	1.072	1.055	–1.60
1	0.258	1.964	1.949	–1.22	1.269	1.222	–3.70
2	0.258	3.142	3.161	5.38	2.424	2.530	4.55
3	0.258	0.272	0.581	–5.08	1.049	0.997	–4.91
1	0.3	2.034	1.954	–3.93	1.270	1.211	–4.65
2	0.3	3.077	3.274	6.40	2.414	2.563	6.17
3	0.3	0.113	0.105	–6.50	1.040	0.964	–7.31
1	0.4	2.142	1.976	–7.77	1.256	0.168	–7.00
2	0.4	2.885	3.133	8.59	2.394	2.559	6.90
3	0.4	–0.228	–0.209	–8.01	1.046	0.962	–8.06

Note: $K_1 = qa^2$.

TABLE 8.17
Bending Stresses of Rectangular Plate with Opposite Sides a Fixed and Opposite Sides b Simply Supported, Linear Thickness Variation and b/a = 1

Point	λ	$\sigma_y(K_2)$ $\times 10^{-2}$	$\sigma_{y_s}(K_2)$ $\times 10^{-2}$	Error (%)	$\sigma_x(K_2)$ $\times 10^{-2}$	$\sigma_{x_s}(K_2)$ $\times 10^{-2}$	Error (%)
1	0.1	0.966	0.962	–0.45	0.748	0.744	–0.58
2	0.1	1.812	1.832	1.10	1.338	1.347	0.67
3	0.1	0.445	0.440	–1.28	0.560	0.554	–1.16
1	0.2	1.253	1.240	–1.04	0.855	0.849	–0.71
2	0.2	1.766	1.855	5.04	1.334	1.370	2.69
3	0.2	0.227	0.222	–2.25	0.487	0.479	–1.60
1	0.258	1.423	1.406	–1.19	0.919	0.885	–3.70
2	0.258	1.726	1.819	5.40	1.330	1.391	4.60
3	0.258	0.117	0.111	–5.11	0.453	0.430	–4.91
1	0.3	1.546	1.485	–3.95	0.966	0.921	–4.65
2	0.3	1.691	1.583	6.40	1.326	1.408	6.18
3	0.3	0.047	0.044	–6.50	0.432	0.400	–7.31
1	0.4	1.839	1.696	–7.78	1.078	1.003	–6.96
2	0.4	1.585	1.721	8.60	1.316	1.407	6.91
3	0.4	–0.087	–0.080	–8.01	3.992	3.670	–8.07

Note: $K_2 = Eqa^2/D_o$.

TABLE 8.18
Comparison of Values of Deflection at Points 1, 2, and 3 for Rectangular Plate with Opposite Sides a Fixed And Opposite Sides b Simply Supported, Linear Thickness Variation and b/a = 1

		Deflection w		
Point	λ	Equivalent system $w(K) \times 10^{-3}$	Finite difference mesh 8×8 $w(K) \times 10^{-3}$	Error (%)
1	0.1	1.232	1.230	–0.111
2	0.1	1.924	1.921	–0.149
3	0.1	1.023	1.021	–0.165
1	0.2	1.370	1.366	–0.324
2	0.2	1.944	1.935	–0.456
3	0.2	0.982	0.977	–0.560
1	0.3	1.531	1.508	–1.150
2	0.3	1.979	1.955	–1.220
3	0.3	0.902	0.884	–1.990
1	0.4	1.715	1.669	–2.709
2	0.4	2.026	1.968	–2.865
3	0.4	0.874	0.845	–3.123
1	0.5	1.922	1.858	–3.333
2	0.5	2.089	2.017	–3.450
3	0.5	0.866	0.835	–3.525

Note: $K = qa^4/D_o$.

methods are compared. For b/a = 1.5, the analogous values of the same quantities are shown in Tables 8.19, 8.20, and 8.21.

For case B with b/a = 1.0, the analogous results are shown in Tables 8.22 through 8.25 and for case C and b/a = 1.0, they are shown in Tables 8.26 through 8.29. Similar observations can be made here as was done for the other cases.

Other cases of rectangular plate problems can be worked out in a similar manner because the method is general.

8.8 INELASTIC ANALYSIS OF THIN PLATES OF UNIFORM AND VARIABLE THICKNESS

When the material of the plate is stressed beyond its elastic limit, the modulus E in Equation 8.7 or Equation 8.10 will become a variable quantity, and its variation along the x and y directions of the plate must be known in order to be able to obtain a solution. The metodology developed in Chapters 6 and 7 regarding the computation of a reduced (or equivalent) modulus for beams, will be extended here to apply for the inelastic analysis of rectangular and circular plates of uniform and variable thickness.

TABLE 8.19
Deflection of Rectangular Plate with Opposite Sides a Fixed and Opposite Sides b Simply Supported, Linear Thickness Variation and b/a = 1.5

Point	λ	$w_o(K)$ $\times 10^{-3}$	$\lambda w_1(K)$ $\times 10^{-4}$	$\lambda^2 w_2(K)$ $\times 10^{-5}$	$\lambda^3 w_3(K)$ $\times 10^{-7}$	$w(K)$ $\times 10^{-3}$	$w_s(K)$ $\times 10^{-3}$	Error (%)
1	0.1	3.267	3.290	3.372	1.828	3.630	3.612	−0.50
2	0.1	5.326	0.0	1.669	0.0	5.343	5.346	0.05
3	0.1	3.267	−3.290	3.372	−1.829	2.972	2.954	−0.61
1	0.2	3.267	6.580	1.349	14.62	4.061	4.024	−0.91
2	0.2	5.326	0.0	6.675	0.0	5.393	5.405	0.22
3	0.2	3.267	−6.580	1.349	−14.62	2.742	2.714	−0.99
1	0.258	3.267	8.488	22.45	31.39	4.343	4.291	−1.19
2	0.258	5.326	0.0	11.11	0.0	5.438	5.457	0.35
3	0.258	3.267	−8.488	22.45	−31.39	2.640	2.605	−1.30
1	0.3	3.267	9.870	30.35	49.35	4.562	4.489	−1.61
2	0.3	5.326	0.0	15.02	0.0	5.476	5.503	0.49
3	0.3	3.267	−9.870	30.35	−49.35	2.479	2.434	−1.81
1	0.4	3.267	1.316	53.96	117.0	5.134	5.006	−2.56
2	0.4	5.326	0.0	26.70	0.0	5.593	5.641	0.86
3	0.4	3.267	−1.316	53.96	−117.0	2.479	2.409	−2.90

Note: $w = w_0 + \lambda w_1 + \lambda^2 w_2 + \lambda^3 w_3$, $K = qa^4/D_o$.

TABLE 8.20
Bending Moments of Rectangular Plate with Opposite Sides a Fixed and Opposite Sides b Simply Supported, Linear Thickness Variation and b/a = 1.5

Point	λ	$M_y(K_1)$ $\times 10^{-2}$	$M_{ys}(K_1)$ $\times 10^{-2}$	Error (%)	$M_x(K_1)$ $\times 10^{-2}$	$M_{xs}(K_1)$ $\times 10^{-2}$	Error (%)
1	0.1	2.916	2.872	–1.51	3.403	3.379	–0.71
2	0.1	4.554	4.603	1.08	5.850	5.868	0.31
3	0.1	2.089	2.030	2.82	3.443	3.411	–0.93
1	0.2	3.217	3.067	–4.66	3.361	3.279	–2.44
2	0.2	4.432	4.629	4.44	5.857	5.929	1.23
3	0.2	1.624	1.565	–3.63	3.488	3.339	–4.27
1	0.258	3.347	3.120	–6.78	3.321	3.198	–3.70
2	0.258	4.324	4.389	5.64	5.863	5.982	2.03
3	0.258	1.363	1.279	–6.15	1.525	1.445	–5.25
1	0.3	3.419	3.134	–8.34	3.283	3.129	–4.69
2	0.3	4.229	4.513	6.72	5.869	6.030	2.74
3	0.3	1.184	1.084	–8.44	3.593	3.349	–6.78
1	0.4	3.515	3.092	–9.25	3.163	2.934	–7.24
2	0.4	3.944	4.242	7.56	5.885	6.171	–4.86
3	0.4	0.828	0.749	–9.55	3.801	3.473	–8.62

Note: $K_1 = qa^2$.

TABLE 8.21
Bending Stresses of Rectangular Plate with Opposite Sides a Fixed and Opposite Sides b Simply Supported, Linear Thickness Variation and b/a = 1.5

Point	λ	$\sigma_y(K_2)$ $\times 10^{-2}$	$\sigma_{y_s}(K_2)$ $\times 10^{-2}$	Error (%)	$\sigma_x(K_2)$ $\times 10^{-2}$	$\sigma_{x_s}(K_2)$ $\times 10^{-2}$	Error (%)
1	0.1	1.775	1.749	−1.46	2.072	2.057	−0.72
2	0.1	2.502	2.529	1.08	3.214	3.224	0.31
3	0.1	1.041	1.011	−2.88	1.715	1.700	−0.93
1	0.2	2.183	2.080	−4.72	2.280	2.224	−2.46
2	0.2	2.435	2.543	4.44	3.218	3.257	1.21
3	0.2	7.376	7.110	−3.61	1.584	1.516	−4.30
1	0.258	2.424	2.260	−6.77	2.405	2.316	−3.70
2	0.258	2.376	2.510	5.66	3.222	3.287	2.02
3	0.258	5.873	5.512	−6.15	1.525	1.410	−5.25
1	0.3	2.601	2.383	−8.38	2.497	2.379	−4.73
2	0.3	2.323	2.480	6.77	3.225	3.313	2.73
3	0.3	0.492	0.451	−8.43	1.493	1.334	−6.79
1	0.4	3.018	2.737	−9.30	2.716	2.519	−7.25
2	0.4	2.167	2.332	7.62	3.223	3.391	4.89
3	0.4	0.316	2.858	−9.55	1.451	1.326	−8.63

Note: $K_2 = Eqa^2/D_o$.

TABLE 8.22
Deflections of Rectangular Plate with Opposite Sides a Simply Supported and Opposite Sides b Fixed, Linear Thickness Variation and b/a = 1

Point	λ	$w_o(K)$ $\times 10^{-3}$	$\lambda w_1(K)$ $\times 10^{-4}$	$\lambda^2 w_2(K)$ $\times 10^{-5}$	$\lambda^3 w_3(K)$ $\times 10^{-7}$	$w(K)$ $\times 10^{-3}$	$w_s(K)$ $\times 10^{-3}$	Error (%)
1	0.1	1.418	1.138	0.9477	1.026	1.541	1.536	−0.32
2	0.1	1.917	0.0	0.2743	0.0	1.920	1.921	0.05
3	0.1	1.418	−1.138	0.9477	−1.026	1.314	1.308	−0.46
1	0.2	1.418	2.275	3.791	8.210	1.684	1.661	−1.37
2	0.2	1.917	0.0	1.097	0.0	1.928	1.934	0.31
3	0.2	1.418	−2.275	3.791	−8.210	1.228	1.206	−1.79
1	0.258	1.418	2.935	6.308	17.62	1.776	1.738	−2.14
2	0.258	1.917	0.0	1.826	0.0	1.936	1.935	0.52
3	0.258	1.418	−2.935	6.308	−17.62	1.185	1.151	2.87
1	0.3	1.418	3.413	8.529	27.71	1.847	1.795	−2.82
2	0.3	1.917	0.0	2.468	0.0	1.942	1.955	0.67
3	0.3	1.418	−3.413	8.529	−27.71	1.159	1.125	−2.92
1	0.4	1.418	4.551	15.160	65.68	2.031	1.967	−3.11
2	0.4	1.917	0.0	4.388	0.0	1.961	1.985	1.22
3	0.4	1.418	−4.551	15.160	−65.68	1.108	1.072	−3.21

Note: $w = w_0 + \lambda w_1 + \lambda^2 w_2 + \lambda^3 w_3,\ K = qa^4/D_o$.

TABLE 8.23
Bending Moments of Rectangular Plate with Opposite Sides a Simply Supported and Opposite Sides b Fixed, Linear Thickness Variation and b/a = 1

Point	λ	$M_y(K_1)$ $\times 10^{-2}$	$M_{y_s}(K_1)$ $\times 10^{-2}$	Error (%)	$M_x(K_1)$ $\times 10^{-2}$	$M_{x_s}(K_1)$ $\times 10^{-2}$	Error (%)
1	0.1	2.325	2.301	−1.03	2.433	2.420	−0.53
2	0.1	2.414	2.441	1.12	3.324	3.332	0.25
3	0.1	2.051	2.019	−1.56	2.640	2.622	−0.68
1	0.2	2.399	2.317	−3.42	2.324	2.281	−1.85
2	0.2	2.342	2.447	4.48	3.321	3.354	0.99
3	0.2	1.878	1.829	−2.61	2.759	2.679	−2.90
1	0.258	2.419	2.295	−5.13	2.255	2.189	−2.93
2	0.258	2.277	2.396	5.23	3.319	3.374	1.66
3	0.258	1.777	1.710	−3.75	2.841	2.698	−5.06
1	0.3	2.422	2.266	−6.44	2.202	2.119	−3.77
2	0.3	2.220	2.382	7.29	3.317	3.391	2.23
3	0.3	1.707	1.609	−5.75	2.909	2.763	−7.08
1	0.4	2.390	2.159	−9.67	2.064	1.899	−7.95
2	0.4	2.051	2.220	8.25	3.311	3.443	3.99
3	0.4	1.567	1.461	−6.75	3.110	2.838	−8.75

Note: $K_1 = qa^2$.

TABLE 8.24
Bending Stresses of Rectagular Plate with Opposite Sides a Simply Supported and Opposite Sides b Fixed, Linear Thickness Variation and b/a = 1

Point	λ	$\sigma_y(K_2)$ $\times 10^{-2}$	$\sigma_{y_s}(K_2)$ $\times 10^{-2}$	Error (%)	$\sigma_x(K_2)$ $\times 10^{-2}$	$\sigma_{x_s}(K_2)$ $\times 10^{-2}$	Error (%)
1	0.1	1.416	1.401	−1.06	1.418	1.474	−0.47
2	0.1	1.327	1.341	1.06	1.826	1.831	0.27
3	0.1	1.022	1.006	−1.57	1.316	1.307	−0.68
1	0.2	1.628	1.572	−3.44	1.577	1.547	−1.90
2	0.2	1.287	1.345	4.51	1.825	1.843	0.99
3	0.2	0.853	0.831	−2.61	1.253	1.217	−2.87
1	0.258	1.752	1.662	−5.14	1.633	1.586	−2.88
2	0.258	1.251	1.316	5.23	1.824	1.854	1.64
3	0.258	0.766	0.737	−3.75	1.225	1.163	−5.06
1	0.3	1.842	1.723	−6.46	1.675	1.612	−3.76
2	0.3	1.220	1.309	7.29	1.823	1.863	2.19
3	0.3	0.709	0.669	−5.75	1.209	1.123	−7.08
1	0.4	2.052	1.854	−9.65	1.772	1.631	−7.97
2	0.4	1.127	1.220	8.25	1.819	1.892	4.01
3	0.4	0.598	0.558	−6.75	1.187	1.083	−8.77

Note: $K_2 = Eqa^2/D_o$.

TABLE 8.25
Comparison of Values of Deflection at Point 1, 2, and 3 for Rectangular Plate with Opposite Sides a Simply Supported and Opposite Sides b Fixed, Linear Thickness Variation and b/a = 1

		Deflection w		
Point	λ	Equivalent system $w(K) \times 10^{-3}$	Finite difference mesh 8×8 $w(K) \times 10^{-3}$	Error (%)
1	0.1	1.541	1.539	–0.123
2	0.1	1.920	1.917	–0.158
3	0.1	1.314	1.312	–0.155
1	0.2	1.684	1.678	–0.355
2	0.2	1.928	1.917	–0.556
3	0.2	1.228	1.219	–0.660
1	0.3	1.847	1.819	–1.540
2	0.3	1.942	1.903	–2.001
3	0.3	1.159	1.132	–2.350
1	0.4	2.031	1.973	–2.876
2	0.4	1.961	1.902	–2.990
3	0.4	1.108	1.072	–3.234
1	0.5	2.237	2.160	–3.433
2	0.5	1.986	1.916	–3.532
3	0.5	1.073	1.034	–3.643

Note: $K = qa^4/D_o$.

8.8.1 RECTANGULAR PLATES

In order to utilize this method for plates, the stress/strain relationship of the material must be known. The stress/strain diagrams for stresses σ_x and σ_y are shown in Figures 8.4b and 8.4a, respectively. Although not required, it is assumed that the compressive yield strength is equal in magnitude to the tensile yield strength, i.e., symmetry. This is not mandarory for the application of the method, but it follows typical engineering practices for steels. The curve AOB, as discussed in Chapter 6, represents the bending stress distribution along the depth h of the plate if h is substituted for Δ.

Let it be assumed that the curvature of the neutral surface produced by the bending moment M_y is r_y. At a distance z from the neutral surface, the unit elongation ε of a fiber can be expressed as

$$\varepsilon = \frac{z}{r_y} \tag{8.92}$$

TABLE 8.26
Deflection of Rectangular Plate with All Sides Fixed, Linear Thickness Variation and b/a = 1

Point	λ	$w_o(K)$ $\times 10^{-3}$	$\lambda^1 w_1(K)$ $\times 10^{-4}$	$\lambda^2 w_2(K)$ $\times 10^{-5}$	$w(K)$ $\times 10^{-3}$	$w_s(K)$ $\times 10^{-3}$	Error (%)
1	0.1	0.758	1.159	0.4823	0.879	0.878	–0.11
2	0.1	1.265	0.0	0.2479	1.268	1.270	0.16
3	0.1	0.758	–1.159	0.4823	0.647	0.646	–0.17
1	0.2	0.758	2.318	1.929	1.009	1.005	–0.39
2	0.2	1.265	0.0	0.920	1.275	1.286	0.86
3	0.2	0.758	–2.318	1.929	0.546	0.542	–0.73
1	0.258	0.758	2.990	3.210	1.089	1.083	–0.55
2	0.258	1.265	0.0	1.650	1.282	1.299	1.33
3	0.258	0.758	–2.990	3.210	0.491	0.485	–1.36
1	0.3	0.758	3.477	4.340	1.149	1.140	–0.78
2	0.3	1.265	0.0	2.231	1.288	1.311	1.79
3	0.3	0.758	–3.477	4.340	0.454	0.445	–2.03
1	0.4	0.758	4.636	7.716	1.299	1.283	–1.23
2	0.4	1.265	0.0	3.966	1.305	1.338	2.52
3	0.4	0.758	–4.636	7.716	0.372	0.360	–3.10

Note: $w = w_0 + \lambda w_1 + \lambda^2 w_2$, $K = qa4/D_o$.

TABLE 8.27
Bending Moments of Rectangular Plate with All Sides Fixed, Linear Thickness Variation and b/a = 1

Point	λ	$M_y(K_1)$ $\times 10^{-2}$	$M_{y_s}(K_1)$ $\times 10^{-2}$	Error (%)	$M_x(K_1)$ $\times 10^{-2}$	$M_{x_s}(K_1)$ $\times 10^{-2}$	Error (%)
1	0.1	1.405	1.402	–0.21	1.371	1.372	0.07
2	0.1	2.283	2.316	1.45	2.289	2.307	0.79
3	0.1	0.724	0.720	–0.62	1.107	1.108	0.09
1	0.2	1.649	1.638	–0.67	1.433	1.434	0.07
2	0.2	2.260	2.307	2.09	2.280	2.325	2.01
3	0.2	0.341	0.377	–2.01	9.052	9.069	0.19
1	0.258	1.758	1.741	–0.97	1.449	1.450	0.07
2	0.258	2.240	2.297	2.56	2.271	2.332	2.65
3	0.258	0.066	0.063	–3.65	0.768	0.771	0.42
1	0.3	1.820	1.799	–1.15	1.451	1.453	0.14
2	0.3	2.222	2.294	3.25	2.263	2.421	6.98
3	0.3	–0.115	–0.109	–4.75	0.659	0.664	0.68
1	0.4	1.917	1.885	–1.67	1.428	1.430	0.14
2	0.4	2.168	2.241	4.75	2.240	2.421	8.08
3	0.4	–0.534	–0.487	–8.75	0.375	0.384	2.43

Note: $K_1 = qa^2$.

TABLE 8.28
Bending Stresses of Rectangular Plate with All Sides Fixed, Linear Thickness Variation and b/a = 1

Point	λ	$\sigma_y(K_2)$ $\times 10^{-2}$	$\sigma_{y_s}(K_2)$ $\times 10^{-2}$	Error (%)	$\sigma_x(K_2)$ $\times 10^{-2}$	$\sigma_{x_s}(K_2)$ $\times 10^{-2}$	Error (%)
1	0.1	0.855	0.853	–0.23	0.835	0.835	0.02
2	0.1	1.255	1.273	1.43	1.258	1.268	0.79
3	0.1	0.361	0.359	–0.64	0.552	0.552	0.04
1	0.2	1.119	1.112	–0.71	0.972	0.973	0.06
2	0.2	1.242	1.267	2.05	1.253	1.291	3.07
3	0.2	0.143	0.133	–6.52	0.411	0.412	0.19
1	0.258	1.273	1.261	–0.94	1.049	1.050	0.10
2	0.258	1.231	1.264	2.66	1.248	1.281	2.66
3	0.258	2.841	2.737	–3.66	0.331	0.332	0.42
1	0.3	1.384	1.368	–1.16	1.103	1.105	0.18
2	0.3	1.221	1.261	3.30	1.243	1.282	3.10
3	0.3	–0.048	–0.043	–4.72	0.274	0.276	0.69
1	0.4	1.646	1.618	–1.70	1.226	1.228	0.16
2	0.4	1.191	1.248	4.80	1.231	1.330	8.08
3	0.4	–0.104	–0.186	–8.75	–0.143	–0.147	2.45

Note: $K_2 = Eqa^2/D_o$.

TABLE 8.29
Comparison of Deflection at Points 1, 2, and 3 for Rectangular Plate with All Sides Fixed, Linear Thickness Variation and b/a = 1

		Deflection w		
Point	**λ**	**Equivalent system w(K) × 10^{-3}**	**Finite difference mesh 8 × 8 w(K) × 10^{-3}**	**Error (%)**
1	0.1	0.879	0.878	–0.113
2	0.1	1.268	1.266	–0.148
3	0.1	0.647	0.646	–0.145
1	0.2	1.009	1.005	–0.412
2	0.2	1.275	1.268	–0.552
3	0.2	0.546	0.542	–0.682
1	0.3	1.149	1.127	–1.880
2	0.3	1.288	1.258	–2.301
3	0.3	0.454	0.442	–2.550
1	0.4	1.299	1.262	–2.877
2	0.4	1.305	1.265	–3.102
3	0.4	0.372	0.358	–3.654
1	0.5	1.458	1.410	–3.323
2	0.5	1.327	1.275	–3.892
3	0.5	0.299	0.286	–4.201

Note: $K = qa^4/D_o$.

Thus,

$$z = \varepsilon r_y \tag{8.93}$$

$$dz = r_y d\varepsilon \tag{8.94}$$

If h_1 and h_2 are the distances from the neutral surface to the lower and upper surfaces of the plate, respectively, we have

$$\varepsilon_1 = \frac{h_1}{r_y} \tag{8.95}$$

$$\varepsilon_2 = \frac{h_2}{r_y} \tag{8.96}$$

and consequently

$$\Delta_y = \varepsilon_1 + \varepsilon_2 = \frac{h_1}{r_y} + \frac{h_2}{r_y} = \frac{h}{r_y} \tag{8.97}$$

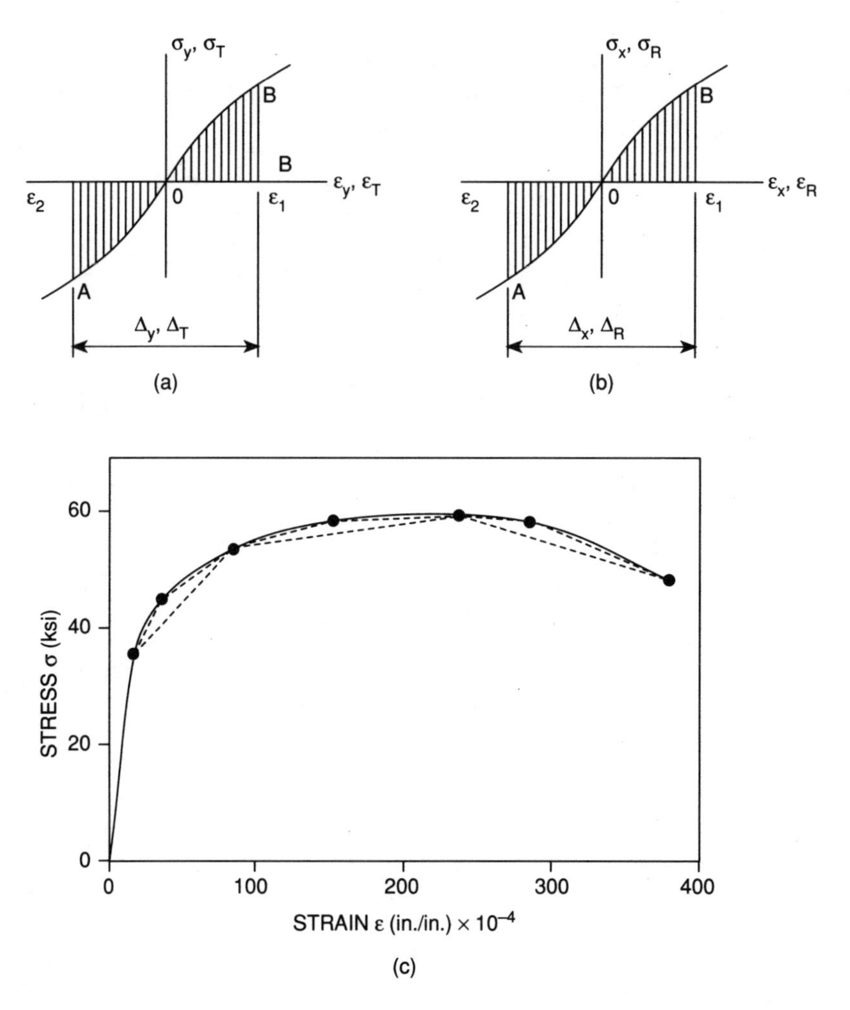

FIGURE 8.4. (a) Stress/strain diagram for σ_y, ε_y and σ_T, ε_T; (b) stress/strain diagram for σ_x, ε_x and σ_R, ε_R; (c) stress/strain diagram of the material of a plate with its shape approximated with four and six straight lines (1 ksi = 6894.757 kPa, 1 in. = 0.0254 m).

Thus the total strains ε_x' and ε_y' and the stresses σ_x and σ_y may be written as follows:

$$\varepsilon_x' = \frac{1}{1-\nu^2}\left(\varepsilon_x + \nu\varepsilon_y\right) \tag{8.98}$$

$$\varepsilon_y' = \frac{1}{1-\nu^2}\left(\varepsilon_y + \nu\varepsilon_x\right) \tag{8.99}$$

$$\sigma_x = \frac{E_x}{1-\nu^2}\left(\varepsilon_x + \nu\varepsilon_y\right) \tag{8.100}$$

$$\sigma_y = \frac{E_y}{1-\nu^2}\left(\varepsilon_y + \nu\varepsilon_x\right) \tag{8.101}$$

Therefore, we can rewrite Equations 8.100 and 8.101 as

$$\sigma_x = E_x \varepsilon'_x \tag{8.102}$$

$$\sigma_y = E_y \varepsilon'_y \tag{8.103}$$

At the cross section under consideration, the bending moment M_y of the plate is

$$M_y = \int_A \sigma_y z dA = \int_{h_2}^{h_1} \sigma_y z dz \tag{8.104}$$

where the width of dA is assumed to be unity. Thus, by using Equations 8.93 and 8.94, we find

$$M_y = r_y^2 \int_{\varepsilon_2}^{\varepsilon_1} \sigma_y \varepsilon_y d\varepsilon_y \tag{8.105}$$

or, by using Equation 8.97 and rearranging, we obtain

$$M_y = \frac{h^3}{\Delta_y^3}\frac{1}{r_y}\int_{\varepsilon_2}^{\varepsilon_1} \sigma_y \varepsilon_y d\varepsilon_y \tag{8.106}$$

On this basis, the reduced rigidity D_{ry} along the y coordinate, may be written as

$$D_{ry} = \frac{h^3}{\Delta_y^3}\cdot\frac{1}{1-\nu^2}\int_{\varepsilon_2}^{\varepsilon_1} \sigma_y \varepsilon_y d\varepsilon_y \tag{8.107}$$

For elastic analysis, Equation 8.106 reduces to $M_y = D_{ry}/r_y$.

In a similar manner, the bending moment M_x and the reduced rigidity D_{rx} along the x coordinate of the plate may be written as

$$M_x = \frac{h^3}{\Delta_x^3}\cdot\frac{1}{r_x}\int_{\varepsilon_2}^{\varepsilon_1} \sigma_x \varepsilon_x d\varepsilon_x \tag{8.108}$$

$$D_{rx} = \frac{h^3}{\Delta_x^3}\cdot\frac{1}{1-\nu^2}\int_{\varepsilon_2}^{\varepsilon_1} \sigma_x \varepsilon_x d\varepsilon_x \tag{8.109}$$

The integrals in Equations 8.109 and 8.107 represent the moment with respect to the vertical axis through the origin 0 of the shaded area in Figures 8.4b and 8.4a, respectively.

By using the well-known curvature moment expressions for rectangular plates and rearranging, we may write

$$\frac{\partial^2 w}{\partial x^2} = \frac{1}{1-\nu^2}\left(\frac{M_x}{D_{rx}} - \nu\frac{M_y}{D_{ry}}\right) \tag{8.110}$$

$$\frac{\partial^2 w}{\partial y^2} = \frac{1}{1-\nu^2}\left(\frac{M_y}{D_{ry}} - \nu\frac{M_x}{D_{rx}}\right) \tag{8.111}$$

In order to facilitate the solution of the above equations, a trial and error solution may be used. For example, by using Figures 8.4b and 8.4a, values of Δ_x and Δ_y may be assumed. This is equivalent to assuming values of ε_1 and ε_2 ($\varepsilon_1 = \varepsilon_2 = \varepsilon$), since $\Delta = 2\varepsilon$. For each assumed value of Δ_x and Δ_y, the required moments $M_{x_{req}}$ and $M_{y_{req}}$ are computed by using Equations 8.108 and 8.106, respectively. The values of D_{rx} and D_{ry} are obtained from Equations 8.109 and 8.107, respectively. Note that the integral in Equations 8.109 and 8.107 represents, respectively, the first moment of the shaded area in Figures 8.4b and 8.4a with respect to a vertical axis through the origin 0. The procedure may be repeated until the required moment $M_{x_{req}}$ and $M_{y_{req}}$ are equal to the actual moments M_x and M_y, respectively, of the plate. The reduced rigidities D_{rx} and D_{ry}, as stated earlier, are obtained from Equations 8.109 and 8.107, respectively.

The deflection w of the rectangular plate along the x and y axis may be obtained from Equations 8.110 and 8.111, respectively, by integration, since M_x, and M_y, D_{ry} and D_{rx} are known. The actual moments M_x and M_y may be obtained from elastic analysis. The method of the equivalent systems, as discussed earlier in this chapter, may be easily applied for this purpose. For statically determinate plates, i.e., simply supported edges, the required moments $M_{x_{req}}$ and $M_{y_{req}}$ that are obtained from inelastic analysis would have to match the corresponding actual moments M_x and M_y obtained from the elastic analysis in order to satisfy the equilibrium requirements. When the plate is statically indeterminate, i.e., all edges fixed, the required moments $M_{x_{req}}$ and $M_{y_{req}}$ will not have the same values as the corresponding moments M_x and M_y that are obtained from elastic analysis, because they depend on both the external load and the distribution of the inelastic moments along the dimensions of the plate.

For statically indeterminate plates, a reasonable solution may be obtained by utilizing the well-established procedures for statically indeterminate problems. Some modifications, however, would have to be made. This may be initiated by first defining the redundants. For example, if the one edge of the plate is fixed and the other three are simply supported, the fixed end moment along the fixed edge may be taken as the redundant. At this point, we may assume that all edges of the plate are simply supported and apply inelastic analysis the way it is discussed in this section. This analysis will yield the rotation along the side of the plate that is supposed to be fixed. The purpose of the redundant moment is to make this rotation zero. Thus, by assuming again that all edges of the plate are simply supported, we apply a moment along the side of the plate that is supposed to be fixed. By following inelastic analysis, the value of this moment should be the one required to counterbalance the rotations produced by the plate loading when all sides are assumed simply supported. Since the rotation along the side of the plate may vary, the fixed end moment along this side may also be variable.

The procedure for statically indeterminate plates is more complicated, but not unrealistic. For statically indeterminate beams, the solution is discussed in Chapters 6 and 7.

8.8.2 CIRCULAR PLATES

The procedure for the inelastic analysis of circular plates of uniform and variable thickness is similar to the one used for rectangular plates. If the radial and tangential directions of the circular plate are designated as R and T, respectively, the procedure yields

$$M_T = \frac{h^3}{\Delta_T^3} \cdot \frac{1}{r_T} \int_{\varepsilon_2}^{\varepsilon_1} \sigma_T \varepsilon_T d\varepsilon_T \tag{8.112}$$

$$M_R = \frac{h^3}{\Delta_R^3} \cdot \frac{1}{r_R} \int_{\varepsilon_2}^{\varepsilon_1} \sigma_R \varepsilon_R d\varepsilon_R \tag{8.113}$$

$$D_{rT} = \frac{h^3}{\Delta_T^3} \cdot \frac{1}{1-\nu^2} \int_{\varepsilon_2}^{\varepsilon_1} \sigma_T \varepsilon_T d\varepsilon_T \tag{8.114}$$

$$D_{rR} = \frac{h^3}{\Delta_R^3} \cdot \frac{1}{1-\nu^2} \int_{\varepsilon_2}^{\varepsilon_1} \sigma_R \varepsilon_R d\varepsilon_R \tag{8.115}$$

where M_T and M_R are the bending moments in the tangential and radial directions, respectively, and D_{rT} and D_{rR} are, respectively, the reduced plate rigidities in the same directions. The symbols r_T and r_R are used to denote the radii of curvature in the tangential and radial directions, respectively.

By using the curvature moment expressions for circular plates and rearranging (see Equations 8.31 and 8.32), we may write

$$\frac{d^2w}{dR^2} = -\frac{1}{1-\nu^2}\left(\frac{M_R}{D_{rR}} - \nu\frac{M_T}{D_{rT}}\right) \tag{8.116}$$

where D_{rR} and D_{rT} are the reduced rigidities in the radial and tangential directions, respectively.

Again, a trial and error procedure similar to the one used for rectangular plates may be applied here. It may be initiated by utilizing Figures 8.4b and 8.4a and assuming values for ε_R and ε_T, which is equivalent to assuming values of $\Delta_R = 2\varepsilon_R$ and $\Delta_T = 2\varepsilon_T$. By using Equations 8.112 and 8.113, the procedure may be repeated for as many times as required until the required moments $M_{T_{req}}$ and $M_{R_{req}}$ are equal to the actual moments M_T and M_R, respectively, of the circular plate. The reduced rigidities D_{rT} and D_{rR} are obtained from Equations 8.114 and 8.115, respectively. For statically indeterminate circular plates, the actual moments M_T and M_R may be obtained using inelastic analysis and computing the redundants in a way similar to the one suggested for rectangular plates. The vertical deflection w of the circular plate may be obtained from Equation 8.116, since M_T, M_R, D_{rT} and D_{rR} can be determined as stated above.

The methodology developed in this section, will be applied in the following sections of this chapter for the inelastic analysis of rectangular and circular plates of uniform and variable thickness.

8.9 INELASTIC ANALYSIS OF SIMPLY SUPPORTED RECTANGULAR PLATES WITH LINEAR, QUADRATIC, AND UNIFORM THICKNESS VARIATIONS

The inelastic analysis method that was developed in the preceding section, will be applied here for the solution of a rectangular plate with linear thickness variation that is loaded as shown in Figure 8.3a. The sides of the plate are a = b = 120.0 in. (3.048 m), the thickness h_o = 8.0 in. (0.2032 m), the parameter $\lambda = 0.3$, and the uniformly distributed load q = 700.0 psi (4,826,329.9 Pa). The material of the plate is mild steel with a stress/strain curve as shown in Figure 8.4c. In the same figure, the approximation of the stress/strain curve with four and six straight lines is also shown. The values of the modulus E and stress σ corresponding to these approximations of the stress/strain curve are shown in Table 8.30. This information is also shown in Figure 6.14 and Table 6.8.

The inelastic analysis is carried out using the four-line approximation of the stress/strain curve of mild steel. The iteration procedure discussed in the preceding section may be initiated by using the stress/strain curve in Figure 8.4c and assuming values of Δ_x and Δ_y, which is equivalent to assuming values of the strains ε_1 and ε_2 (see also Figures 8.4b and 8.4a). Because of symmetry we have $\varepsilon_1 = \varepsilon_2 = \varepsilon$, and $\Delta = 2\varepsilon$. Then, for each value of Δ_x amd Δ_y, we compute the required moments $M_{x_{req}}$ and $M_{y_{req}}$ by using Equations 8.108 and 8.106,

TABLE 8.30
Values of E and σ for 4-Line and 6-Line Approximation of the Stress/Strain Curve

1 Modulus E (psi)	3 4-Line approximation	3 6-Line approximation
E_1	94.366×10^5	94.366×10^5
E_2	28.289×10^4	40.351×10^4
E_3	3.785×10^4	21.053×10^4
E_4	-7.9710×10^4	3.785×10^4
E_5	—	-1.942×10^4
E_6	—	-11.561×10^4
Stress σ (psi)		
σ_1	33.5×10^3	33.5×10^3
σ_2	55×10^3	45×10^3
σ_3	61×10^3	55×10^3
σ_4	—	61×10^3
σ_5	—	60×10^3

Note: 1 psi = 6.895 kPa.

respectively, and the values of the reduced rigidities D_{rx} and D_{ry} from Equations 8.109 and 8.107, respectively. The procedure may be repeated as required until $M_{x_{req}}$ and $M_{y_{req}}$ are equal to the actual moment M_x, and M_y, respectively, of the rectangular plate. The actual moments M_x and M_y may be obtained by using the method of equivalent systems for plates in the same way as it was done in Section 8.6.

By following the above procedure, the values of ε_x, ε_y, $M_{x_{req}}$, $M_{y_{req}}$, D_{rx}, and D_{ry} are determined along the y axis at x = 0, and they are shown in Table 8.31. With known M_x, M_y, D_{rx}, and D_{ry}, the deflections and rotations at any point of the plate may be determined from Equations 8.110 and 8.111 by integration. Simpson's rule may be used for this purpose. For points along the y axis with x = 0, the deflections and rotations are determined in this manner, and they are shown in Table 8.32, columns 5 and 4, respectively. From this table, we observe that the maximum vertical deflection w = 1.9245 in. (0.0489 m), and it occures at a point of the plate where x = 0 and y = –12.0 in. (0.3048 m). The maximum rotation is 0.069279 rads or 3.9714°, and it occurs at y = –60.0 in. (1.524 m), as shown in the same table.

In Figure 8.5a, the variation of the vertical deflection w at the quarter points 1, 2, and 3 along the y axis, with distributed load q, is shown. We note again here that w increases rapidly for loads q > 750 psi (5,171,067.75 Pa), which suggests that at this level of loading the plate is very deep into the inelastic range, and any further increase in the load may result in a catastrophic failure. This information is very useful to the practicing design engineer who wants to establish criteria for ultimate or permissible design loads. This information

TABLE 8.31
Reduced Rigidities and Required Bending Moments for Simply Supported Rectangular Plate with Linear Thickness Variation

y (in.)	h (in.)	$\varepsilon_x \times 10^{-4}$	$\varepsilon_y \times 10^{-4}$	$M_{x_{req}} = M_x \times 10^4$(in.-lb)	$M_{y_{req}} = M_y \times 10^4$(in.-lb)	$D_{rx} \times 10^8$	$D_{ry} \times 10^8$
−60	5.60	0.0	0.0	0.0	0.0	1.5176	1.5176
−54	5.84	5.229	17.871	−6.242	−11.459	1.7212	1.7212
−48	6.08	10.976	29.012	−12.573	−22.681	1.9422	1.9422
−42	6.32	16.085	37.262	−18.821	−29.613	2.1814	2.0233
−36	6.56	15.775	55.100	−26.403	−36.151	2.4395	1.8034
−30	6.80	12.525	74.275	−30.344	−40.976	2.6972	1.6247
−24	7.04	16.400	76.525	−35.420	−44.223	2.8827	1.7393
−18	7.28	24.625	62.000	−39.809	−45.989	3.0506	2.1954
−12	7.52	31.525	45.825	−43.374	−46.518	3.2781	2.8792
−6	7.76	34.725	34.400	−46.086	−45.985	3.6124	3.6228
0	8.00	34.550	27.400	−47.828	−44.575	4.0705	4.2964
+6	8.24	32.550	23.450	−48.689	−42.786	4.6114	4.8295
+12	8.48	29.100	23.575	−48.610	−44.121	5.1851	5.2696
+18	8.72	26.544	20.720	−47.297	−37.695	5.7282	5.7298
+24	8.96	27.189	17.055	−49.257	−34.981	6.2161	6.2161
+30	9.20	24.137	14.737	−41.777	−32.151	6.7291	6.7291
+36	9.44	20.234	12.788	−37.072	−29.044	7.2696	7.2696
+42	9.68	15.816	10.854	−30.887	−25.261	7.8382	7.8382
+48	9.92	10.932	8.518	−22.940	−20.066	8.4358	8.4358
+54	10.16	5.638	5.178	−12.830	−12.255	9.0630	9.0630
+60	10.40	0.0	0.0	0.0	0.0	9.7206	9.7206

Note: 1 in. = 0.0254 m, 1 in.-lb = 0.113 Nm.

TABLE 8.32
Deflections and Rotations Along the y Axis at x = 0 for Rectangular Plate of Quadratic and Linear Thickness Variations Along the y Axis

	Quadratic Plate		Tapered Plate	
y (in.)	Slope × 10^{-2}	w (in.)	Slope × 10^{-2}	w (in.)
–60	4.4086	0.0	6.9279	0.0
–54	4.3645	0.2635	6.7346	0.4113
–48	4.2492	0.5222	6.2336	0.8013
–42	4.0716	0.772	5.5326	1.1549
–36	3.8272	1.0092	4.5852	1.4596
–30	3.5028	1.2295	3.2972	1.6973
–24	3.0978	1.4278	1.8353	1.8513
–18	2.5932	1.599	0.5484	1.9216
–12	1.9289	1.7354	–0.4111	1.9245
–6	1.0532	1.8256	–1.099	1.8784
0	–2.1684	1.8577	–1.6128	1.7966
6	–1.0532	1.8256	–2.024	1.6872
12	–1.9289	1.7354	–2.3906	1.5547
18	–2.5932	1.599	–2.6971	1.4017
24	–3.0978	1.4278	–2.9396	1.2324
30	–3.5028	1.2295	–3.1456	1.0498
36	–3.8272	1.0092	–3.3228	0.8556
42	–4.0716	0.772	–3.4714	0.6517
48	–4.2492	0.5222	–3.5906	0.4397
54	–4.3645	0.2635	–3.6736	0.2216
60	–4.4086	0.0	–3.7057	0.0

Note: 1 in. = 0.0254 m.

would be particularly useful in the design of engineering structures that are designed to resist the effects of blust and earthquake.

If $\lambda = 0$, the rectangular plate in Figure 8.3a becomes a flat plate of uniform thickness $h_o = 8.0$ in. (0.2032 m). The vertical deflections w at the quarter points 1, 2, and 3 of this plate are plotted in Figure 8.6 by increasing the load q. We note again that for values of q > 850 psi (5,860,543.45 Pa), the slope of the curves increases rapidly, even with small increases of the vertical load q beyond this value. This indicates that the ultimate capacity of the plate to resist load and deformation is reached. For the same plate problem, along the y direction of the plate and at x = 0, the variation of the rigidity functions

$$g(x) = \frac{D_{rx}}{D_x} \tag{8.117}$$

$$g(y) = \frac{D_{ry}}{D_y} \tag{8.118}$$

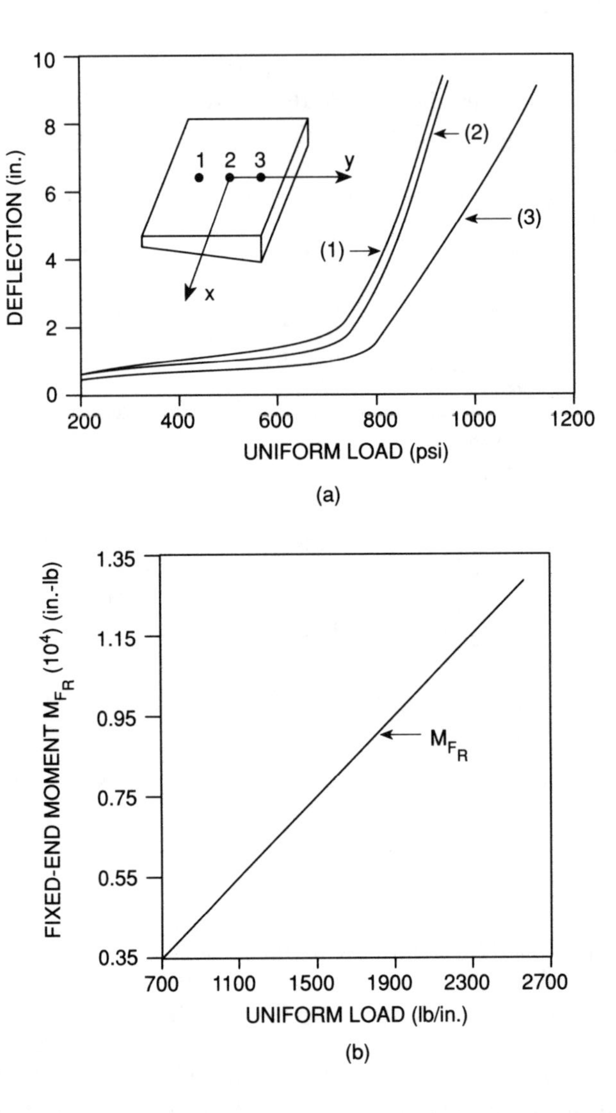

FIGURE 8.5. (a) Vertical deflection variation at points 1, 2, and 3 for rectangular plate of linear thickness variation; (b) variation of the fixed end moment M_{F_R} of a clamped circular plate with quadratic thickness variation (1 in. = 0.0254m, 1 psi = 6894.757 Pa, 1 lb/in. = 175.1268 N/m).

in the x and y directions, respectively, and for q = 800 psi (5,515,805.6 Pa) is shown in Figure 8.7. The smallest value of the rigidity ratios is at the center of the plate, and the plate is elastic, i.e., g(x) = g(y) = 1.0, for the interval of y in the range of about ± 35 in. ≤ y ≤ ± 60 in.

Now we examine the case where the above rectangular plate with b = a = 120 in. (3.048 m) has a quadratic thickness variation in the y direction as shown in Figure 8.2b. The thickness h_o = 8.0 in. (0.2032 m), λ = 0.3, and Poisson ratio ν = 0.3. The plate is loaded with a uniform vertical load q = 900 psi (6,205,281.3

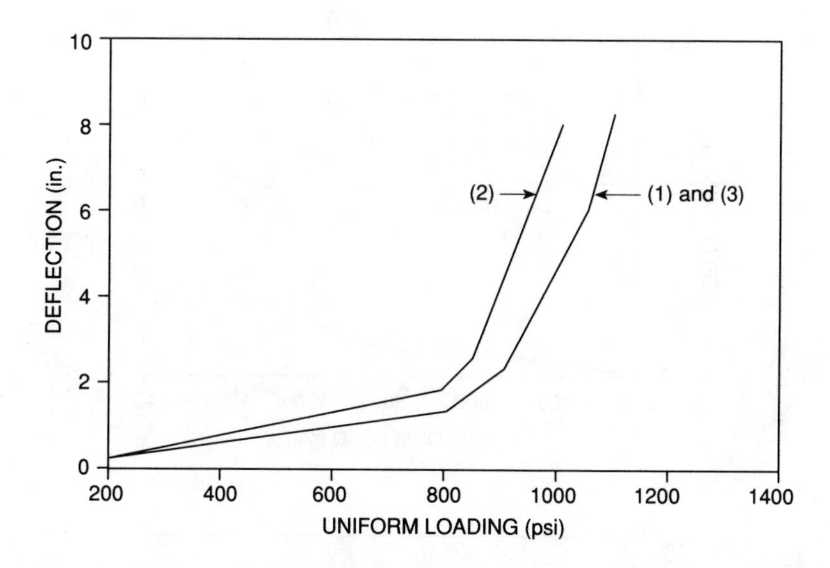

FIGURE 8.6. Variation of vertical deflection at the quarter points 1, 2, and 3 of the rectangular plate (b/a = 1) of uniform thickness (1 in. = 0.0254 m, 1 psi = 6894.757 Pa).

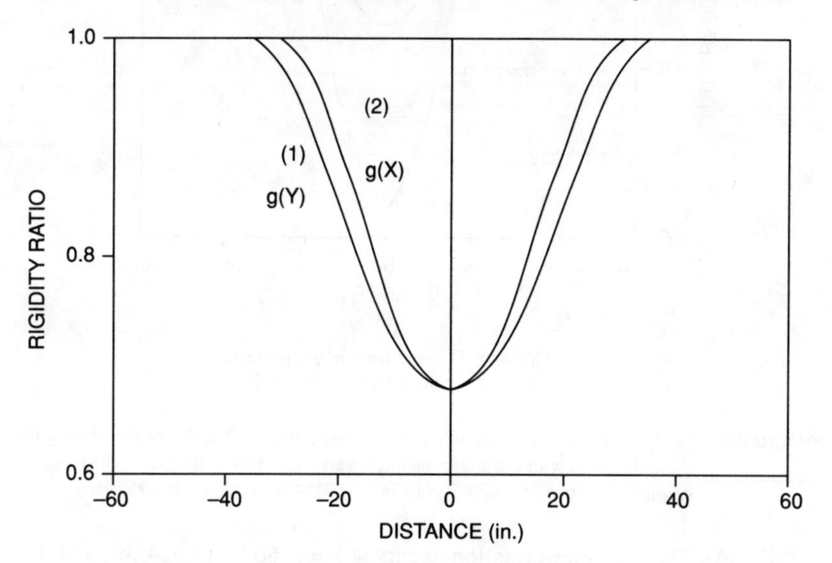

FIGURE 8.7. Variation of plate rigidity functions g(x) and g(y) for a rectangular plate (b/a = 1.0) of uniform thickness and q = 800 psi (5,515,805.6 Pa) (1 in. = 0.0254 m).

Pa). By using the four-line approximation of the stress/strain curve and applying the iteration procedure discussed above, the values of ε_x, ε_y, $M_{x_{req}}$, $M_{y_{req}}$, D_{rx}, and D_{ry} are determined along the y axis at x = 0, and they are shown in Table 8.33. The deflections and rotations at points along the y axis at x = 0 are shown in Table 8.32. From the results in this table, we observe that maximum vertical deflection occurs at the center of the plate (y = 0), and it is equal to 1.8577 in.

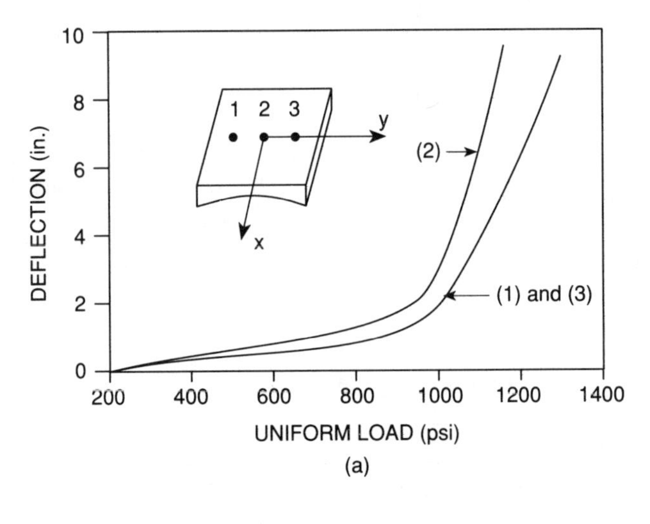

(a)

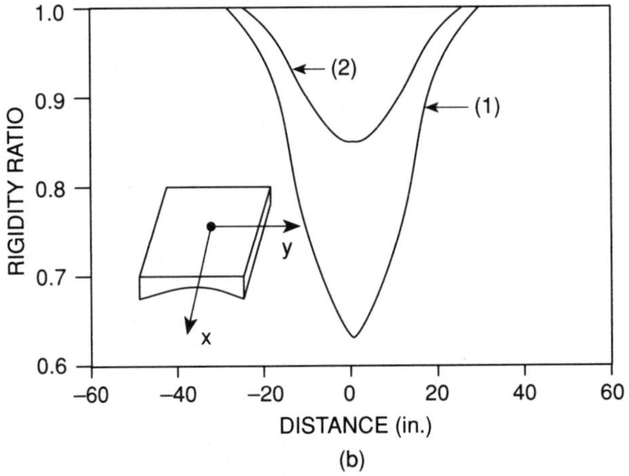

(b)

Curve 1. Rigidity ratio in y direction.
Curve 2. Rigidity ratio in x direction.

FIGURE 8.8. (a) Variation of the deflection at the quarter points 1, 2, and 3 of the rectangular plate (b/a = 1) with quadratic thickness variation in the y direction; (b) variation of plate rigidities g(x) and g(y) (1 in. = 0.0254 m, 1 psi = 6894.757 Pa).

(0.0472 m). The maximum rotation occurs at $x = \pm 60$ in. (1.524 m), and it is equal to 0.044086 rads or 2.5272°.

The deflections at the quarter points 1, 2, and 3 along the y axis at $x = 0$ are plotted in Figure 8.8a with increasing values of the load q. Again we note that the slope of these curves for values of $q > 1000$ psi (6,894,757 Pa) increases rapidly, even with small increases of q beyond this value, indicating that the ultimate capacity of the plate to resist load and deformation is reached. The variation of the rigidity ratios g(x) and g(y) for $q = 900$ psi (6,205,281.3 Pa) are shown by the curves 2 and 1, respectively, in Figure 8.8b. This figure illustrates that the changes of the rigidity ratio g(y) are far more severe

TABLE 8.33
Reduced Rigidities and Required Bending Moments for Simply Supported Rectangular Plate with Quadratic Thickness Variation

y (in.)	h (in.)	$\varepsilon_x \times 10^{-4}$	$\varepsilon_y \times 10^{-4}$	$M_{x_{req}} = M_x \times 10^5$ (in.-lb)	$M_{y_{req}} = M_y \times 10^5$ (in.-lb)	$D_{rx} \times 10^8$	$D_{ry} \times 10^8$
−60	10.40	0.0	0.0	0.0	0.0	9.7206	9.7206
−54	9.94	6.373	6.918	−1.4438	−1.5091	8.4972	8.4972
−48	9.54	12.109	11.632	−2.4515	−2.3991	7.4936	7.4936
−42	9.18	17.205	16.024	−3.2033	−3.0830	6.6766	6.6766
−36	8.86	21.662	20.699	−3.7849	−3.6933	6.0184	6.0184
−30	8.60	24.850	24.850	−4.5378	−4.2490	5.4965	5.4965
−24	8.38	25.900	28.175	−4.5629	−4.7295	5.0669	5.0193
−18	8.22	28.875	35.700	−4.8405	−5.1315	4.5712	4.3242
−12	8.10	30.975	46.900	−5.0164	−5.4165	4.1023	3.5508
−6	8.02	31.625	59.900	−5.1167	−5.5869	3.7667	2.9396
0	8.00	31.450	66.550	−5.1490	−5.6462	3.6453	2.7047
+6	8.02	31.625	59.900	−5.1167	−5.5869	3.7667	2.9396
+12	8.10	30.975	46.900	−5.0164	−5.4165	4.1023	3.5508
+18	8.22	28.875	35.700	−4.8405	−5.1315	4.5712	4.3242
+24	8.38	25.900	28.175	−4.5629	−4.7295	5.0669	5.0193
+30	8.60	24.850	24.850	−4.5378	−4.2490	5.4965	5.4965
+36	8.86	21.662	20.699	−3.7849	−3.6933	6.0184	6.0184
+42	9.18	17.205	16.024	−3.2033	−3.0830	6.6776	6.6766
+48	9.54	12.109	11.632	−2.4515	−2.3991	7.4936	7.4936
+54	9.94	6.373	6.918	−1.4438	−1.5091	8.4972	8.4972
+60	10.40	0.0	0.0	0.0	0.0	9.7206	9.7206

Note: 1 in. = 0.0254 m, 1 in.-lb = 0.113 Nm.

compared to the changes of g(x), with the lowest values occuring at the center of the plate.

8.10 INELASTIC ANALYSIS OF CLAMPED CIRCULAR PLATES OF QUADRATIC THICKNESS VARIATION

Inelastic analysis of circular plates with quadratic thickness variation is discussed in this section. The cross section along a diameter of the plate will have the shape shown in Figure 8.2a, and its quadratic thickness variation h is given by the expression

$$h = h_o\left[1 + \lambda\left(\frac{R}{a}\right)^2\right] \tag{8.119}$$

where a is the radius of the circular plate. The radius a = 60 in. (1.524 m), h_o = 8 in. (0.2032 m), $\lambda = 0.3$, and the plate is loaded with a uniformly distributed load q = 2200 psi (15,168,465.4 Pa). The material of the circular plate is mild steel.

Since the geometry of the circular plate and its loading are both symmetrical, a diametral section of the plate is sufficient for calculating deflections and stresses. The problem, however, is statically indeterminate because the edge of the circular plate is clamped. Therefore, the procedure discussed in Section 8.8 for statically indeterminate plates must be used. The purpose of the uniform fixed end moment M_{F_R} around the edge of the plate is to make the rotation at the edge of the circular plate equal to zero. This condition must be satisfied during inelastic analysis, as it was done for the fixed/fixed beam in Section 7.5.

By utilizing the trial-and-error procedure discussed in the preceding two sections and satisfying the boundary condition of zero rotation at the edge of the circular plate, the values of ε_R, ε_T, $M_{T_{req}}$, $M_{R_{req}}$, D_{rT}, D_{rR}, and vertical deflection w are determined and shown in Table 8.34. From this table we observe that the maximum vertical deflection occurs at the center of the plate, and it is equal to 1.1407 in. (0.029 m). The rigidity functions

$$g(R) = \frac{D_{rR}}{D_R} \tag{8.120}$$

$$g(T) = \frac{D_{rT}}{D_T} \tag{8.121}$$

TABLE 8.34
Values of ε_R, ε_T, , D_{rT}, D_{rR}, and Deflection w for a Clamped Circular Plate of Quadratic Thickness Variation

R (in.)	h (in.)	$\varepsilon_R \times 10^{-3}$	$\varepsilon_T \times 10^{-3}$	$M_{T_{req}} = M_T \times 10^5$ (in.-lb)	$M_{R_{req}} = M_R \times 10^5$ (in.-lb)	$D_{rT} \times 10^8$	$D_{rR} \times 10^8$	Deflection w(in.)
–60	10.40	25.094	–5.463	3.8603	11.0390	9.7206	2.2271	0.0
–54	9.94	4.890	–0.299	1.9969	7.7652	8.4972	7.3192	0.05735
–48	9.54	3.341	–0.784	3.4302	4.8815	7.4936	7.4936	0.16869
–42	9.18	1.525	–0.302	–1.1057	2.3510	6.6766	6.6766	0.30555
–36	8.86	–0.430	1.862	–2.3534	0.1748	6.0184	6.0184	0.45822
–30	8.60	0.544	2.499	–3.4033	–1.6543	5.4965	5.4965	0.61664
–24	8.38	1.778	2.697	–4.3128	–3.1426	5.0927	5.0927	0.77056
–18	8.22	2.358	3.365	–4.9186	–4.3065	4.5152	4.7813	0.91026
–12	8.10	3.378	4.515	–5.3917	–5.1190	3.5927	3.9849	1.02700
–6	8.02	5.285	5.968	–5.6742	–5.6078	2.7428	2.8937	1.10970
0	8.00	6.703	6.703	–5.7692	–5.7692	2.4101	2.4101	1.14070
6	8.02	5.285	5.968	–5.6742	–5.6078	2.7428	2.8937	1.10970
12	8.10	3.378	4.515	–5.3917	–5.1190	3.5927	3.9849	1.02700
18	8.22	2.358	3.365	–4.9186	–4.3065	4.5152	4.7813	0.91026
24	8.38	1.778	2.697	–4.3128	–3.1426	5.0927	5.0927	0.77056
30	8.60	0.544	2.499	–3.4033	–1.6543	5.4965	5.4965	0.61664
36	8.86	–0.430	1.862	–2.3534	0.1748	6.0184	6.0184	0.45822
42	9.18	1.525	–0.302	1.1057	2.3510	6.6766	6.6766	0.30555
48	9.54	3.341	–0.784	3.4302	4.8815	7.4936	7.4936	0.16869
54	9.94	4.890	–0.299	1.9969	7.7652	8.4972	7.3192	0.05735
60	10.40	25.094	–5.463	3.8603	11.0390	9.7206	2.2271	0.0

Note: 1 in. = 0.0254 m, 1 in-lb = 0.113 Nm.

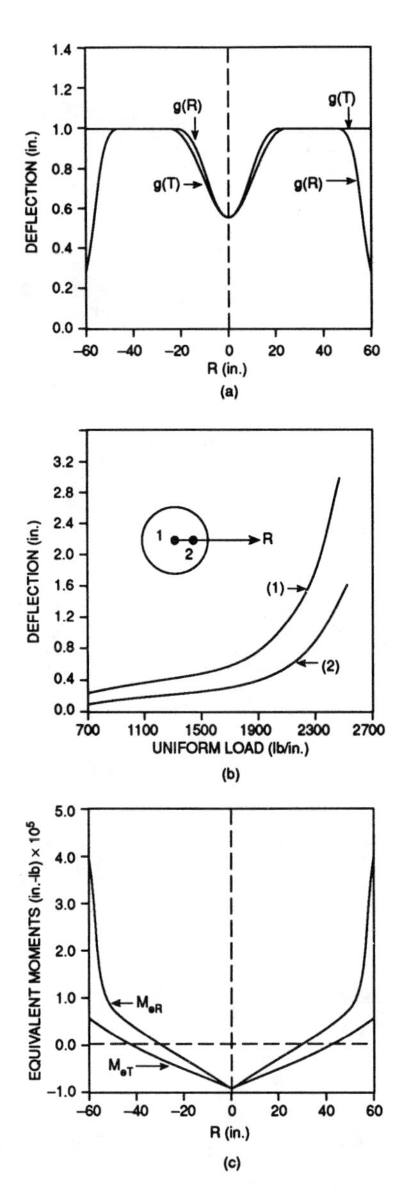

FIGURE 8.9. (a) Variation of the rigidity functions g(R) and g(T) for a clamped circular plate; (b) vertical deflection variation for points 1 and 2 with increasing load q; (c) variation of the tangential and radial equivalent moments M_{eT} and M_{eR}, respectively, of the clamped circular plate (1 in. = 0.0254 m, 1 lb/in. = 175.1268 N/m, 1 in.-lb = 0.113 Nm).

along a diameter of the plate for q = 2200 psi. (15,168,465.4 Pa) are shown plotted in Figure 8.9a. This shows that most of the yielding of the plate occurs at the center and edge of the plate, as should be expected.

The variation of the vertical deflection w of the circular plate with increasing load q is shown in Figure 8.9b. Curve 1 gives the variation of w at point 1 located at the center of the plate, and curve 2 gives the variation of w at point 2 located at a distance equal to a/2 from the center of the plate. Again, a critical value of the load q exists which defines the ultimate capacity of the plate in resisting load and deformation.

The curves in Figure 8.9c provide a plot of the parameters

$$M_{eR} = \frac{M_R}{g(R)} \tag{8.122}$$

$$M_{eT} = \frac{M_T}{g(T)} \tag{8.123}$$

where g(R) and g(T) are given by Equations 8.120, and 8.121, respectively. In the above equations, M_{eR} and M_{eT} may be thought of as equivalent moments in the radial and tangential directions, respectively.

The above rsults were obtained using the four-line approximation of the stress/strain curve in Figure 8.4c. The variation of the fixed end moment M_{F_R} of the clamped circular plate with increasing load q is shown in Figure 8.5b. This curve becomes nonlinear for values of q larger than about 1500 lb/in. (262,690.2 N/m).

Inelastic analysis of circular plates with other boundary conditions and applied loading may be carried out in a similar manner. The methodology is general.

PROBLEMS

8.1 A rectangular plate with linearly varying thickness in the y direction is loaded with a triangular loading q over the entire area of the plate, as shown in Figure 8.10. By using the procedure of Section 8.2, derive an equivalent system consisting of flat plates such as the one represented by Equation 8.21. Discuss in detail what each differential equation of the equivalent system represents, and solve the first differential equation by assuming that all sides of the plate are simply supported.

8.2 For the rectangular plate and loading of Problem 8.1, determine a simplified equivalent system by following the procedure discussed in Section 8.3.

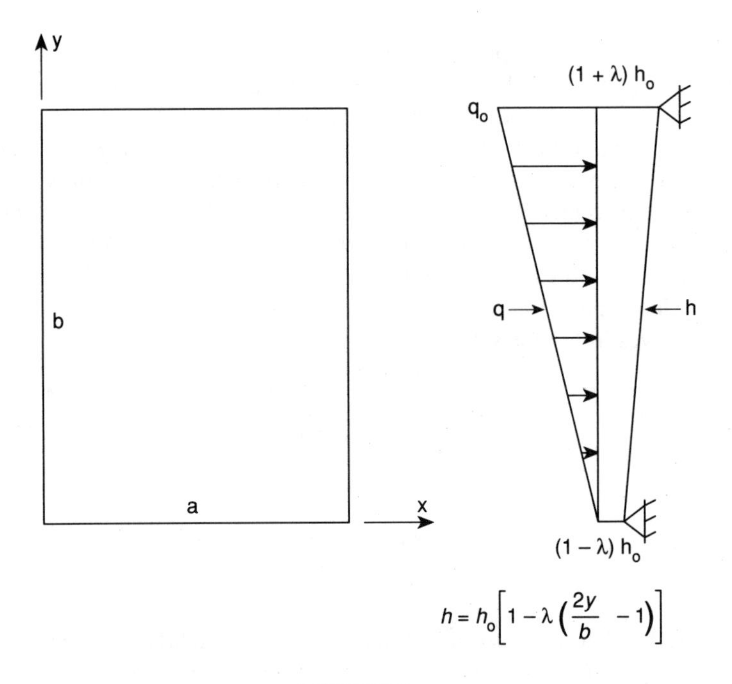

FIGURE 8.10.

8.3 By using the equivalent system derived in Problem 8.1 and assuming that a = b = 200 in. (5.08 m), h_o = 10 in. (0.254 m), Poisson ratio $\nu = 0.3$, $E = 30 \times 10^6$ psi ($206{,}842.71 \times 10^6$ Pa), and $\lambda = 0.2$, determine the vertical deflection at the center of the plate and the angular rotations at y = 0 and y = b.
Answer: $w_{center} = 0.0568 q_o$ (use units of psi for q_o).

8.4 For the rectangular plate in Problem 8.3, determine the vertical deflection at the center of the plate and the angular rotations at y = 0 and y = b by using a simplified equivalent system, and compare the results.

8.5 For the rectangular plate in Problem 8.3, determine the bending moments and the bending stresses at the center of the plate. Determine the value of the loading q_o so that the maximum bending stress at the center of the plate is equal to 30,000 psi (206.8428×10^6 Pa).

8.6 Solve Problem 8.3 by assuming that the two sides of the square plate are fixed and the other two sides are simply supported, and compare the results.

8.7 Solve Problem 8.3 by assuming that all sides of the square plate are fixed, and compare the results.

8.8 Solve Problem 8.3 by assuming that side b = 200 in. (5.08 m) and side a = 100 in. (2.54 m), and compare the results.

8.9 Solve Problem 8.1 by assuming that the rectangular plate has a quadratic thickness variation in the y direction.

8.10 The simply supported rectangular plate in Problem 8.9 has side b = 300 in. (7.62 m) and side a = 200 in. (5.08 m).The material of the plate is steel with modulus of elasticity $E = 30 \times 10^6$ psi ($206{,}842.71 \times 10^6$ Pa). The thickness h_o of the plate is 8.0 in. (0.2032 m), $\lambda = 0.4$, and Poisson ratio $\nu = 0.3$. If the plate is loaded with a uniformly distribution load q = 150 psi (1,034,213.55 Pa), determine the vertical deflection at the center of the plate using equivalent flat plates.
Answer: $w_{center} = 1.022$ in. (0.026 m).

8.11 For Problem 8.10, determine the bending moments M_x and M_y and the maximum bending stresses $\sigma_{x_{max}}$ and $\sigma_{y_{max}}$ at the center of the plate.
Answer: $M_x = 390{,}240$ in.-lb (44,091.2 Nm), $M_y = 279{,}720$ in.-lb (31,604.1 Nm), $\sigma_{x_{max}} = 4573.6$ psi (31,533,860.6 Pa), $\sigma_{y_{max}} = 3277.3$ psi (22,596,187.1 Pa).

8.12 Solve Problem 8.10 by assuming that side a and its opposite side are fixed and the remaining two sides are simply supported, and compare the results. Assume linear thickness variation in the y direction. Also, determine the bending moments and the maximum bending stresses at the center of the plate.
Answer: $w_{center} = 0.9543$ in. (0.0242 m), $M_x = 353{,}100$ in.-lb (39,894.9 Nm), $M_y = 236{,}640$ in.-lb (26,736.7 Nm), $\sigma_{x_{max}} = 4124.43$ psi (28,436,942.6 Pa), and $\sigma_{y_{max}} = 2{,}773.08$ psi (19,119,712.7 Pa).

8.13 Solve Problem 8.12 by assuming that all sides are fixed, and compare the results.

8.14 By following the procedure for inelastic analysis of plates discussed in Section 8.8, reproduce the results for ε_x, ε_y, $M_{x_{req}}$, $M_{y_{req}}$, D_{rx}, and D_{ry} that are shown in Table 8.31. Discuss the results.

8.15 By following the procedure for inelastic analysis of plates discussed in Section 8.8, reproduce the results for the vertical deflections and rotations shown in Table 8.32. Discuss the results.

8.16 By following the procedure for inelastic analysis of plates discussed in Section 8.8, reproduce the results for ε_x, ε_y, $M_{x_{req}}$, $M_{y_{req}}$, D_{rx}, and D_{ry} that are shown in Table 8.31 using the six-line approximation of the stress/strain curve shown in Figure 8.4c, and compare the results.

8.17 Solve Problem 8.14 by assuming that b/a = 1.5.

8.18 Solve Problem 8.15 for b/a = 1.5.

8.19 By following the procedure for inelastic analysis of plates discussed in Section 8.8, reproduce the results for ε_x, ε_y, $M_{x_{req}}$, $M_{y_{req}}$, D_{rx}, and D_{ry} that are shown in Table 8.33.

8.20 By following the procedure for inelastic analysis of circular plates discussed in Section 8.8 reproduce the results for ε_x, ε_y, $M_{x_{req}}$, $M_{y_{req}}$, D_{rx}, and D_{ry} that are shown in Table 8.34. Discuss the results.

Chapter 9

INELASTIC ANALYSIS OF FLEXIBLE BARS WITH UNIFORM AND VARIABLE STIFFNESS

9.1 INTRODUCTION

In the first three chapters the nonlinear problem of flexible bars was investigated by assuming that the elastic limit of the material is not exceeded, and consequently the modulus of elasticity E remains elastic. The large deformations were obtained using the Euler-Bernoulli nonlinear differential equation in conjunction with equivalent pseudolinear and/or equivalent simplified nonlinear systems in order to simplify the solution of the complex nonlinear problem. In Chapters 6 and 7, the material of the member was stressed well beyond its elastic limit, all the way to failure, and consequently the modulus E will vary along the length of the member. The deformations, however, were kept small, and application of large deformation theory was not required.

In this chapter, the flexible member will be stressed well beyond the elastic limit of its material, and therefore the modulus E will vary along its length. Consequently, the problems to be solved in this chapter include flexible members that are subjected to large deformations and acted upon by loading conditions that stress the flexible member heavily into the inelastic range. Such loading conditions establish limits of such loads or define ultimate loads, which makes it possible to determine the ultimate capacity of the flexible member in resisting load, stress, and deformation. The flexible member can be of uniform or variable thickness, and it can be subjected to complicated loading conditions. Simplified methods of analysis are introduced in order to reduce the complexity of the problem. Axial restraints, or axial loads, are not taken into consideration in this chapter. This problem requires special attention, and it is thoroughly investigated in Chapter 10.

9.2 EQUIVALENT SYSTEMS FOR INELASTIC ANALYSIS OF FLEXIBLE BARS

The inelastic analysis of flexible members with uniform or variable cross-sectional dimensions along their length, and variable modulus E, may be carried out by using the concept of simplified equivalent nonlinear systems of constant stiffness and the concept of pseudolinear equivalent systems of constant stiffness as discussed in Chapters 1, 2, and 3. If the flexible member is subjected to complicated combined loading conditions, a simplified nonlinear equivalent system of constant stiffness is first obtained. The simplified nonlinear system is then solved by pseudolinear analysis, as discussed in Chapters 1 to 3. The purpose of the simplified nonlinear system is to simplify the mathematical model of the initial problem and reduce the complexity of its solution. The simplified system is then solved using pseudolinear equivalent systems, which further simplify the solution. In other words, two types of simplifications may be applied if the original nonlinear problem is very complicated. This approach solves complicated problems in a more relaxed and relatively easy way. Athough the utilization of such equivalent systems was thoroughly discussed in earlier chapters, some of the aspects are also brought up here for better clarity, because now the member is subjected to large deformations and the modulus E varies along the length of the member.

The general nonlinear differential equation of the elastic line of a flexible member with variable moment of inertia I_x and variable modulus of elasticity E_x may be expressed here as

$$\frac{y''}{\left[1+(y')^2\right]^{3/2}} = -\frac{M_x}{E_x I_x} \tag{9.1}$$

where M_x is the bending moment at a cross section along the length of the member.

If both E_x and I_x are permitted to vary along the length of the member, the variable stiffness $E_x I_x$ may be expressed as

$$E_x I_x = E_1 I_1 g(x) f(x) \tag{9.2}$$

where g(x) represents the variation of E_x with respect to a reference value E_1 and f(x) represents the variation of I_x with respect to a reference value I_1. If E and I are both constant, then f(x) = g(x) = 1.00.

By substituting Equation 9.2 into Equation 9.1, we find

$$\frac{y''}{\left[1+(y')^2\right]^{3/2}} = -\frac{M_e}{E_1 I_1} \tag{9.3}$$

where

$$M_e = \frac{M_x}{g(x)f(x)} \tag{9.4}$$

It was shown in Chapter 1 that M_e in Equation 9.4 represents the moment variation along the length of an equivalent beam of constant stiffness $E_1 I_1$, which has the same elastic line, length, and boundary conditions, as the initial variable stiffness member represented by Equation 9.1. In other words, Equation 9.3 is the nonlinear differential equation of an equivalent beam of constant stiffness $E_1 I_1$ and moment M_e. With known M_e, the shear force V_e and load W_e

of the nonlinear equivalent member, may be obtained by differentiation, i.e.,

$$V_e = \frac{d}{dx}(M_e) = \frac{d}{dx}\left[\frac{M_x}{g(x)f(x)}\right] \tag{9.5}$$

$$W_e = -\frac{d}{dx}(V_e) = -\frac{d^2}{dx^2}\left[\frac{M_x}{g(x)f(x)}\right] \tag{9.6}$$

Equations 9.3, 9.4, 9.5, and 9.6 completely define the constant stiffness equivalent nonlinear system, provided that g(x) and f(x) are known.

9.2.1 EQUIVALENT PSEUDOLINEAR SYSTEMS OF CONSTANT STIFFNESS

The nonlinear differential equation of a flexible bar in bending, Equation 9.1, may be rewritten here as

$$y'' = -\frac{M_e Z_e}{E_1 I_1} \tag{9.7}$$

where

$$Z_e = \left[1 + (y')^2\right]^{3/2} \tag{9.8}$$

or as

$$y'' = -\frac{M_e'}{E_1 I_1} \tag{9.9}$$

where

$$M_e' = M_e Z_e = \frac{M_x Z_e}{g(x)f(x)} \tag{9.10}$$

The quantity M_e' in Equation 9.10 represents the variation of the moment along the x axis of an equivalent pseudolinear system of constant stiffness $E_1 I_1$ that has the same elastic line and the boundary conditions as the initial variable stiffness member represented by Equation 9.1. The elastic line and boundary conditions are also indentical with those of the equivalent nonlinear system represented by Equation 9.3. With known M_e', the equivalent shear force V_e' and equivalent load W_e', can be obtained by differentiation as indicated above. Linear analysis may be used to solve the pseudolinear equivalent system represented by Equation 9.9, provided that the M_e' is known.

The purpose in this section is to perform inelastic analysis of members with various types of loading conditions and moment of inertia variations, using simplified equivalent nonlinear systems of constant stiffness E_1I_1 and equivalent pseudolinear systems of constant stiffness E_1I_1, coupled with equivalent linear analysis. In other words, if the initial problem is complicated, it may be converted first into a simpler nonlinear equivalent system of constant stiffness E_1I_1, represented by Equation 9.3. Pseudolinear analysis represented by Equation 9.9 may then be used to solve the equivalent nonlinear system represented by Equation 9.3. Exact, as well as accurate approximate, solutions are obtained and the results are compared, when applicable, with the results obtained using other methods of analysis. In many cases pseudolinear analysis may be applied directly to the initial problem, as it is illutrated in the following sections of the chapter.

In order to start the inelastic analysis for flexible beams, the methodology concerning the evaluation of the modulus function g(x) in Equation 9.2 must be formulated. The procedure regarding the computation of g(x) is the same as the one discussed in Chapter 6, and it is based on the determination of a reduced modulus E_r as discussed in this chapter. In order to compute E_r, the stress/strain curve of the material of the member must be known. For example, if the material is Monel, its stress/strain curve is as shown in Figure 6.3a, and its shape may be approximated with two, three, or six straight lines, depending on how accurate the solution is required. For the indicated straight-line approximations, the values of the modulus E and stress σ are shown in Tables 6.1 and 6.2, respectively.

For members that are free of any axial restraints, or axial tensile and compressive loads, the methodology regarding the computation of E_r, and consequently the computation of the modulus function

$$g(x) = \frac{E_r}{E} \tag{9.11}$$

where E is the reference elastic modulus, is thoroughly discussed in Section 6.2. For members with axial restraints, or for members that are also subjected to axial compressive or tensile loads, the determination of g(x) is based on the computation of an equivalent modulus E_e, as discussed in Section 7.2. This procedure takes into consideration the shifting of the neutral axis at cross sections along the length of the member, which results from the presence of the axial restraint or axial force.

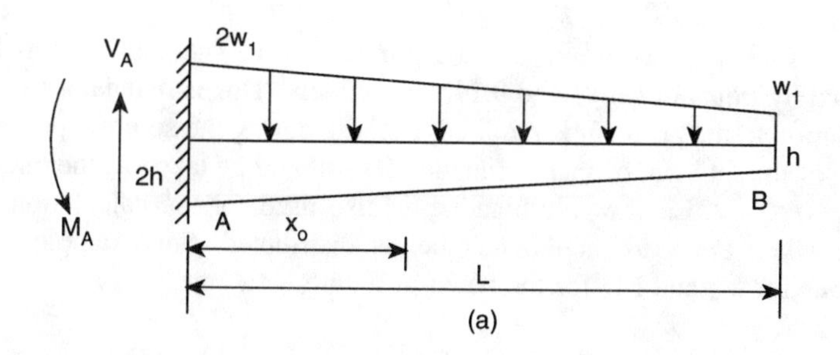

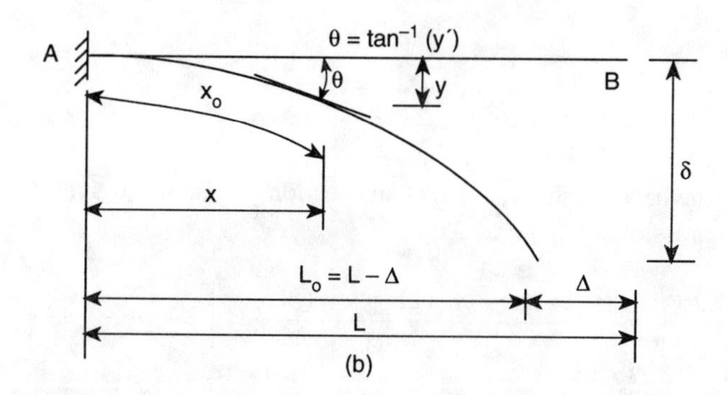

FIGURE 9.1. (a) Flexible cantilever bar loaded with a trapezoidal loading; (b) large deformation configuration of the flexible bar.

9.3 INELASTIC ANALYSIS OF FLEXIBLE BARS BY USING EQUIVALENT PSEUDOLINEAR SYSTEMS

The procedure here may be illustrated by considering the tapered cantilever bar in Figure 9.1a, which is loaded as shown. In Figure 9.1b we note that for every length x_o there corresponds a length x, where $0 \le x \le L_o$ and $L_o = L - \Lambda$. Thus, we may write

$$x_o = x + \Lambda(x) \tag{9.12}$$

where $\Lambda(x)$ represents the horizontal displacement of x_o. We also know that

$$x_o = \int_0^x \left[1 + (y')^2\right]^{1/2} dx \tag{9.13}$$

and

$$L = \int_0^{L-\Lambda} \left[1 + (y')^2\right]^{1/2} dx \tag{9.14}$$

Thus, correspondence between x and x_o and $L_o = L - \Lambda$ and L, can be established by Equations 9.13 and 9.14, respectively. This is mandatory if an exact solution to the problem is required, because during the solution process the value of the moment of inertia function f(x) should be taken as the one at x_o, where $0 \leq x_o \leq L$. The coordinate x_o is also used for the calculation of moments when the load on the member is distributed. For example, the expressions for V_A and M_A for the beam in Figure 9.1a are

$$V_A = \frac{3w_1L}{2} \tag{9.15}$$

$$M_A = \frac{2w_1LL_o}{3} \tag{9.16}$$

Thus, the bending moment M_x at any section of distance $0 \leq x \leq L_o$ is given by the expression

$$M_x = \frac{2w_1LL_o}{3} - \frac{3w_1Lx}{2} + \frac{w_1(2L - x_o)x_ox}{2L} + w_1\left[2 - \frac{(2L - x_o)}{L}\right]\frac{x_ox}{3} \tag{9.17}$$

In Equation 9.17 the expression for the coordinate x_o is given by Equation 9.12, and the distributed vertical load w at any $0 \leq x_o \leq L$ is given by the equation

$$w = \frac{w_1(2L - x_o)}{L} \tag{9.18}$$

The moment of inertia I_x at any $0 \leq x_o \leq L$ is given by the expression

$$I_x = \frac{bh^3}{12}\left[\frac{2L - x_o}{L}\right]^3 \tag{9.19}$$

where b is the width of the member. If the moment of inertia I_B at the free end of the member is taken as the reference value of I_x, then we have

$$f(x) = \left[\frac{2L - x_o}{L}\right]^3 \tag{9.20}$$

and

$$I_B = \frac{bh^3}{12} \tag{9.21}$$

The cross-sectional shape of the member is assumed to have a rectangular shape.

In order to determine a pseudolinear system of constant stiffness E_1I_1, the rotation y' must be known. This rotation can be determined from Equation 9.9 by integration, i.e.,

$$y' = \frac{1}{E_1I_1}\int_o^x M_e' dx + C_1 \tag{9.22}$$

where C_1 is the constant of integration. In order to carry out the integration of Equation 9.22, the horizontal displacement Λ at the free end of the beam and the function g(x) must be known. Thus, inelastic analysis should be initiated for the computation of g(x), and nonlinear analysis must be utilized for the computation of Λ.

The methodology regarding the computation of the large deflections and rotations by inelastic analysis may be carried out by employing trial and error procedures and the equivalent pseudolinear system represented by Equation 9.9. The required integrations may be carried out with sufficient accuracy by using Simpson's rule. The following step-by-step procedure is suggested here for convenience:

Step 1: The first approximation of $g(x) = E_r/E_1$, where E_r is the reduced modulus, may be obtained by assuming that $g(x) = 1$. Proceed with the integration of Equation 9.9, as shown in Equation 9.22, to determine y'. The boundary condition of zero rotation at the fixed end of the member may be used to determine the constant of integration C_1. Since M_e' in Equation 9.22 depends upon $L_o = (L - \Lambda)$, the value of the horizontal displacement Λ of the free end of the bar must first be determined. The value of Λ can be determined from Equation 9.14 by applying a trial and error procedure, i.e., assume a value of Λ and determine L from Equation 9.14 in conjunction with Equation 9.22. The procedure may be repeated until the correct length L of the bar is obtained. With known Λ, the values of y' can be determined from Equation 9.22, and the values of M_x at any $0 \leq x \leq L_o$ can be determined from Equation 9.17. The procedure converges rather fast.

Step 2: With known M_x, we can start the procedure to determine E_r, and consequently $g(x) = E_r/E_1$, by using Timoshenko's method as explained in Section 6.2. Since the problem is statically determinate, we have $M_{req} = M_x$.

Step 3: By using the values of y' from Step 1 we determine Z_e from Equation 9.8. With known Z_e and using the values of g(x) from Step 2, we determine M_e' from Equation 9.10.

Step 4: By using M_e' from Step 1 and g(x) from Step 2, we determine a new y' from Equation 9.22 which incorporates the inelastic behavior of the member. The boundary condition of zero rotation at the fixed end of the beam may be used to determine C_1.

Step 5: By utilizing Equation 9.14 and the new y' obtained from Step 4, a new horizontal displacement Λ and a new $L_o = L - \Lambda$ may be obtained. Thus, a new $M_{req} = M_x$ may be determined from Equation 9.17, and a new g(x) may be calculated as discussed in Step 2.

Step 6: By using the new y' from Step 4, a new M_e' may be obtained from Equation 9.10 and a new y' may be calculated by using Equation 9.22 and the g(x) from Step 5.

The procedure may be repeated until the last y' is almost identical to the one obtained from the preceding trial. Usually four to seven repetitions are sufficient. The utilization of pseudolinear analysis, i.e., M_e', greatly facilitates the solution and convergence of the trial and error procedure. With known y', the large deflections may be obtained by either (1) integrating Equation 9.22 once and satisfying the boundary condition of zero deflection at the fixed end or (2) using the moment area method, since M_e' is known. The second procedure is the most convenient (see Chapters 1, 2, and 3).

In order to illustrate the above procedure, consider the flexible tapered cantilever beam in Figure 9.1a and let it be assumed that L = 1000 in. (25.4 m), w_1 = 3.0 lb/in. (525.38 N/m), h = 3.0 in. (0.0762 m), and width b = 3.0 in. (0.0762 m). The material of the beam is Monel with a stress/strain curve as shown in Figure 6.3a. The values of E and σ for a two-, three-, and six-line approximation of this stress/strain curve are shown in Table 6.1 and 6.2, respectively. The three-line approximation is used here.

By applying the step-by-step procedure stated above and repeating it about four times, the final values of y', g(x), Z_e, M_e', and strain Δ are obtained, and they are shown in Table 9.1. The moment diagram M_e' of the equivalent pseudolinear system of constant stiffness E_1I_1, where $E_1 = 22 \times 10^6$ psi (151.68×10^6 kPa) and I_1 = 6.75 in.^{4} (2.8096×10^{-6} m^4), is shown in Figure 9.2a. The approximation of its shape with three straight line segments as shown in the same figure leads to the pseudolinear equivalent system of constant stiffness E_1I_1 shown in Figure 9.2b. By using the pseudolinear system in Figure 9.2b and applying the well-known moment area method, the rotations y' and deflections y along the length L_o of the member may be easily obtained. Since the pseudolinear system yields y', the actual rotation $\theta = \tan^{-1}(y')$. Table 9.2 shows the values of the vertical displacement δ, horizontal displacement Λ, and rotation θ at the free end of the beam for various values of the distributed load w_1. The column notation "Exact PES", which means exact pseudolinear equivalent system, is used to denote the results obtained by using the pseudolinear equivalent system. For example, for w_1 = 3.0 lb/in. (525.38 N/m), we find δ = 646.57 in. (16.42 m), Λ = 263.90 in. (6.70 m), and $\theta = 52.03°$. For the results obtained above, the

TABLE 9.1
Final Values of y', $g(x)$, Z_e, Δ, and M_e'

x (in.)	f(x)	Δ	g(x)	y'	Z_e	$M_{req}=M_x$ (in.-kip)	$M_e' = \frac{M_x Z_e}{f(x)\,g(x)}$ (in.-kip)
0.0	8.0	4.6214×10^{-2}	0.1609	0.0	1.0	1472.20	1143.80
36.8	7.56	2.8364×10^{2}	0.2466	2.3658×10^{-1}	1.0848	1334.00	776.01
73.6	7.12	1.4914×10^{-2}	0.4395	3.8983×10^{-1}	1.2356	1201.20	474.04
110.4	6.69	9.1136×10^{-3}	0.6710	4.8703×10^{-1}	1.3749	1073.90	329.23
147.2	6.25	6.8136×10^{-3}	0.8331	5.6109×10^{-1}	1.5061	952.90	275.72
184.0	5.83	5.6136×10^{-3}	0.9327	6.2626×10^{-1}	1.6409	834.49	253.38
220.8	5.41	4.8636×10^{-3}	0.9850	6.8793×10^{-1}	1.7862	730.90	244.96
257.6	5.00	4.3538×10^{3}	1.0	7.4875×10^{-1}	1.9468	630.28	245.33
294.4	4.60	3.9185×10^{-3}	1.0	8.0992×10^{-1}	2.1279	536.73	248.12
331.2	4.22	3.4866×10^{-3}	1.0	8.7156×10^{-1}	2.3310	450.35	249.03
368.0	3.84	3.0592×10^{-3}	1.0	9.3310×10^{-1}	2.5552	371.27	247.12
404.8	3.48	2.6381×10^{-3}	1.0	9.9368×10^{-1}	2.7983	299.60	241.25
441.6	3.13	2.2256×10^{-3}	1.0	1.0522	3.0551	235.49	230.20
478.4	2.79	1.8249×10^{-3}	1.0	1.1072	3.3171	179.05	212.83
515.3	2.47	1.4404×10^{-3}	1.0	1.1570	3.5725	130.38	188.37
552.1	2.17	1.0782×10^{-3}	1.0	1.1999	3.8064	89.55	156.84
588.9	1.89	7.4635×10^{-4}	1.0	1.2343	4.0035	56.57	119.53
625.7	1.64	4.5567×10^{-4}	1.0	1.2590	4.1506	31.33	79.43
662.5	1.40	2.2060×10^{-4}	1.0	1.2739	4.2418	13.67	41.39
669.3	1.19	6.0220×10^{-5}	1.0	1.2803	4.2820	3.35	12.05
736.1	1.0	0.0	1.0	1.2815	4.2888	0.0	0.0

Note: 1 in. = 0.0254 m, 1 in.-kip = 112.9848 Nm.

TABLE 9.2
Values of δ, Λ, and θ at the Free End of the Member vs. Values of the Distributed Load w_1 Using the Exact PES and the Simplified NES

w_1 (lb/in.)	δ Exact PES (in.)	δ Simplified NES (in.)	Difference %	Λ Exact PES (in.)	Λ Simplified NES (in.)	Difference %	θ Exact PES (°)	θ Simplified NES (°)	Difference %
1	184.66	184.6	−0.03	20.9	20.80	−0.48	15.63	15.67	0.26
2	361.41	361.99	0.16	80.1	81.08	1.22	30.52	31.02	1.63
3	646.57	639.53	−1.09	263.9	257.20	−2.54	52.03	51.95	−0.15
4	779.18	778.25	−0.12	414.3	413.98	−0.08	64.01	63.98	−0.05
5	841.07	850.74	1.15	516.1	519.90	0.74	70.84	71.59	1.06
6	881.7	881.52	−0.02	587.1	584.10	−0.51	75.26	75.45	0.25

Note: 1 in. = 0.0254 m, 1 lb/in. = 175.1268 N/m.

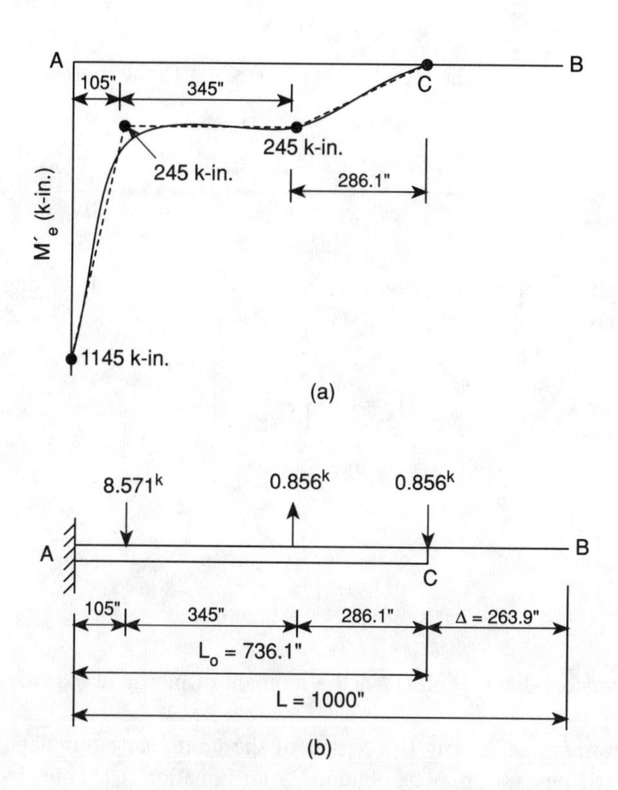

FIGURE 9.2. (a) Moment diagram M_e' of the pseudolinear equivalent system with its shape approximated with three straight lines; (b) equivalent pseudolinear system of constant stiffness E_1I_1 (1 in. = 0.0254 m, 1 kip-in. = 112.9848 Nm, 1 kip = 4.448 kN).

utilization of Equation 9.9 and the application of the preceding step-by-step procedure was carried out by subdividing each time the length L_o into 40 equal segments and carrying out the required integrations by using Simpson's rule. This procedure yields almost exact results.

9.4 INELASTIC ANALYSIS OF FLEXIBLE BARS BY USING SIMPLIFIED NONLINEAR EQUIVALENT SYSTEMS

This method of analysis is particularly useful for flexible beam problems where the moment of inertia of the member and its loading conditions are complicated, and in many ways arbitrary. The application of this methodology is illustrated by again using the variable stiffness cantilever beam loaded as shown in Figure 9.1a. More complicated beam problems are discussed later in the chapter. In order to compare results, the geometric dimensions and loading values for the problem in Figure 9.1a are considered to be the same as the ones used to carry out the inelastic analysis in the preceding section by using the pseudolinear system represented by Equation 9.9.

Along the length of the member, at any $0 \le x \le L$, the moment M_x is given by the equation

$$M_x = \frac{2w_1L^2}{3} - \frac{3w_1Lx}{2} + \frac{w_1x^2}{2L}(2L - x) + \frac{w_1x^3}{3L} \tag{9.23}$$

The straight undeformed configuration is used to derive M_x. Along the same axis we have

$$I_x = \frac{bh^3}{12}\left[\frac{2L - x}{L}\right]^3 \tag{9.24}$$

and

$$f(x) = \left[\frac{2L - x}{L}\right]^3 \tag{9.25}$$

The reference value $I_1 = bh^3/12$ is the moment of inertia at the free end of the beam.

The moment M_e at any $0 \le x \le L$ of the nonlinear equivalent system of constant stiffness EI_1, may be obtained from Equation 9.4. Thus, by substituting Equations 9.23 and 9.25 into Equation 9.4, we have

$$\begin{aligned} M_e &= \frac{M_x}{f(x)} \\ &= \frac{L^2}{6(2L - x)^3}\left[4w_1L^3 - 9w_1L^2x + 3w_1x^2(2L - x) + 2w_1x^3\right] \end{aligned} \tag{9.26}$$

Note that g(x) in Equation 9.4 is taken to be equal to one, because inelastic analysis is not yet initiated.

By assuming the same values for w_1 and L as for the pseudolinear analysis in the preceding section, the plot of M_e is shown in Figure 9.3a. The approximation of its shape with one straight line as shown leads to the simplified nonlinear equivalent system of constant stiffness EI_1, shown in Figure 9.3b. Accurate large deflections and rotations may be obtained here by using the simplified nonlinear system in Figure 9.3b in place of the original variable stiffness nonlinear system in Figure 9.1a. The solution of the simplified nonlinear problem in Figure 9.3b may now be obtained by either (1) using simple nonlinear existing methods of analysis, if such methods are readily available or (2) by applying pseudolinear analysis and the step-by-step procedure stated in the preceding section.

The inelastic analysis here was carried out by using the simplified nonlinear equivalent system in Figure 9.3b and applying pseudolinear analysis as discussed in Section 9.3. In other words, we apply the step-by-step procedure stated in this section. This procedure produces the final values for g(x), Δ, y', Z_e, and M_e' shown in Table 9.3. Thus, with known M_e', the values of the large displacement of the member may be determined as stated earlier by either integrating Equation 9.9 once or by using the pseudolinear system of constant stiffness E_1I_1 and applying the moment area method. The second approach was used here. Table 9.2, under column heading "Simplified NES" shows values of the vertical displacement δ, horizontal displacement Λ, and rotation θ at the free end of the member for various values of the distributed load w_1. In the same table, the results obtained using the exact PES are shown for comparison purposes. For practical purposes the difference is considered to be negligible, but if greater accuracy is required, more straight lines may be used to approximate the M_e diagram as shown by the dashed lines in Figure 9.3c. On this basis, the simplified nonlinear equivalent system will be loaded with more than one concentrated load. The inelastic levels of maximum stress σ at the fixed end of the beam vs. given values of w_1 are shown in Figure 9.3d. The solid line represents the results obtained by the pseudolinear system, and the black dots are the results obtained by the simplified nonlinear equivalent system. It should be pointed out here that solving the simplified nonlinear equivalent system of constant stiffness is a much simpler task.

9.5 INELASTIC ANALYSIS OF A PRISMATIC CANTILEVER BEAM WITH CONCENTRATED LOAD AT THE FREE END

Let it be assumed that the prismatic cantilever beam in Figure 9.4a is loaded with a concentrated load P at the free end as shown. The beam is assumed to be uniform throughout its length so that f(x) = 1. The large deformation configuration of the beam is shown in Figure 9.4b, with Λ representing the horizontal displacement of the free end. The bending moment M_x at any $0 \le x \le (L - \Lambda)$ is given by the equation

$$M_x = -P(L - x - \Lambda) \tag{9.27}$$

By substituting Equation 9.27 into Equation 9.9, we find

$$\begin{aligned} y'' &= \frac{PZ_e}{E_1 Ig(x)}(L - x - \Lambda) \\ &= -\frac{M_e'}{E_1 I} \end{aligned} \tag{9.28}$$

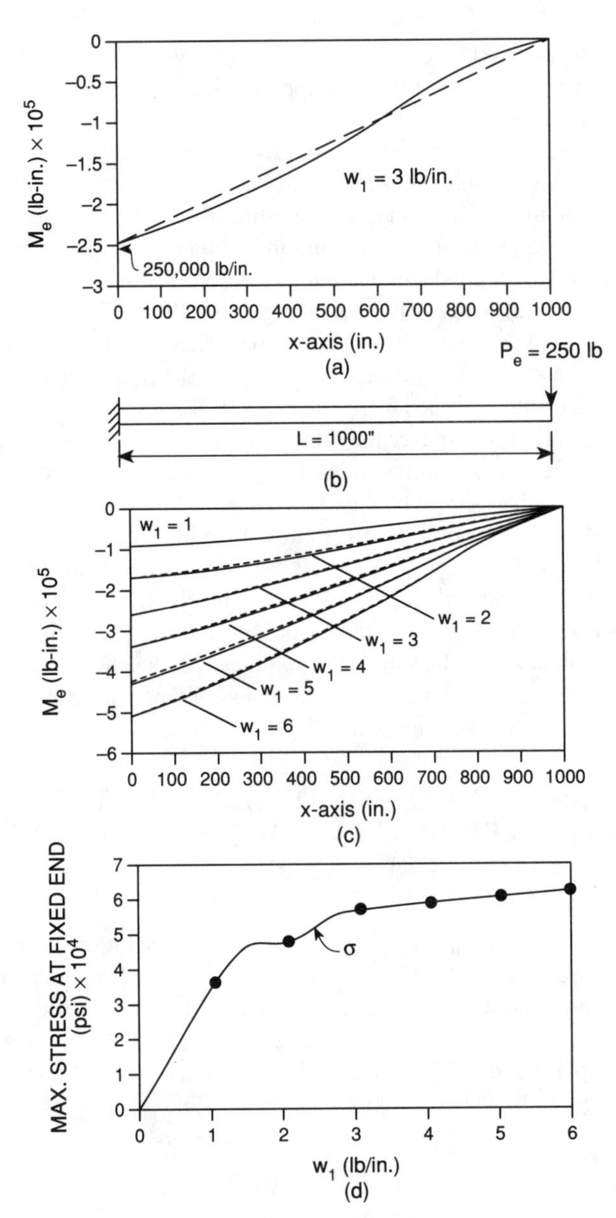

FIGURE 9.3. (a) Moment diagram M_e of the simplified NES with its shape approximated with one straight line; (b) simplified NES of constant stiffness EI_1; (c) M_e diagrams for various values of w_1 with their shape approximated with three straight lines; (d) maximum normal stress at the fixed end of the bar with increasing w_1 (1 in. = 0.0254 m, 1 lb/in. = 175.1268 N/m, 1 psi = 6894.757 Pa).

TABLE 9.3
Final Values of Δ, y', $g(x)$, Z_e, and M_e'

x(in.)	f(x)	Δ	g(x)	y'	Z_e	M_e (in.-kip)	$M_e' = \frac{M_e Z_e}{g(x)}$ (in.-kip)
0.0	1.0	4.9114×10^{-2}	1.5278×10^{-1}	0.0	1.0	185.71	1215.60
37.1	1.0	2.8314×10^{-2}	2.4701×10^{-1}	0.2481	1.0929	176.42	780.50
74.3	1.0	1.3714×10^{-2}	4.7354×10^{-1}	0.3978	1.2441	167.14	439.11
111.4	1.0	8.3136×10^{-3}	7.2155×10^{-1}	0.4879	1.3755	157.85	300.68
148.6	1.0	6.3136×10^{-3}	8.7459×10^{-1}	0.5568	1.4976	148.57	254.27
185.7	1.0	5.2636×10^{-3}	9.5959×10^{-1}	0.6177	1.6224	139.28	235.16
222.9	1.0	4.6136×10^{-3}	9.9577×10^{-1}	0.6756	1.7568	130.00	228.83
260.0	1.0	4.1650×10^{-3}	1.0	0.7329	1.9056	120.71	230.03
297.1	1.0	3.7395×10^{-3}	1.0	0.7905	2.0723	111.43	230.90
334.3	1.0	1.6644×10^{-3}	1.0	0.8483	2.2568	102.14	230.51
371.4	1.0	2.9335×10^{-3}	1.0	0.9057	2.4587	92.86	228.30
408.6	1.0	2.5543×10^{-3}	1.0	0.9622	2.6764	83.57	223.67
445.7	1.0	2.1921×10^{-3}	1.0	1.0173	2.9068	74.28	215.93
482.8	1.0	1.8476×10^{-3}	1.0	1.0699	3.1448	64.99	204.41
520.0	1.0	1.5216×10^{-3}	1.0	1.1191	3.3836	55.71	188.51
557.1	1.0	1.2152×10^{-3}	1.0	1.1637	3.6142	46.43	167.80
594.3	1.0	9.2894×10^{-4}	1.0	1.2026	3.8257	37.14	142.10
631.4	1.0	6.6375×10^{-4}	1.0	1.2344	4.0068	27.86	111.61
688.6	1.0	4.2023×10^{-4}	1.0	1.2580	4.1460	18.57	76.99
705.7	1.0	1.9888×10^{-4}	1.0	1.2726	4.2338	9.29	39.31
742.8	1.0	0.0	1.0	1.2775	4.2639	0.0	0.0

Note: 1 in. = 0.0254 m, 1 in.-kip = 112.9848 Nm.

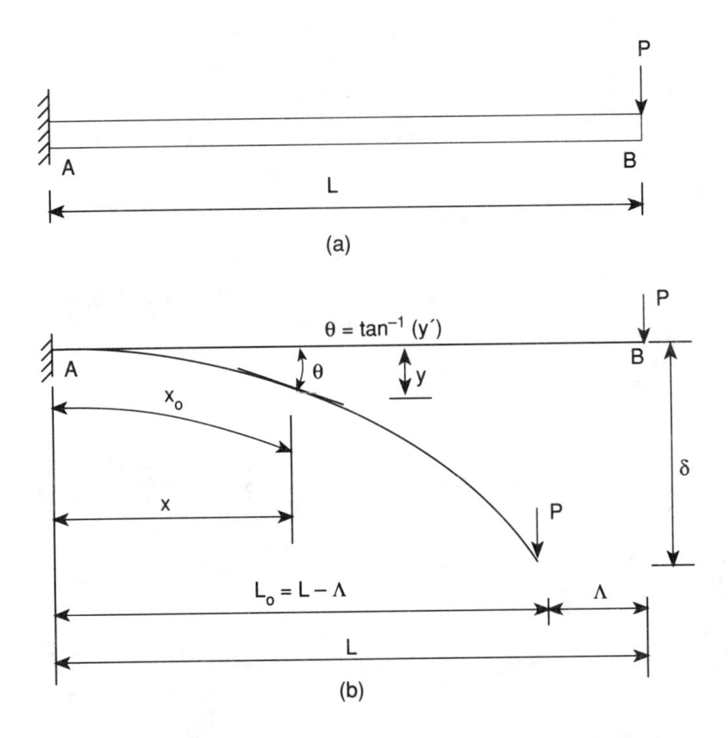

FIGURE 9.4. (a) Prismatic cantilever beam loaded with a concentrated load P at the free end; (b) large deformation configuration of the flexible cantilever beam.

where, from Equations 9.10 and 9.28, we have

$$M'_e = -\frac{PZ_e}{g(x)}(L - x - \Lambda) \tag{9.29}$$

By integrating Equation 9.28 once, we may write

$$y' = \frac{1}{E_1 I}\int_0^x M'_e dx + C_1 \tag{9.30}$$

where C_1 is the constant of integration.

In order to be able to carry out the integration in Equation 9.30, the horizontal displacement Λ at the free end of the beam and the variation g(x) of $E_x = E_1 g(x)$ must be known. Thus, inelastic analysis should be initiated for the computation of g(x), and nonlinear analysis must be used for the computation of Λ. The inelastic analysis can be carried out now by using the step-by-step procedure discussed in Section 9.3. The material of the flexible beam is Monel, and the three-line approximation of the stress/strain curve of Monel is used in order to carry out the inelastic analysis.

From step 1, a first approximation of $g(x) = E_r/E_1$, where E_r is the reduced modulus, may be obtained by assuming that g(x) in Equation 9.28 is equal to one. One can then proceed with the integration of Equation 9.28 to determine y′. For example, with g(x) = 1, the integration of Equation 9.28 yields

$$y' = \frac{k^2\left(Lx - \frac{x^2}{2} - \Lambda x\right)}{\left\{1 - \left[k^2\left(Lx - \frac{x^2}{x} - \Lambda x\right)\right]^2\right\}^{1/2}} \tag{9.31}$$

where

$$k^2 = \frac{P}{EI} \tag{9.32}$$

The boundary condition of zero rotation at the fixed end is used here to determine the constant of integration.

In order to determine y′ from Equation 9.31, the value of Λ must be known. This value can be determined from Equation 9.14 by trial and error. This procedure may be initiated assuming a value of Λ and determining L by using Equation 9.14. The procedure may be repeated until the correct length L of the member is obtained. With known Λ, the values of y′ can be determined from Equation 9.14, and the values of the actual moment M_x at any $0 \le x \le L_o$, where $L_o = L - \Lambda$, can be determined from Equation 9.27.

In order to obtain numerical results, let it be assumed that P = 800 lb (3560 N), L = 600 in. (15.24 m), $E_l = 22 \times 10^6$ psi ($151{,}684.65 \times 10^6$ Pa), and I = 6.75 in^4 (2.8096×10^{-6} m^4). With g(x) = 1, the first trial values of y′ and Λ are obtained from Equations 9.31 and 9.14 using the indicated trial and error procedure in Step 1. This yields Λ = 92.7 in. (2.3268 m). Thus values of y′ at any $0 \le x \le Lo$, where $L_o = L - \Lambda = 600 - 92.7 = 507.3$ in. (12.7332 m), may be obtained from Equation 9.31, and the values of the actual moment M_x may be determined from Equation 9.27. The vertical deflection at the free end, as a result of the above analysis, is 290.97 in. (7.3033 m), and the rotation θ at the same end is $\theta = \tan^{-1}(y') = \tan^{-1}(0.96292) = 43.9178°$.

Since, for statically determinate problems, the required moment M_{req} is equal to the actual moment M_x, the procedure explained in Step 2 may be initiated to determine the reduced modulus E_r along the length L_o of the member, and consequently the function $g(x) = E_r/E_1$. Table 9.4 gives a summary of the calculations for the strain Δ, g(x), and the required moment $M_{req} = M_x$ for various locations x along the length L_o of the bar. The bar has a square cross section with sides equal to 3 in. (0.0753 m). By following the procedure in Step 2, the values of Z_e and M_e' are determined; they are shown in the same table.

TABLE 9.4
Values of g(x), y′, Z_e, M_{req}=M_x, and M_e' Obtained from the First Trial

1	2	3	4	5	6	7
x (in.)	f(x)	$g(x) = E_r/E_1$	y′	$Ze = [1 + (y')^2]^{3/2}$	$M_{req} = M_x$ (in.-kip)	$M_e' = \frac{Z_e M_x}{f(x)\,g(x)}$ (in.-kip)
0.0	1.0	0.07268	0.0	1.0	405.87	5584.30
25.40	1.0	0.11711	0.0678	1.0	385.58	3315.30
50.70	1.0	0.16840	0.1329	1.0268	365.28	2226.90
76.10	1.0	0.25424	0.1961	1.0586	344.99	1436.50
101.50	1.0	0.43640	0.2579	1.1020	324.70	819.01
126.80	1.0	0.65648	0.3185	1.1569	304.40	536.25
152.20	1.0	0.81315	0.3183	1.2235	284.11	427.79
177.60	1.0	0.91641	0.4373	1.3019	263.82	374.64
202.60	1.0	0.91328	0.4956	1.3922	243.52	348.97
228.30	1.0	0.99840	0.5530	1.4946	223.23	333.92
253.70	1.0	1.0	0.6094	1.6094	202.94	326.40
279.00	1.0	1.0	0.6644	1.7326	182.64	316.45
304.40	1.0	1.0	0.7173	1.8654	162.35	302.84
329.80	1.0	1.0	0.7675	2.0037	142.05	284.63
355.10	1.0	1.0	0.8142	2.1432	121.76	260.96
380.50	1.0	1.0	0.8563	2.2187	101.47	231.21
405.90	1.0	1.0	0.8928	2.4035	81.17	195.10
431.20	1.0	1.0	0.9346	2.5107	60.88	152.85
456.60	1.0	1.0	0.9447	2.5933	40.59	105.26
482.00	1.0	1.0	0.9583	2.6455	20.29	53.69
507.30	1.0	1.0	0.9629	2.6634	0.0	0.0

Note: 1 in. = 0.0254 m, 1 in.-kip = 112.9848 Nm.

By following the methodology stated in Steps 4, 5, and 6 and repeating the procedure about four times, the final values of y', $g(x)$, Z_e, M_e', and strain Δ are obtained. They are shown in Table 9.5. Note in this table that $0 \leq x \leq L_0$ with $L_0 = 442.0$ in. (11.0942 m). In Table 9.4, where elastic analysis is used for the first trial values of y', the value of $L_0 = 507.3$ in. (12.7332 m), which is 14.77% larger than the correct value of 442.0 in. (11.0942 m) obtained by inelastic analysis.

The moment diagram M_e' of the equivalent system of constant stiffness E_1I_1 is plotted in Figure 9.5a. Its approximation with four straight line segments leads to the constant stiffness equivalent system shown in Figure 9.5b. By using the equivalent system in Figure 9.5b and applying the well-known moment area method, the deflection y along the length L_0 of the bar and the rotation y' can be easily determined. Since the equivalent system yields y', the actual rotations $\theta = \tan^{-1}(y')$. At the free end of the bar, i.e., x = 442.0 in. (11.0942 m), the equivalent system in Figure 9.5b yields the deflection y = 393.674 in. (9.8812 m) and $\theta = \tan^{-1}(y') = \tan^{-1}(1.436) = 55.1471°$. The values obtained by utilizing Equation 9.30 and integrating once are y = 384.12 in. (9.7566 m) and $\theta = \tan^{-1}(1.352) = 53.512°$. Better agreement may be obtained by using more straight lines to approximate the M_e' diagram in Figure 9.5a. The utilization of Equation 9.30 and the step-by-step procedure was performed by subdividing each time L_0 into 40 equal segments, which should yield almost exact results.

Figure 9.5c shows various graphs representing variations in horizontal and vertical free end displacements Λ and δ, respectively, vs. load P for comparison purposes. For example, curve 1 shows δ vs. P by using small deflection theory and inelastic analysis. In curve 2 large deflection theory coupled with inelastic analysis was used, while in curve 3 large deflection theory with elastic analysis was used. Curves 4 and 5 show variations of Λ vs. P for large deflection inelastic and large deflection elastic analysis, respectively. In all five curves in Figure 9.5c, the solution was simplified by pseudolinear equivalent systems. It should be noted that for large deformations and large loads P, large deformation theory combined with inelastic analysis must be used in order to obtain accurate results (see curves 2 and 4 in Figure 9.5c). The curve in Figure 9.6a represents the variation of g(x) at the fixed end with load P, and the curve in Figure 9.6b shows the variation of the maximum normal stress at the same end with load P.

The results obtained by using pseudolinear equivalent systems were also compared with those obtained using the 4th order Runge-Kutta method. In order to apply the 4th order Runge-Kutta method, the stress/strain curve of Monel was curve-fitted by the equation $\sigma_{max} = 103992\varepsilon^{0.156988}$, which is represented by the curve shown in Figure 9.6c. In the same figure, the three-line approximation that was used for the pseudolinear equivalent systems is also shown for comparison. This is the best curve fitting we could obtain. The 3-line approximation is considered to be a better approximation of the shape of the stress/strain curve.

In Figure 9.7 the results obtained for end displacements δ and Λ and end rotation θ, by using large deformation theory with inelastic analysis and

TABLE 9.5
Final Results from the Last Trial for Strain Δ, g(x), y′, Z_e, M_{req}, and M_e'

1	2	3	4	5	6	7	8
x	f(x)	Strain Δ	g(x)	y′	$Z_e = [1 + (y')^2]^{3/2}$	$M_{req} = M_x$ (in.-kip)	$M_e' = \frac{M_x Z_e}{f(x)\, g(x)}$ (in.-kip)
0.0	1.0	3.4164×10^{-1}	0.20910	0.0	1.0	353.59	1691.1
22.1	1.0	2.7214×10^{-1}	0.32220	0.20641	1.0646	335.91	110.0
44.2	1.0	1.2614×10^{-2}	0.50972	0.34162	1.1800	318.23	736.7
66.3	1.0	8.8136×10^{-3}	0.68926	0.43722	1.2998	300.55	567.1
88.4	1.0	6.9636×10^{-3}	0.82111	0.51524	1.4233	282.87	490.6
110.5	1.0	5.8636×10^{-3}	0.91227	0.58488	1.5547	265.19	451.3
132.6	1.0	5.1636×10^{-3}	0.96664	0.65082	1.6986	247.51	434.2
154.7	1.0	4.6636×10^{-3}	0.99407	0.71507	1.8582	229.83	429.0
176.8	1.0	4.2859×10^{-3}	1.0	0.77909	2.0315	212.15	432.3
198.9	1.0	3.9288×10^{-3}	1.0	0.84369	2.2403	194.47	435.7
221.0	1.0	3.5716×10^{-3}	1.0	0.90860	2.4673	176.79	436.2
243.1	1.0	3.2145×10^{-3}	1.0	0.97329	2.7183	159.12	432.5
265.2	1.0	2.8573×10^{-3}	1.0	1.03700	2.9908	141.44	423.0
287.3	1.0	2.5001×10^{-3}	1.0	1.09880	3.2803	123.76	406.0
309.4	1.0	2.1430×10^{-3}	1.0	1.15730	3.5788	106.08	379.6
331.5	1.0	1.7858×10^{-3}	1.0	1.21120	3.8750	88.40	342.5
353.6	1.0	1.4286×10^{-3}	1.0	1.25860	4.1538	70.72	293.8
375.7	1.0	1.0715×10^{-3}	1.0	1.29790	4.3980	53.04	233.3
397.8	1.0	7.1432×10^{-4}	1.0	1.32750	4.5893	35.36	162.3
419.9	1.0	3.5716×10^{-4}	1.0	1.34580	4.7116	17.68	83.3
442.0	1.0	0.0	1.0	1.35200	4.7537	0.0	0.0

Note: 1 in. = 0.0254 m, 1 in.-kip = 112.9848 Nm.

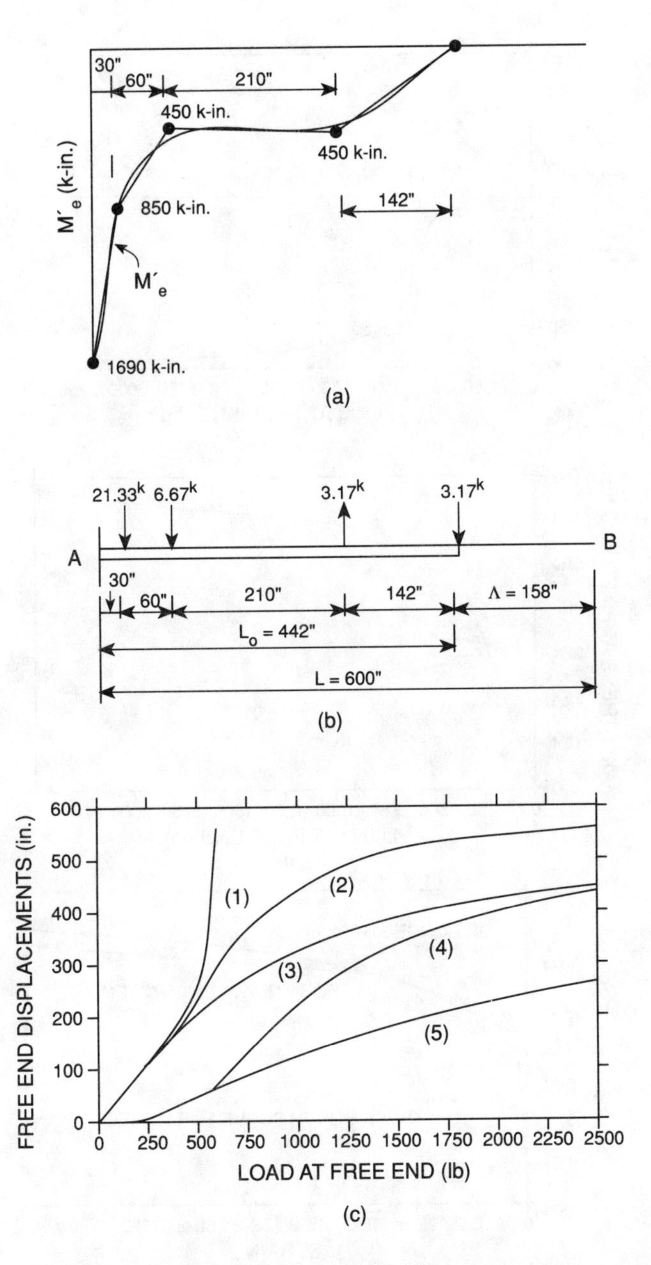

FIGURE 9.5. (a) Moment diagram M_e' of the pseudolinear equivalent system with its shape approximated with four straight lines; (b) equivalent pseudolinear system of constant stiffness E_1I; (c) graphs of free end displacements vs. end load P (1 in. = 0.0254 m, 1 kip-in. = 112.9848 Nm, 1 kip = 4448.222 N).

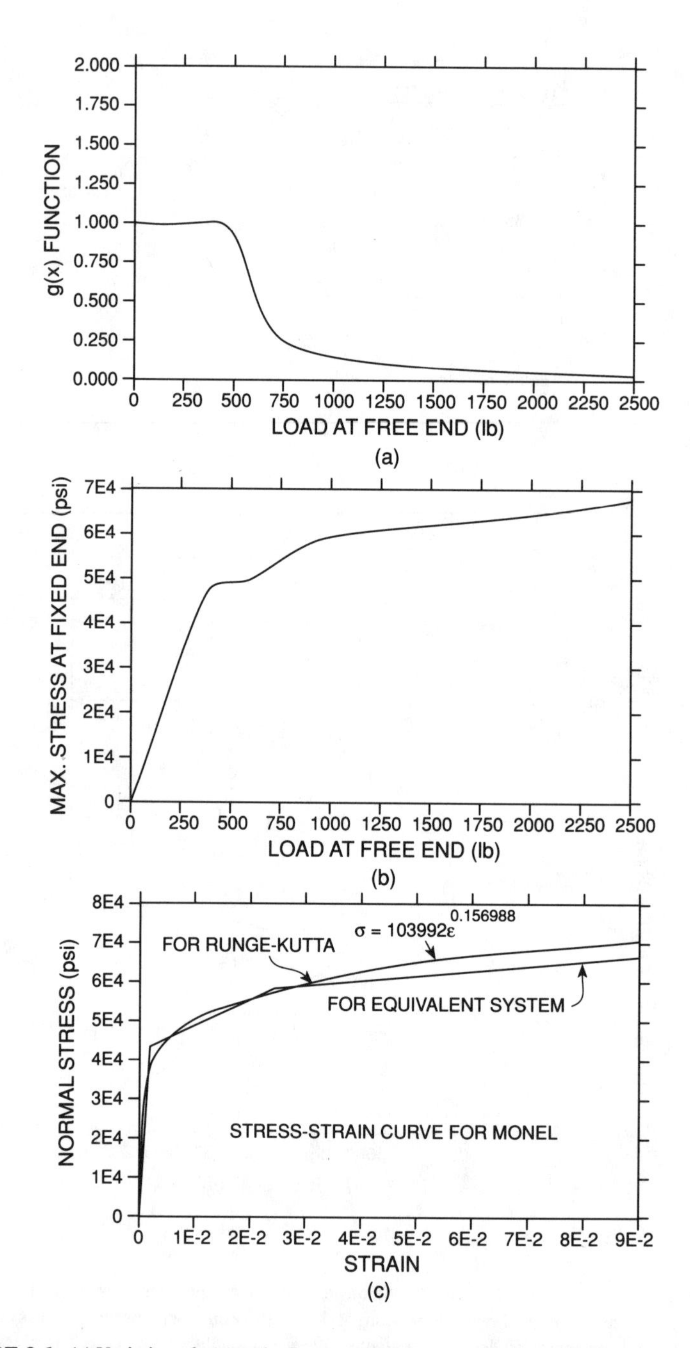

FIGURE 9.6. (a) Variation of g(x) at the fixed end with end load P; (b) variation of the maximum normal stress at the fixed end with load P; (c) equivalent systems and Runge-Kutta approximations of the stress/strain curve of Monel (1 in. = 0.0254 m, 1 lb = 4.448 N, 1 psi = 6894.757 Pa).

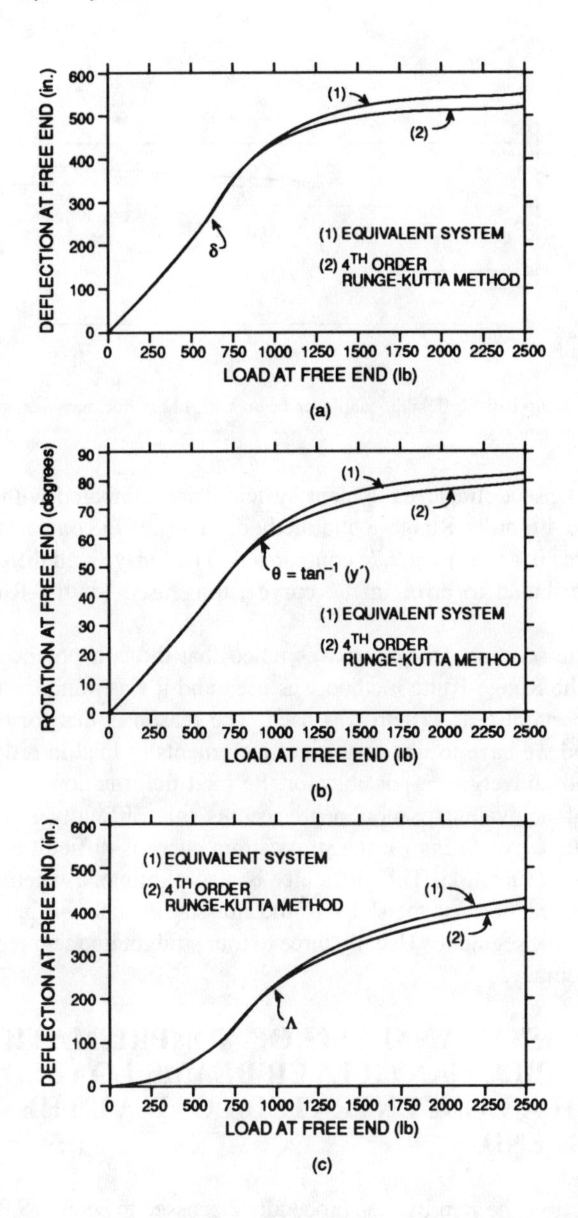

FIGURE 9.7. (a) Curves of end deflection δ vs. load P; (b) curves of end rotation θ vs. load P; (c) curves of end displacement Λ vs. load P (1 in. = 0.0254m, 1 lb = 4.4482 N).

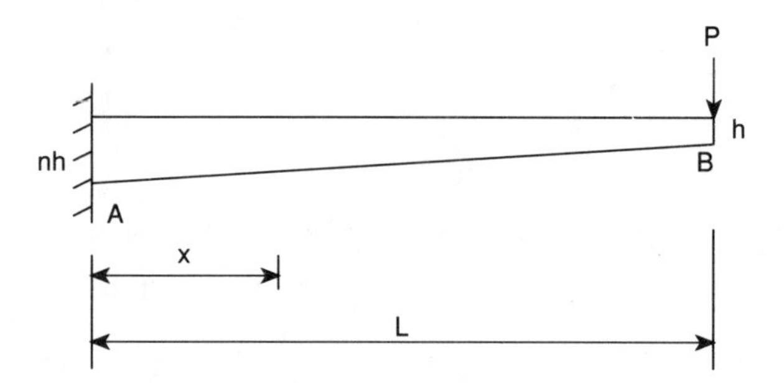

FIGURE 9.8. Nonprismatic flexible cantilever beam with linear thickness variation along its length.

utilization of pseudolinear equivalent systems, are compared with those obtained by the 4th order Runge-Kutta method. Figure 9.7a compares δ, Figure 9.7b compares θ, and Figure 9.7c compares Λ. The fairly small differences are primarily attributed to error in the curve fitting used for the Runge-Kutta method.

During the solution process, it was noted that the convergence was very slow when the Runge-Kutta method was used and it was rather fast when the equivalent pseudolinear system was used. The reason is that for the Runge-Kutta method we have to use very small increments of load in order to make the next load convergence possible for the load deformation curves. In the utilization of equivalent pseudolinear systems this difficulty is eliminated. Also, accurate curve fitting for the stress/strain curve is difficult to obtain for the Runge-Kutta method. This difficulty is also eliminated when equivalent systems are used, because the shape of the stress/strain curve is approximated with straight line segments. Usually three to four straight line segments provide excellent accuracy.

9.6 INELASTIC ANALYSIS OF NONPRISMATIC FLEXIBLE CANTILEVER BEAMS LOADED WITH A CONCENTRATED LOAD AT THE FREE END

In this section, the step-by-step procedure discussed in Section 9.3 is applied to a nonprismatic cantilever beam loaded at the free end with vertical concentrated load P, as shown in Figure 9.8. The thickness h varies linearly along the length of the member as shown in the same figure. This problem is similar to the one in the preceding section, with the exception of the thickness variation. The step-by-step procedure is identical to the one used in the preceding section, with the difference that the moment of inertia function f(x) is no longer equal to one and its variation along the length of the member must be taken into consideration. Therefore, correspondence between x and x_0 should be maintained during the process if an exact solution to the problem is desired. This correspondence can be maintained by making use of Equations 9.12, 9.13, and 9.14. That is, for every length $0 \le x \le L_0$, there corresponds a length $0 \le x \le L$, and the value of the function f(x) at any $0 \le x \le L_0$ should be taken as the one corresponding to $0 \le x \le L$. If the correspondence between x and x_0 is not maintained and one selects to express $I_x = I_1 f(x)$ as a function of $0 \le x \le L_0$, the error would be well within the 10% for cantilever beams and well within 5% for beams supported at both ends.

Consider, for example, the cantilever bar in Figure 9.8 and let it be assumed that n = 2, h = 3 in. (0.0762 m), width b = 3 in. (0.0762 m), and L = 600 in. (15.24 m). The inelastic analysis of the preceding section was carried out by subdividing the length L_0 in each trial into 40 segments and retaining the correspondence between x and x_0 in the computation of f(x). Table 9.6 shows the variation of the free end displacements δ, Λ, and θ with load P. The results are compared with those obtained by the 4th order Runge-Kutta method. Fairly close agreement in results is observed. In Figure 9.9a, curve 1 shows the variation of δ with P using inelastic analysis, and curve 2 shows the variation of δ by assuming that the bar remains elastic during deformation. The same comparison is made for Λ in curves 3 and 4, with curve 3 representing the inelastic analysis. In Figure 9.6b the variation of g(x) at the fixed end vs. load P is shown. In Figure 9.10, the free end displacements δ, θ, and Λ are compared with the results obtained using the 4th order Runge-Kutta method. For practical applications the results in Figure 9.10 are considered to be in close agreement.

When the vertical load on the flexible member is distributed, correspondence between x and x_0 in the computation of the bending moment at any $0 \le x \le L_0$ should also be maintained. The inelastic analysis for the solution of the cantilever beam problems in Sections 9.3 and 9.4 takes into consideration the correspondence between x and x_0. For beam cases such as the ones in Sections 9.3 and 9.4, where the moment of inertia is variable and the load is distributed, the error tends to be rather small for practical applications when the correspondence between x and x_0 is not maintained. The reason is that the error introduced when the funtion f(x) is considered is opposite in sign compared to the error introduced when the distributed load is considered and the net total error tends to be smaller.

9.7 INELASTIC ANALYSIS OF NONPRISMATIC FLEXIBLE CANTILEVER BEAMS SUBJECTED TO A UNIFORMLY DISTRIBUTED LOADING

Consider a tapered nonprismatic cantilever bar loaded by a uniformly distributed load w as shown in Figure 9.11a. Its large deflection configuration is shown in the same figure. The moment M_x at ant $0 \le x \le L_0$ from the fixed end is given by the expression

$$M_x = \frac{wLL_0}{2} - wLx + \frac{wx_0x}{2} \tag{9.33}$$

TABLE 9.6
Values of End Displacement δ, Λ, and θ with Increasing End Load P

1	2	3	4	5	6	7	8	9	10
Load P (lb)	Equivalent systems δ(in.)	4th order Runge–Kutta δ(in.)	Difference (%)	Equivalent systems Λ (in.)	4th order Runge–Kutta Λ (in.)	Difference (%)	Equivalent systems θ(°)	4th order Runge–Kutta θ(°)	Difference (%)
1000	96.37	97.04	0.70	8.40	8.41	0.12	16.85	16.91	0.36
1500	139.44	140.82	0.99	21.10	21.12	0.09	24.45	24.58	0.53
2000	189.04	191.40	1.25	40.00	40.90	2.25	32.77	33.18	1.25
2500	273.06	277.22	1.52	83.10	85.50	2.89	44.55	45.64	2.45
3000	352.61	360.19	2.15	139.50	141.49	1.43	55.03	56.78	3.18
3500	404.01	407.61	0.89	188.20	189.83	0.87	62.16	61.67	–0.79
4000	433.99	430.10	–0.90	228.00	228.23	0.10	67.07	66.23	–1.25
4500	453.61	447.96	–1.25	259.60	257.35	–0.87	71.32	69.78	–2.16
5000	467.54	455.57	–2.56	287.70	282.26	–1.89	75.30	72.85	–3.25
5500	479.42	463.84	–3.25	309.76	302.14	–2.26	78.76	75.41	–4.25
6000	487.75	468.98	–3.85	327.63	316.86	–3.29	81.77	77.79	–4.87

Note: 1 in. = 0.0254 m, 1 lb = 4.448 N.

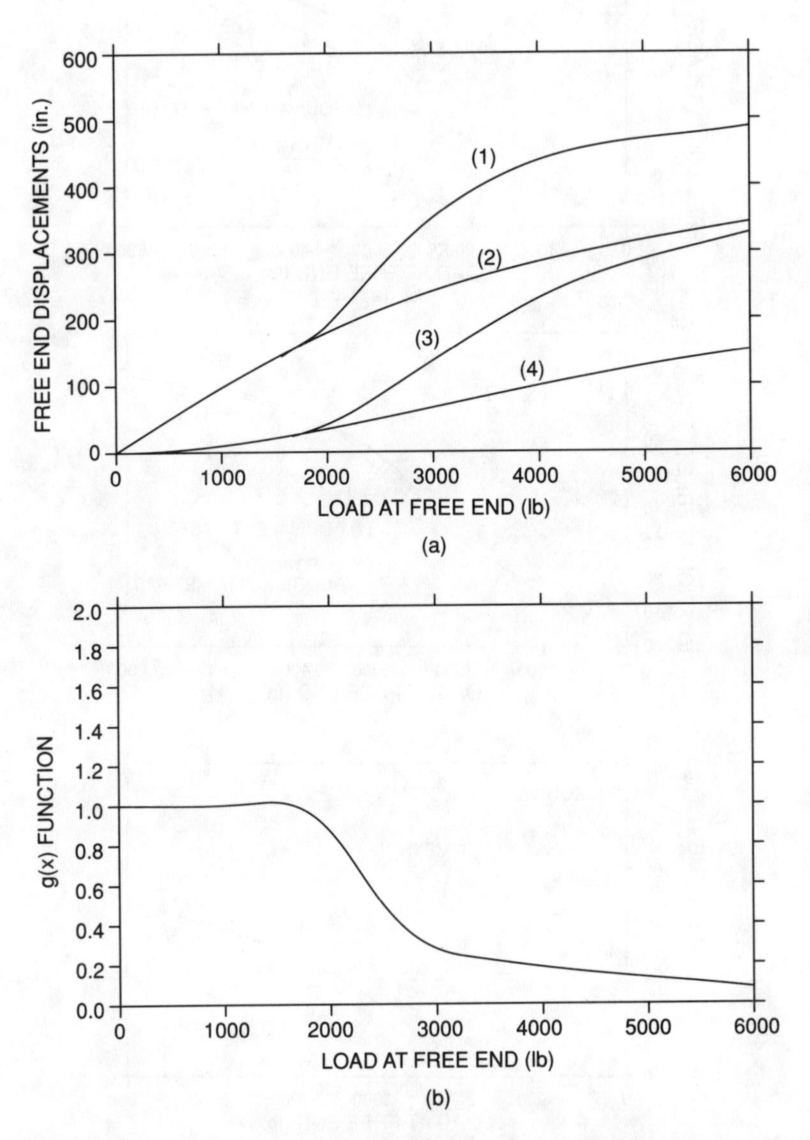

FIGURE 9.9. (a) Curves of free end displacements vs. load P; (b) variation of g(x) at the fixed end of the member with load P (1 in. = 0.0254 m, 1 lb = 4.448 N).

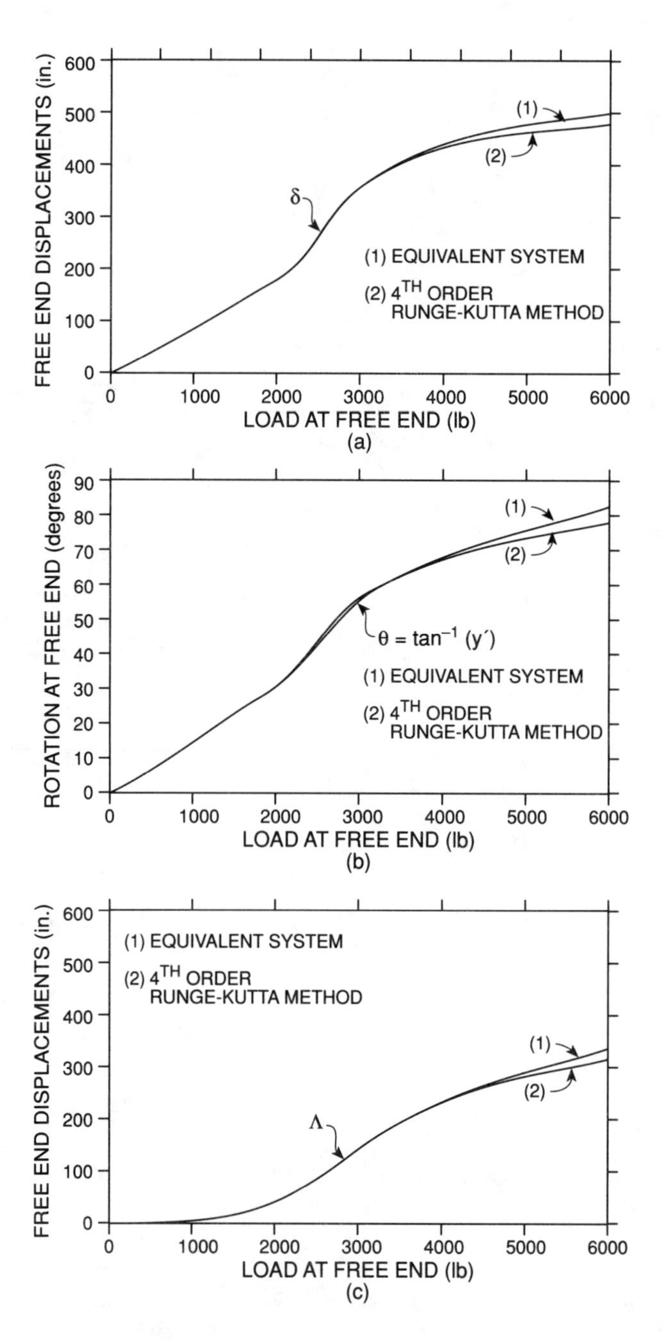

FIGURE 9.10. (a) Curves of end displacement δ vs. P; (b) curves of end rotation θ vs. load P; (c) curves of end displacement Λ vs. load P (1 in. = 0.0254 m, 1 lb = 4.448 N).

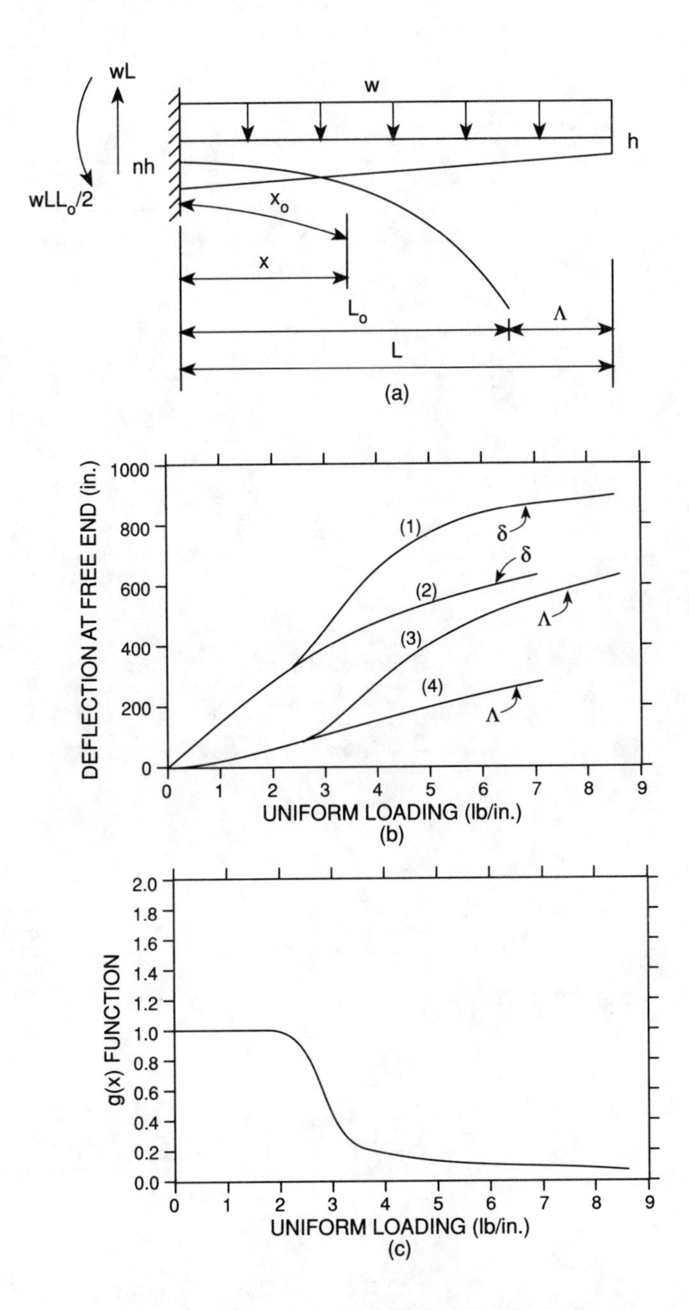

FIGURE 9.11. (a) Tapered nonprismatic cantilever beam loaded with a distributed load w; (b) variation of δ and Λ with increasing load w; (c) variation of g(x) at the fixed end of the member with increasing load w (1 in. = 0.0254 m, 1 lb/in. = 175.1268 N/m).

TABLE 9.7
Final Values for Strain Δ, g(x), y′, Z_e, M_{req} and M_e'

1	2	3	4	5	6	7	8
x(in.)	f(x)	Strain Δ	g(x)	y′ (radians)	Z_e	$M_{req} = M_x$ (in.-kip)	$M_e' = \frac{Z_e M_x}{f(x)\,g(x)}$ (in.-kip)
0.0	8.0	3.0964×10^{-2}	0.2281	0.0	1.0	1398.40	766.35
40.0	7.53	1.8664×10^{-2}	0.3593	1.6745×10^{-1}	1.0427	1275.10	491.54
79.9	7.07	1.1414×10^{-2}	0.5559	2.7567×10^{-1}	1.1174	1156.40	329.01
119.9	6.61	8.1636×10^{-3}	0.7317	3.5351×10^{-1}	1.1955	1042.40	257.38
159.8	6.15	6.5636×10^{-3}	0.8537	4.1824×10^{-1}	1.2766	933.11	226.15
199.8	5.74	5.6136×10^{-3}	0.9327	4.7683×10^{-1}	1.3635	828.87	211.63
239.7	5.32	4.9136×10^{-3}	0.9823	5.5970×10^{-1}	1.4585	729.80	203.22
279.7	4.91	4.4636×10^{-3}	0.9993	5.8699×10^{-1}	1.5643	636.05	203.34
319.6	4.52	4.0506×10^{-3}	1.0	6.4179×10^{-1}	1.6835	547.77	204.22
359.5	4.13	3.6488×10^{-3}	1.0	6.9681×10^{-1}	1.8169	465.14	204.49
399.5	3.76	3.2428×10^{-3}	1.0	7.5163×10^{-1}	1.9641	388.36	202.70
439.5	3.41	2.8340×10^{-3}	1.0	8.0559×10^{-1}	2.1237	317.66	197.98
479.4	3.07	2.4243×10^{-3}	1.0	8.5776×10^{-1}	2.2924	253.29	189.35
519.4	2.74	2.0169×10^{-3}	1.0	9.0697×10^{-1}	2.4649	195.55	175.83
559.4	2.43	1.6162×10^{-3}	1.0	9.5179×10^{-1}	2.6337	144.73	156.66
599.3	2.14	1.2291×10^{-3}	1.0	9.9065×10^{-1}	2.7895	101.13	131.63
639.3	1.87	8.6494×10^{-4}	1.0	1.0221	2.9218	65.05	101.49
679.2	1.62	5.3734×10^{-4}	1.0	1.0449	3.0218	36.73	68.40
719.2	1.39	2.6508×10^{-4}	1.0	1.0589	3.0846	16.37	36.23
759.1	1.19	7.4089×10^{-5}	1.0	1.0650	3.1124	4.11	10.78
799.1	1.0	0.0	1.0	1.0661	3.1175	0.0	0.0

Note: 1 in. = 0.0254 m, 1 kip-in. = 112.9848 Nm.

TABLE 9.8
Values of End Displacement δ, Λ, and θ, with Increasing Distributed Load w

1 Load W (lb/in.)	2 Equivalent systems δ(in.)	3 4th order Runge–Kutta δ(in.)	4 Difference (%)	5 Equivalent systems Λ(in.)	6 4th order Runge–Kutta Λ(in.)	7 Difference (%)	8 Equivalent systems θ(°)	9 4th order Runge–Kutta θ(°)	10 Difference (%)
1.0	150.08	150.98	0.60	12.4	12.55	1.21	12.90	13.15	1.94
1.5	219.14	219.57	0.20	27.7	27.95	0.90	18.93	19.09	0.86
2.0	281.85	282.98	0.40	48.4	48.87	0.97	24.53	24.84	1.26
2.5	349.89	351.96	0.59	75.9	76.73	1.07	30.39	31.17	2.57
3.0	450.59	452.61	0.49	126.2	128.58	1.89	38.18	39.54	3.56
3.5	565.39	569.40	0.71	200.9	206.92	0.51	46.83	48.65	3.89
4.0	652.16	646.31	–0.90	273.6	272.23	–0.50	53.75	53.88	0.24
4.5	713.35	706.31	–0.99	336.5	333.18	–0.99	59.03	57.91	–1.89
5.0	757.4	747.97	–1.25	390.0	385.13	–1.25	63.13	60.95	–3.45
5.5	791.49	780.21	–1.43	435.9	444.09	–1.88	66.44	64.07	–3.57
6.0	817.50	793.87	–2.89	475.5	469.56	–1.25	69.13	66.44	–3.88
6.5	838.10	811.91	–3.12	510.1	495.56	–2.89	71.38	68.51	–4.02
7.0	854.72	830.02	–2.90	540.4	524.24	–2.99	73.27	70.24	–4.14
7.5	864.98	840.85	–2.79	567.3	549.61	–3.12	76.71	72.96	–4.89
8.0	874.83	848.59	–2.99	591.3	578.76	–2.12	78.90	74.96	–4.99
8.5	885.61	857.30	–3.20	613.2	594.89	–2.99	80.21	75.99	–5.26

Note: 1 in. = 0.0254 m, 1 lb/in. = 175.1268 N/m.

where x_o is taken along the deformed length of the member as shown in Figure 9.11a. Since $x_o = x + \Lambda(x)$ and $L_o = L - \Lambda$, Equation 9.33 may be written as

$$M_x = \frac{wL}{2}(L - \Lambda) - wLx + \frac{wx}{2}[x + \Lambda(x)] \tag{9.34}$$

where Λ is the horizontal displacement of the free end of the beam and $\Lambda(x)$ is the horizontal displacement at any $0 \le x \le L_o$.

The inelastic analysis here may be carried out as in the preceding sections by following the indicated step-by-step procedure. The moment M_x at any x, however, would have to be determined by using Equation 9.34. Correspondence between x and x_o during the solution process can be maintained through Equations 9.12 and 9.13 as discussed in preceding sections.

Let it be assumed in Figure 9.11a that L = 1000 in. (25.4 m), n = 2, h = 3 in. (0.0762 m), and width b = 3 in. (0.0762 m). The inelastic analysis, for various values of the uniformly distributed load w, was carried out as in the preceding sections and by subdividing L_o in each trial into 40 segments. Table 9.7 shows the final values of g(x), f(x), Δ, y', Z_e, $M_{req} = M_x$, and $M_e' = Z_e M_x / f(x)g(x)$ for a load w = 3.5 lb/in. (613.19 N/m). With known M_e', deflections and rotations at any $0 \le x \le L_o$ can be determined as discussed in the preceding sections. In Table 9.8 the values of δ, Λ, and θ at the free end of the bar, for various values of the distributed load w, are shown. The results obtained by the 4th order Runge-Kutta method are also shown in the same table for comparison purposes. For practical applications, the results are in close agreement.

Figure 9.11b shows curves of end displacements vs. load w for both elastic and inelastic analyses. For example, curve 1 is the plot of δ vs. w for large deflection inelastic analysis, while curve 2 shows the displacement δ obtained by large deflection elastic analysis. The same comparisons are made for Λ by using curves 3 and 4, where inelastic analysis was used for the results in curve 3. Note that for loads w larger than 2.5 lb/in. (437.82 N/m), inelastic analysis must be used if accurate values for δ and Λ are desired. Figure 9.11c shows the variation of the function g(x) at the fixed end of the bar with increasing load w.

9.8 INELASTIC ANALYSIS OF TAPERED SIMPLY SUPPORTED BEAMS

Consider the flexible variable stiffness simply supported beam loaded with a concentrated load P at the center of the bar as shown in Figure 9.12a. Its large deformation configuration is shown in Figure 9.12b. In this figure, the coordinate x_P locates the position of the load P after deformation, and x_D locates the point where the deflection y is maximum and rotation θ is zero. The horizontal movement of the end B is Λ. The bending moment M_x at any $0 \le x \le x_P$ is

$$M_x = \frac{Px}{2} \qquad 0 \le x \le x_P \tag{9.35}$$

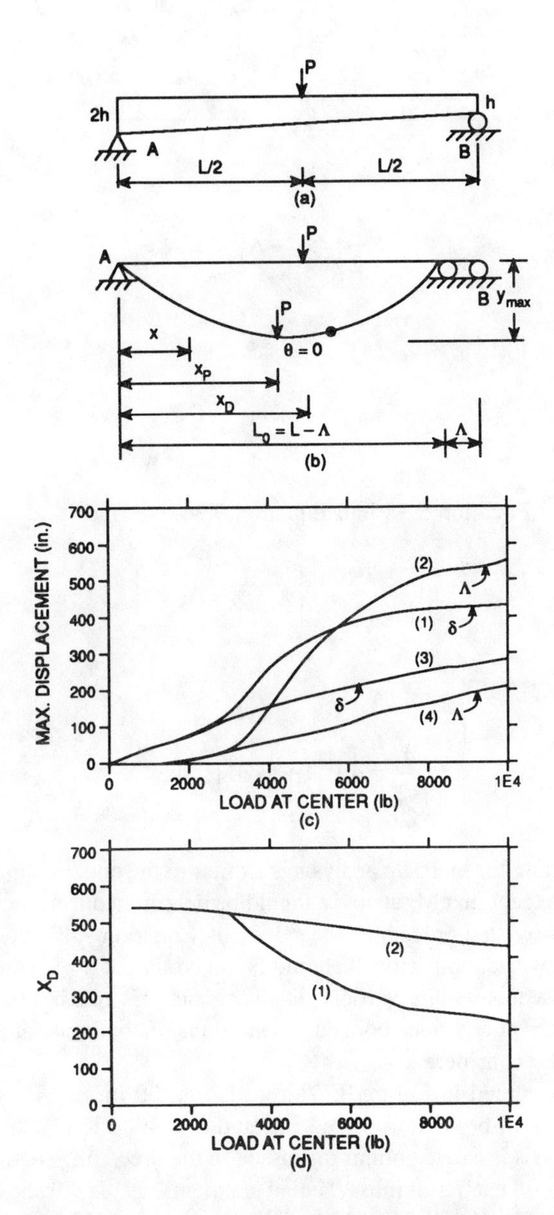

FIGURE 9.12. (a) Simply supported tapered flexible member loaded with concentrated load P at midspan; (b) large deformation configuration of the flexible member; (c) curves of maximum displacements vs. increasing load P; (d) variation of x_D with increasing load P (1 in. = 0.0254 m, 1 lb = 4.448 N).

and the bending moment at any $x_P \leq x \leq L_o$ is

$$M_x = \frac{Px}{2} - P(x - x_P) \qquad x_P \leq x \leq L_o \tag{9.36}$$

We know that

$$\frac{L}{2} = x_P + \Lambda_{x=x_P} \tag{9.37}$$

or

$$x_P = \frac{L}{2} - \Lambda_{x=x_P} \tag{9.38}$$

By subsituting Equation 9.38 into Equation 9.36, we find

$$M_x = \frac{Px}{2} - P\left[x - \frac{L}{2} + \Lambda_{x=x_P}\right] \tag{9.39}$$

We also have

$$\frac{L}{2} = \int_o^{x_P} \left[1 + (y')^2\right]^{1/2} dx \tag{9.40}$$

The procedure for inelastic analysis is similar to the one used in the preceding sections, except that M_x at any x should be determined from Equation 9.39 when the step-by-step procedure is carried out. Equation 9.40 may be used to determine x_P by trial and error. It should be noted that at $x = x_D$ the rotation $\theta = 0$, and consequently the vertical displacement y is maximum. These are important observations since boundary conditions of continuity at these points may need to be considered.

Let it be assumed in Figure 9.12a that L = 1000 in. (25.4 m), h = 3 in. (0.0762 m), width b = 3 in. (0.0762 m), and P = 4000 lb (17.792 kN). The inelastic analysis is carried out as discussed in the preceding sections, and by subdividing L_o in each trial into 40 equal segments. Table 9.9 shows the final values of g(x), f(x), Δ, y', Z_e, $M_{req} = M_x$, and $M_e' = Z_e M_x / f(x)g(x)$. With known M_e', deflections and rotations at any x may be determined as discussed earlier. That is, either by integrating y' or by using the pseudolinear system of constant stiffness E_1I_1 and applying the moment area method. Table 9.9 shows that the smallest value of g(x) = 0.1769, located at x = 411.6 in. (10.4546 m), and the largest rotation θ is at the right support of the bar, i.e., $\theta = \tan^{-1}y' = \tan^{-1}(1.1099) = 47.98°$.

TABLE 9.9
Final Values for Strain Δ, g(x), y', Z_e, M_{req}, and M_e'

1	2	3	4	5	6	7	8
x(in)	f(x)	Strain Δ	g(x)	y' (rads)	$Z_e = [1 + (y')^2]^{3/2}$	$M_{req} = M_x$ (kip-in.)	$M_e' = \frac{Z_e M_x}{f(x)\,g(x)}$ (kip-in.)
0.0	8.0	0.0	1.0	1.8379×10^{-1}	2.0529	0.0	0.0
41.2	7.39	4.3832×10^{-4}	1.0	7.8379×10^{-1}	2.0437	82.31	22.77
82.3	6.81	9.2548×10^{-4}	1.0	7.7085×10^{-1}	2.0147	164.62	48.69
123.5	6.27	1.4672×10^{-3}	1.0	7.5342×10^{-1}	1.9647	246.93	77.39
164.5	5.76	2.0697×10^{-3}	1.0	7.2773×10^{-1}	1.8937	329.20	108.23
205.8	5.29	2.7393×10^{-3}	1.0	6.9329×10^{-1}	1.8037	411.55	140.37
246.9	4.85	3.4826×10^{-3}	1.0	6.4986×10^{-1}	1.6983	493.86	172.97
288.1	4.44	4.3064×10^{-3}	1.0	5.9743×10^{-1}	1.5826	576.17	205.20
329.2	4.07	5.5636×10^{-3}	0.9367	5.3474×10^{-1}	1.4599	658.48	251.81
370.4	3.73	1.0014×10^{-2}	0.6212	4.4607×10^{-1}	1.3139	740.79	419.97
411.6	3.42	4.1364×10^{-2}	0.1769	2.0658×10^{-1}	1.0651	823.10	1447.00
452.7	3.15	2.7264×10^{-2}	0.2555	-1.1865×10^{-1}	1.0211	740.79	940.17
493.9	2.88	1.5164×10^{-2}	0.4330	-3.3310×10^{-1}	1.1712	658.48	617.62
535.0	2.62	9.1636×10^{-3}	0.6681	-4.7841×10^{-1}	1.3630	576.17	448.39
576.2	2.36	6.6136×10^{-3}	0.8496	-5.9309×10^{-1}	1.5733	493.86	386.65
617.3	2.11	5.2636×10^{-3}	0.9596	-6.9692×10^{-1}	1.8143	411.55	368.42
658.5	1.87	4.3636×10^{-3}	1.0	-7.9866×10^{-1}	2.1006	329.24	368.37
699.6	1.63	3.5967×10^{-3}	1.0	-9.0155×10^{-1}	2.4459	246.93	369.75
740.8	1.41	2.6470×10^{-3}	1.0	-9.9967×10^{-1}	2.8320	164.62	351.05
781.9	1.20	1.4758×10^{-3}	1.0	−1.0776	3.7812	82.31	218.92
823.1	1.0	0.0	1.0	−1.1099	3.3378	0.0	0.0

Note: 1 in. = 0.0254 m, 1 kip-in. = 112.9848 Nm.

Figure 9.12c shows curves of maximum vertical displacement δ and maximum horizontal displacement Λ of the right support of the bar. Curve 1 shows the variation of δ with load P when large deformation inelastic analysis is used, while curve 3 shows the same variation of δ when large deformation elastic analysis is used. Curves 2 and 4 make the same comparisons for Λ, with curve 2 representing the large deformation inelastic analysis. Note in curve 2 the large increases in Λ with increasing P.

The variation with load of the location x_D of the maximum vertical displacement is shown in Figure 9.12d. The large deformation inelastic analysis is represented by curve 1, while curve 2 shows the large deformation elastic analysis results. Again, it is noted that the deviation between elastic and inelastic analysis becomes progressively larger for loads P larger than about 3000 lb (13,345 N). The results are also compared with the results obtained by the 4th order Runge-Kutta method. Figure 9.13a compares the maximum vertical displacement with increasing load P. The same comparison regarding the horizontal displacement Λ of the right support is given in Figure 9.13b, while Figure 9.13c compares the variation with load of the maximum rotation θ occuring at the right support. For practical purposes, the results are considered to be very close.

9.9 INELASTIC ANALYSIS OF FLEXIBLE TAPERED SIMPLY SUPPORTED BEAMS BY USING SIMPLIFIED NONLINEAR EQUIVALENT SYSTEMS

In this section, the methodology discussed in Section 9.4 is applied to simply supported beams of variable stiffness. Consider, for example, the simply supported variable stiffness beam loaded as shown in Figure 9.14a. Loading and stiffness are considered to be symmetrical from the center of the beam to the end supports. This is not mandatory, since these two quantities can vary in any arbitrary manner along the length of the beam. They are chosen to be symmetrical in order to make it easier to compare the results with those obtained using the original system.

By utilizing Equation 9.4 and $g(x) = 1$, the moment diagrams M_e of the simplified equivalent system of constant stiffness EI_1, for various values of the distributed load w_1, are shown in Figure 9.14b. The reference moment of inertia I_1 is the one at the center of the member where the depth h = 3.0 in. (0.0762 m) and width b = 3.0 in. (0.0762 m). The depth increases linearly to 4.5 in. (0.1143 m) at the end supports and the width remains constant. The approximation of the shape of the M_e diagram with two straight lines as shown in Figure 9.14b leads to a simplified equivalent nonlinear system of constant stiffness EI_1 that is loaded with a concentrated load at the center of the member. For the case of w_1 = 2.5 lb/in. (438.00 N/m), the simplified equivalent nonlinear system is as shown in Figure 9.14c. The solution for inelastic analysis of the

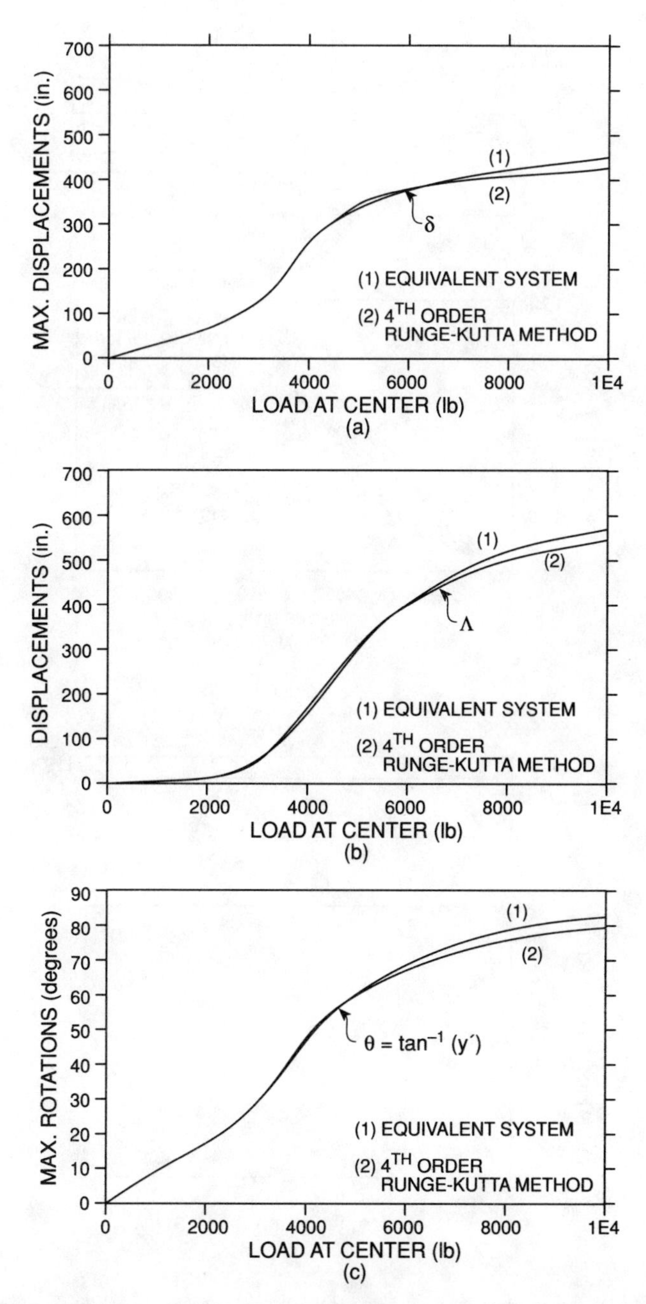

FIGURE 9.13. (a) Curves of maximum displacement δ vs. load P; (b) variation of the horizontal displacement Λ vs. load P; (c) variation of the rotation θ at the right support with increasing load P (1 in. = 0.0254 m, 1 lb = 4.448 N).

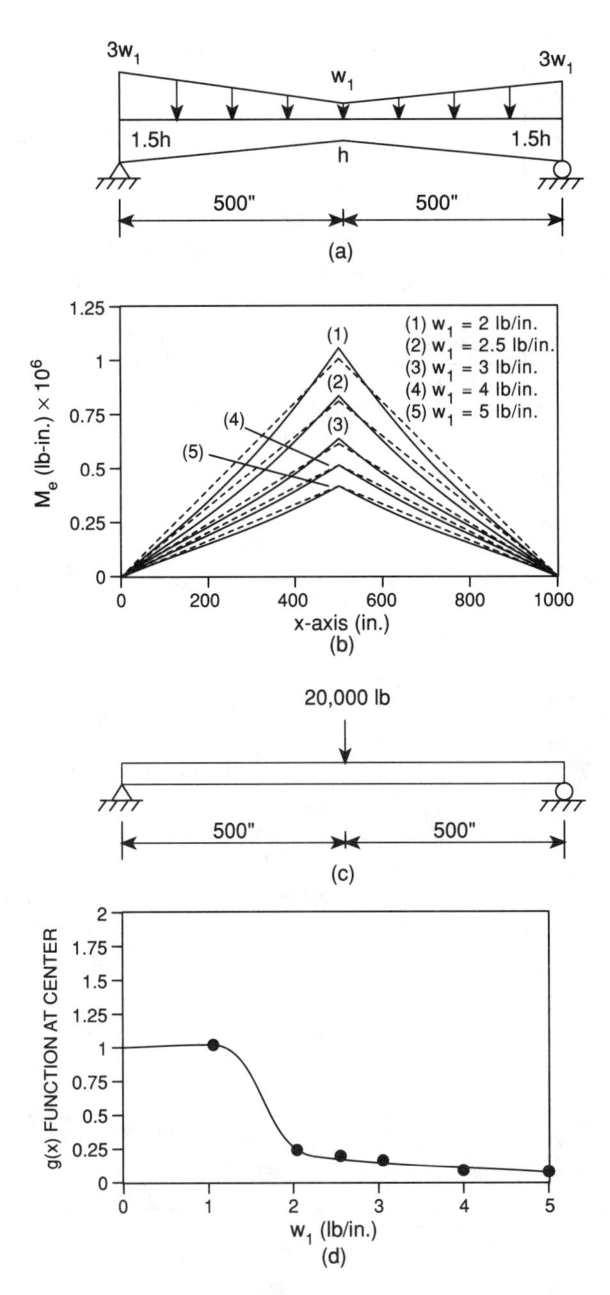

FIGURE 9.14. (a) Flexible simply supported beam of variable stiffness; (b) moment diagrams M_e of the equivalent nonlinear system for various loads w_1 and with their shape approximated with two straight lines; (c) simplified equivalent nonlinear system of constant stiffness EI_1; (d) variation of g(x) at midspan with increasing load w_1 (1 in. = 0.0254 m, 1 in.-lb = 0.11298 Nm, 1 lb/in. = 175.1268 N/m).

problem in Figure 9.14c is much easier than the solution of the original problem in Figure 9.14a.

The inelastic analysis of the problem in Figure 9.14c is carried out in the same way as in the preceding section, by using the indicated step-by-step procedure. The three-line approximation of the stress/strain curve of Monel with $E_1 = 22 \times 10^6$ psi (151.68×10^6 kPa) is used to carry out the inelastic analysis. The final values of Δ, y', $g(x)$, Z_e, and M_e' for the simplified nonlinear system in Figure 9.14c are shown in Table 9.10. With known M_e' and y', the horizontal and vertical displacements at any $0 \leq x \leq (L - \Lambda)$ may be determined as explained in the preceding section. The horizontal displacement Λ of the roller support, the vertical displacement δ at the center of the beam, and the rotation θ of its end supports for various values of the distributed load w_1 are shown in Table 9.11 under the column heading "Simplified NES".

Inelastic analysis was also used to solve directly the original problem in Figure 9.14a by using Equation 9.9 and following the step-by-step procedure as discussed in preceding sections. In this case, the final values of Δ, y', Z_e, M_e, and M_e' are shown in Table 9.12. With known y' and M_e', pseudolinear analysis, as before, can be used to determine horizontal and vertical displacements along the length $L_0 = (L - \Lambda)$ of the member. In Table 9.11 under column heading "Exact PES", the horizontal displacement Λ of the roller support, vertical displacement δ of the beam's center, and end rotations θ are shown. Note that the values obtained by the simplified system in Figure 9.14c, for all practical purposes, are closely identical with the values calculated using the initial system in Figure 9.14a and applying pseudolinear analysis.

Figure 9.14d shows the variation of $g(x)$ at the center of the member vs. the distributed load w_1. The solid curve represents the results obtained from the pseudolinear analysis of the original beam in Figure 9.14a, and the black dots are the results obtained by using a simplified system in Figure 9.14c. These results are in close agreement. Figures 9.15a, 9.15b, and 9.15c provide similar comparisons with curve 1 representing the results obtained by using large deflection theory and inelastic analysis, while curve 2 gives the results obtained by large deflection theory and elastic analysis. Note that for loads w_1 larger than 1.5 lb/in. (262.69 N/m), the gap between these two curves becomes wider. Again here the solid lines represent the results obtained by the original system in Figure 9.14a, and the black dots represent the results computed by using a simplified system such as the one in Figure 9.14c.

9.10 INELASTIC ANALYSIS OF FLEXIBLE BARS WITH ELABORATE LOADING CONDITIONS AND STIFFNESS VARIATIONS

In order to point out the advantage in using the simplified nonlinear equivalent system for inelastic analysis, the variable stiffness cantilever bar loaded as shown in Figure 9.16a is considered here. The loading and moment of inertia

TABLE 9.10
Final Values of y', $g(x)$, Z_e, Δ, and M_e'

x(in.)	f(x)	Δ	g(x)	y'	Z_e	M_e (in.-kip)	$M_e' = \frac{M_e Z_e}{g(x)}$ (in.-kip)
0.0	1.0	0.0	1.0	1.3878	5.0052	0.0	0.0
35.8	1.0	1.0417×10^{-3}	1.0	1.3666	4.8559	35.83	174.00
71.7	1.0	1.9966×10^{-3}	1.0	1.3065	4.4536	71.67	319.18
107.5	1.0	2.8695×10^{-3}	1.0	1.2166	3.9058	107.50	419.88
143.3	1.0	3.6669×10^{-3}	1.0	1.1079	3.3242	143.34	476.47
179.3	1.0	4.4136×10^{-3}	0.99981	0.9896	2.7847	179.17	501.00
215.0	1.0	5.2636×10^{-3}	0.95959	0.8668	2.3177	215.00	518.14
250.8	1.0	6.8136×10^{-3}	0.83322	0.7355	1.9129	250.84	576.31
286.7	1.0	1.0164×10^{-2}	0.61350	0.5812	1.5473	286.67	723.14
322.5	1.0	2.0064×10^{-2}	0.33665	0.3611	1.2018	322.51	1151.30
358.3	1.0	3.8014×10^{-2}	0.19042	0.0	1.0	358.34	1881.80
394.2	1.0	2.0064×10^{-2}	0.33665	–0.3611	1.2018	322.51	1151.30
430.0	1.0	1.0164×10^{-2}	0.61350	–0.5812	1.5473	286.67	723.14
465.8	1.0	6.8136×10^{-3}	0.83322	–0.7355	1.9129	250.84	576.31
501.7	1.0	5.2636×10^{-3}	0.95959	–0.8668	2.3177	215.00	518.14
537.5	1.0	4.4136×10^{-3}	0.99981	–0.9896	2.7847	179.17	501.00
573.3	1.0	3.6669×10^{-3}	1.0	–1.1079	3.3247	143.34	476.47
609.2	1.0	2.8695×10^{-3}	1.0	–1.2166	3.9058	107.50	419.88
645.0	1.0	1.9966×10^{-3}	1.0	–1.3065	4.4536	71.67	319.18
680.8	1.0	1.0417×10^{-3}	1.0	–1.3666	4.8559	35.83	174.00
716.7	1.0	0.0	1.0	–1.3878	5.0052	0.0	0.0

Note: 1 in. = 0.0254 m, 1 in.-kip = 112.9848 Nm.

TABLE 9.11
Values of δ, Λ, and θ vs. Increasing Values of the Distributed Load w_1 by using the Exact PES and the Simplified NES

w_1 (lb/in.)	δ Exact PES (in.)	δ Simplified NES (in.)	Difference %	Λ Exact PES (in.)	Λ Simplified NES (in.)	Difference %	θ Exact PES (°)	θ Simplified NES (°)	Difference %
1.0	101.59	101.20	–0.38	21.40	21.2	–0.93	17.04	17.01	–0.18
2.0	259.55	254.15	–2.08	162.50	160.0	–1.23	41.10	41.95	2.07
2.5	336.72	331.70	–1.49	287.20	283.3	–1.36	53.29	54.23	1.76
3.0	378.38	374.62	–0.97	384.50	379.2	–1.38	60.99	61.97	1.61
4.0	414.89	412.56	–0.56	540.90	532.3	–1.59	70.80	71.80	1.41
5.0	439.35	435.78	–0.81	620.50	608.7	–1.90	75.35	75.98	0.84

Note: 1 in. = 0.0254 m, 1 lb/in. = 175.1268 N/m.

TABLE 9.12
Final Values of Δ, g(x), y′, Z_e, and M_e'

x(in.)	f(x)	Δ	g(x)	y′	Z_e	M_e (in.-kip)	$M_e' = \frac{M_e Z_e}{g(x) f(x)}$ (in.-kip)
0.0	3.38	0.0	1.0	1.3410	4.6770	0.0	0.0
35.6	2.99	7.9212×10^{-4}	1.0	1.3260	4.5773	81.37	124.59
71.3	2.64	1.5688×10^{-3}	1.0	1.2819	4.2937	148.30	241.28
106.9	2.33	2.3313×10^{-3}	1.0	1.2121	3.8762	202.57	337.61
142.6	2.05	3.0815×10^{-3}	1.0	1.1223	3.3930	246.06	407.49
178.2	1.81	3.8222×10^{-3}	1.0	1.0188	2.9065	280.61	451.53
213.8	1.59	4.5636×10^{-3}	0.9972	0.9074	2.4595	307.87	475.55
249.5	1.41	5.7136×10^{-3}	0.9246	0.7883	2.0627	329.24	520.14
285.1	1.25	8.4636×10^{-3}	0.7116	0.6495	1.6941	345.99	658.62
320.8	1.11	2.0114×10^{-2}	0.3359	0.4343	1.2955	359.51	1243.90
356.4	1.0	4.8964×10^{-2}	0.1532	0.0	1.0	371.17	2424.00
392.1	1.11	2.0114×10^{-2}	0.3359	−0.4343	1.2955	359.51	1243.90
427.7	1.25	8.4636×10^{-3}	0.7116	−0.6495	1.6941	345.99	658.62
463.3	1.41	5.7136×10^{-3}	0.9246	−0.7883	2.0627	329.24	520.14
499	1.59	4.5636×10^{-3}	0.9972	−0.9074	2.4595	307.87	475.55
534.6	1.81	3.8222×10^{-3}	1.0	−1.0188	2.9065	280.61	451.53
570.3	2.05	3.0815×10^{-3}	1.0	−1.1223	3.3930	246.06	407.49
605.9	2.33	2.3313×10^{-3}	1.0	−1.2121	3.8762	202.57	337.61
641.5	2.64	1.5688×10^{-3}	1.0	−1.2819	4.2937	148.30	241.28
677.2	2.99	7.9212×10^{-4}	1.0	−1.3260	4.5773	81.37	124.59
712.8	3.38	0.0	1.0	−1.3410	4.6770	0.0	0.0

Note: 1 in. = 0.0254 m, 1 in.-kip = 112.9848 Nm.

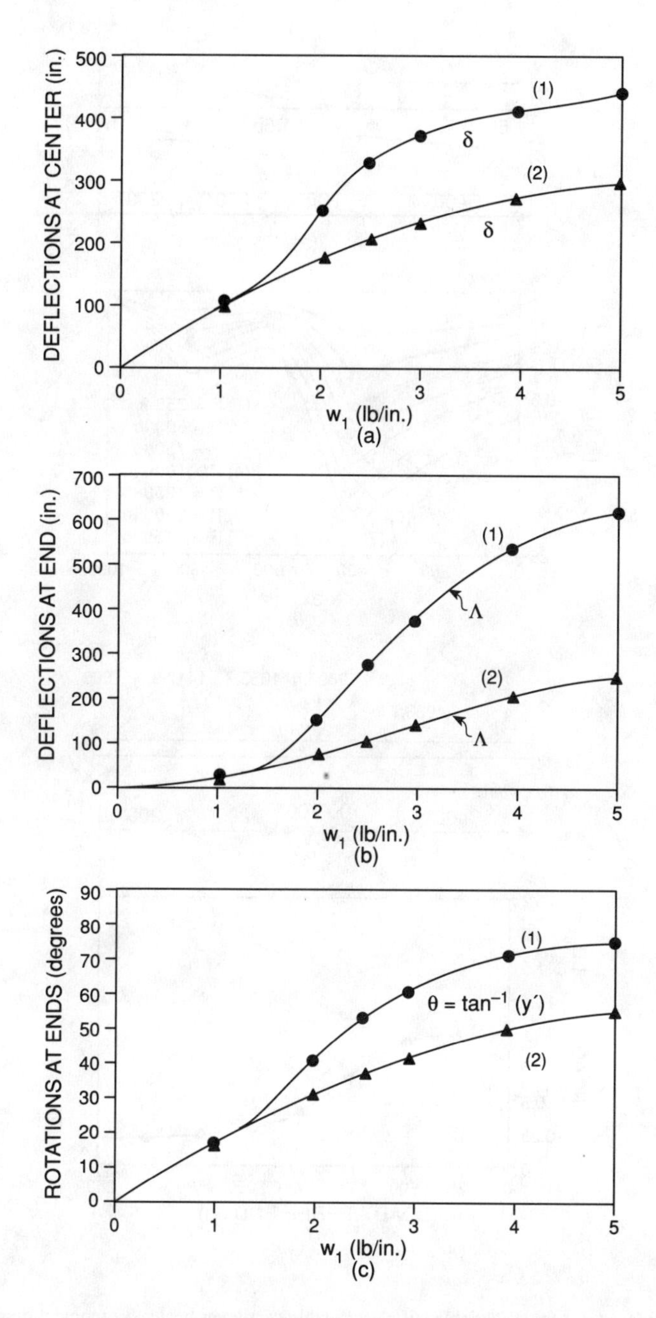

FIGURE 9.15. (a) Vertical deflection δ at midspan with increasing w_1; (b) horizontal displacement Λ of the roller support with increasing w_1; (c) end rotation θ with increasing w_1 (1 in. = 0.0254 m, 1 lb/in = 175.1268 N/m).

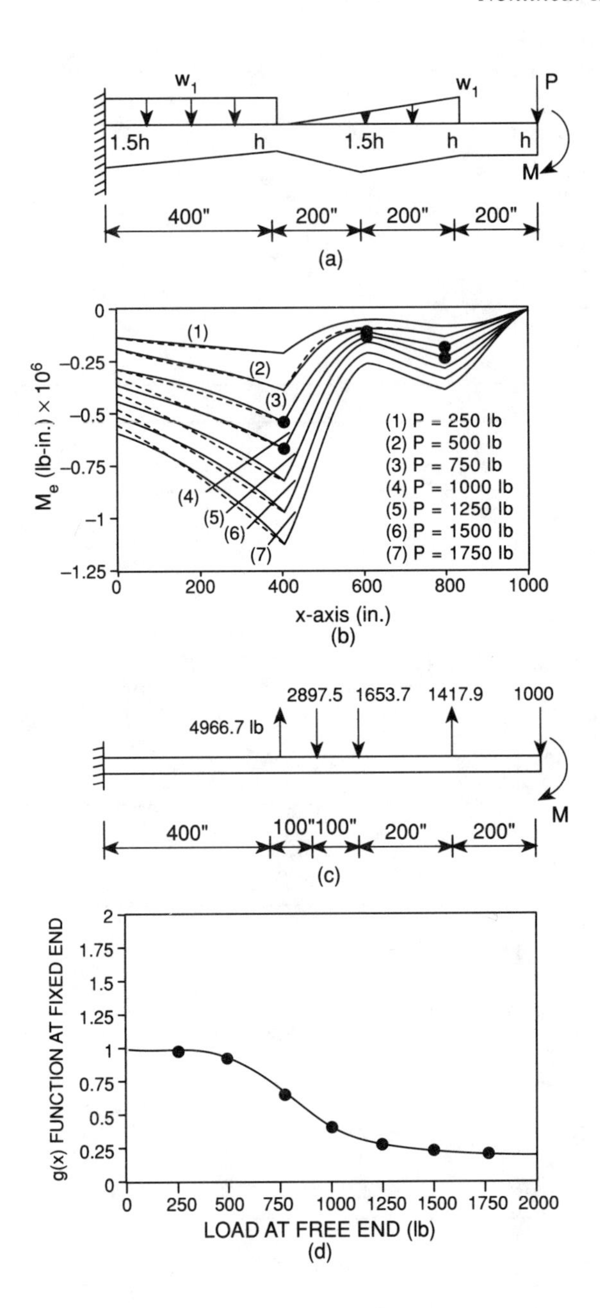

FIGURE 9.16. (a) Initial variable stiffness flexible cantilever beam; (b) moment diagrams M_e vs. load P approximated with five straight lines; (c) simplified nonlinear equivalent system of constant stiffness EI_1; (d) variation of g(x) at the fixed end with increasing load P (1 in. = 0.0254 m, 1 in.-lb = 0.11298 Nm, 1 lb = 4.448 N).

variations are arbitrary enough to illustrate the advantage. The material of the member is Monel.

The simplified nonlinear equivalent system of constant stiffness EI_1 may be determined in the same way as in preceding sections by using Equation 9.4 with g(x) = 1. The plots of this equation for various values of the concentrated load P, but keeping w_1 = 1 lb/in. (175.1268 N/m) and M = 10,000 in.-lb (1130.0 Nm), are shown in Figure 9.16b. The width b = 3.0 in. (0.0762 m) and h = 3.0 in. (0.0762 m). The approximation of the shape of these diagrams with five straight lines, as shown by the dashed lines, yields simplified nonlinear equivalent systems of constant stiffness EI_1 loaded with concentrated loads. For example, when P = 1000 lb (4.448 kN) the simplified nonlinear equivalent system is as shown in Figure 9.16c. Inelastic analysis may now be used to solve this simplified nonlinear system. By applying the step-by-step procedure as explained in preceding sections, the final values of the strain Δ, function g(x), y′, Z_e, and

$$M'_e = \frac{M_x Z_e}{g(x) f(x)} = \frac{M_e Z_e}{f(x)}$$

are shown in Table 9.13. With known M_e' and y′, pseudolinear analysis may be applied for the solution of the problem in Figure 9.16c in order to determine its large deformations.

By retaining the values of w_1 and M as given above and permitting P to vary from 250 lb (1112.5 N) to 1750 lb (7787.5 N), the calculated values of the vertical displacement δ, horizontal displacement Λ, and rotation θ at the free end of the beam are shown in Table 9.14. In this table, the notation "Simplified NES" is used in the columns to desingate these results. In the same table, under the column notation "Exact PES", the values of δ, Λ, and θ are also listed, which are obtained by solving the original problem in Figure 9.16a by pseudolinear analysis as discussed in preceding sections, which involves utilization of Equations 9.9 and 9.10. This procedure, however, is more tedious than using the simplified nonlinear equivalent system in Figure 9.16c. The maximum difference in results, as it can be seen in Table 9.14, is only about 2%, which is almost negligible for practical applications. If better accuracy is required, it can be obtained by using more straight lines to approximate the shape of M_e and M_e' diagrams. Table 9.15 shows the final values of Δ, g(x), y′, Z_e, M_e, and M_e', which are obtained when the pseudolinear analysis was used to solve the original problem in Figure 9.16a.

Figure 9.17a shows the variation of the end vertical displacement δ with increasing vertical load P, while keeping the value of w_1 constant. The solid line represents the results from the pseudolinear analysis, while the black dots give the results obtained by the simplified equivalent nonlinear system. For comparison purposes, curve 1 gives the results from large deformation and inelastic analysis, and curve 2 represents the results obtained by using large

TABLE 9.13
Final Values of y', $g(x)$, Z_e, Δ, and M_e'

x (in.)	f(x)	Δ	g(x)	y'	Z_e	M_e (in.-kip)	$M_e' = \frac{M_e Z_e}{g(x)}$ (in.-kip)
0.0	1.0	2.0164×10^{-2}	0.33514	0.0	1.0	223.03	665.40
29.00	1.0	1.3964×10^{-2}	0.46602	0.1047	1.0169	220.04	480.21
58.00	1.0	1.2264×10^{-2}	0.52241	0.1954	1.0593	222.29	450.67
87.00	1.0	1.0914×10^{-2}	0.57770	0.2820	1.1251	224.56	437.37
116.00	1.0	9.8136×10^{-3}	0.63166	0.3672	1.2148	226.97	436.63
145.10	1.0	8.9136×10^{-3}	0.68309	0.4533	1.3326	229.60	448.03
174.10	1.0	8.1636×10^{-3}	0.73171	0.5432	1.4864	232.50	472.29
203.10	1.0	7.6136×10^{-3}	0.77094	0.6401	1.6906	236.15	518.14
232.10	1.0	7.1136×10^{-3}	0.80919	0.7478	1.9688	243.23	637.11
261.10	1.0	6.7636×10^{-3}	0.83729	0.8731	2.3669	246.31	695.81
290.10	1.0	6.5136×10^{-3}	0.85789	1.0271	2.9787	254.27	882.04
319.10	1.0	6.4136×10^{-3}	0.86622	1.2332	4.0353	266.13	1238.70
348.10	1.0	4.7636×10^{-3}	0.98996	1.5130	5.9773	220.72	1336.30
377.20	1.0	3.1914×10^{-3}	1.0	1.7469	8.1525	132.34	1078.90
406.20	1.0	2.1361×10^{-3}	1.0	1.9301	10.2663	78.57	806.67
435.20	1.0	1.4383×10^{-3}	1.0	2.6632	12.0529	47.71	575.08
464.20	1.0	1.4633×10^{-3}	1.0	2.1940	14.0099	54.80	767.79
493.20	1.0	1.4900×10^{-3}	1.0	2.3755	17.1069	64.64	1105.70
522.20	1.0	1.3742×10^{-3}	1.0	2.6413	22.5250	68.02	1532.20
551.20	1.0	7.8812×10^{-4}	1.0	2.9072	29.1328	39.01	1136.50
580.20	1.0	2.0202×10^{-4}	1.0	3.0550	33.3651	10.00	333.65

Note: 1 in. = 0.0254 m, 1 in.-kip = 112.9848 Nm.

TABLE 9.14
Values of δ, Λ, and θ at the Free End of the Member vs. Various Values of the Load P Using Exact PES and Simplified NES

P (lb)	δ Exact PES (in.)	δ Simplified NES (in.)	Difference %	Λ Exact PES (in.)	Λ Simplified NES (in.)	Difference %	θ Exact PES (°)	θ Simplified NES (°)	Difference %
250	389.93	387.23	−0.69	98.8	98.8	0.0	35.73	35.55	−0.50
500	542.88	538.36	−0.83	207.9	204.3	−1.73	52.02	51.61	−0.79
750	663.15	656.67	−0.98	327.6	321.3	−1.92	64.51	63.96	−0.85
1000	743.91	736.03	−1.06	428.3	419.8	−1.98	72.45	71.88	−0.79
1250	796.96	787.18	−1.23	503.8	493.6	−2.02	77.39	76.77	−0.80
1500	838.24	822.45	−1.88	561.5	549.3	−2.17	80.65	79.98	−0.83
1750	859.64	851.34	−0.97	606.9	594.9	−1.98	82.93	82.39	−0.65

Note: 1 in. = 0.0254 m, 1 lb = 4.448 N.

TABLE 9.15
Final Values of Δ, g(x), y′, Z_e, and M_e'

x(in.)	f(x)	Δ	g(x)	y′	Z_e	M_e (in.-kip)	$M_e' = \frac{M_e Z_e}{g(x)f(x)}$ (in.-kip)
0.0	3.38	1.7362×10^{-2}	0.38349	0.0	1.0	741.71	573.00
28.6	3.14	1.5064×10^{-2}	0.43560	0.10483	1.0167	696.39	517.76
57.2	2.91	1.3064×10^{-2}	0.49427	0.20070	1.0617	651.90	480.78
85.7	2.69	1.1464×10^{-2}	0.55386	0.29117	1.1313	608.25	461.45
114.3	2.48	1.0164×10^{-2}	0.61350	0.37921	1.2257	565.48	455.59
142.9	2.27	9.1636×10^{-3}	0.66808	0.46757	1.3487	523.65	465.33
171.5	2.07	8.3136×10^{-3}	0.72155	0.55904	1.5085	482.79	487.08
200.1	1.87	$7,7136 \times 10^{-3}$	0.76358	0.65701	1.7189	442.98	532.42
228.7	1.68	7.2136×10^{-3}	0.80136	0.76593	2.0047	404.31	602.09
257.2	1.49	6.8136×10^{-3}	0.83322	0.89189	2.4107	366.91	711.67
285.8	1.30	6.5636×10^{-3}	0.85374	1.04580	3.0314	330.95	901.28
314.4	1.12	6.4136×10^{-3}	0.86622	1.25000	4.0981	296.70	1252.90
343.0	1.14	4.9136×10^{-3}	0.98232	1.53080	6.0988	261.53	1419.00
371.6	1.66	3.2767×10^{-3}	1.0	1.77900	8.4854	227.29	1162.60
400.2	2.38	2.1885×10^{-3}	1.0	1.97550	10.8452	193.36	879.38
428.7	3.37	1.4390×10^{-3}	1.0	2.11840	12.8471	160.05	610.47
457.3	2.35	1.4605×10^{-3}	1.0	2.25430	14.9932	127.76	815.38
485.9	1.51	1.4884×10^{-3}	1.0	2.44380	18.4089	96.96	1182.20
514.5	1.0	1.3569×10^{-3}	1.0	2.72340	24.4283	67.17	1640.70
543.1	1.0	7.7945×10^{-4}	1.0	3.00480	31.7632	38.58	1225.50
571.7	1.0	2.0202×10^{-4}	1.0	3.16240	36.4715	10.00	364.72

Note: 1 in. = 0.0254 m, 1 in.-kip = 112.9848 Nm.

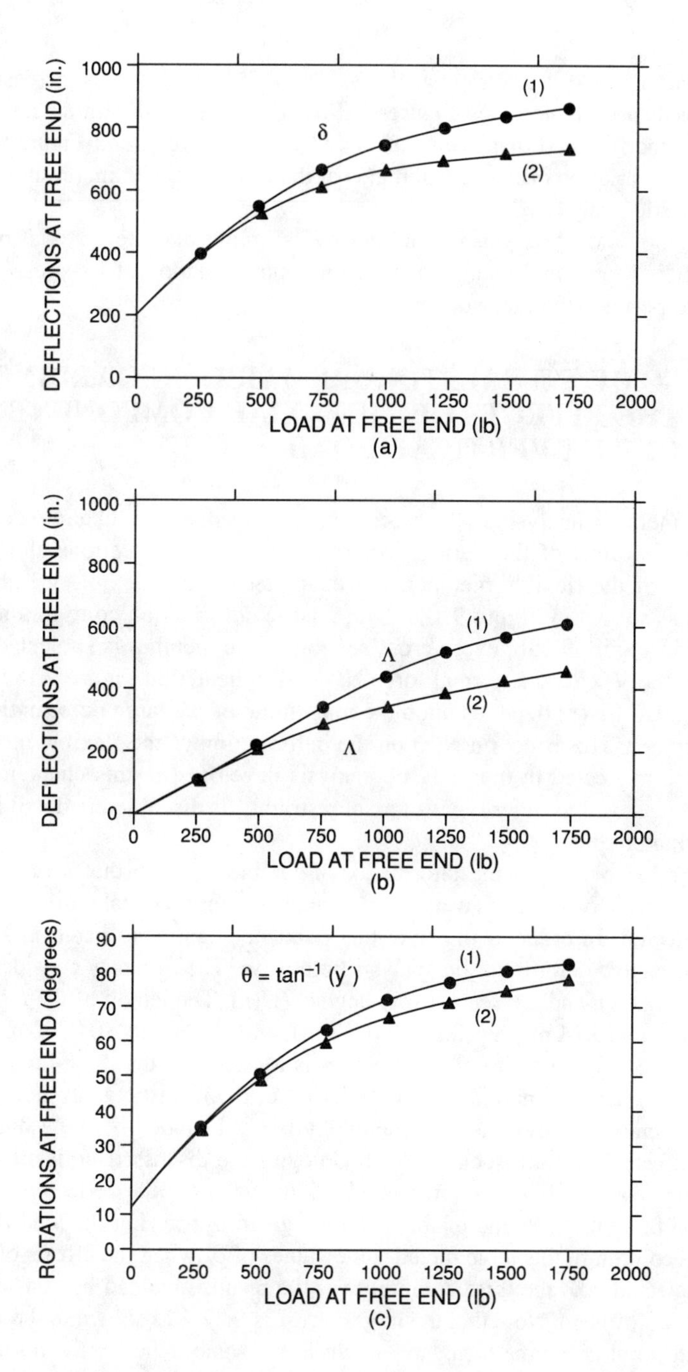

FIGURE 9.17. (a) End vertical displacement with increasing load P; (b) end horizontal displacement Λ with increasing load P; (c) end rotation θ with increasing load P (1 in. = 0.0254 m, 1 lb = 4.448 N).

deformation theory and elastic analysis. The difference becomes increasingly wider as the load P increases. Figures 9.17b and 9.17c provide similar comparisons for the free end displacements Λ and θ, respectively, and Figure 9.16d shows the variation of the function g(x) at the fixed end with increasing P for the inelastic analysis.

The above analysis was carried out by assuming that the material of the member is Monel and using the three-line approximation of its stress/strain curve to perform the inelastic analysis.

9.11 SOME INTERESTING REMARKS REGARDING THE EFFECTS OF THE AXIAL COMPONENT OF THE VERTICAL LOAD

The inelastic analysis in the preceding sections does not take into consideration the effect of the normal force that acts at cross sections along the length L of the flexible member. For example, when a flexible member is loaded as shown in Figure 9.18a, with a large deformation configuration as shown in Figure 9.18b, every cross section of the member is subjected to a shear force V(x) and a normal force N(x), as indicated in Figure 9.18c. The magnitude of N(x) depends upon the magnitude of the large deformation of the member. The effect on N(x) on the deformation of the flexible member has been neglected in the inelastic analysis developed in this chapter. This problem, as well as other types of axial restraints, is discussed in detail in the following chapter.

The effect of N(x) on the deformation of flexible members that are subjected to transversed loads only is usually very small, and for practical purposes it can be neglected. In order to illustrate this point, the cantilever beam in Figure 9.18a is analyzed for inelastic analysis both ways. The first way considers the effects of N(x), and the second way neglects N(x). The length of the member L = 600 in. (15.24 m), h = 3 in. (0.0762 m), width b = 3 in. (0.0762 m), and taper n = 2. The material of the member is Monel, and the three-line stress/strain curve approximation is used to carry out the inelastic analysis.

The detailed analysis and methodology is not included here, because the axial effects and the associated methodologies are discussed in detail in the following chapter. However, in Table 9.16, the values of the deflection δ_B and rotation θ_B at the free end of the beam are given by varying the load P. The second column in this table includes the values of δ_B when the effects of N(x) are neglected, and the third column gives the results obtained by considering the effects of N(x). Note that the largest error is 0.82% as shown in the fourth column. Similar comparisons are shown in the same table for the horizontal displacement Λ_B and rotation θ_B at the free end B. For Λ_B, the maximum error is 2.69% as shown in column 7, and for θ_B the maximum error is 0.99% as indicated in the tenth column. The problem of axial restraints is discussed in greater detail in the following chapter.

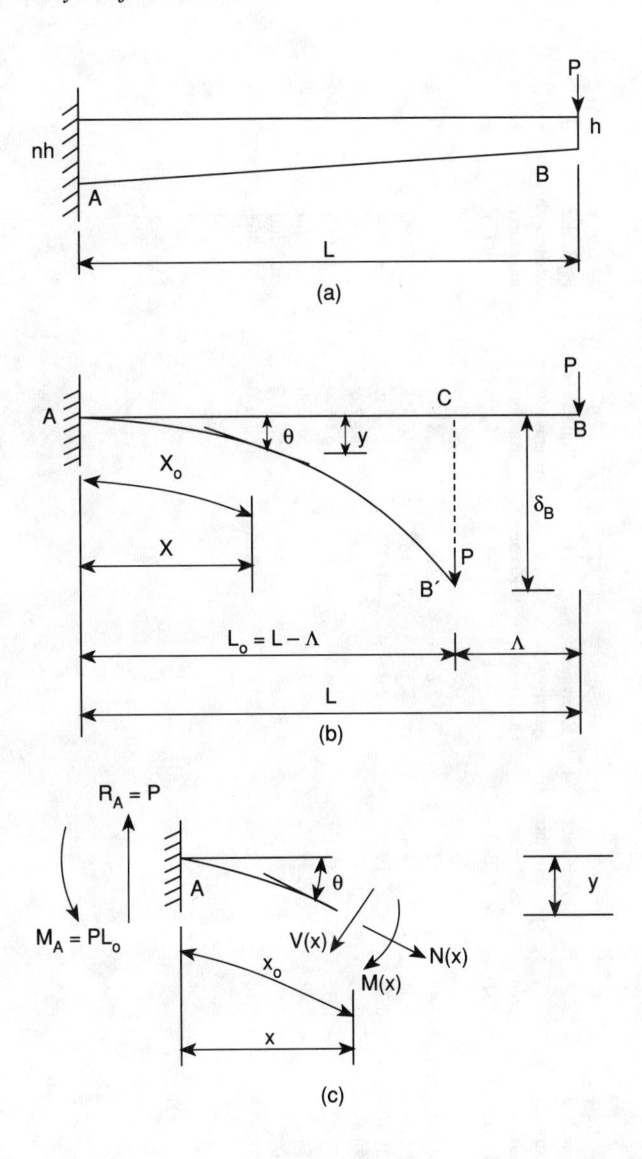

FIGURE 9.18. (a) Original nonprismatic flexible member loaded with a vertical load P at the free end; (b) large deformation configuration of the member; (c) free body diagram.

PROBLEMS

9.1 The uniform cantilever beam in Figure 9.4 is loaded at the free end with a concentrated load P = 1500 lb (6672.3 N). The length L = 600 in. (15.24 m) and moment of inertia I = 6.75 in.^{4} (2.8096×10^{-6} m^4). The material of the member is Monel. By using the three-line approximation of the stress/strain curve of Monel and applying pseudolinear

TABLE 9.16
Comparison of Values of δ_B, Λ_B, and θ_B at the Free End of the Beam wth Increasing P by Considering and Not Considering the Axial Component Effects

1	2	3	4	5	6	7	8	9	10
P (lb)	Equivalent systems without axial force δ_B (in.)	Equivalent systems with axial force δ_B (in.)	Error (%)	Equivalent systems without axial force Λ_B (in.)	Equivalent systems with axial force Λ_B (in.)	Error (%)	Equivalent systems without axial force θ_B (°)	Equivalent systems with axial force θ_B (°)	Error (%)
500	49.19	49.41	0.45	2.2	2.2	0.0	8.611	8.683	0.83
1000	96.37	96.71	0.35	8.4	8.5	1.18	16.852	16.990	0.81
1500	139.44	139.10	–0.24	19.2	19.0	–1.05	24.449	24.584	0.55
2000	189.04	189.49	0.24	40.0	40.9	2.20	32.768	33.090	0.97
2500	273.06	274.02	0.35	83.1	85.4	2.69	44.549	44.990	0.99
3000	352.61	354.27	0.47	139.5	141.9	1.69	55.030	55.550	0.94
3500	404.01	407.36	0.82	188.5	190.6	1.10	62.160	62.760	0.96
4000	433.99	433.39	0.14	228.0	230.8	1.21	67.070	67.461	0.60

Note: 1 in. = 0.0254 m, 1 lb = 4.448 N.

analysis, reproduce the quantities shown in Table 9.5, and compare the results. Show all details of the inelastic analysis used to produce the table. Also plot the M_x and M_e' moment diagrams.

9.2 By approximating the shape of the M_e' diagram in Problem 9.1 with one straight line, determine an equivalent system of constant stiffness EI that is loaded with an equivalent concentrated load at the free end. Also, by using the equivalent system and applying the moment area method, determine the vertical deflection at the free end. Compare and check the result with the result obtained in the graph of Figure 9.7a.

9.3 Repeat Problem 9.2 by approximating the shape of the M_e' diagram with two straight lines, and compare the results.

9.4 The tapered cantilever beam in Figure 9.8 has a rectangular cross section, and it is loaded with a concentrated load P = 2500 lb (11,120.56 N) at the free end. The material of the member is Monel, the length L = 600 in. (15.24 m), h = 3 in. (0.0762 m), width b = 3 in. (0.0762 m), and taper n = 2. By using the three-line approximation of the stress/strain curve of Monel and applying pseudolinear analysis, determine the vertical and horizontal displacements δ and Λ, respectively, at the free end, as well as the rotation θ at the same end.
Answer: δ = 273.06 in. (6.9357 m), Λ = 83.1 in. (2.1107 m), and θ = 44.55°.

9.5 Solve Problem 9.4 by using the bilinear approximation of the stress/strain curve of Monel, and compare the results.

9.6 Reproduce the results in Table 9.1 by using the bilinear approximation of the stress/strain curve of Monel, and compare the results.

9.7 Solve Problem 9.6 by using the three-line approximation of the stress/strain curve of Monel and w_1 = 4.0 lb/in. (700.5 N/m).

9.8 By using the M_e' diagram from Problem 9.7 and approximating its shape with one straight line, determine an equivalent system of constant stiffness E_1I_1 and use it to determine the displacements δ, Λ, and θ at the free end. Compare the results with the results obtained in Table 9.2.

9.9 Reproduce the results shown in Table 9.3 when w_1 = 4.0 lb/in. (700.5 N/m). Use a simplified nonlinear equivalent system of constant stiffness as discussed in Section 9.4 for the inelastic analysis.

9.10 By utilizing the M_e' diagram from Problem 9.9 and approximating its shape with one straight line, determine the analogous equivalent system and use it to determine the rotation and vertical displacement at the free end. Compare the results with the results shown in Table 9.2.

9.11 Reproduce the results shown in Table 9.9 by using the bilinear approximation of the stress/strain curve of Monel, and compare the results. Discuss the results.

9.12 Reproduce curve 1 in Figure 9.12c using the bilinear approximation of the stress/strain curve of Monel, and compare the results.

9.13 For the tapered simply supported beam loaded as shown in Figure 9.14a, determine a simplified nonlinear equivalent system of constant stiffness EI_1, such as the one in Figure 9.14c, when $w_1 = 4.0$ lb/in. (700.5 N/m). All other parameters are as discussed in Section 9.9.

9.14 By using the simplified nonlinear equivalent system obtained in Problem 9.13 and the three-line approximation of the stress/strain curve of Monel, determine the vertical displacement at midspan and the rotation at the ends of the member. Compare the results with the results obtained in Table 9.11.

9.15 Solve Problem 9.14 by directly using the original problem in Figure 9.14a and applying pseudolinear analysis. Compare the results.

9.16 Solve Problem 9.14 using the bilinear approximation of the stress/strain curve of Monel, and compare the results.

Chapter 10

INELASTIC ANALYSIS OF VARIABLE STIFFNESS FLEXIBLE BARS WITH AXIAL RESTRAINTS

10.1 INTRODUCTION

In Chapter 9 flexible members of either uniform or variable stiffness were subjected to transverse loading conditions, which caused the material of the member to be stressed well beyond its elastic limit and all the way to failure. The flexible members examined were free of axial restraints, and the effect of the normal cross-sectional component of the transversed load during the large deformation of the flexible member was neglected.

The analysis in this chapter takes into consideration the effect of the cross-sectional normal component of the transversed loading of the member during its large deformation. The flexible member is also loaded by both axial and transversed loadings, and the thickness of the member can vary in any arbitrary manner. In order to perform the inelastic analysis, a step-by-step procedure is also suggested in order to simplify the solution process and methodology. Interesting cases of loading and boundary conditions are investigated in detail. The solution, however, is general, and it can be extended to apply to other types of flexible beam problems.

10.2 INELASTIC ANALYSIS OF FLEXIBLE CANTILEVER BARS BY INCLUDING THE EFFECT OF THE AXIAL COMPONENTS OF LOADING

By following the discussion in Section 9.2, we can write the nonlinear differential equation of a flexible bar in bending as follows:

$$y'' = -\frac{M'_e}{E_1 I_1} \tag{10.1}$$

where

$$M'_e = \frac{M_x Z_e}{f(x)g(x)} \tag{10.2}$$

$$Z_e = \left[1 + (y')^2\right]^{3/2} \tag{10.3}$$

In the above equations, M_x is the moment at any $0 \le x \le (L - \Lambda)$ of the original problem, f(x) is the function representing the variation of I_x with respect to a reference value I_1, and g(x) is a function representing the variation of the modulus E_x with respect to a reference value E_1.

Equation 10.1 is also the differential equation of a pseudolinear equivalent system of constant stiffness E_1I_1. It has the same deflection curve, length, and boundary conditions as the original nonlinear variable stiffness E_xI_x bar. The moment M_e' of the equivalent pseudolinear system is given by Equation 10.2. Since g(x) in Equation 10.2 may also represent the variation of the modulus of elasticity along the length of the member when its material is stressed beyond its elastic limit, Equation 10.2 may be used for both elastic and inelastic analysis when g(x) is known. The procedure regarding the inelastic analysis of flexible members by taking into consideration axial effects may be illustrated by the following hypothetical example.

Let it be assumed that the nonprismatic cantilever beam in Figure 10.1a is loaded with a concentrated load P at the free end as shown. Its large deformation configuration is shown in Figure 10.1b, where the symbol Λ is used to represent the horizontal displacement of the free end. In this figure we note that for every length $0 \le x_o \le L$ there corresponds a length $0 \le x \le L_o$, where $L_o = (L - \Lambda)$. Thus we may write the expression

$$x_o = x + \Lambda(x) \tag{10.4}$$

where $\Lambda(x)$ represents the horizontal displacement of x_o. We also know that

$$x_o = \int_0^x \left[1 + (y')^2\right]^{1/2} dx \tag{10.5}$$

$$L = \int_0^{L-\Lambda} \left[1 + (y')^2\right]^{1/2} dx \tag{10.6}$$

By integrating Equation 10.1 once, we may write

$$y' = -\frac{1}{E_1I_1}\int_0^x M_e' dx + C_1 \tag{10.7}$$

where C_1 is the constant of integration.

In order to carry out the integration in Equation 10.7, the horizontal displacement Λ at the free end of the beam, as well as $\Lambda(x)$ and g(x), must be known. Thus inelastic analysis should be initiated for the computation of g(x), and nonlinear analysis must be used for the computation of Λ and $\Lambda(x)$. The whole procedure for computing the large deflections and rotations by inelastic

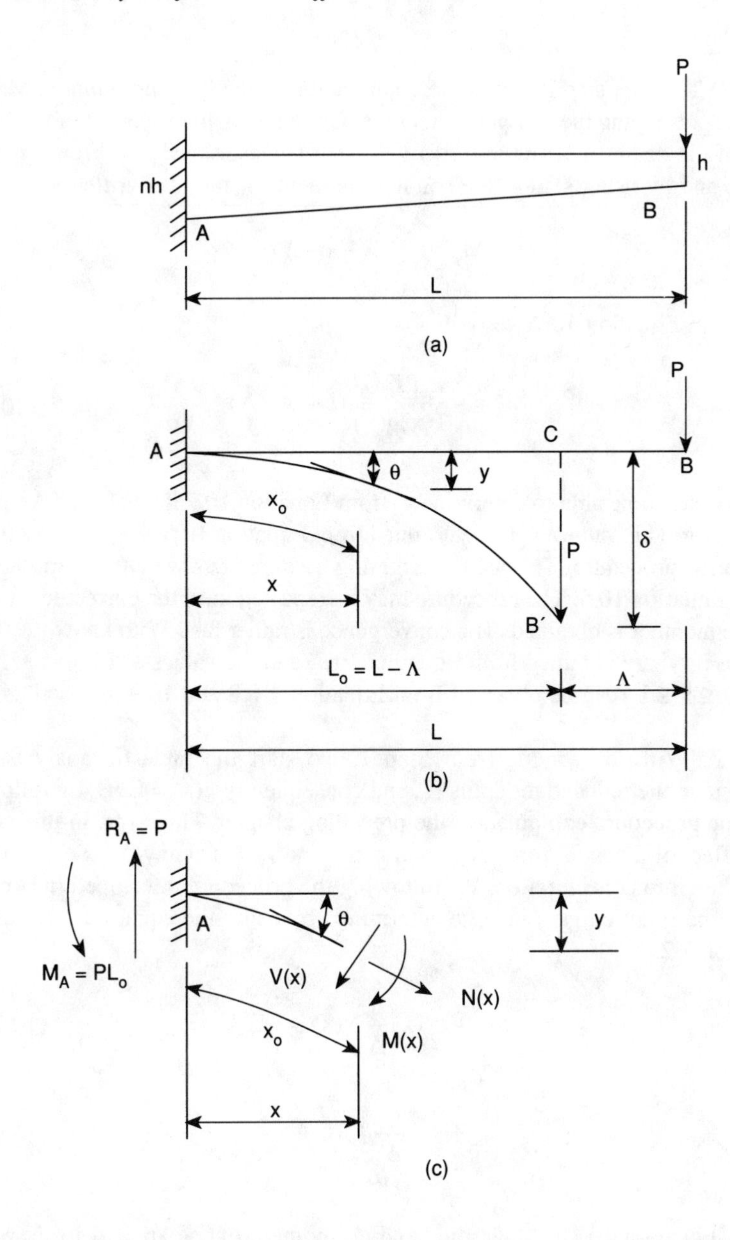

FIGURE 10.1. (a) Initial nonprismatic cantilever beam loaded with a concentrated load P at the free end; (b) large deformation configuration of the flexible member; (c) free body diagram.

analysis may be carried out by trial and error and utilization of equivalent pseudolinear systems represented by Equation 10.1. Simpson's rule may be used to carry out the required integration. The global trial and error procedure may be initiated by following a step-by-step procedure as shown below.

Step 1: The first step involves the computation of the bending moment M_x at any $0 \le x \le L_o$ and the computation of the horizontal displacement Λ of the free end of the beam by assuming g(x) = 1. That is elastic analysis. From Figure 10.1b and by using statics, the bending moment M_x may be written as

$$M_x = -P(L - x - \Lambda) \tag{10.8}$$

and from Equation 10.2 we write

$$M_e' = -\frac{PZ_e}{f(x)g(x)}(L - x - \Lambda) \tag{10.9}$$

In order to be able to determine y′ from Equation 10.7 the value of Λ must be known. This value can be determined from Equation 10.6 by applying a trial and error procedure. That is, by assuming a value of Λ we can determine L from Equation 10.6. The procedure may be repeated until the correct length L of the member is obtained. The convergence is rather fast. With known Λ, the values of y′ are obtained from Equation 10.7, and the values of M_x and M_e' at any $0 \le x \le L_o$ can be obtained from Equations 10.8 and 10.9, respectively.

Step 2: With known M_x from Step 1, we start the inelastic analysis to determine the reduced modulus E_e, and consequently $g(x) = E_e/E_1$ by following the procedure explained in the preceding chapter. However, in this step the effect of the axial force N(x) and axial strain $\varepsilon_o(x)$ at any $0 \le x \le L_o$ must be taken into consideration. By following the procedure developed in Chapter 7, the function g(x) may be determined by using the equations (see also Section 7.2.

$$M_x + \varepsilon_o(x)\frac{h}{\Delta}N(x) = \frac{I}{r}E_e \tag{10.10}$$

$$E_e = \frac{12}{\Delta^3}\int_{\varepsilon_2}^{\varepsilon_1} \sigma\varepsilon d\varepsilon \tag{10.11}$$

In Equation 10.10, M_x is the bending moment of the applied loads with respect to the centroidal axis, and $\varepsilon_o(x)hN(x)/\Delta$ is the additional moment due to the presence of the axial force N(x) (see Figure 10.1c regarding N[x]). The quantity E_e, as stated in Section 7.2, may be thought of as an equivalent modulus that corrects for the effect of the shifting of the neutral axis resulting from the presence of the axial component force.

The values of the axial strain $\varepsilon_o(x)$, axial force N(x), and shear force V(x), Figure 10.1c, may be determined from the static equations:

$$V(x) = P\cos\theta \tag{10.12}$$

$$N(x) = P\sin\theta \tag{10.13}$$

$$\varepsilon_o(x) = \frac{N(x)}{A_o\alpha(x)E} \tag{10.14}$$

where $\alpha(x)$ in Equation 10.14 gives the variation of the cross-sectional area A along the length of the member with respect to the reference area A_o at the free end, and E is the modulus of elasticity which may be either elastic or inelastic.

In order to be able to determine the equivalent reduced modulus E_e from Equation 10.11, the position of the neutral axis at sections along the length of the member must be known. This is accomplished by trial and error as discussed in Section 7.3. For example, by assuming values for ε_1 and ε_2 (Figure 7.2a), we adjust Δ until the sum of the two terms on the left side of Equation 10.10 matches the value obtained from the term on the right side of the equation. We may assume as many values of ε_1 and Δ as is necessary to match the two sides of Equation 10.10. The procedure converges rather fast. The value of the equivalent modulus E_e is obtained from Equation 10.11. Note that the axial force along the length of the member is variable, because ε and Δ are variable. With known E_e, the function $g(x) = E_e/E_1$ would also be known.

Step 3: By using the values of y' from Step 1, we can determine Z_e from Equation 10 3. Thus, with known Z_e and using the values of g(x) obtained from Step 2, we determine the moment M_e' of the pseudolinear system of constant stiffness E_1I_1 from the equation

$$M_e' = \frac{\left[M_x + \varepsilon_o(x)\frac{h}{\Delta}N(x)\right]Z_e}{f(x)g(x)} \tag{10.15}$$

Step 4: By using M_e' from Step 3 and g(x) from Step 2, we can use Equation 10.7 to obtain a new y' that incorporates inelastic behavior of the member and the effect of the axial component force N(x). The boundary condition of zero rotation at the fixed end may be used to determine the constant of integration C_1.

Step 5: By using the new y' that is obtained from Step 4, a new horizontal displacement Λ and a new $L_o = (L - \Lambda)$ may be obtained from the equation

$$L + \Delta L = \int_0^{L+\Delta L-\Lambda} \left[1+(y')^2\right]^{1/2} dx + \int_0^{L+\Delta L} \varepsilon_o(x)\cos\theta dx \qquad (10.16)$$

In Equation 10.16 ΔL represents the elongation of the length L due to the presence of the axial force N(x). Since this quantity is usually small, it may be neglected, and Equation 10.16 becomes Equation 10.6. By using Equation 10.16 or Equation 10.6 and applying a trial and error procedure by assuming values of Δ, the correct value of Λ may be obtained when Equation 10.16, or Equation 10.6, is satisfied. Thus, a new g(x) may now be calculated as discussed in Step 2.

Step 6: By using the new y' obtained from Step 4, a new M_e' may be determined by using Equation 10.15. Thus, a new y' may be obtained from Equation 10.7 using the new g(x) obtained in Step 5.

The procedure may be repeated for as many times as required until the last y' is almost identical to the one obtained from the preceding trial. Usually 5 to 10 repetitions are sufficient. The whole procedure may be easily computerized for convenience, and Simpson's rule may be used to carry out the required integrations. The utilization of pseudolinear analysis, i.e., M_e', facilitates a great deal the solution and convergence of the above trial and error procedure, and consequently the solution of such types of problems.

Now that the correct values of y' are known by using the above procedure, the large deflections may be obtained by either (1) integrating Equation 10.7 once and determining the constant of integration by satisfying the condition of zero deflection at the fixed end or (2) using the moment area method, since M_e' is known. You may also refer to the preceding chapter for details.

As an illustration regarding the application of the above methodology, let it be assumed in Figure 10.1 that L = 600 in. (15.24 m), P = 4000 lb (17.792 kN), h = 3 in. (0.0762 m), n = 2, and constant width b = 3 in. (0.0762 m). The material of the member is Monel, and the three-line approximation of the stress/strain curve is used in the analysis. By following the methodology described in the preceding six steps, the values of f(x), ε_1, ε_2, Δ, E_e, g(x), Z_e, M_{req}, and M_e' at any $0 \leq x \leq L_o$ where L_o = 369.2 in. (9.3777 m), are given in Table 10.1. If better accuracy is desired the procedure may be repeated. The moment diagram M_e' of the equivalent system of constant stiffness E_1I_B, where $E_1 = 22 \times 10^6$ psi (151.69×10^6 kPa) and $I_B = 6.75$ in.^{4} (2.81×10^{-6} m^4), is shown plotted in Figure 10.2a. The approximation of its shape with four straight line segments is shown in the same figure, which leads to the equivalent system of constant stiffness E_1I_B shown in Figure 10.2b. By using the equivalent system and applying the well-known moment area method for linear analysis, we find that the deflection δ_C at the free end C is 435.689 in. (11.0665 m). The solution obtained by integrating Equation 10.7 and using Simpson's rule to facilitate the integration yields $\delta_C = \delta_B = 433.39$ in. (11.0081 m), a difference of only 0.53%.

TABLE 10.1
Values of f(x), ε_1, ε_2, Δ, g(x), Z_e, M_{req}, and M_e' for a Nonprismatic Cantilever Beam Loaded with a Concentrated Load P at the Free End

1	2	3	4	5	6	7	8	9	10
x (in.)	f(x)	$\varepsilon_1 \times 10^{-3}$	$\varepsilon_2 \times 10^{-3}$	$\Delta \times 10^{-3}$	$E_e \times 10^6$ (psi)	g(x)	Z_e	$M_{req} \times 10^6$ (in.-lb)	$M_e \times 10^6$ (in.-lb)
0.0	8.0	24.832	24.832	49.664	3.3382	0.15173	1.0	1.4882	1.7259
18.5	7.63	21.914	21.932	43.846	3.7465	0.17029	1.0338	1.4159	1.1257
36.9	7.27	18.927	18.957	37.884	4.3061	0.19573	1.1281	1.3437	1.0646
55.4	6.91	15.871	15.907	31.777	5.0955	0.23161	1.2781	1.2713	1.0148
73.8	6.55	12.919	12.957	25.876	6.2361	0.28346	1.4805	1.1988	0.9561
92.3	6.18	10.171	10.207	20.378	7.8401	0.35637	1.7276	1.1261	0.8832
110.8	5.81	7.874	7.907	15.781	9.8888	0.44449	2.0087	1.0532	0.8101
129.2	5.44	6.203	6.232	12.435	12.1580	0.55263	2.3182	0.9801	0.7564
147.7	5.06	5.031	5.057	10.087	14.3390	0.65178	2.6596	0.9068	0.7311
166.2	4.69	4.208	4.232	8.439	16.3510	0.74324	3.0440	0.8334	0.7275
184.6	4.32	3.634	3.657	7.290	18.1090	0.82315	3.4855	0.7597	0.7447
203.1	3.96	3.184	3.207	6.391	19.5650	0.88930	4.0042	0.6859	0.7806
221.5	3.60	2.834	2.857	5.691	20.7470	0.94306	4.6247	0.6119	0.8334
240.0	3.24	2.559	2.582	5.140	21.4940	0.97701	5.3871	0.5376	0.9148
258.5	2.89	2.308	2.332	4.640	21.9480	0.99762	6.3439	0.4631	1.0194
276.9	2.55	2.087	2.112	4.199	22.0480	1.00220	7.5143	0.3884	1.1492
295.4	2.22	1.845	1.872	3.717	22.0460	1.00210	9.1737	0.3135	1.2908
313.9	1.90	1.553	1.582	3.134	22.0430	1.00200	11.2160	0.2383	1.4021
332.3	1.60	1.188	1.219	2.407	22.0410	1.00190	13.6580	0.1628	1.3910
350.8	1.31	0.717	0.751	1.467	22.0540	1.00250	16.0600	0.0870	1.2841
369.2	1.0	—	—	—	—	—	17.2220	0.0	0.0

Note: 1 in. = 0.0254 m, 1 psi = 6.895 kPa, 1 in.-lb = 0.11298 Nm.

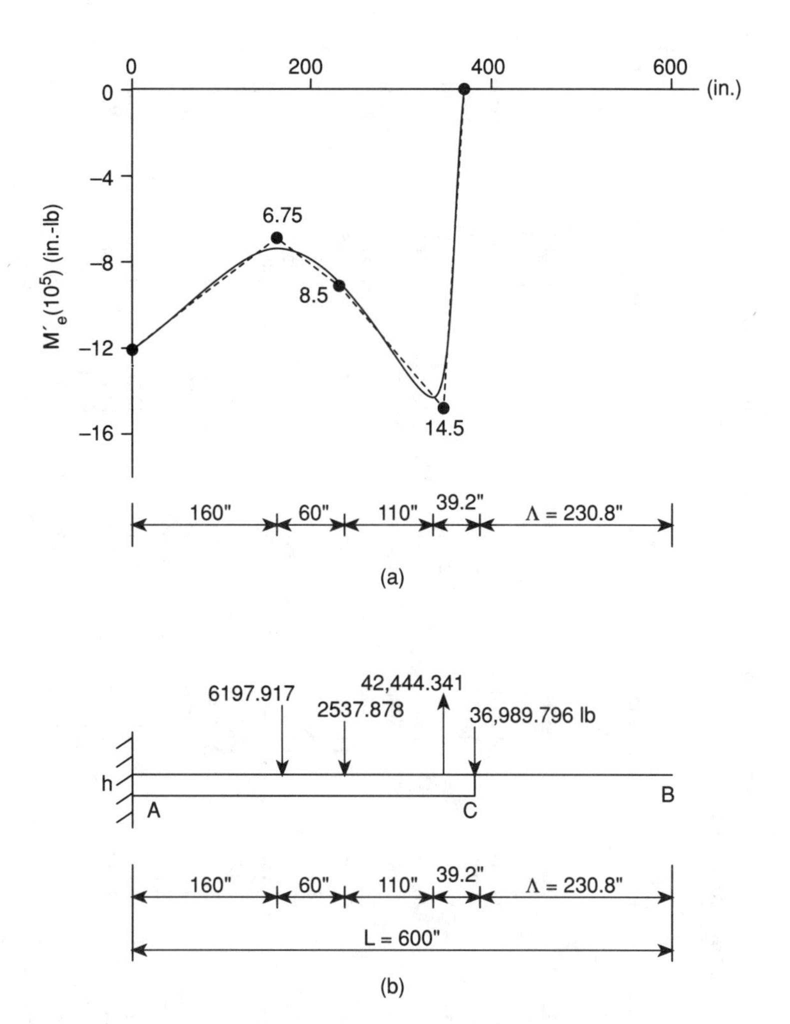

FIGURE 10.2. (a) Moment diagram M_e' of the pseudolinear system of constant stiffness with its shape approximated with four straight lines; (b) pseudolinear equivalent system of constant stiffness (1 in. = 0.0254 m, 1 lb = 4.448 N, 1 in.-lb = 0.11298 Nm).

In the above example, the step-by-step procedure was carried out by subdividing each time the length L_0 into 40 equal segments and carrying out the integrations by using Simpson's rule. The correspondence between x_0 and x is established by using Equations 10.4 and 10.5. This approach facilitates the convergence of the trial and error procedure and yields very accurate results. In Step 1, however, since L_0 is not known, the length L of the member is subdivided into 40 segments and a trial $L_0 = (L - \Lambda)$ is established as explained in this step. In Table 10.1 the final length $L_0 = 369.2$ in. (9.3777 m). In the same table, column 2 gives the values of f(x) for each x_0 that corresponds to each x given in column 1. Equations 10.4 and 10.5 are used to find the x_0 that corresponds to each value of $0 \leq x \leq L_0$. In Table 10.2 the values of the deflection δ_B and rotation θ_B at the free end of the member are given for various values of the load P. The second column in this table gives the values of δ_B by neglecting the effects of the axial force N(x), and in the third column the effects of N(x) are included. Note that the largest error when the effects of N(x) are neglected is 0.82% as shown in column 4. Similar comparisons are made for the horizontal displacement Λ_B of the free end B, and the maximum error is 2.69% as shown in column 7. For the rotation θ_B at the free end B, the maximum error is 0.99%, as shown in column 10. For most practical purposes, these results indicate that the effect of N(x) may be neglected as far as large displacements and rotations are concerned.

The deflection curves of the flexible cantilever bar for P = 3000 lb (13.334 kN) and 4000 lb (17.792 kN) are shown in Figure 10.3a. The dotted line represents the results obtained by neglecting the effects of the axial force N(x). The variation of N(x) is shown in Figure 10.3b and Figure 10.3c shows the shifting of the neutral axis at cross sections $0 \leq x \leq L_0$. The symbol h_1 indicates the distance from the bottom of the bar to neutral axis location. Note that as $x \to L_o$ the position of the neutral axis approaches infinity.

10.3 INELASTIC ANALYSIS OF VARIABLE THICKNESS FLEXIBLE CANTILEVER BEAMS LOADED WITH TRANSVERSED AND AXIAL COMPRESSIVE LOADS AT THE FREE END

In this section we consider a tapered cantilever beam that is loaded with a vertical concentrated load P at the free end B, and with an axial compressive load Q at the same end, as shown in Figure 10.4a. The large deformation configuration of the member is shown in the same figure. The material of the member is Monel, and the three-line approximation of its stress/strain curve will be used here to carry out the inelastic analysis. The length L of the member is equal to 600 in. (15.24 m). It is further assumed that P = 3500 lb (15.568 kN) and $Q = 0.4Q_{cr}$, where Q_{cr} is the critical Euler load for a prismatic cantilever beam of length L = 600 in. (15.24 m), h = 3 in. (0.0762 m), b = 3 in. (0.0762 m), and $E = 22 \times 10^6$ psi (151.69×10^6 kPa).

By using the free body diagram in Figure 10.4b and applying equations of statics, we find

$$M(x) = P(L_o - x) - Qy \tag{10.17}$$

$$V(x) = Q\sin\theta + P\cos\theta \tag{10.18}$$

$$N(x) = Q\sin\theta - P\sin\theta \tag{10.19}$$

TABLE 10.2
Comparison of Values of δ_B, Λ_B, and θ_B at the Free End Of The Beam with Increasing P by Considering and Not Considering the Axial Component Effects

1	2	3	4	5	6	7	8	9	10
P (lb)	Equivalent systems without axial force δ_B (in.)	Equivalent systems with axial force δ_B (in.)	Error (%)	Equivalent systems without axial force Λ_B (in.)	Equivalent systems with axial force Λ_B (in.)	Error (%)	Equivalent systems without axial force θ_B (°)	Equivalent systems with axial force θ_B (°)	Error (%)
500	49.19	49.41	0.45	2.2	2.2	0.0	8.611	8.683	0.83
1000	96.37	96.71	0.35	8.4	8.5	1.18	16.852	16.99	0.81
1500	139.44	139.1	−0.24	19.2	19.0	−1.05	24.449	24.584	0.55
2000	189.04	189.49	0.24	40.0	40.9	2.20	32.768	33.09	0.97
2500	273.06	274.02	0.35	83.1	85.4	2.69	44.549	44.99	0.99
3000	352.61	354.27	0.47	139.5	141.9	1.69	55.03	55.55	0.94
3500	404.01	407.36	0.82	188.5	190.6	1.10	62.16	62.76	0.96
4000	433.99	433.39	0.14	228.0	230.8	1.21	67.07	67.461	0.60

Note: 1 in. = 0.0254 m, 1 lb = 4.448 N.

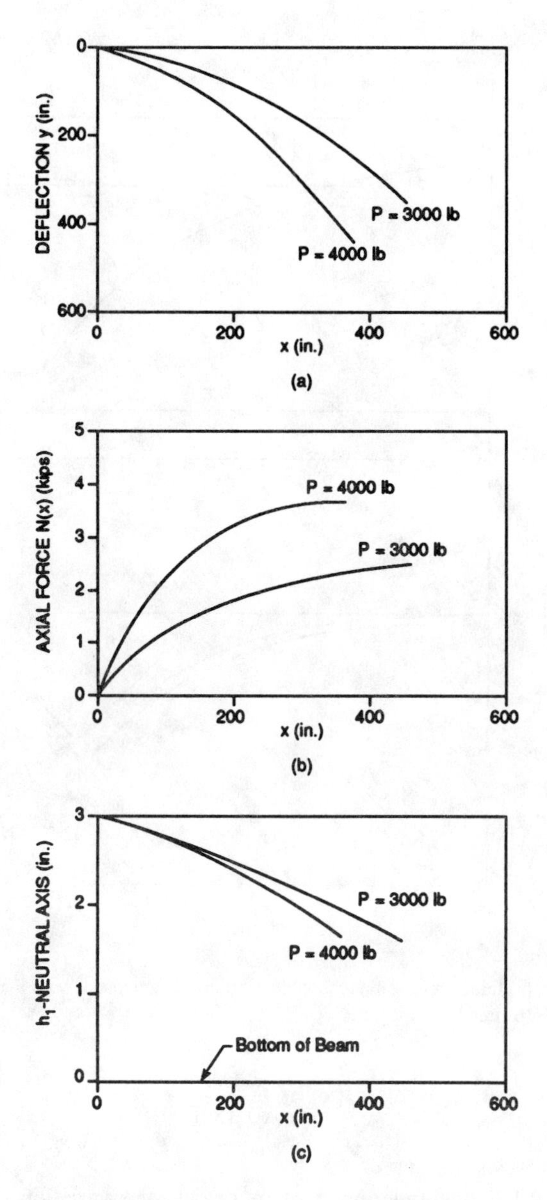

FIGURE 10.3. (a) Vertical deflection curves for P = 3000 and 4000 lb; (b) variation of the normal force N(x) with x when P = 3000 and 4000 lb; (c) shifting of the neutral axis position along the length of the member for P = 3000 and 4000 lb (1 in. = 0.0254 m, 1 lb = 4.448 N).

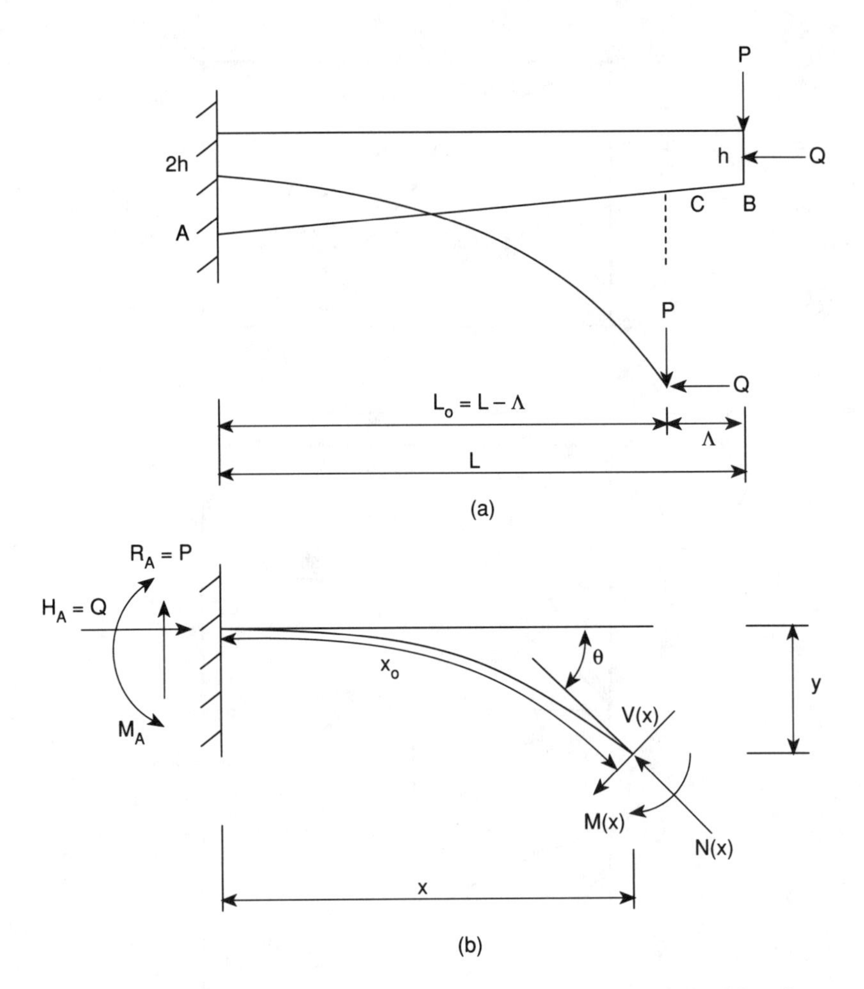

FIGURE 10.4 . (a) Initial nonprismatic flexible cantilever beam loaded with P and Q as shown; (b) free body diagram.

$$\varepsilon_o(x) = \frac{N(x)}{A_o \alpha(x) E} \tag{10.20}$$

In Equation 10.20 the function $\alpha(x)$ represents the variation of the cross-sectional area $A_x = A_o\alpha(x)$ with respect to a reference area A_o. In this case, A_o is the cross-sectional area at the free end B of the member.

The inelastic analysis for this problem may be carried out by using the step-by-step procedure discussed in the preceding section. However, for this case, some modifications should be made in Step 1 in order to initiate the procedure. In Equation 10.17 we note that the bending moment M(x) at any $0 \le x \le L_o$ is dependent upon the large vertical displacement y of the member, which is unknown. In order to overcome this difficulty, Step 1 may be initiated by assuming $g(x) = 1$ and use in Equation 10.17 the y obtained by solving the problem in Figure 10.4a when $Q = 0$. In other words, we start with a trial deflection y and carry out the step-by-step procedure as discussed in the preceding section. This procedure will yield a new deflection y which most probably will be different than the one assumed. By using the new deflection, the step-by-step procedure may be repeated to determine a new y. This repetition may continue until the deflections y of the last trial are approximately equal to the ones obtained in the preceding trial. The rate of convergence for this procedure is satisfactory and the results are accurate.

By applying the above methodology, the values of ε_1, Δ, E_e, $g(x)$, Z_e, M_x, M_e', y', and y, for various values of $0 \leq x \leq L_o$, are shown in Table 10.3. The procedure was carried out by subdividing each time the length L_o into 40 segments as explained in the preceding section and carring out the required integrations using Simpson's rule. We observe in Table 10.3 that the smallest value of the function $g(x)$ is 0.1657, and it occurs at the fixed end of the member. This indicates that the member is heavily stressed at its fixed end and it progressed deep into the inelastic range. From the same table we note that L_o = 365.5 in. (9.2837 m). The maximum vertical deflection at the free end is 438.22 in. (11.1308 m), the horizontal displacement Λ = 234.5 in. (5.9563 m), and the rotation θ at the same end is 68.938°. Other inportant values regarding these quantities may be found in the same table (see, for example, columns 5, 6, 9, 10, and 11). Column 9 shows values of the moment M_e' of the pseudolinear equivalent system of constant stiffness EI_B, where I_B is the moment of the inertia at the free end B of the member. The actual moments M_x are given in column 8 of the table. If better accuracy is required, the procedure may be repeated a few additional times.

Figure 10.5a shows the variation of the vertical deflection δ at the free end of the beam with increasing load P, and for Q/Q_{cr} = 0.4, 0.6, and 0.8. For the same values of Q/Q_{cr}, the horizontal displacement Λ and rotation θ at the free end are plotted in Figures 10.5b and 10.5c, respectively, for increasing values of the load P. In all cases, the curves become progressively steeper as Q/Q_{cr} increases.

10.4 INELASTIC ANALYSIS OF FLEXIBLE NONPRISMATIC SIMPLY SUPPORTED BEAMS WITH VERTICAL LOADS AND AXIAL COMPRESSIVE FORCES AT THE END SUPPORTS

Consider the nonprismatic simply supported beam that is subjected to an axial compressive force Q and a uniformly distributed load w as shown in Figure 10.6a. Its large deformation configuration is shown in the same figure. The coordinate x_D locates the point on the beam where the deflection y is maximum and rotation θ is zero. The horizontal displacement of the end B to

TABLE 10.3
Values of f(x), ε_1, Δ, E_e, g(x), Z_e, M_x, M_e', y', and y for a Nonlinear Cantilever Beam Loaded with a Vertical Load P and a Horizontal Compressive Load Q at the Free End for $Q = 0.4\ Q_{cr}$

1	2	3	4	5	6	7	8	9	10	11
x (in.)	f(x)	$\varepsilon_1 \times 10^{-3}$	$\Delta \times 10^{-3}$	$E_e \times 10^6$ (psi)	g(x)	Z_e	$M_x \times 10^6$ (in.-lb)	$M_e' \times 10^6$ (in.-lb)	y'	y (in.)
0.0	8.0	22.369	44.726	3.645	0.1657	1.0	1.4657	1.1058	0.0	0.0
18.3	7.64	20.405	40.812	3.982	0.1810	1.027	1.4028	1.0415	0.1318	1.21
36.5	7.28	18.244	36.501	4.438	0.2017	1.103	1.3390	1.0051	0.2576	4.78
54.8	6.93	15.936	31.893	5.067	0.2303	1.227	1.2741	0.9801	0.3797	10.60
73.1	6.57	13.557	27.138	5.941	0.2701	1.401	1.2081	0.9538	0.4988	18.64
91.4	6.21	11.230	22.487	7.144	0.3247	1.624	1.1411	0.9186	0.6147	28.81
109.6	5.85	9.056	18.137	8.744	0.3974	1.891	1.0732	0.8732	0.7244	41.05
127.9	5.48	7.257	14.539	10.716	0.4871	2.200	1.0042	0.8276	0.8291	55.25
146.2	5.11	5.884	11.791	12.790	0.5814	2.551	0.9343	0.8023	0.9293	71.32
164.5	4.74	4.861	9.743	14.794	0.6724	2.953	0.8635	0.8	1.0277	89.20
182.7	4.37	4.137	8.294	16.688	0.7585	3.422	0.7918	0.8168	1.1270	108.89
201.0	4.01	3.563	7.144	18.398	0.8363	3.980	0.7192	0.8538	1.2298	130.42
219.3	3.65	3.138	6.295	19.749	0.8977	4.657	0.6455	0.9182	1.3386	153.88
237.6	3.29	2.788	5.594	20.922	0.9510	5.495	0.5708	1.0023	1.4565	179.40
255.8	2.94	2.487	4.994	21.778	0.9899	6.561	0.4950	1.1177	1.5865	207.19
274.1	2.59	2.236	4.493	22.030	1.0014	7.959	0.4179	1.2843	1.7336	237.50
292.4	2.25	1.984	3.990	22.056	1.0025	9.841	0.3394	1.4789	1.9034	270.71
310.7	1.92	1.681	3.387	22.053	1.0024	12.365	0.2592	1.6634	2.0971	307.24
328.9	1.60	1.293	2.613	22.051	1.0023	15.572	0.1773	1.7172	2.3069	347.47
347.2	1.30	0.776	1.581	22.056	1.0026	18.951	0.0934	1.3557	2.5008	391.45
365.5	1.0	—	—	—	—	20.673	0.0	0.0	2.5967	438.22

Note: 1 in. = 0.0254 m, 1 psi = 6.895 kPa, 1 in.-lb = 0.11298 Nm.

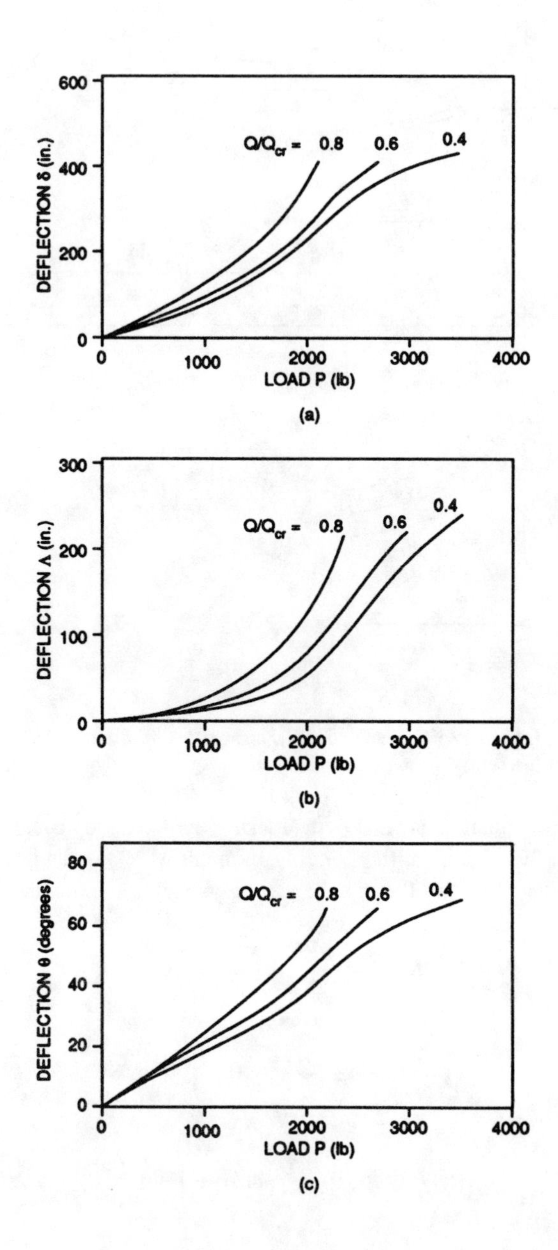

FIGURE 10.5. (a) Variation of vertical free end displacement δ with increasing load P and $Q/Q_{cr} = 0.4, 0.6$, and 0.8; (b) variation of horizontal free end displacement Λ with load P and $Q/Q_{cr} = 0.4, 0.6$, and 0.8; (c) variation of free end rotation θ with load P and $Q/Q_{cr} = 0.4, 0.6$, and 0.8 (1 in. = 0.0254 m, 1 lb = 4.448 N).

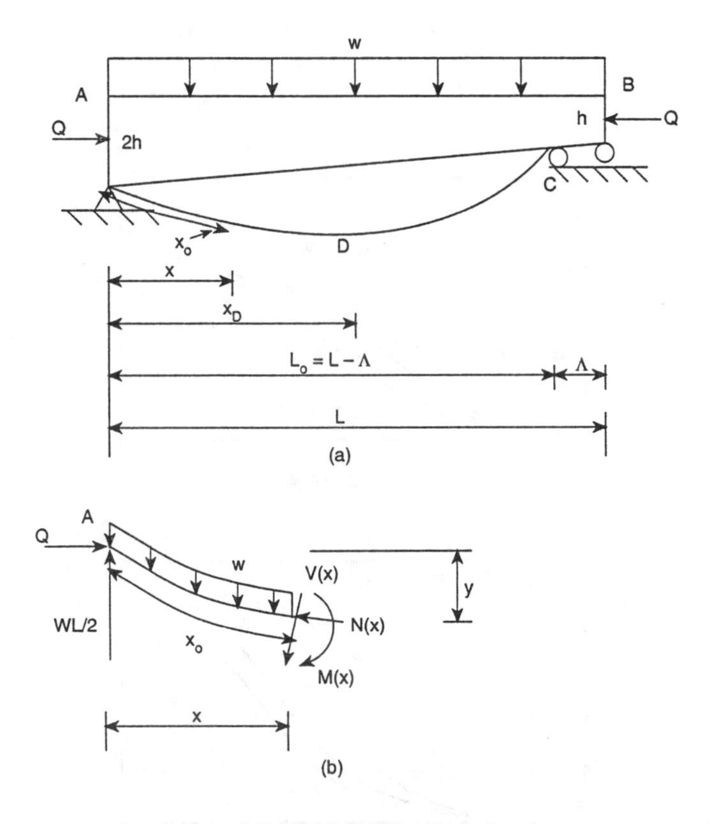

FIGURE 10.6. (a) Initial nonprismatic simply supported flexible beam loaded with a distributed load w and an axial compressive force Q; (b) free body diagram.

position C is designated as Λ. By considering the free body diagram in Figure 10.6b and applying statics, the bending moment M(x), shear force V(x), and normal force N(x) at any $0 \leq x \leq L_o$, are given by the expressions

$$M(x) = \frac{wLx}{2} - \frac{wx_0x}{2} + Qy \tag{10.21}$$

$$V(x) = Q\sin\theta + \frac{wL}{2}\cos\theta - \int_0^{x_0} w\cos\theta\, dx_o \tag{10.22}$$

$$N(x) = Q\cos\theta - \frac{wL}{2}\sin\theta + \int_0^{x_0} \sin\theta\, dx_o \tag{10.23}$$

Also,

$$\varepsilon_0(x) = \frac{N(x)}{A_0\alpha(x)E} \tag{10.24}$$

In the above equations, x_o may be obtained by using Equations 10.4 and 10.5 and Λ by using Equation 10.6 as discussed in the preceding two sections.

The step-by-step procedure discussed in the preceding sections may be also used here to determine the inelastic large deflections and rotations of the member. In order to start with Step 1 of this procedure and determine M_e' in this step, the large displacement y in Equation 10.21 must be known. Since this deflection is not known, we may initiate the procedure with Step 1 by utilizing the displacement y of the simply supported beam obtained by assuming Q = 0 and neglecting the axial restraint from the axial component of the load w. We may also assume that g(x) = 1 throughout the length of the member, which means elastic analysis. We know that this is not the correct deflection, but it serves as a first trial deflection to complete the step-by-step procedure and obtain a new deflection y that can be used to repeat the procedure. This procedure may be repeated until the deflection y obtained from the last trial closely approximates the one obtained from the preceding trial. It converges fairly fast and the results obtained are very accurate.

For the problem in Figure 10.6a let it be assumed that L = 1000 in. (25.4 m), h = 3 in. (0.0762 m), width b = 3 in. (0.0762 m), w = 6.5 lb/in. (1138.267 N/m), and Q = 0.2 Q_{cr}, where Q_{cr} is the critical Euler load for a prismatic simply supported beam of length L = 1000 in. (25.4 m) and h = b = 3 in. (0.0762 m). The material of the member is Monel, and the three-line approximation of its stress/strain curve with $E_1 = 22 \times 10^6$ psi (151.69×10^6 kPa) is used in the analysis. By applying the methodology discussed above, the final values of ε_1, Δ, E_e, g(x), Z_e, M_x, M_e', y′, and y for various values of $0 \leq x \leq L_o$ are shown in Table 10.4. The procedure was carried out as in the examples of the preceding sections by subdividing each time the length L_o into 40 segments and carrying out the required integrations using Simpson's rule.

From Table 10.4, column 11, we note that the maximum deflection occurs at x = 474.8 in. (12.06 m), and it is equal to 289.15 in. (7.3444 m). From the same table, the rotation θ_B of the end B is maximum, with a value of $\theta_B = \tan^{-1}y' =$ 59.60°. At support A, $\theta_A = 40.10°$. The smallest value of the function g(x) = 0.2817 and occurs at x = 514.3 in. (13.06 m). The values of the moment M_e' of the equivalent pseudolinear system of constant stiffness E_1I_B are shown in column 9 of Table 10.4. The values of the function f(x) are shown in the second column of the table. They are the values at each x_o that corresponds to the value of x in the first column of the table. Equations 10.4 and 10.5 are used to determine the correspondence between x and x_o. The actual moments M_x of the original beam in Figure 10.6a are given in column 8 of the same table. It would be interesting to study Table 10.4 in detail and draw useful conclusions.

The variation of the maximum deflection δ of the member with increasing load w and Q/Q_{cr} = 0.2, 0.4, and 0.6 is shown in Figure 10.7a. Similar curves are plotted for the horizontal displacement Λ of the end B of the member and for the rotation θ_B at the end B, and they are shown in Figures 10.7b and 10.7c, respectively. The variation of the axial normal force N(x) along the length of the member for w = 6.5 lb/in. (1138.324 N/m) and Q/Q_{cr} = 0.2 is shown in

TABLE 10.4

Values of f(x), ε_1, Δ, E_e, g(x), Z_e, M_x, M_e', y', and y, for a Nonprismatic Simply Supported Beam Loaded with a Uniformly Distributed Load w and an Axial Force Q at the Ends for $Q = 0.2\ Q_{cr}$

1	2	3	4	5	6	7	8	9	10	11
x(in.)	f(x)	$\varepsilon_1 \times 10^{-3}$	$\Delta \times 10^{-3}$	$E_e \times 10^7$ (psi)	g(x)	Z_e	$M_x \times 10^5$ (in.-lb)	$M_e' \times 10^5$ (in.-lb)	y'	y(in.)
0.0	8.0	—	—	—	—	2.2344	0.0	0.0	0.8421	0.0
39.6	7.40	0.3472	0.7028	2.2044	1.0020	2.1323	1.3218	0.38	0.8371	33.24
79.1	6.84	0.7012	1.4098	2.2058	1.0027	2.0851	2.5142	0.7646	0.8104	66.08
118.7	6.31	1.0557	2.1177	2.20	1.0036	2.0092	3.5788	1.1359	0.7802	98.12
158.3	5.81	1.4077	2.8205	2.20	1.0046	1.9105	4.5138	1.4767	0.7409	128.97
197.8	5.38	1.7532	3.5103	2.2123	1.0056	1.7961	5.3176	1.7742	0.7182	158.26
237.4	4.93	2.0875	4.1778	2.2144	1.0066	1.6743	5.9888	2.0214	0.6676	185.69
276.9	4.54	2.4552	4.9120	2.2045	1.0020	1.5521	6.5260	2.2358	0.6109	210.99
316.5	4.17	2.9813	5.9632	2.0209	0.9186	1.4310	6.9280	2.5819	0.5468	233.92
356.1	3.84	3.9576	7.9145	1.7228	0.7831	1.3096	7.1932	3.1339	0.4716	254.09
395.6	3.54	5.8846	11.7660	1.2950	0.5886	1.1869	7.3191	4.1750	0.3134	270.88
435.2	3.26	9.4128	18.8200	0.8855	0.4025	1.0706	7.3018	5.9652	0.1544	283.16
474.8	3.0	12.9650	25.9220	0.6705	0.3048	1.0020	7.1357	7.8214	−0.0560	289.15
514.3	2.76	14.2620	28.5190	0.6198	0.2817	1.0510	6.8150	9.2185	−0.1716	286.95
553.9	2.52	12.6290	25.2610	0.7039	0.32	1.2900	6.3349	10.1340	−0.4305	275.10
593.5	2.28	8.5730	17.1550	1.0058	0.4572	1.7681	5.6930	9.6736	−0.6976	252.76
633.0	2.02	4.8215	9.6533	1.5475	0.7034	2.4556	4.8900	8.4395	−0.9381	220.32
672.6	1.76	3.0452	7.5029	1.8162	0.8255	2.8737	4.4288	8.7383	−1.1561	178.86
712.2	1.50	2.1530	4.3208	2.2393	1.0178	4.5825	2.8067	8.4063	−1.3735	128.85
751.7	1.25	1.3196	2.6596	2.2411	1.0187	6.1492	1.5248	7.3862	−1.5900	70.16
791.3	1.0	—	—	—	—	7.0938	0.0	0.0	−1.7046	0.0

Note: 1 in. = 0.0254 m, 1 psi = 6.895 kPa, 1 in.-lb = 0.11298 Nm.

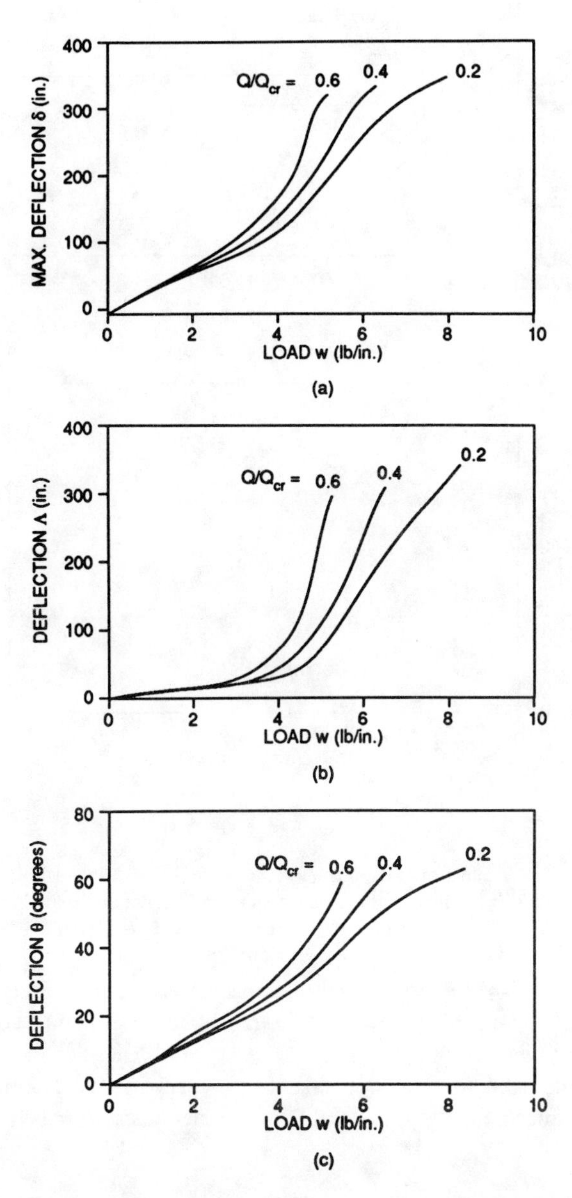

FIGURE 10.7. (a) Variation of maximum vertical displacement δ with increasing load w and $Q/Q_{cr} = 0.2$, 0.4, and 0.6; (b) variation of horizontal displacement Λ of support B with load w and $Q/Q_{cr} = 0.2$, 0.4, and 0.6; (c) variation of rotation θ_B of support B with load w and $Q/Q_{cr} = 0.2$, 0.4, and 0.6 (1 in. = 0.0254 m, 1 lb/in. = 175.1268 N/m).

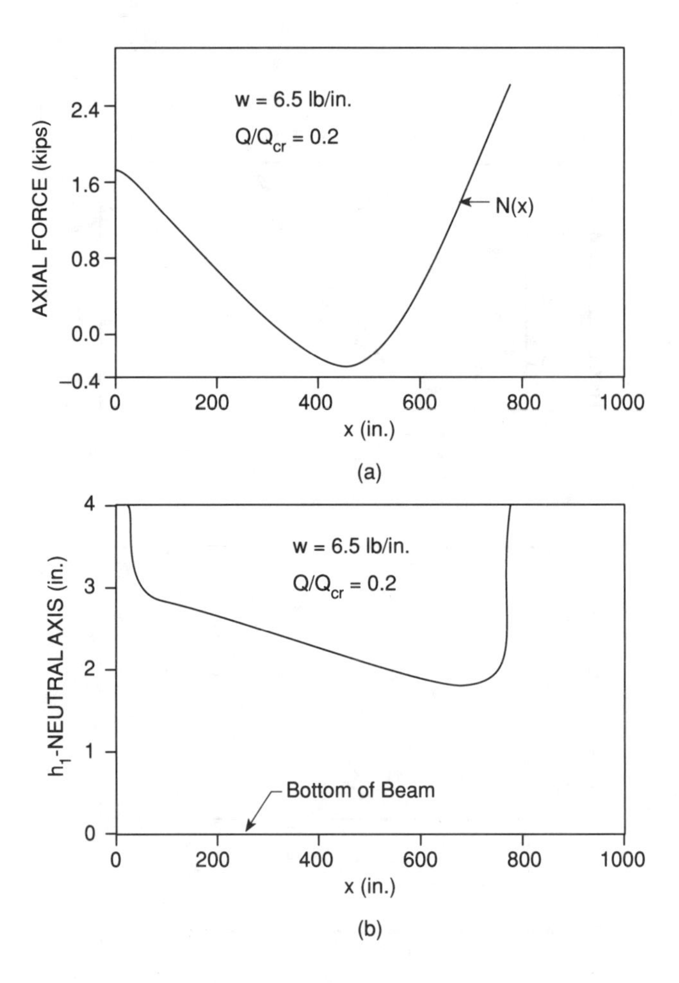

FIGURE 10.8 . (a) Variation of axial force N(x) with x for w = 6.5 lb/in and $Q/Q_{cr} = 0.2$; (b) shifting h_1 of the neutral axis measured from bottom of beam for w = 6.5 lb/in. and $Q/Q_{cr} = 0.2$ (1 in. = 0.0254 m, 1 lb = 4.448 N, 1 lb/in. = 175.1268 N/m).

Figure 10.8a. The shifting of the neutral axis at cross sections along the length of the flexible member for w = 6.5 lb/in. (1138.324 N/m) and $Q/Q_{cr} = 0.2$ is illustrated in Figure 10.8b.

From Table 10.4 we also note that $L_0 = 791.3$ in. (20.099 m), and the horizontal displacement Λ of the end B of the member is $\Lambda = 1000 - 791.3 = 208.7$ in. (5.301 m).

10.5 GENERAL DISCUSSION AND REMARKS

The analysis in this chapter showed that inelastic analysis of nonprismatic flexible bars that includes the effect of axial restraints may be carried out using the concept of pseudolinear equivalent systems of constant stiffness and an equivalent modulus E_e. The variation of E_x along the length of the bar is caused by the applied load when it is large enough to stress the material of the bar beyond its elastic limit. The member becomes increasingly inelastic with increasing load, until its ultimate capacity to resist load and deformation is reached. This progressive deterioration of the flexible member's ability to resist load and deformation may be used in practice to establish specific performance criteria in a given situation.

In the analysis, the effects of N(x), which includes the components of the vertical load during the large deformation of the member, have also been taken into consideration. These effects, in terms of displacements and rotations, are often small, and they could be neglected in practical situations (see, e.g., the results in Table 10.2 where values of displacements and rotations are given for both cases). The flexible nonprismatic bar was also assumed to be loaded with vertical as well as axial concentrated compressive loads at the ends of the member. The results show that the deflections and rotations approach infinity as the axial load Q approaches the critical buckling load of the member.

When the axial load effects are considered in the inelastic analysis of flexible bars, coaxiality between centroidal and neutral axes no longer exists. At sections along the length of the member, the neutral axis will shift to different positions, depending upon the relative values of the normal stresses due to bending and axial force Q. The purpose of the equivalent modulus E_e used in the pseudolinear analysis is to correct for the shifting of the neutral axis so that we can always assume coaxiality between centroidal and neutral axes when the pseudolinear equivalent system is used.

The use of equivalent systems, together with the subdivision of the length L_0 into segments as discussed in this chapter, facilitates the application of the suggested step-by-step procedure and its convergence and assures very accurate results.

The methodology discussed in this chapter is general, and it may be used for stress/strain curve variations of other types of materials. The approximation of the shapes of these curves with three to four straight line segments is usually sufficient for practical applications.

Some additional discussion regarding the moment M_e' of the equivalent pseudolinear system of constant stiffness E_1I_1 would be helpful here in order to better comprehend its significance. This is illustrated by using the values of M_e' from Table 10.4 and plotting the M_e' diagram shown in Figure 10.9a. By approximating its shape with six straight lines as shown, we can obtain the pseudolinear equivalent system of constant stiffness E_1I_1, which is loaded with five concentrated loads as shown in Figure 10.9b. The equivalent system in Figure 10.9b is now linear, and it replaces the original nonlinear problem in Figure 10.6a. Therefore, deflections and rotations of the original nonlinear problem may be determined using the pseudolinear system in Figure 10.9b and applying linear analysis. For example, by using the moment area method, we find that y_C' at the end C of the pseudolinear system in Figure 10.9b is –1.707, and therefore the rotation $\theta_C = \tan^{-1} y_C' = 59.637°$. In Table 10.4, the value of

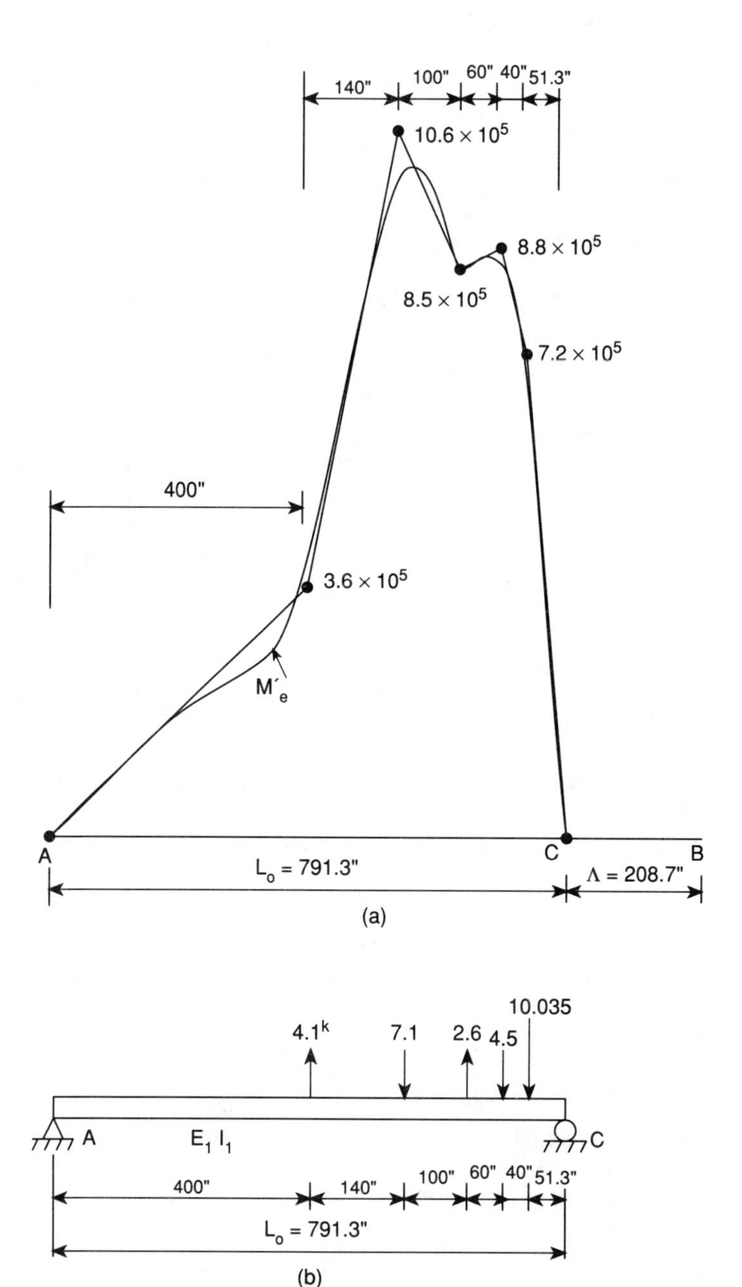

FIGURE 10.9. (a) M_e' moment diagram of the pseudolinear equivalent system of constant stiffness E_1I_1 with its shape approximated with six straight lines; (b) pseudolinear equivalent system of constant stiffness E_1I_1 loaded with five concentrated loads (1 in. = 0.0254 m, 1 kip = 4448.222 N, 1 in.-lb = 0.11298 Nm).

y_C' is −1.7046 and $\theta_C = \tan^{-1} y_C' = 59.602°$. The difference is only 0.06%, which shows that the approximation of the M_e' diagram in Figure 10.9a with six straight lines as shown, gives excellent accuracy.

It should also be noted here that the effect of the axial force Q in Figure 10.6a is taken into consideration when the M_e' diagram in Figure 10.9a was determined, and consequently the pseudolinear system in Figure 10.9b is free of axial effects and it can be solved directly for deflections and rotations using linear analysis. The shifting of the neutral axis is also taken into consideration in the derivation of the pseudolinear system, and thus coaxiality between neutral and centroidal axes can be maintained when the pseudolinear system is used.

PROBLEMS

10.1 The tapered flexible cantilever beam in Figure 10.1a is loaded with a concentrated load P = 3000 lb (13,345 N) at the free end. The length L = 600 in. (15.24 m), h = 3 in. (0.0762 m), width b = 3 in. (0.0762 m), and taper n = 2. The material of the member is Monel. Using the three-line approximation of the stress/strain curve of Monel and applying the step-by-step procedure discussed in Section 10.2, determine in two ways the vertical and horizontal displacements δ_B and Λ_B, respectively, at the free end B and the rotation θ_B at the same end. First consider the effect of N(x) Figure 10.1c, and in the second case neglect N(x), and compare the results.
Answer: δ_B = 352.61 in., Λ_B = 139.5 in., and θ_B = 55.03°.

10.2 Solve Problem 10.1 when P = 2000 lb (8896.4 N).
Answer: δ_B = 189.49 in., Λ_B = 40.9 in., and θ_B = 33.09°.

10.3 Solve Problem 10.1 when P = 4000 lb (17,793 N).
Answer: δ_B = 433.39 in., Λ_B = 230.8 in., and θ_B = 67.46°.

10.4 By using the values of M_e' shown in Table 10.1, plot the moment diagram M_e' of the equivalent pseudolinear system of constant stiffness E_1I_1. Then, by approximating the shape of M_e' with straight lines, derive the pseudolinear equivalent system of constant stiffness E_1I_1 and use it to determine the rotation and deflection at the free end. Compare these values with the values shown in Table 10.2.

10.5 Using the values of M_e' shown in Table 10.3, plot the moment diagram M_e' of the equivalent pseudolinear system of constant stiffness E_1I_1. By approximating the shape of M_e' with straight lines, obtain a pseudolinear system of constant stiffness E_1I_1 and use it to determine the vertical deflection and rotation at the free end. Compare the results with the values obtained in Section 10.3.

10.6 The flexible cantilever beam in Figure 10.4a is loaded with a load P = 2000 lb (8896.4 N) and an axial load $Q = 0.6\ Q_{cr}$, where Q_{cr} is the critical Euler load of a prismatic beam of length L = 600 in. (15.24 m), h = b = 3 in. (0.0762 m), and $E = 22 \times 10^6$ psi (151.69×10^6 kPa). Following the procedure discussed in Section 10.3, reproduce a table similar to Table 10.3 and discuss the results. Use the three-line approximation of the stress/strain curve of Monel.

10.7 Using the results obtained in Problem 10.6, determine a pseudolinear equivalent system of constant stiffness E_1I_1, where $E_1 = 22 \times 10^6$ psi (151.69×10^6 kPa) and $I_1 = 6.75$ in.^{4} (2.8096×10^{-6} m^4), and use it to determine the vertical deflection and rotation at its free end.

10.8 Using the results shown in Table 10.4, plot M_x, M_e', g(x), E_e, and y, and discuss the results. Compare the graphs of M_x and M_e' by puting them in the same figure and make interesting observations.

10.9 The tapered simply supported flexible member in Figure 10.6a, is loaded with a distributed load w = 4.0 lb/in. (700.5 N/m) and an axial load $Q = 0.4Q_{cr}$, where Q_{cr} is the Euler critical load for a prismatic simply supported beam of length L = 1000 in. (25.4 m) and h = b = 3 in. (0.0762 m). The material of the member is Monel. Using the three-line approximation of the stress/strain curve of Monel and following the procedure discussed in Section 10.4, reproduce a table similar to Table 10.4 and discuss the results.

10.10 Using the results from Problem 10.9, determine a pseudolinear equivalent system of constant stiffness E_1I_1, where $E_1 = 22 \times 10^6$ psi (151.69×10^6 kPa) and $I_1 = 6.75$ in.^{4} (2.8096×10^{-6} m^4), and use it to obtain the vertical deflection and rotation at its free end.

Chapter 11

HYSTERETIC MODELS AND ANALYSIS OF UNIFORM AND VARIABLE STIFFNESS MEMBERS UNDER CYCLIC LOADING

11.1 INTRODUCTION AND BACKGROUND INFORMATION

Historically, the problem of inelastic vibration received considerable attention by many researchers and practicing engineers. Bleich[57] and Bleich and Salvadori[58] proposed an approach based on normal modes for the inelastic analysis of beams under transient and impulsive loads. This approach is theoretically sound, but it can be applied only to situations where the number of possible plastic hinges is determined beforehand and where the number of load reversals is negligible. Baron et al.[59] and Berge and DaDeppo[60] solved the required equation of motion by using methods that are based on numerical integration. This, however, involved concentrated kink angles which are used to correct for the amount by which the deflection of the member surpasses the actual elastic-plastic point. The methodology is simple, but the actual problem may become very complicated, because multiple correction angles and several hinges may appear simultaneously. Lee and Symonds[61] proposed the method of rigid plastic approximation for the deflection of beams, which is valid only for a single possible yield with no reversals. Toridis and Wen[62] used lumped mass and flexibility models to determine the response of beams.

In all the models developed in the above references the precise location of the point of reversal of loading is essential. A hysteretic model where the location of the loading reversal point is not required, and where the reversal is automatically accounted for, was first suggested by Bonc[63] for a spring-mass system, and it was later extended by Wen[64] and Iyender and Dash.[65] In recent years, Sues et al.[66] have provided a solution for a single degree of freedom model for a degrading inelastic model. This work was later extended by Fertis and Lee,[67] and they developed a model that adequately describes the dynamic structural response of variable and uniform stiffness members. In their work, the material of the member can be stressed well beyond its elastic limit, thus causing the modulus E to vary along the length of the member.

The above discussion, is not intended to provide a complete historical treatment of the subject, and I wish to apologize for any unintentional omition of the work of other investigators. It provides, however, some incite regarding the state of the art and how the ideas regarding this very important subject have been initiated.

The work in this chapter includes the dynamic analysis of variable and/or uniform stiffness members that are subjected to cyclic dynamic loadings. The analysis takes into consideration the restoring force behavior of such members by using appropriate hysteretic restoring force models, which adequately describe the dynamic response of such continuous structural systems. Since the material of the member is stressed beyond its elastic limit, the analysis also considers the resulting variation of the modulus E along the length of the member. Appropriate mathematical equivalent systems are also used, as discussed in the preceding chapters, in order to simplify the mathematical complexity of the nonlinear problem. Various beam cases and dynamic loadings are included, and dynamic deflections for a large number of cycles are obtained. The derived differential equations are solved using the finite difference method in combination with the 4th order Runge-Kutta method.

11.2 THEORETICAL FORMULATION OF THE PROBLEM

In this section, the theory regarding the dynamic analysis of members that are subjected to dynamic cyclic loadings is formulated. The derived differential equations take into consideration the restoring force behavior of such members by using appropriate hysteretic restoring force models. The values for the hysteresis loop shape parameters and identification of the hysteretic restoring force have been determined by extending the methodology for single-degree-of-freedom systems, which was developed by Sues et al.[66] This methodology is extended to apply for systems with continuous mass and elasticity, and to cases where the modulus E and cross-sectional dimensions along the length of the member are variable. The material of the member is stressed beyond its elastic limit, and the method of the equivalent systems as developed by Fertis and his collaborators and discussed in the preceding chapters, will be used to simplify the mathematical complexity of the problem.

The basic idea in the model by Sues et al.[66] is to split the restoring force into hysteretic and elastic portions. The hysteretic component is expressed in terms of a nonlinear equation, where its solution is bounded by unity in both loading and unloading phases.

By considering a model that involves a single-degree-of-freedom oscillator with mass m and viscous damping coeffecient c and subjected to a dynamic excitation q(t), we may write the following differential equation that describes this system:

$$m\ddot{u} + c\dot{u} + P(u,t) = q(t) \tag{11.1}$$

where u is the displacement and $\ddot{u}$ and $\dot{u}$ are, respectively, the acceleration and velocity of the mass m.

The restoring force P(u,t) is given by the equation

$$P(u,t) = \alpha ku + (1-\alpha)k\bar{z} \tag{11.2}$$

where α is the slope coefficient of the load displacement curve, αk is the slope for the nonlinear part after yield, and $\bar{z}$ is the restoring displacement. By using Equation 11.2, the differential equation of motion given by Equation 11.1 may be written as

$$m\ddot{u} + c\dot{u} + \alpha k u + (1-\alpha)k\bar{z} = q(t) \tag{11.3}$$

The nonlinear differential equation governing $\bar{z}$, is given by Sues et al.[66] and it is written as

$$\bar{z} = B^{*}\dot{u} - \beta^{*}|\dot{u}||\bar{z}|^{n^{*}-1}\bar{z} - \gamma\dot{u}|\bar{z}|^{n^{*}} \tag{11.4}$$

where B^*, β^*, γ^*, and n^* are parameters that control the shape of the hysteresis loop. The resulting hysteretic loops for the case $n^* = 1$ and for constant amplitude cyclic motion of displacement u and various combinations of β^* and γ^* are shown in Figure 11.1. The sharpness of the transition from the linear to nonlinear range is governed by the parameter n^*, with the hysteresis approaching the bilinear behavior as n^* approaches infinity.

The above model for single-degree-of-freedom systems can be extended to apply for beams where E, I, and mass along the length of the member can vary. The general differential equation of motion for a beam may be written as

$$\left[\rho A(x) + \sum_{i=1}^{R} Q_i \delta(x - a_i)\right]\ddot{y} + c\dot{y} + \frac{\partial^2 M}{\partial x^2} = q_1(x,t) + q_2(x) + \sum_{j=1}^{n} F_j(x,t) + \sum_{l=1}^{m} F_l(x) \tag{11.5}$$

where y(x,t), M(x,t), and $q_1(x,t)$ are the displacement, bending moment, and external loading at any location x and time t. In the same equation, ρ is the mass density, A(x) is the cross-sectional area at any x, Q_i is the mass of a concentrated weight, δ is the Dirac delta function, c is the viscous damping coefficient, $q_2(x)$ is the applied distributed static load, $F_j(x,t)$ is a concentrated dynamic load at a position x, and F_l is the concentrated static load at a position x. With the help of Equation 11.2, we may write

$$M(x,t) = \alpha k \frac{\partial^2 y}{\partial x^2} + (1-\alpha)kz \tag{11.6}$$

where k may be expressed as

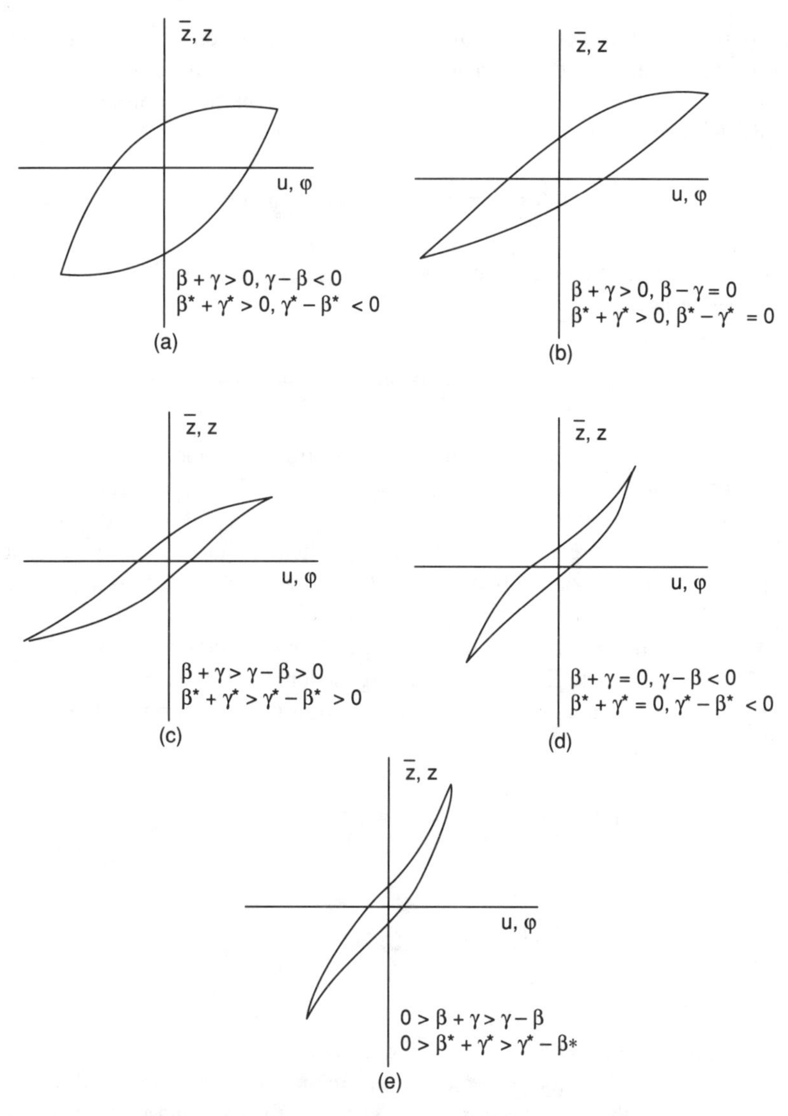

FIGURE 11.1. Hysteretic shapes for combinations of β, β^*, γ, and γ^* with $n^* = n = 1$.

$$k = k_o f(x) \quad \text{(a)}, \quad k_o = E_o I_o \quad \text{(b)} \qquad (11.7)$$

In Equations 11.6 and 11.7, the modulus of elasticity E_o before yield and I_o are reference values for the variables E and I, respectively, k_o is a reference stiffness of the variable stiffness k = EI of the member, z is the restoring curvature, f(x) gives the variation of the stiffness, and α is the slope coeffecient of the moment curvature curve.

By using the expression

$$\frac{\partial^2 M}{\partial x^2} = \alpha k_o \left[\frac{\partial^2 f(x)}{\partial x^2} \frac{\partial^2 y}{\partial x^2} + z \frac{\partial f(x)}{\partial x} \frac{\partial^3 y}{\partial x^3} + f(x) \frac{\partial^4 y}{\partial x^4} \right]$$
$$+(1-\alpha) k_o \left[z \frac{\partial^2 f(x)}{\partial x^2} + 2 \frac{\partial f(x)}{\partial x} \frac{\partial z}{\partial x} + f(x) \frac{\partial^2 z}{\partial x^2} \right] \tag{11.8}$$

and substituting into Equation 11.5, we obtain

$$\left[\rho A(x) + \sum_{i=1}^{R} Q_i \delta(x - a_i) \right] \ddot{y} + c\dot{y}$$
$$+\alpha k_o \left[\frac{\partial^2 f(x)}{\partial x^2} \frac{\partial^2 y}{\partial x^2} + z \frac{\partial f(x)}{\partial x} \frac{\partial^3 y}{\partial x^3} + f(x) \frac{\partial^2 y}{\partial x^2} \right]$$
$$+(1-\alpha) k_o \left[z \frac{\partial^2 f(x)}{\partial x^2} + 2 \frac{\partial f(x)}{\partial x} \frac{\partial z}{\partial x} + f(x) \frac{\partial^2 z}{\partial x^2} \right] \tag{11.9}$$
$$= q_1(x,t) + q_2(x) + \sum_{j=1}^{n} F_j(x,t) + \sum_{l=1}^{m} F_l(x)$$

The nonlinear differential equation governing z may now be written as

$$\dot{z} = B \frac{\partial^3 y}{\partial x^2 \partial t} - \beta \left| \frac{\partial^3 y}{\partial x^2 \partial t} \right| |z|^{n-1} z - \gamma \frac{\partial^3 y}{\partial x^2 \partial t} |z|^n \tag{11.10}$$

In Equation 11.10, B, β, γ, and n are paremeters that control the shape of the hysteresis loop for the beam. In this case, the hysteretic shapes for constant amplitude cyclic motions of curvature ϕ vs. restoring curvature z for various combinations of β and γ are shown in Figure 11.1. The sharpness of the transition from linear to nonlinear range is governed by the parameter n. The hysteresis approaches the bilinear behavior of moment curvature relationships as n approaches infinity. By dividing Equation 11.10 by $\partial^3 y/\partial x^2 \partial t$, we obtain

$$\frac{\partial z}{\left(\frac{\partial^2 y}{\partial x^2}\right)} = B - \beta \frac{\left|\frac{\partial^3 y}{\partial x^2 \partial t}\right|}{\frac{\partial^3 y}{\partial x^2 \partial t}} |z|^{n-1} z - \gamma |z|^n \tag{11.11}$$

By solving Equation 11.11 as a function of y, the nature and versatility of the model may be revealed. The quantity z is a transformed curvature variable, such that the restoring moment given by Equation 11.6 exhibits smooth hysteretic behavior. The maximum value of z may be obtained by setting Equation 11.11 to zero. For positive $\partial^3 y/\partial x^2 \partial t$ and z, it yields

$$z_{max} = \left[\frac{B}{\beta + \gamma}\right]^{1/n} \tag{11.12}$$

The yield level M_{yield} for the moment curvature relationship is

$$M_{yield} = (1 - \alpha) k z_{max} \tag{11.13}$$

Other important physical properties worth noting are the initial stiffness k_i at any point along the length of the member before yield and the postyield stiffness k_f at corresponding points. They are as follows:

$$k_i = \alpha k + (1 - \alpha) k B \tag{11.14}$$

and

$$k_f = \alpha k \tag{11.15}$$

Note that the ratio of postyield to initial stiffness reduces to the value α when $B = 1$, revealing the physical significance of α.

It is convenient to nondimensionalize the dependent and independent variables in Equation 11.9. Therefore, we let M_{ε_o} and φ_{ε_o} be the yield moment and yield curvature, respectively, at a reference cross section. On this basis, the reference initial stiffness slope k_o is

$$k_o = \frac{M_{\varepsilon_o}}{\varphi_{\varepsilon_o}} \tag{11.16}$$

We also define

$$M^* = \frac{M}{M_{\varepsilon_o}} \tag{11.17}$$

$$\varphi^* = \frac{\varphi}{\varphi_{\varepsilon_o}} \tag{11.18}$$

$$z^* = \frac{z}{\varphi_{\varepsilon_o}} \tag{11.19}$$

$$\xi = \frac{x}{L} \tag{11.20}$$

$$y^* = \frac{y}{\delta_{\varepsilon_o}} \tag{11.21}$$

where δ_{ε_o} is the reference yield displacement.

On this basis, Equations 11.9 and 11.10 may be written as

$$\begin{aligned}
&\ddot{y}^* + \frac{1}{\left[\rho A(\xi) + \sum_{i=1}^{R} Q_i \delta(\xi - a_i)\right]} \left\{ \left(\frac{2nc}{\omega_n}\right) \dot{y}^* \right. \\
&+ \left(\frac{\alpha k_o}{L^4}\right)\left(\frac{2\pi}{\omega_n}\right)^2 \left[\frac{\partial^2 f(\xi)}{\partial \xi^2}\frac{\partial^2 y^*}{\partial \xi^2} + 2\frac{\partial f(\xi)}{\partial \xi}\frac{\partial^3 y^*}{\partial \xi^3} + f(\xi)\frac{\partial^4 y^*}{\partial \xi^4}\right] \\
&\left. + \left(\frac{\varphi_{\varepsilon_o}}{\delta_{\varepsilon_o}}\right)\left(\frac{2\pi}{\omega_n}\right)^2 \left(\frac{1-\alpha}{L^2}\right) k_o \left[z^* \frac{\partial^2 f(\xi)}{\partial \xi^2} + 2\frac{\partial f(\xi)}{\partial \xi}\frac{\partial z^*}{\partial \xi} + f(\xi)\frac{\partial^2 z^*}{\partial \xi^2}\right] \right\} \\
&= \left(\frac{2\pi}{\omega_n}\right)\left(\frac{1}{\delta_{\varepsilon_o}}\right) \frac{1}{\left[\rho A(\xi) + \sum_{i=1}^{R} Q_i \delta(\xi - a_i)\right]} \left\{ q_1(\xi, T^*) + q_2(\xi) \right. \\
&\left. + \sum_{j=1}^{n} F_j(\xi, T^*) + \sum_{l=1}^{m} F_l(\xi) \right\}
\end{aligned} \tag{11.22}$$

$$\dot{z}^* = \frac{\delta_{\varepsilon_o} B}{L^2 \varphi_{\varepsilon_o}} \frac{\partial^3 y^*}{\partial \xi^2 \partial T^*} - \frac{\beta \varphi_{\varepsilon_o}^{n-1} \delta_{\varepsilon_o}}{L^2} \left| \frac{\partial^3 y^*}{\partial \xi^2 \partial T} \right| \left| z^* \right|^{n-1} z^*$$

$$- \frac{\gamma \varphi_{\varepsilon_o}^{n-1} \delta_{\varepsilon_o}}{L^2} \left(\frac{\partial^3 y^*}{\partial \xi^2 \partial T^*} \right) \left| z^* \right|^n \tag{11.23}$$

In the above two equations L is the length of the member, ω_n is the natural frequency of vibration of mode n for the purely elastic case, and $T^* = t/\tau_n$, where $\tau_n = 2\pi/\omega_n$.

Equations 11.22 and 11.23 are highly nonlinear and it seems unlikely that closed form solutions are possible, even for thc simple inputs. One way to solve these equations is to use the finite difference method in conjunction with the 4th order Runge-Kutta method. The utilization of these two methods becomes much easier if equivalent systems are used as shown in the following section.

If Δ_x is taken as the step size and the member is subdivided into n elements, Equations 11.22 and 11.23 may be written in finite difference forms as shown below, i.e., by using central differencies, we find

$$\ddot{y}_i^* = -\frac{1}{\left[\rho A(\xi) + \sum_{i=1}^{R} Q_i \delta(\xi - a_i)\right]} \left\{ \left[\frac{2nc}{\omega_n}\right] \dot{y}^* \right.$$

$$+ \left[\frac{\alpha k_o}{L^4}\right] \left[\frac{2\pi}{\omega_n}\right]^2 \left[\frac{\partial^2 f(\xi)}{\partial \xi^2} \frac{\left(y_{i+1}^* + y_{i-1}^* - 2y_i^*\right)}{\Delta_x^2} \right.$$

$$+ \frac{\partial f(\xi)}{\partial \xi} \frac{\left(y_{i+2}^* - 2y_{i+1}^* + 2y_{i-1}^* - y_{i-2}^*\right)}{\Delta_x^4}$$

$$\left. + f\left(\xi \frac{\left(y_{i+2}^* - 4y_{i+1}^* + 6y_i^* - 4y_{i-1}^* + y_{i-2}^*\right)}{\Delta_x^3} \right) \right]$$

$$+ \left[\frac{\varphi_{\varepsilon_o}}{\delta_{\varepsilon_o}}\right] \left[\frac{2\pi}{\omega_n}\right]^2 \left[\frac{1-\alpha}{L^2}\right] k_o \left[z^* \frac{\partial^2 f(\xi)}{\partial \xi^2} + \frac{\partial f(\xi)}{\partial \xi} \frac{z_{i+1}^* - z_{i-1}^*}{\Delta_x} \right.$$

$$\left. + f(\xi) \frac{\left(z_{i+1}^* + z_{i-1}^* - 2z_i^*\right)}{\Delta_x^2} \right] - \left[\frac{2\pi}{\omega_n}\right]^2 \left[\frac{1}{\delta_{\varepsilon_o}}\right] \left[q_1(\xi, T^*) + q_2(\xi) \right.$$

$$\left. \left. + \sum_{j=1}^{n} F_j(\xi, T^*) + \sum_{l=1}^{m} F_l(\xi) \right] \right\} \tag{11.24}$$

$$\dot{z}_i^* = \frac{\delta_{\varepsilon_o} B}{L^2 \varphi_{\varepsilon_o}} \frac{\left(\dot{y}_{i+1}^* - 2\dot{y}_i^* + \dot{y}_{i-1}^*\right)}{\Delta_x^2} - \frac{\beta \varphi_{\varepsilon_o}^{n-1} \delta_{\varepsilon_o}}{L^2} \left| \frac{\left(\dot{y}_{i+1}^* - 2\dot{y}_i^* + \dot{y}_{i-1}^*\right)}{\Delta_x^2} \right| \left|z^*\right|^{n-1} z^*$$

$$- \frac{\gamma \varphi_{\varepsilon_o}^{n-1} \delta_{\varepsilon_o}}{L^2} \frac{\left(\dot{y}_{i+1}^* - \dot{y}_i^* + \dot{y}_{i-1}^*\right)}{\Delta_x^2} \left|z^*\right|^n \tag{11.25}$$

In the above two equations we assumed that damping is purily due to hysteresis, i.e., c = 0. These two equations can be solved by the 4th order Runge-5Kutta method.

In the application of the Runge-Kutta method, at a time *t*, Equation 11.24 may be written in the form

$$\begin{aligned} \ddot{y}_{i,t}^* = F\Big(&A(\xi), Q_i, \omega_n, q_1, q_2, F_j, F_1, y_{i+2,t}^*, \\ &y_{i+1,t}^*, y_{i,t}^*, y_{i-1,t}^*, y_{i-2,t}^*, z_{i-1,t}^*, z_{i,t}^*, z_{i+1,t}^*, \\ &f(\xi), \frac{\partial f(\xi)}{\partial \xi}, \frac{\partial^2 f(\xi)}{\partial \xi^2}\Big) \end{aligned} \tag{11.26}$$

At $T^* = 0$ we have

$$\dot{y}_{i,o}^* = 0 \tag{11.27}$$

$$y_{i,o}^* = 0 \tag{11.28}$$

For the next step, $\dot{y}_{i,t+1}^*$ and $y^*_{i,t+1}$ can be expressed as

$$\dot{y}_{i,t+1}^* = \dot{y}_{i,t}^* + \frac{1}{6}\left[A_1 + 2A_2 + 2A_3 + A_4\right] \tag{11.29}$$

$$y_{i,t+1}^* = y_{i,t}^* + (\Delta t)\dot{y}_{i,t}^* + \frac{(\Delta t)}{6}\left[A_1 + A_2 + A_3\right] \tag{11.30}$$

where Δt in Equation 11.30 is the time step increment and

$$A_1 = (\Delta t)F\Big(A(\xi), Q_i, \omega_n, q_1, q_2, F_j, F_1, y^*_{i+2,t}, y^*_{i+1,t}, y^*_{i,t}, y^*_{i-1,t}, y^*_{i-2,t}, z^*_{i-1,t}, z^*_{i,t}, z^*_{i-1,t}, f(\xi), \frac{\partial f(\xi)}{\partial \xi}, \frac{\partial^2 f(\xi)}{\partial \xi^2}\Big) \tag{11.31}$$

$$A_2 = (\Delta t)F\Big(A(\xi), Q_i, \omega_n, q_1, q_2, F_j, F_1, y^*_{i+2,t}, y^*_{i+1,t}, y^*_{i,t}, \left(y^*_{i,t} + \frac{(\Delta t)}{2}\dot{y}^*_{i,t}\right), y^*_{i-1,t}, y^*_{i-2,t}, z^*_{i+1,t}, z^*_{i,t}, z^*_{i-1,t}, f(\xi), \frac{\partial f(\xi)}{\partial \xi}, \frac{\partial^2 f(\xi)}{\partial \xi^2}\Big) \tag{11.32}$$

$$A_3 = (\Delta t)F\Big(A(\xi), Q_i, \omega_n, q_1, q_2, F_j, F_1, y^*_{i+2,t}, y^*_{i+1,t}, \left(y^*_{i,t} + \frac{(\Delta t)}{2}\dot{y}^*_{i,t} + \frac{(\Delta t)}{4}A_1\right), y^*_{i-1,t}, y^*_{i-2,t}, z^*_{i+1,t}, z^*_{i,t}, z^*_{i-1,t}, f(\xi), \frac{\partial f(\xi)}{\partial \xi}, \frac{\partial^2 f(\xi)}{\partial \xi^2}\Big) \tag{11.33}$$

$$A_4 = (\Delta t)F\Big(A(\xi), Q_i, \omega_n, q_1, q_2, F_j, F_1, y^*_{i+2,t}, y^*_{i+1,t}, \left(y^*_{i,t} + (\Delta t)\dot{y}^*_{i,t} + \frac{(\Delta t)}{2}A_2\right), y^*_{i-1,t}, y^*_{i-2,t}, z^*_{i+1,t}, z^*_{i,t}, z^*_{i+1,t}, f(\xi), \frac{\partial f(\xi)}{\partial \xi}, \frac{\partial^2 f(\xi)}{\partial \xi^2}\Big) \tag{11.34}$$

By following the same procedure at a time t, Equation 11.25 may be written in the form

$$\dot{z}^*_{i,t} = F\left(\varphi_{\varepsilon_0}, \delta_{\varepsilon_0}, n, B, \beta, \gamma, y^*_{i+1,t}, y^*_{i,t}, y^*_{i-1,t}, z^*_{i,t}\right) \tag{11.35}$$

For the next time step, $z^*_{i,t+1}$ can be expressed as

$$z^*_{i,t+1} = z^*_{i,t} + \frac{1}{6}\left(A_1^* + 2A_2^* + 2A_3^* + A_4^*\right) \tag{11.36}$$

where

$$A_1^* = (\Delta t)F\left(\varphi_{\varepsilon_o}, \delta_{\varepsilon_o}, n, B, \beta, \gamma, y^*_{i+1,t}, y^*_{i,t}, y^*_{i-1,t}, z^*_{i,t}\right) \tag{11.37}$$

$$A_2^* = (\Delta t)F\left(\varphi_{\varepsilon_o}, \delta_{\varepsilon_o}, n, B, \beta, \gamma, y^*_{i+1,t}, y^*_{i,t}, y^*_{i-1,t}, \left(z^*_{i,t} + \frac{A_1}{2}\right)\right) \tag{11.38}$$

$$A_3^* = (\Delta t)F\left(\varphi_{\varepsilon_o}, \delta_{\varepsilon_o}, n, B, \beta, \gamma, y^*_{i+1,t}, y^*_{i,t}, y^*_{i-1,t}, \left(z^*_{i,t} + \frac{A_2}{2}\right)\right) \tag{11.39}$$

$$A_4^* = (\Delta t)F\left(\varphi_{\varepsilon_o}, \delta_{\varepsilon_o}, n, B, \beta, \gamma, y^*_{i+1,t}, y^*_{i,t}, y^*_{i-1,t}, \left(z^*_{i,t} + A_3\right)\right) \tag{11.40}$$

11.3 UTILIZATION OF EQUIVALENT SYSTEMS TO SIMPLIFY THE SOLUTION

The mathematical model developed in Section 11.2 may be simplified by using equivalent systems as discussed in earlier chapters. This may be accomplished by writting Equation 11.5 in the following manner:

$$\left[\sum_{i=1}^{R} Q_{e_i}\delta(x - a_i)\right]\ddot{y} + c\dot{y} + \frac{\partial^2 M_e}{\partial x^2} = \sum_{j=1}^{n} F_{e_j}(t)\delta(x - a_j) + \sum_{l=1}^{m} F_{e_l}(x) \tag{11.41}$$

Equation 11.41 represents the differential equation of motion of an equivalent system of constant stiffness EI_o, which has the same length L and boundary conditions as the initial variable stiffness member. The distributed mass of the initial system is replaced by the equivalent concentrated masses Q_{e_i} (i = 1,2,3,...). M_e is the moment of the equivalent system at any x and time t. F_{e_j} (j = 1,2,3,...) are equivalent concentrated dynamic loads that replace the dynamic loading of the initial variable stiffness member, and F_{e_l} (l = 1,2,3,...) are equivalent static loads that replace the static loading on the initial variable stiffness member.

The quantities Q_{e_i}, M_e, F_{e_j}, and F_{e_l} may be determined by using the method of the equivalent systems as developed in the preceding chapters. For example,

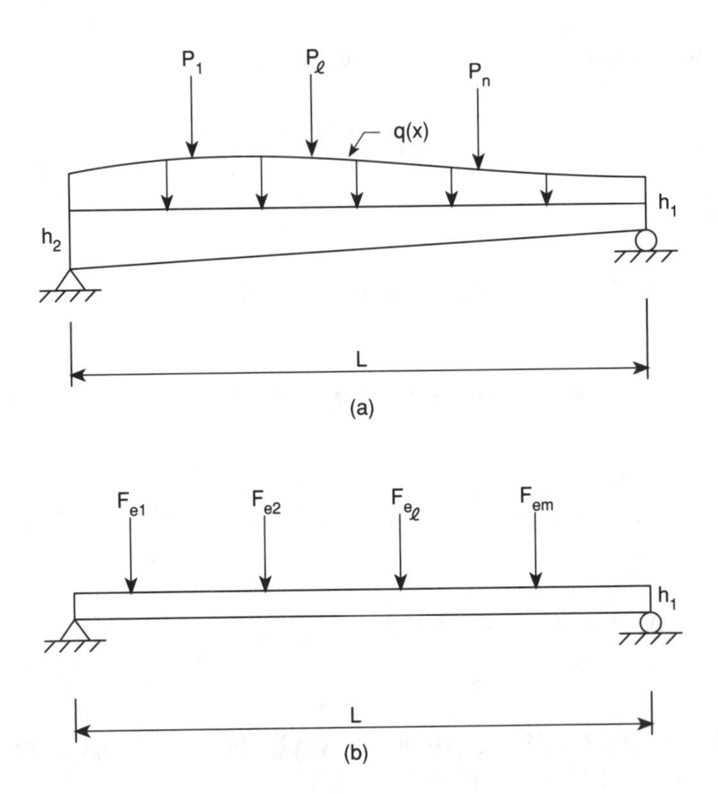

FIGURE 11.2. (a) Initial variable stiffness member loaded with distributed static load q(x) and concentrated static loads P_l, l = 1,2,3,...; (b) equivalent system of constant stiffness EI_o loaded with equivalent concentrated static loads F_{e_l}.

the static loading F_{el} (l = 1,2,3,...) on the equivalent system in Figure 11.2b replaces the static loadings q(x) and P_l (l = 1,2,3, ...) acting on the initial system in Figure 11.2a. This is appropriate because the method of the equivalent systems as shown in Chapter 1 proves that the moment M_e of the equivalent system of constant stiffness EI_o is given by the expression

$$M_e = \frac{M_x}{f(x)} \tag{11.42}$$

In the above equation, M_x is the bending moment at any x of the initial variable stiffness member in Figure 11.2a, and

$$f(x) = \frac{I_x}{I_o} \tag{11.43}$$

represents the variation of the moment of inertia I_x with respect to the constant moment of inertia I_o of the equivalent system in Figure 11.2b.

If the modulus of elasticity E is also variable, then

$$M_e = \frac{M_x}{f(x)g(x)} \tag{11.44}$$

where

$$g(x) = \frac{E_x}{E_o} \tag{11.45}$$

is a modulus function that represents the variation of the modulus E_x of the member in Figure 11.2a with respect to the constant modulus E_o of the equivalent system in Figure 11.2b. If the shape of the M_e diagram is approximated with straight lines as discussed in preceding chapters, then the equivalent system of constant stiffness E_oI_o will be loaded with concentrated weights as shown in Figure 11.2b. If E and I are uniform, then f(x) = g(x) = 1. This problem is thoroughly discussed in the preceding chapters and in the work by Fertis and his colaborators listed in the bibliography.

In a similar manner, if the weight w(x) of the member and other possible attached weights W_i (i = 1,2,3,...) are applied as loads as shown in Figure 11.3a, the equivalent system of constant stiffness EI_o or E_oI_o will be loaded with concentrated weights R_{e_i} (i = 1,2,3,...) as shown in Figure 11.3b. This is equivalent to lumping the weight of the member and its attached weights at discrete points along the length of the equivalent system by using the method of the equivalent systems. If the equivalent weights R_{e_i} are divided by the acceleration of gravity g, they yield the equivalent concentrated masses Q_{e_i} (i = 1,2,3,...), which are attached on the equivalent system as shown in Figure 11.3c.

The same analogy can be applied for the dynamic loads $q(x,t) = q_o(x)f(x)$ and $F_j(t) = F_{oj}f(t)$ (j = 1,2,3,...) that are applied to the initial variable stiffness member in Figure 11.4a. We assume here that the time function f(t) is the same for all dynamic loads. Again, it would be reasonable to apply $q_o(x)$ for F_{oj} (j = 1,2,3,...) as static loads on the initial variable stiffness member as shown in Figure 11.4b. Then, proceeding with the method of the equivalent systems as stated earlier, the equivalent system of constant stiffness EI_o or E_oI_o in Figure 11.4c will be loaded with concentrated loads F_{e_j} (j = 1,2,3,...). The time function for each of these loads is the same and it is equal to f(t).

It will be convenient in the analysis if the concentrated quantities F_{e_i}, Q_{e_i} and F_{e_j} on the equivalent systems in Figures 11.2b, 11.3c, and 11.4c, respectively, are acting at the same corresponding points on the equivalent system. This can be easily accomplished by dividing the initial variable stiffness member into n

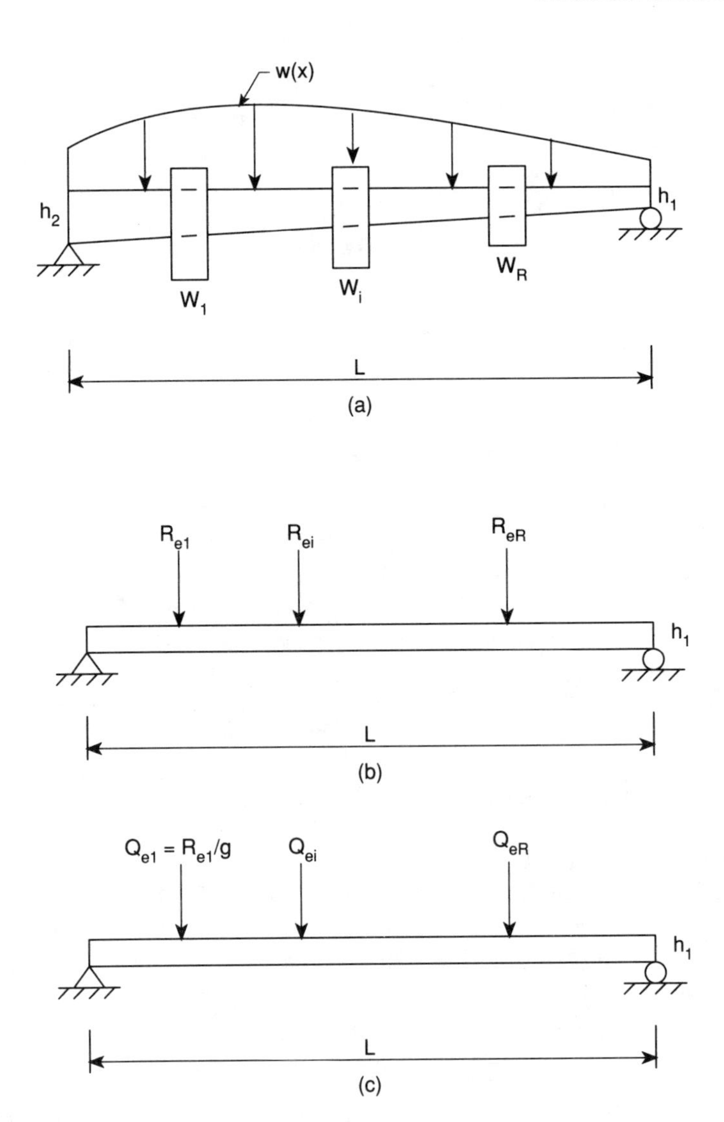

FIGURE 11.3. (a) Initial variable stiffness member loaded with weights w(x) and W_i; (b) equivalent constant stiffness system loaded with equivalent concentrated weights R_{e_i}; (c) equivalent constant stiffness system with equivalent attached concentrated masses $Q_{e_i} = R_{e_i}/g$.

segments. On this basis, all M_e diagrams may be subdivided into n straight lines, and consequently all equivalent quantities F_{e_1}, Q_{e_i}, and F_{e_j} will have common points of application on the equivalent system. Once the equivalent system of constant stiffness is established, it can be used in place of the initial variable stiffness member to determine moment curvature relationships, displacements, and other pertinent quantities resulting from the action of the various loadings.

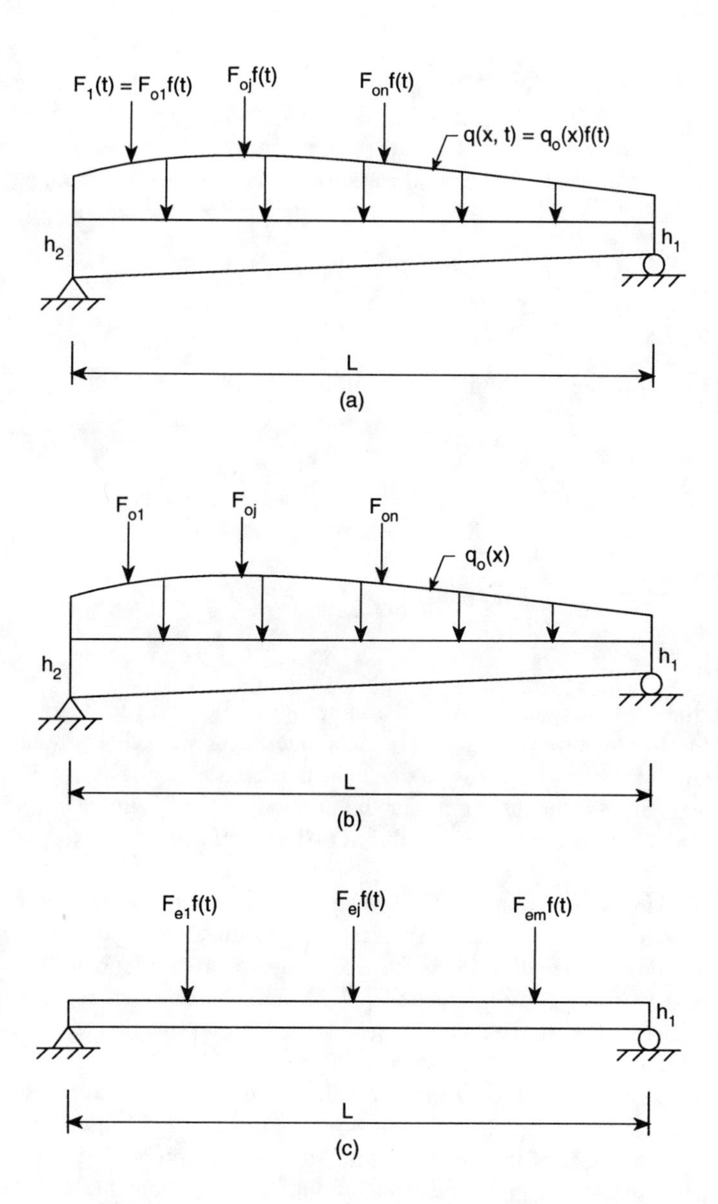

FIGURE 11.4. (a) Initial variable stiffness member subjected to dynamic loads q(x,t) and $F_j(t)$; (b) initial stiffness member subjected to static loads $q_o x$ and F_{oj}; (c) Equivalent system of constant stiffness loaded with equivalent concentrated dynamic loads F_{ej} with time function f(t).

By using the equivalent system, the equivalent bending moment $M_e(x,t)$ at any section x and time t may be expressed as

$$M_e(x,t) = \alpha k_o \frac{\partial^2 y}{\partial x^2} + (1-\alpha) k_o z_e \tag{11.46}$$

where k_o, as in Equation 11.7b, is the reference stiffness E_oI_o of the member. Again, the quantity z_e in the above equation is an equivalent quantity that reveals the nature and versatility of the equivalent systems model. By proceeding as in Equation 11.10, the nonlinear differential equation governing z_e may be written as

$$\dot{z}_e = B\frac{\partial^3 y}{\partial x^2 \partial t} - \beta\left|\frac{\partial^3 y}{\partial x^2 \partial t}\right|\left|z_e\right|^{n-1} z_e - \gamma\frac{\partial^3 y}{\partial x^2 \partial t}\left|z_e\right|^n \tag{11.47}$$

By substituting Equation 11.46 into Equation 11.41, we obtain

$$\left[\sum_{i=1}^{R} Q_{e_i}\delta(x-a_i)\right]\ddot{y} + c\dot{y} + \alpha k_o\frac{\partial^2 y}{\partial x^2} + (1-\alpha)k_o\frac{\partial^2 z_e}{\partial x^2} = \sum_{j=1}^{n} F_{e_j}(t)\delta(x-a_j) + \sum_{l=1}^{m} F_{e_l} \tag{11.48}$$

Equation 11.48 may be solved in the same way as Equation 11.9 by using the FDM in conjunction with the 4th order Runge-Kutta method. The solution of Equation 11.48, however, has proven to be much easier and faster, and it saves a great deal of computer time. For the problems solved here, about 45% in computer time savings may be attributed to the use of equivalent systems in the analysis.

The reduction in computer time when equivalent systems are used may be attributed to two basic reasons. The first reason is that the number of variables in Equations 11.47 and 11.48 are reduced compared to the number of variables involved with the use of Equations 11.22 and 11.23, which represents the original variable stiffness member. Second, due to the reduction of the number of variables in Equations 11.47 and 11.48, the utilization of equivalent systems does not create convergence problems. In addition, equivalent systems may utilize larger time intervals in the convergence process, and consequently they reduce the number of time steps required to solve the problem. For example, in order to establish convergence in the solution of Equations 11.22 and 11.23, a nondimensional time increment $\Delta T^* = 0.01$ should be used. Only $\Delta T^* = 0.05$ would have to be used for convergence when the equivalent system represented by Equations 11.47 and 11.48 is used. This is typical of all cases that have been examined here.

11.4 APPLICATION OF THE THEORY TO STRUCTURAL PROBLEMS

The theory developed in the preceding two sections will be used here to determine the dynamic response of structural members that are subjected to dynamic cyclic loadings. Two methodologies are used here in order to accomplish this purpose, and the results obtained from the two methods are compared. The first methodology involves the solution of Equations 11.9 and 11.10, or Equations 11.22 and 11.23. These sets of equations represent the initial variable stiffness member, and they are solved by using the FDM in conjunction with the 4th order Runge-Kutta method, as discussed in Section 11.2. The second methodology requires the solution of Equations 11.47 and 11.48, which represent an equivalent system of constant stiffness EI_o, or E_oI_o, which is derived as discussed in Section 11.3. Equations 11.47 and 11.48 are solved by using the finite difference method to derive the finite difference equations, which in turn are solved by using the 4th order Runge-Kutta method. In other words, the first methodology provides the direct solution of the initial problem, while the second methodology solves the differential equations of an equivalent system of uniform stiffness that replaces the initial complicated variable stiffness problem. Since the solution of the equivalent system is much simpler, and very accurate for practical applications, its utilization for practical applications is strongly recommended.

In order to initiate the solution procedure, certain assumptions are made regarding the parameters B, β, γ, and n in Equations 11.23 and 11.47. In the applications discussed in this chapter, we assume that the material is steel, and we let $B = n = 1$. On this basis, Equation 11.12 yields

$$\beta = \gamma = \frac{0.5}{\varphi_{\varepsilon_i}} \tag{11.49}$$

where φ_{ε_i} is the yield curvature at any section along the length of the member. The assumption of 0.5 in Equation 11.49 means that the unloading in the moment curvature curve is linear and of the same stiffness k as in the preyield condition.

When the depth of the member is variable, φ_{ε_i} in Equation 11.49 must be known at every cross section along the length of the member. This complication may be easily resolved by expressing the thickness h(x) of the member at any x, as $h(x) = h_o \bar{h}(x)$, where h_o is a reference thickness value and $\bar{h}(x)$ is a function representing the variation of h(x). For elastic or inelastic analysis, the curvature 1/r, where r is the radius of curvature, may be expressed as discussed in Chapter 6 by the equation

$$\frac{1}{r} = \frac{\Delta}{h} = \varphi \tag{11.50}$$

where $\Delta = \varepsilon_1 + \varepsilon_2$ is the distance between strains ε_1 and ε_2 in the strees/strain curve in Figure 11.5a. Thus, if φ_{ε_o} is the yield curvature at the reference cross section of thickness h_o, we may write

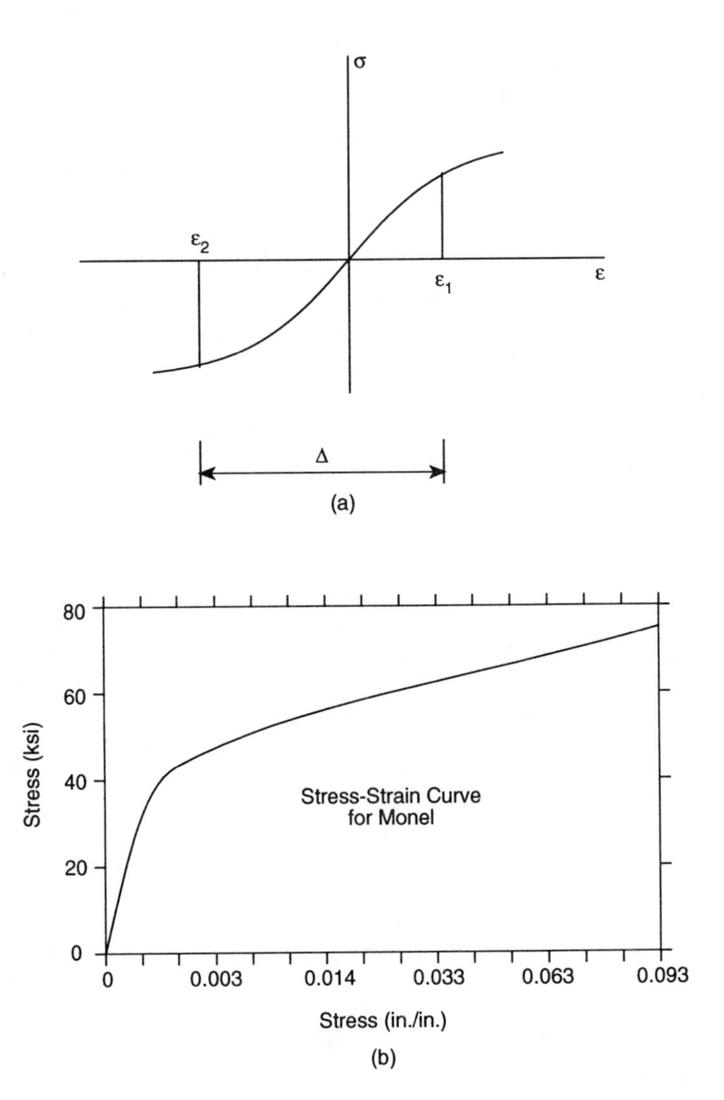

FIGURE 11.5. (a) General stress/strain curve; (b) stress/strain curve of Monel.

$$\frac{\varphi_{\varepsilon_i}}{\varphi_{\varepsilon_o}} = \frac{\Delta_{\varepsilon_i}}{\Delta_{\varepsilon_o}} \cdot \frac{1}{\bar{h}(x)} \tag{11.51}$$

Since φ_{ε_i} and φ_{ε_o} are both yield curvatures, their corresponding Δ_{ε_i} and Δ_{ε_o} are equal; therefore,

$$\varphi_{\varepsilon_i} = \frac{\varphi_{\varepsilon_o}}{\bar{h}(x)} \tag{11.52}$$

Equation 11.52 indicates that it is necessary to calculate only the yield curvature φ_{ε_0} at the reference cross section of thickness h_0. With known φ_{ε_0}, the yield curvature φ_{ε_i} at any other cross section i may be determined from Equation 11.52. The reference curvature φ_{ε_0} may be computed by applying the reduced modulus concept as discussed in Chapters 6 and 7. On this basis, the general expression for the curvature ϕ for either elastic or inelastic range may be expressed as

$$\varphi = \frac{M}{E_r I} \tag{11.53}$$

where the reduced modulus E_r may be obtained from the expression

$$E_r = \frac{12}{\Delta^3}\int_{\varepsilon_2}^{\varepsilon_1} \sigma\varepsilon d\varepsilon \tag{11.54}$$

Equations 11.53 and 11.54 establish the moment curvature relationship for a given material and its stress/strain curve, which is required in order to define the hysteretic model.

The numerical examples in the following sections illustrate the application of the theory for both methodologies.

11.5 NONPRISMATIC CANTILEVER BEAMS SUBJECTED TO SINUSOIDAL DYNAMIC LOADINGS

The methodologies developed in the preceding sections of this chapter are used here in order to determine the dynamic response of cantilever beams that are acted upon by concentrated and/or distributed cyclic dynamic loadings. The cross-sectional dimensions of the member can vary in any arbitrary manner along its length, and the applied loading can be large enough so that the material of the member can be stressed well beyond its elastic limit. Such a loading condition causes the modulus E to vary along the length of the member.

As a first application, we assume that the nonprismatic cantilever beam in Figure 11.6a is loaded with a concentrated sinusoidal dynamic load at the free end as shown, i.e., the concentrated dynamic load P(L,t) is of the form

$$P(L,t) = P\sin\omega_f t \tag{11.55}$$

In nondimensional form, Equation 11.55 becomes

$$P(1,T^*) = P\sin(2\pi\omega_f / \omega_1)T^* \tag{11.56}$$

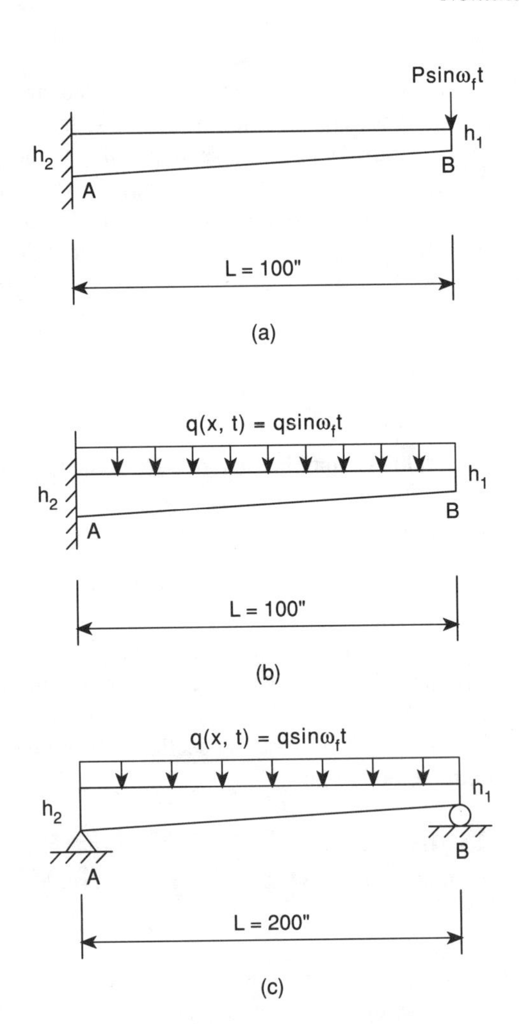

FIGURE 11.6. (a) Variable stiffness cantilever beam loaded with a concentrated dynamic load P sin $\omega_f t$; (b) variable stiffness cantilever beam loaded with a distributed load q sin $\omega_f t$; (c) variable stiffness simply supported beam loaded with a distributed load q sin $\omega_f t$ (1 in. = 0.0254 m).

In the above equations, ω_f is the forced frequency, P is the magnitude of the load, and ω_1 is the fundamental linear elastic frequency of the member in radians per second (rps). It is further assumed that P = 40 kips (177,928.9 N), the length L of the member is 100 in.(2.54 m), depth h_1 = 8 in. (0.2032 m), depth h_2 = 12 in. (0.3048 m), and the constant width b = 6 in. (0.1524 m). The material of the member is Monel, with a stress/strain curve as shown in Figure 11.5b or Figure 6.3a. The values of the modulus E and normal stress σ for a bilinear, trilinear, or six-line approximation of the stress/strain curve of Monel are shown in Tables 6.1 and 6.2, respectively.

By applying the reduced modulus concept as discussed in the preceding sections of this chapter and in Chapter 6, the moment curvature curves at

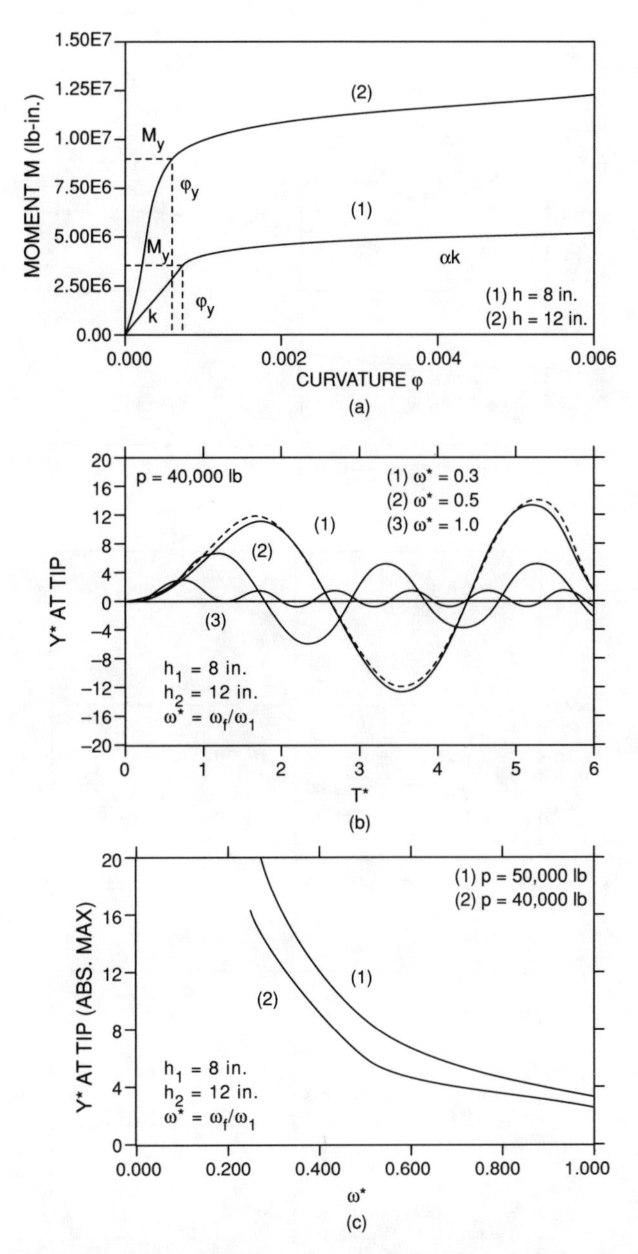

FIGURE 11.7. (a) Moment curvature curves; (b) Y^* vs. T^* curves for $\omega^* = 0.3$, 0.5, and 1.0; (c) Y^* at free end vs. ω^* curves for two values of the concentrated load P at the free end (1 in. = 0.0254 m, 1 lb = 4.448 N).

depths $h_1 = 8$ in. (0.2032 m), and $h_2 = 12$ in. (0.3048 m) are determined, and they are shown in Figure 11.7a. The solution of Equations 11.26 through 11.40, which represents the direct solution of the nonprismatic member in Figure 11.6a, is carried out numerically using a digital computer, in order to obtain

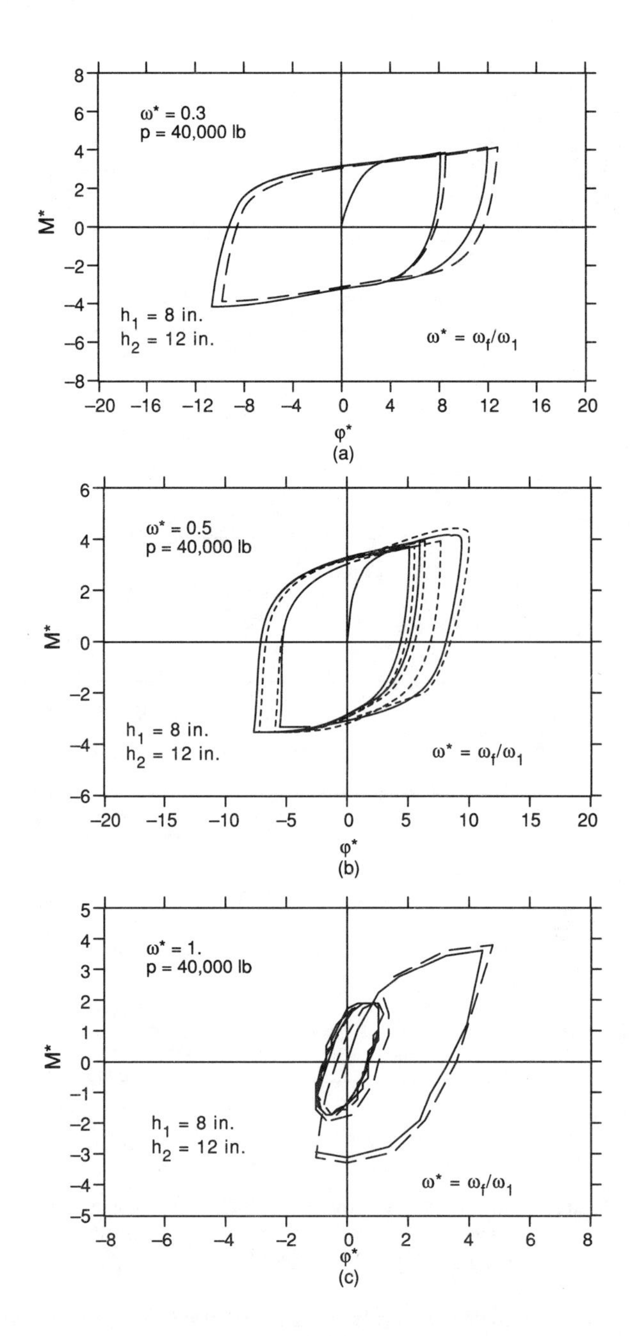

FIGURE 11.8. (a) Moment curvature curve for $\omega^* = 0.3$; (b) moment curvature curve for $\omega^* = 0.5$; (c) moment curvature curve for $\omega^* = 1.0$ (1 in. = 0.0254 m, 1 lb = 4.448 N).

deflections, bending moments, and curvature. In a similar manner, the solution of Equations 11.47 and 11.48, which represent an equivalent system of constant stiffness E_0I_0 is obtained.

Hysteresis loops of root M^* vs. curvature ϕ^* are constructed for both cases and they are shown in Figures 11.8a, 11.8b, and 11.8c for values of $\omega^* = 0.3$, 0.5, and 1.0, respectively. The solid lines in these figures represent the solution of Equations 11.22 through 11.40 of the original problem in Figure 11.6a, and the dotted lines provide the results obtained from the solution of Equations 11.47 and 11.48, which represent an equivalent system of constant stiffness E_0I_0, where I_0 is the moment of inertia at the free end of the member. The quantities M^* and ϕ^* are defined by Equations 11.17 and 11.18, respectively, and ω^* represents the ratio of the forced frequency ω_f to the fundamental frequency ω_1 for the purely elastic case.

Figure 11.7b shows the variation of the deflection magnification factor Y^* at the free end of the member with increasing T^* for values of $\omega^* = 0.3, 0.5$, and 1.0, where $T^* = t/\tau_1$ and $\tau_1 = 2\pi/\omega_1$. Figure 11.7c shows the variation of Y^* with increasing ω^* for values of P = 40,000 lb (177,929 N) and 50,000 lb (222,411 N). Figure 11.7b shows that the largest value of Y^* is obtained when $\omega^* = 0.3$, and from Figure 11.7c we note that Y^* approaches infinity when ω^* approaches the value of 0.25. In the same figures we also note that the results obtained by the equivalent system are in close agreement, for practical purposes, with the results obtained using the mathematical model of the original system.

It is also noted that an increase in ω^* narrows the hysteresis loop. Also, the magnitudes of the curvature and moment are decreasing with increasing ω^*, and eventually the system approaches the steady-state response. Again, from Figure 11.7c, we observe the drastic effect of the load P on the hysteresis curve. That is, an appreciable increase of the load P would appreciably alter the size of the hysteresis loop.

As another illustration of the application of the theory, let it be assumed that a nonprismatic cantilever beam is loaded by a uniformly distributed sinusoidal dynamic load q(x,t) as shown in Figure 11.6c, i.e.,

$$q(x,t) = q \sin \omega_f t \tag{11.57}$$

In nondimensional form, the above equation takes the form

$$q(\xi, T^*) = q \sin(2\pi\omega_f / \omega_1)T^* \tag{11.58}$$

where $\xi = x/L$.

It is further assumed here that the dimensions of the member, as well as other related material parameters, are the same as the ones in the preceding problem, and that q = 1000 lb/in. (175,126.8 N/m). By following the methodology discussed earlier, the moment curvature curves for $\omega^* = 0.3, 0.5$, and 1.0 are shown in Figures 11.9a, 11.9b, and 11.9c, respectively. These figures

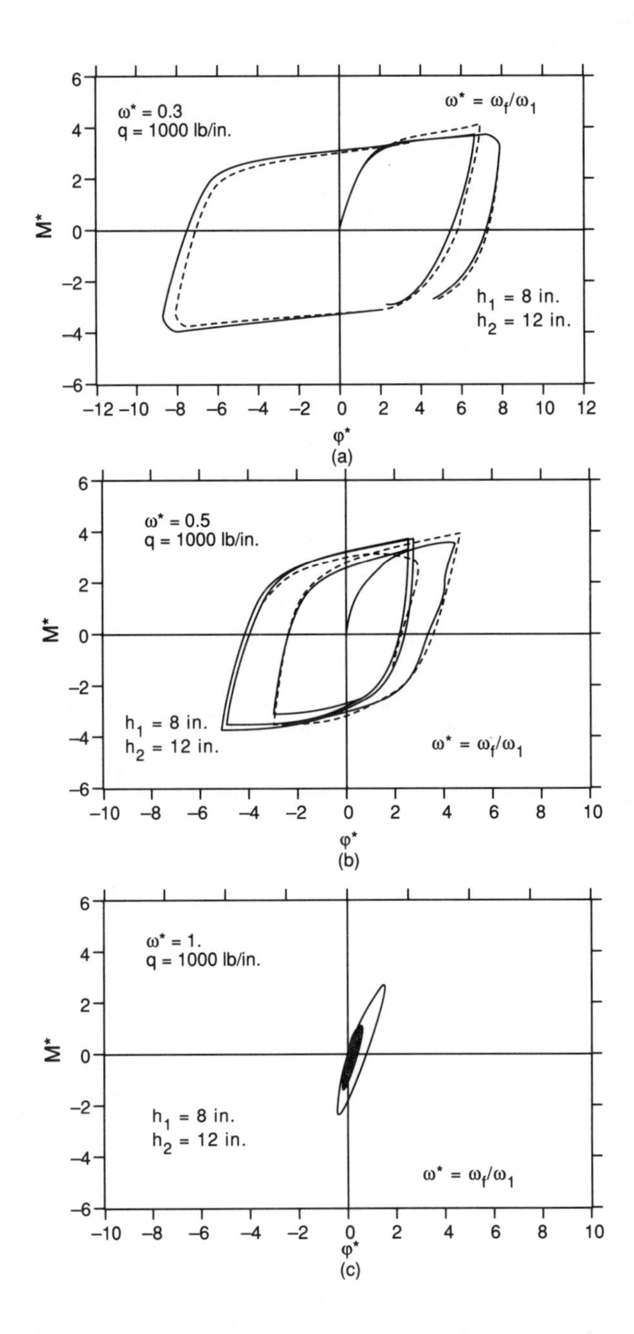

FIGURE 11.9. (a) Moment curvature curve for $\omega^* = 0.3$; (b) moment curvature curve for $\omega^* = 0.5$; (c) moment curvature curve for $\omega^* = 1.0$ (1 in. = 0.0254 m, 1 lb/in. = 175.1268 N/m).

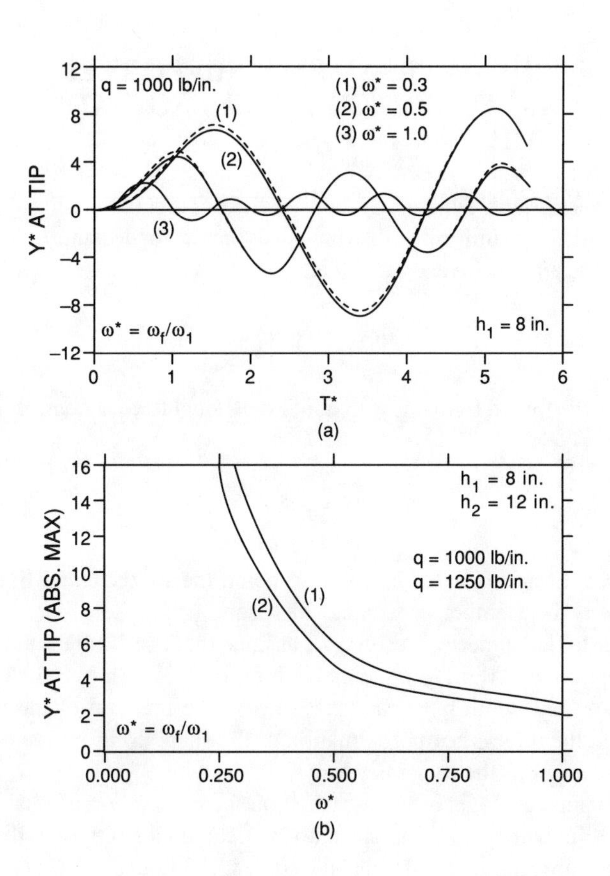

FIGURE 11.10. (a) Y^* at free end vs. T^* curves for values of $\omega^* = 0.3$, 0.5, and 1.0; (b) Y^* at free end vs. ω^* curves for two values of the uniformly distributed load q (1 in. = 0.0254 m, 1 lb/in. = 175.1268 N/m).

indicate that the size of the hysteresis loop decreases with increasing ω^*, as it was also observed for the case of loading in Figure 11.6a. However, for $\omega^* = 1.0$, the reduction is rather dramatic when it is compared with the hysteresis loop for $\omega^* = 0.3$. The preceding case of loading did not exhibit such large reduction for the case $\omega^* = 1.0$.

Figure 11.10a shows the variation of the deflection magnification factor Y^* at the free end of the member with increasing T^* for values of $\omega^* = 0.3$, 0.5, and 1.0. We note again here that the largest values of Y^* are associated with $\omega^* = 0.3$. In all cases above, the solid lines represent the results obtained using the mathematical model of the initial system in Figure 11.6b, and the dashed lines represent the results obtained by the equivalent system of constant stiffness E_oI_o, where I_o is the moment of inertia at the free end of the member. Figure 11.10b shows the variation of Y^* with increasing ω^* for q = 1000 lb/in. (175,126.8 N/m) and 1250 lb/in. (218,908.5 N/m). Again, we note here the dramatic increase in Y^* as ω^* approaches the value of about 0.25.

11.6 NONPRISMATIC SIMPLY SUPPORTED BEAMS SUBJECTED TO SINUSOIDAL DYNAMIC LOADINGS

In this section the dynamic response of a nonprismatic simply supported beam loaded by a uniformly distributed dynamic load $q\sin\omega_f t$ as shown in Figure 11.6c, will be investigated, i.e.,

$$q(x,t) = q\sin\omega_f t \tag{11.59}$$

where ω_f is the forced frequency. In nondimensional form, Equation 11.59 may be written as

$$q(\xi, T^*) = q\sin(2\pi\omega^*) \tag{11.60}$$

where $\xi = x/L$ and $\omega^* = \omega_f/\omega_1$, with ω_1 being the fundamental linear elastic frequency of the member in radians per second (rps).

In order to get numerical results, we assume that q = 1000 lb/in. (175,126.8 N/m), length L = 200 in. (5.08 m), h_1 = 8 in. (0.2032 m), h_2 = 12 in. (0.3048 m), and constant width b = 6 in. (0.1524 m). The material of the member is Monel, and the three-line approximation of the shape of its stress/strain curve is used to carry out the dynamic analysis.

The dynamic analysis is again carried out here by solving Equations 11.22 through 11.40, which represent the mathematical model of the original variable stiffness member in Figure 11.6c, and also by using Equations 11.47 and 11.48, which represent the mathematical model of an equivalent system of constant stiffness E_oI_o, where E_o is the reference value of the modulus E_x and I_o is the reference value of the moment of inertia I_x. In the analysis here, the reference value I_o is taken as the moment of inertia at the right support B of the member in Figure 11.6c. The results obtained from both solutions are compared. The procedure followed is similar to the one used in the preceding section.

Figures 11.11a through 11.11c show the moment curvature curves for ω^* = 0.3, 0.5, and 1.0, respectively. Again, we note here that as ω^* increases, the shape of the M^* vs. ϕ^* loop at midspan of the member narrows down as it was observed for the beam problems in Figures 11.6a and 11.6b. However, the change is not so dramatic as the one noticed in Figure 11.9, which represents the beam case in Figure 11.6b. Figure 11.12a shows the variation of the amplitude magnification factor Y^* at midspan with increasing T^* for ω^* = 0.3, 0.5, and 1.0. The largest values of Y^* are again obtained for the case with ω^* = 0.3. Figure 11.12b gives the variation of Y^* at midspan vs. ω^* for q = 800 lb/in. (140,101.44 N/m) and 1000 lb/in. (175,126.8 N/m). We note again here the rapid increase in Y^* at midspan as ω^* increases. As ω^* approaches the value of about 0.25, Y^* approaches infinity.

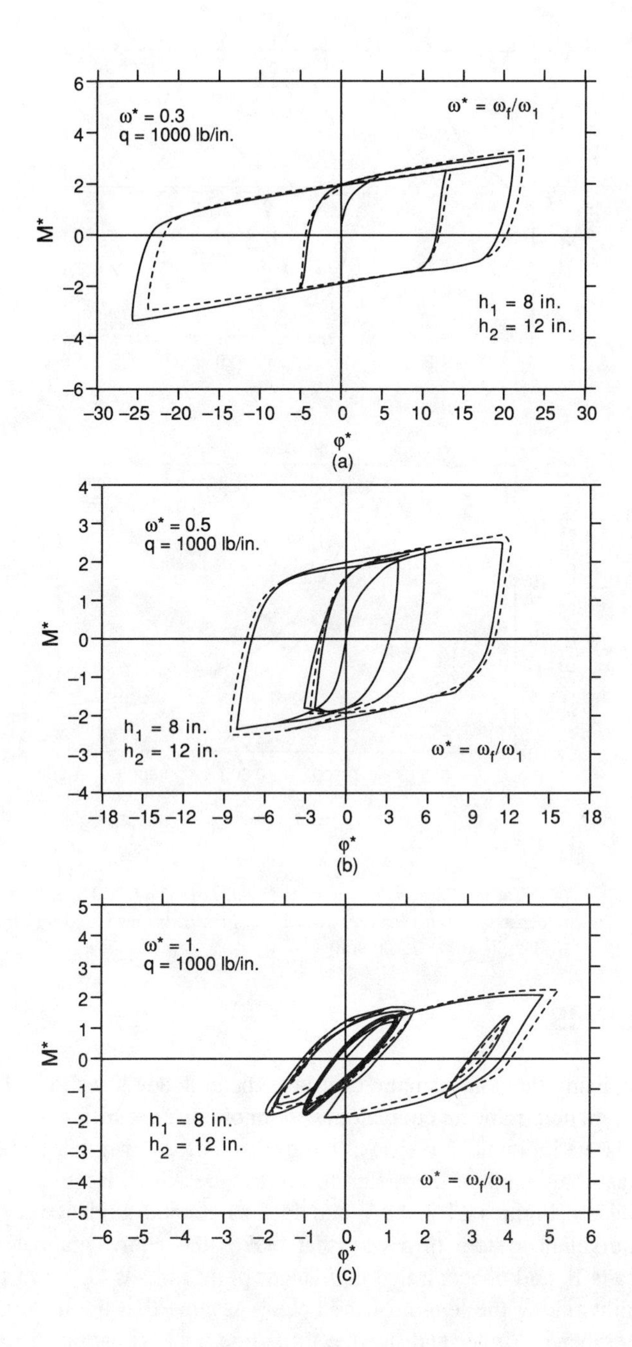

FIGURE 11.11. (a) Moment curvature curve for $\omega^* = 0.3$; (b) moment curvature curve for $\omega^* = 0.5$; (c) moment curvature curve for $\omega^* = 1.0$ (1 in. = 0.0254 m, 1 lb/in. = 175.1268 N/m).

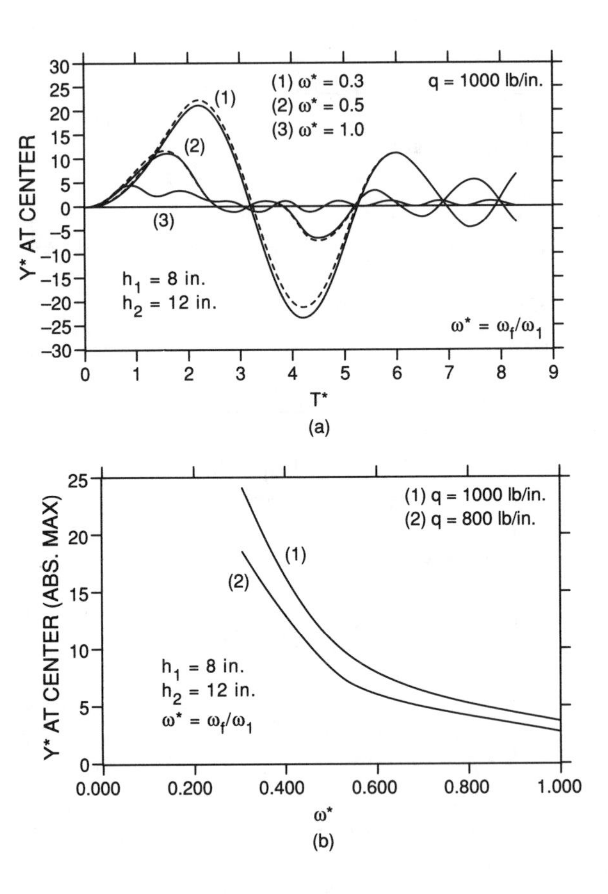

FIGURE 11.12. (a) Y^* at midspan vs. T^* curves for values of $\omega^* = 0.3$, 0.5, and 1.0; (b) Y^* at midspan vs. ω^* curves for two values of the distributed load q (1 in. = 0.0254 m, 1 lb/in. = 175.1268 N/m).

PROBLEMS

11.1 By using the nonprismatic cantilever beam loaded as shown in Figure 11.6b, determine an equivalent system of constant stiffness E_0I_0, where I_0 is the moment of inertia at the free end of the member and E_0 is the elastic modulus of the material. Follow the discussion in Section 11.3 and use Figures 11.2, 11.3, and 11.4 as general guidance. Derive the equivalent system in a way that makes the equivalent concentrated loads F_e and concentrated equivalent point masses Q_e act at the same points along the length of the beam. Assume that the material of the member is Monel and use the three-line approximation of the stress/strain curve. Assume that q = 1000 lb/in. (175,126.8 N/m), L = 100 in. (2.54 m), h_1 = 8 in. (0.2032 m), h_2 = 12 in. (0.3048 m), and constant width b = 6 in.(0.1524 m).

11.2 Repeat Problem 11.1 by assuming that q = 1250 lb/in. (218,908.5 N/m).

11.3 Repeat Problem 11.1 by assuming that the nonprismatic cantilever beam is loaded with a concentrated dynamic load $P\sin\omega_f t$ at the free end, where P = 40,000 lb (177,929 N).

11.4 Using the equivalent system obtained in Problem 11.1, determine the M^* vs. ϕ^* hysteresis loop for $\omega^* = 0.3$, and compare the results with the results obtained in Figure 11.9a.

11.5 Using the equivalent system obtained in Problem 11.1, determine the M^* vs. ϕ^* hysteresis loop for $\omega^* = 0.5$, and compare the results with the results obtained in Figure 11.9b.

11.6 Repeat Problem 11.4 for $\omega^* = 0.75$, and compare the results.

11.7 Using the equivalent system obtained in Problem 11.3, determine the M^* vs. ϕ^* hysteresis loop for $\omega^* = 0.3$, and compare the results with the results obtained in Figure 11.8a.

11.8 Repeat Problem 11.7 for $\omega^* = 0.5$, and compare the results with the ones obtained in Figure 11.8b.

11.9 Repeat Problem 11.7 for $\omega^* = 0.75$, and compare the results.

11.10 Repeat Problem 11.3 when P = 50,000 lb (222,411 N).

11.11 Using the equivalent system derived in Problem 11.1, determine the Y^* at the free end vs. ω^* curve, and compare the results with the results obtained in Figure 11.10b. Discuss the results and point out important findings.

11.12 Using the equivalent system derived in Problem 11.1, determine the Y^* at the free end vs. T^* curve for $\omega^* = 0.3$, and compare the results with the ones obtained in Figure 11.10a.

11.13 Repeat Problem 11.12 for $\omega^* = 0.5$.

11.14 Using the equivalent system derived in Problem 11.3, determine the Y^* at the free end vs. ω^* curve, and compare the results with the ones obtained in Figure 11.7c.

11.15 Repeat Problems 11.3 and 11.14 with P = 30,000 lb (133,446.7 N), and compare the results.

11.16 Using the nonprismatic simply supported beam loaded as shown in Figure 11.6c, determine an equivalent system of constant stiffness E_0I_0, where I_0 is the moment of inertia at the right support B of the member and E_0 is the elastic modulus of the material. Follow the discussion in Section 11.3 and use Figures 11.2, 11.3, and 11.4 as general guidance. Assume that the material of the member is Monel, and use the three-line approximation of its stress/strain curve. The loading q = 1000 lb/in. (175,126.8 N/m), L = 200 in. (5.08 m), h_1 = 8 in.(0.2032 m), h_2 = 12 in. (0.3048 m), and b = 6 in. (0.1524 m).

11.17 Solve Problem 11.16 by assuming that q = 800 lb/in. (140,101.4 N/m) and 700 lb/in. (122,588.8 N/m).

11.18 Using the equivalent system obtained in Problem 11.16, determine the M^* vs. ϕ^* hysteresis loop for $\omega^* = 0.3$, and compare the results with the ones obtained in Figure 11.11a.

11.19 Using the two equivalent systems obtained in Problem 11.17, determine the respective M^* vs. ϕ^* hysteresis loop for $\omega^* = 0.3$, and compare the results.

11.20 Solve Problem 11.18 for $\omega^* = 0.5$, and compare the results with the results obtained in Figure 11.11b.

11.21 Using the equivalent system derived in Problem 11.16, determine the Y^* at the free end vs. ω^* curve, and compare the results with the results shown in Figure 11.12b.

11.22 Using the two equivalent systems obtained in Problem 11.17, determine the corresponding Y^* at midspan vs. ω^* curves and compare the results. Also, discuss the results by making interesting observations.

Chapter 12

NONLINEAR VIBRATION AND INSTABILITIES OF ELASTICALLY SUPPORTED BEAMS

12.1 INTRODUCTION

The work and methodologies included in this chapter are intended to examine the linear and nonlinear vibrations, as well as static and flutter instabilities of elastically supported uniform beams. In particular, we wish to examine the nonlinear vibration response of such beam problems when initial transversed displacement conditions are present, and when the ends of the member are subjected to compressive restraints. Kounadis[68,69] and Fertis and Lee[29,70] recognized the inherent peculiarities in the static and dynamic response of such problems, and they have investigated the existance of static and flutter instabilities associated with the static and vibration behavior of such problems.

The cases examined in this chapter are uniform beams that are elastically supported at the ends by vertical and horizontal springs. Axial restraints are applied to such members in the form of axial compressive forces, or by initial precompression of the horizontal springs. The vertical restraints are applied in the form of an initial static transversed displacement y_{st}. The vibration analysis takes into consideration the rotatory and shear effects and their influence regarding the development of flutter instabilities, but damping is neglected.

The methodologies developed in this chapter may be used to study the response of many practical structural and mechanical engineering problems. A very interesting practical application, just to name one, would be the vibration response of suspension bridges entering a nonlinear mode where the effects of forces are not proportional, or even predictable. An upward lurch from a strong wind or an earthquake could kick a bridge into a nonlinear condition that needs to be investigated. Another interesting application of such methodologies would be the space station. Such space stations are subjected to various dynamic loads during shuttle docking, solar tracking, altitude adjustment, and other functional conditions. Accurate predictions regarding the natural frequencies and mode shapes of the space station components, including the solar arrays, is an important requirement in determining the structural adequacy of the components and designing an appropriate dynamic control system (see the work by Bosela et al.[71-73,76]

12.2 UNIFORM BEAMS SUPPORTED BY A VERTICAL SPRING AT EACH END AND SUBJECTED TO AXIAL COMPRESSIVE LOADS

The first problem to consider here is a uniform beam that is supported by vertical springs of spring constant k_v and subjected to an axial compressive

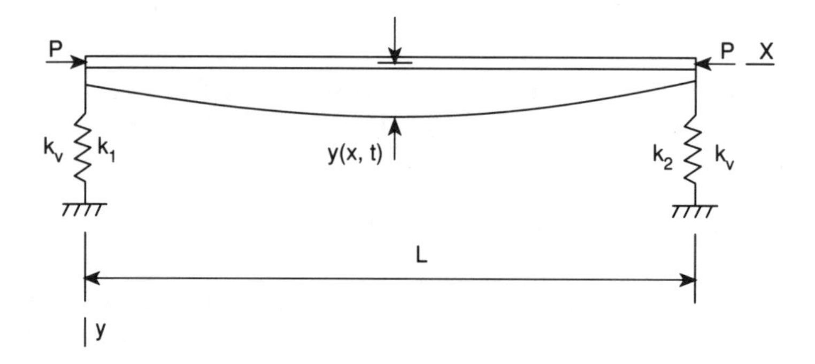

FIGURE 12.1. Uniform elastically supported beam subjected to compressive loads at its ends.

force P as shown in Figure 12.1. Its vibrational motion from the initial straight configuration position of the member is represented by the vertical displacement y(x,t), which is a function of both x and time t.

By using the principle of virtual work, we can write the following variational equation for the motion of the member:

$$\int_0^L \rho A\ddot{y}\delta y dx + \int_0^L \rho I\ddot{\theta}\delta dx + \delta(U + \Pi) = 0 \qquad (12.1)$$

In the above equation, U and Π are functionals of potential energy of the internal and external loads, respectively, A is the cross-sectional area, ρ is the material density, θ(x,t) is the slope of the deflection curve produced by bending, and δ is the Dirac delta function.

The functionals U and Π may be expressed as

$$U = \frac{1}{2}\int_0^L \left[EI(\theta')^2 + K'AG(y' - \theta)^2\right]dx + \frac{1}{2}k_1 y^2(0,t) + k_2 y^2(L,t) \qquad (12.2)$$

$$\Pi = -P\int_0^L (y')^2 dx \qquad (12.3)$$

Where G is the shear modulus and K′ is the Timoshenko shear coefficient. In Equation 12.2 the spring constants k_1 and k_2 of the vertical end springs are assumed to have different values in order to make the problem somewhat more general. By carrying out the variation of the functionals U and Π, integrating by parts, and then using Equation 12.1, we can write the following differential equations of motion:

$$K'AG(y'' - \theta') - \rho A\ddot{y} - Py' = 0 \qquad (12.4)$$

$$EI\theta''' - \rho I\ddot{\theta}' + m\ddot{y} + Py'' = 0 \tag{12.5}$$

Since the compressive force P is considered to be normal to the cross section of the member, we can assume that $y' \cong \theta$ and $Py' = P\theta$. On this basis, Equation 12.4 takes the form

$$K'AG(y' - \theta') - \rho A\ddot{y} - P\theta = 0 \tag{12.6}$$

and Equation 12.5 can be transformed into the following equation

$$\frac{EI(K'AG)}{(K'AG+P)}y'''' - \frac{EI\rho}{(K'AG+P)}\left[A + \frac{k'AG}{E}\right]\ddot{y}'' + m\ddot{y} + Py'' + \frac{I\rho^2 A}{(K'AG+P)}\ddot{\ddot{y}} = 0 \tag{12.7}$$

The general solution of Equation 12.7 is of the form

$$y(x,t) = \phi(x)e^{i\omega t} \tag{12.8}$$

where ω is the free frequency of vibration.

By substituting Equation 12.8 into Equation 12.7 and carrying out the required manipulations, the following equation is obtained:

$$\frac{d^4\phi(\xi)}{d\xi^4} + \lambda^2\frac{d^2\phi(\xi)}{d\xi^2} - \gamma^2\phi(\xi) = 0 \tag{12.9}$$

where $\xi = x/L$ is a dimensionless independent variable and

$$\lambda^2 = \frac{\rho\omega^2 L^2}{K'G} + \frac{\rho\omega^2 L^2}{E} + \frac{PL^2(k'AG+P)}{EI(K'AG)} \tag{12.10}$$

$$\gamma^2 = \frac{m\omega^2(K'AG+P)}{EI(K'AG)}\left[1 - \frac{\rho^2\omega^2 IA}{(K'AG+P)m}\right] \tag{12.11}$$

In the above equations, m is the mass per unit length of the member and I is its cross-sectional moment of inertia.

The general solution of Equation 12.9 is

$$\phi(\xi) = C_1 \cosh \beta\xi + C_2 \sinh \beta\xi + C_3 \cos \overline{\beta}\xi + C_4 \sin \overline{\beta}\xi \tag{12.12}$$

where

$$\beta = \left[-\frac{\lambda^2}{2} + \left(\frac{\lambda^4}{4} + \gamma^2 \right)^{1/2} \right]^{1/2} \tag{12.13}$$

$$\overline{\beta} = \left[\frac{\lambda^2}{2} + \left(\frac{\lambda^4}{4} + \gamma^2 \right)^{1/2} \right] \tag{12.14}$$

By utilizing Equations 12.8 and 12.12 and applying the boundary conditions for rotation and vertical deflection at the ends of the member, we obtain four linear algebraic equations involving the four constants C_1, C_2, C_3, and C_4. In matrix form, these four equations are written as follows:

$$\begin{bmatrix} \phi & (0) \\ \theta & (0) \\ \phi & (1) \\ \theta & (1) \end{bmatrix} = \begin{bmatrix} 1 & 0 & 1 & 0 \\ 0 & A_1 & 0 & A_2 \\ ch & sh & c & s \\ A_1 sh & A_1 ch & -A_2 s & P_2 c \end{bmatrix} \begin{bmatrix} C_1 \\ C_2 \\ C_3 \\ C_4 \end{bmatrix} \tag{12.15}$$

where

$$A_1 = \frac{1}{L}(\beta + \rho * \beta^3 + \rho * \beta\lambda^2) \tag{12.16}$$

$$A_2 = \frac{1}{L}(\overline{\beta} - \rho * \overline{\beta} + \rho * \overline{\beta}\lambda^2) \tag{12.17}$$

$$ch = \cosh \beta \tag{12.18}$$

$$sh = \sinh \beta \tag{12.19}$$

$$c = \cos \overline{\beta} \tag{12.20}$$

$$s = \sin \overline{\beta} \tag{12.21}$$

$$\rho^* = \frac{E}{K'AG + P} \cdot \frac{1}{\left[1 - \rho\omega^2 \dfrac{I}{K'AG + P}\right] L^2} \tag{12.22}$$

We can also use the boundary conditions

$$V(0) + k_v y(0) = 0 \tag{12.23}$$

$$V(L) - k_v y(L) = 0 \tag{12.24}$$

$$M(0) = 0 \tag{12.25}$$

$$M(L) = 0 \tag{12.26}$$

in order to derive a system of four equations in terms of the constants C_1, C_2, C_3, and C_4. In matrix form, these four equations are written as

$$\begin{bmatrix} V^* & (0) \\ M & (0) \\ -V^* & (1) \\ -M & (1) \end{bmatrix} = \begin{bmatrix} 0 & \overline{A} & 0 & \overline{B} \\ \overline{C} & 0 & \overline{D} & 0 \\ \overline{A}sh & \overline{A}ch & -\overline{B}s & \overline{B}c \\ \overline{C}ch & \overline{C}sh & Dc & \overline{D}s \end{bmatrix} \begin{bmatrix} C_1 \\ C_2 \\ C_3 \\ C_4 \end{bmatrix} \tag{12.27}$$

where

$$\overline{A} = \overline{M}(\beta^3 + \lambda^2\beta) \tag{12.28}$$

$$\overline{B} = \overline{M}(-\overline{\beta}^3 + \lambda^2\overline{\beta}) \tag{12.29}$$

$$\overline{C} = \overline{E}\left(\beta^2 + \frac{\rho\omega^2 L^2}{K'G}\right) \tag{12.30}$$

$$\overline{D} = \overline{E}\left(\overline{\beta}^2 + \frac{\rho\omega^2 L^2}{K'G}\right) \tag{12.31}$$

$$\overline{E} = -\frac{EIK'AG}{(K'AG + P)} \tag{12.32}$$

$$\overline{M} = \frac{\left[\dfrac{EIK'AG}{K'AG + P}\right]}{\left[1 - \dfrac{\rho\omega^2 I}{(K'AG + P)}\right]} \tag{12.33}$$

Also, $V^*(0)$ and $V^*(1)$ are the sums of the shear forces due to bending and springs at $\xi = 0$ and $\xi = 1$, respectively, and M(0) and M(1) are the bending moments at $\xi = 0$ and $\xi = 1$, respectively.

Vibration analysis can be performed by using Equation 12.27 and applying the boundary conditions given by Equations 12.23 to 12.26. On this basis, the left-hand side of Equation 12.27 goes to zero, and a nontrivial solution may be obtained by setting the square determinant on the right side of Equation 12.27 equal to zero. This establishes the frequency equation as a function of the horizontal compressive force. Therefore, by keeping the other parameters constant and varying the axial compressive force, its influence on the natural frequencies of vibration and rigid body motion may be determined. The Bisection method[74] may be used to determine the eigenfrequencies of the lateral vibration and rigid body motion.

As an application regarding the above methodology, let it be assumed that the beam in Figure 12.1 has a length L = 100 in. (2.54 m), depth h = 8 in. (0.2032 m), width b = 6 in. (0.1524 m), modulus of elasticity $E = 30 \times 10^6$ psi ($206{,}843 \times 10^6$ Pa), moment of inertia I = 256 in.^{4} (100×10^{-6} m^4), shear coefficient K′ = 2/3, and Poisson ratio $\nu = 0.25$. The mass of the member per cubic inch is 0.0088 lb-sec^2/in. The spring constant k_v of the vertical springs at the end supports is assumed to be the same for both supports, and values of k_v = 10, 100, 500, 1000, 2000, and 5000 lb/in. are considered in the analysis (1 lb/in. = 175.1268 N/m).

By applying the procedure discussed earlier in this section, the first two frequencies of vibration, in radians per second, are obtained, and the results are tabulated in Tables 12.1 and 12.2. For each case of spring constant k_v, the axial compressive load was permitted to vary while other parameter of the member remained the same. For example, in the first column of Table 12.1, the axial load P was allowed to vary from zero to 355 lb (1579.12 N). It is important to note here that when the load P = 355 lb (1579.12 N), the first and second frequencies ω_1 and ω_2, respectively, coincide to the value of $\omega_1 = \omega_2 = 0.689$ rps. This indicates that at P = 355 lb (1579.12 N) the elastically supported member experiences a state of flutter instability. Since ω_1 and ω_2 are, respectively, the frequencies corresponding to vertical translational and rotational rigid body type vibrational motions, we can say that the system becomes unstable when the translational and rotational motions coincide. Similar instabilities are developed for the other cases of spring constants k_v as shown in Tables 12.1 and 12.2. We also note that as the value of the spring constant k_v increases, it will require higher values of the axial load P in order for ω_1 and

Table 12.1
Natural Frequencies for Various Values of the Spring Constant k_v

$k_v = 10$ lb/in.			$k_v = 100$ lb/in.			$k_v = 500$ lb/in.		
P (lb)	ω_1 (rps)	ω_2 (rps)	P (lb)	ω_1 (rps)	ω_2 (rps)	P (lb)	ω_1 (rps)	ω_2 (rps)
0	0.689	1.189	0	2.177	3.758	0	4.877	8.402
100	0.689	1.063	500	2.177	3.565	2,000	4.877	8.059
200	0.689	0.921	1,000	2.177	3.362	4,000	4.877	7.701
300	0.689	0.752	1,500	2.177	3.144	6,000	4.877	7.325
355	0.689	0.689	2,000	2.177	2.911	8,000	4.877	6.929
			2,500	2.177	2.658	10,000	4.877	6.509
			3,000	2.177	2.377	12,000	4.877	6.059
			3,350	2.177	2.177	14,000	4.877	5.574
						16,000	4.877	5.024
						16,550	4.877	4.877

Note: 1 lb/in. = 175.1268 N/m.

Table 12.2
Natural Frequencies for Various Values of the Spring Constant k_v

k_v = 1000 lb/in.			k_v = 2000 lb/in.			k_v = 5000 lb/in.		
P (lb)	ω_1 (rps)	ω_2 (rps)	P (lb)	ω_1 (rps)	ω_2 (rps)	P (lb)	ω_1 (rps)	ω_2 (rps)
0	6.876	11.882	0	9.713	16.546	0	11.882	20.572
5,000	6.876	11.272	12,000	9.713	15.761	10,000	11.882	19.876
10,000	6.876	10.628	24,000	9.713	14.647	20,000	11.882	19.153
15,000	6.876	9.941	30,000	9.713	14.058	30,000	11.882	18.402
20,000	6.876	9.294	39,000	9.713	13.122	40,000	11.882	17.620
25,000	6.876	8.402	45,000	9.713	12.461	50,000	11.882	16.800
30,000	6.876	7.516	51,000	9.713	11.762	60,000	11.882	15.939
35,501	6.876	6.876	57,000	9.713	11.019	70,000	11.882	15.028
			63,000	9.713	10.222	80,000	11.882	14.058
			66,065	9.713	9.713	95,564	11.882	11.882

Note: 1 lb/in. = 175.1268 N/m.

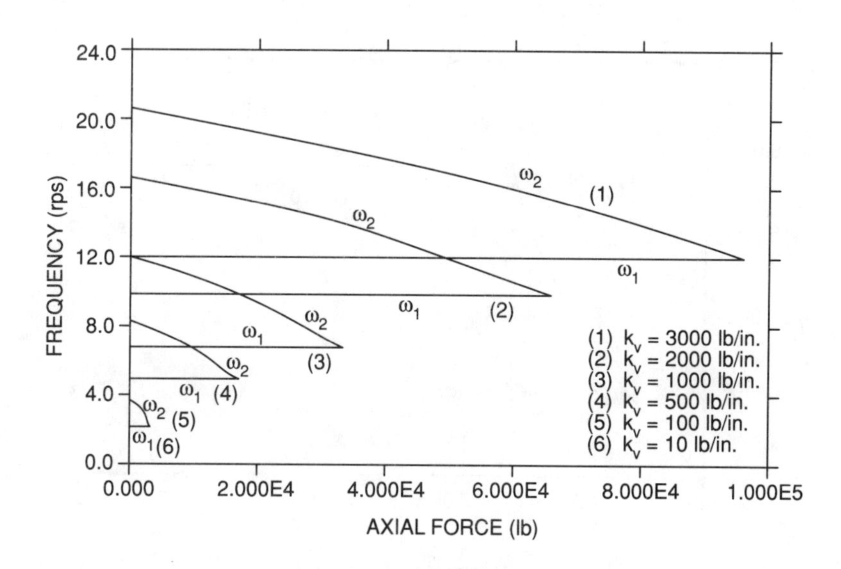

FIGURE 12.2. Natural frequencies of vibration ω_1 and ω_2 with increasing axial compressive force P.

ω_2 to become equal. Therefore, there will be a value of k_v that establishes the limit of such flutter instability. If we name this value as the critical spring constant $k_{v_{cr}}$, then flutter instabilities may occur when $k_v < k_{v_{cr}}$. That is, for every $k_v < k_{v_{cr}}$, there exists an axial load $P_{flutter}$ that makes the frequencies ω_1 and ω_2 equal. This can be easily observed by examining the results in Figure 12.2.

In Figure 12.2 we observe that as k_v increases, larger values of axial load P are required to produce flutter instability, and consequently ω_1 to become equal to ω_2. The point where ω_1 and ω_2 coincide may be referred to as the point of double instability. In Figure 12.3 the axial flutter loads corresponding to spring constants k_v are plotted. We note in this figure that when $k_v < 1.52 \times 10^5$ lb/in. (266.19×10^5 N/m), flutter instability occurs, thus forming the region of flutter instability. When the spring constant $k_v > 1.52 \times 10^5$ lb/in. (266.19×10^5 N/m), the critical axial load P_{cr} remains constant with increasing k_v, and static instability with $\omega_1 = 0$ governs, thus establishing the region of static instability.

In Figure 12.4 the axial compressive load P is assumed to be constant and equal to 1 million lb (4.448222×10^6 N), and the length of the member was permitted to vary as shown. The cases examined are $k_v = 25{,}000$ lb/in. (4378.17×10^3 N/m), $k_v = 50{,}000$ lb/in. (8756.34×10^3 N/m), and $k_v = 100{,}000$ lb/in. ($17{,}512.68 \times 10^3$ N/m). We note in this figure that for $k_v = 25{,}000$ lb/in. the double point of instability ($\omega_1 = \omega_2$) occurs when the length L = 123.18 in. (3.1288 m), and it will occur at L = 60.5 in. (1.5367 m) and L = 31 in. (0.7874 m) when $k_v = 50{,}000$ lb/in. and 100,000 lb/in., respectively. We also note that ω_1 becomes zero when L = 275 in. (6.985 m), which establishes the static critical length L_{cr} for static instability.

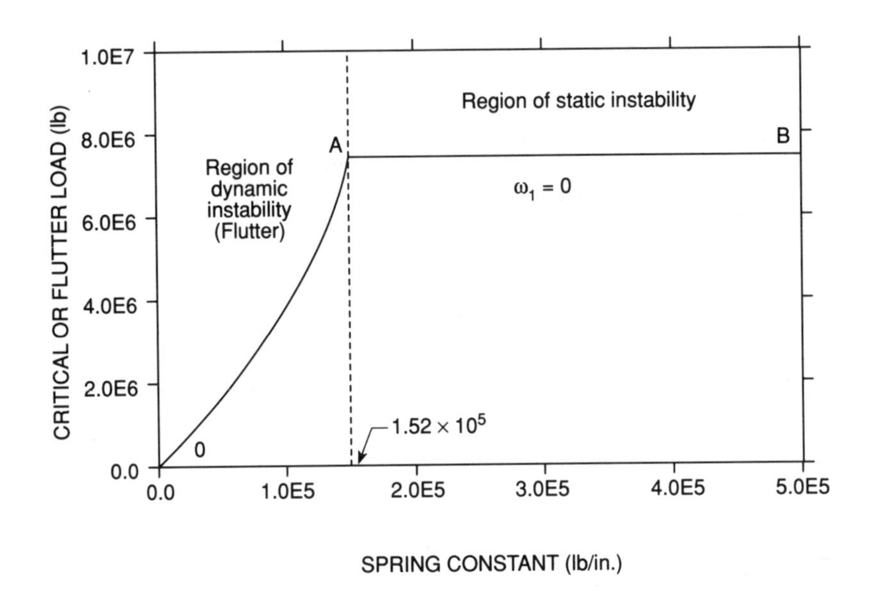

FIGURE 12.3. Regions of flutter and static instabilities.

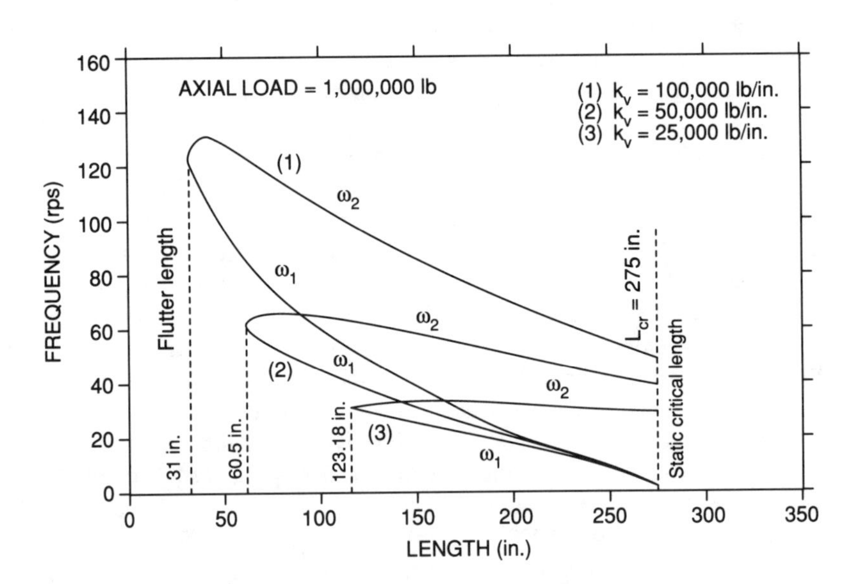

FIGURE 12.4. Natural frequencies of vibration ω_1 and ω_2 with increasing beam length.

In Figure 12.5, the frequencies ω_1 and ω_2 are plotted by varying the length L of the member up to 275 in. (6.985 m). The spring constant k_v is assumed to be constant and equal to 50,000 lb/in. (8756.34×10^3 N/m). The axial force P $= 10^6$ lb (4448×10^3 N), and the mass m per unit of length is 0.4224 lb-sec²/in. In this case, Figure 12.5 shows that the frequencies ω_1 and ω_2 will coincide

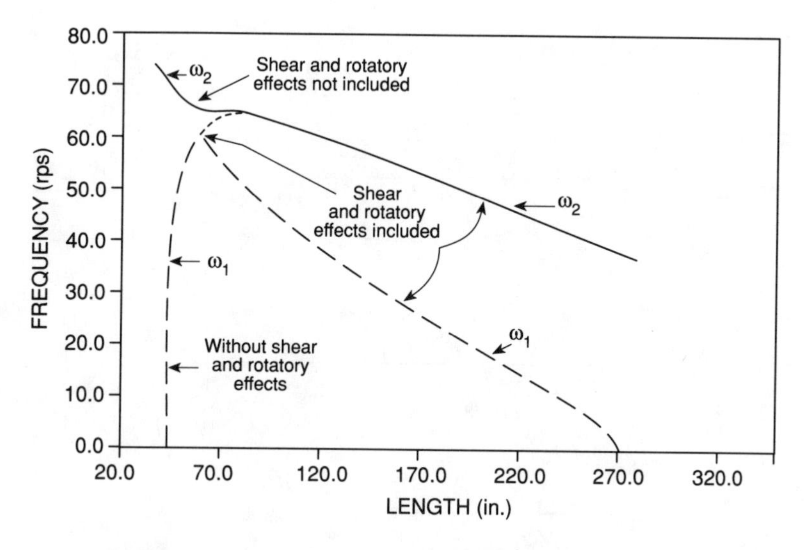

FIGURE 12.5. Effect of shear and rotatory inertia (1 in. = 0.0254 m).

only when the shear and rotatory effects are taken into consideration. Therefore, for this case, it can be concluded that the existence of flutter instability will not be recognized if the shear and rotatory effects are not included in the analysis. Similar observations were made when the spring constant k_v was increased to 100,000 lb/in. ($17{,}512.68 \times 10^3$ N/m), which indicates that shear and rotatory effects become an important parameter when the vertical springs at the end supports reach a certain level of rigidity.

12.3 UNIFORM BEAMS SUPPORTED BY VERTICAL AND HORIZONTAL SPRINGS AT THE ENDS AND SUBJECTED TO AXIAL AND VERTICAL RESTRAINTS

The work in this section deals with the static and vibration analysis of a uniform member that is elastically supported at each end by a vertical and a horizontal spring of constants k_v and k_h, respectively. The member is assumed to be subjected to an axial compressive restraint consisting of an initial precompression x_o of the horizontal springs and an initial transversed static displacement y_{st}. These two restraints may also be thought of as initial imperfections, since they can be very small, or they may be the result of temperature changes and other possible conditions arising from the utilization of the structural system. Rotatory and shear effects are also taken into consideration in the analysis, but damping is neglected.

Consider the elastically supported beam in Figure 12.6a, where x_o is the horizontal restaint, x_1 is the undeformed length of the spring k_h, and y_{st} is an initial transversed static displacement. The total vertical displacement from the initial straight position of the member is denoted by $y(x,t)$. The derivation of

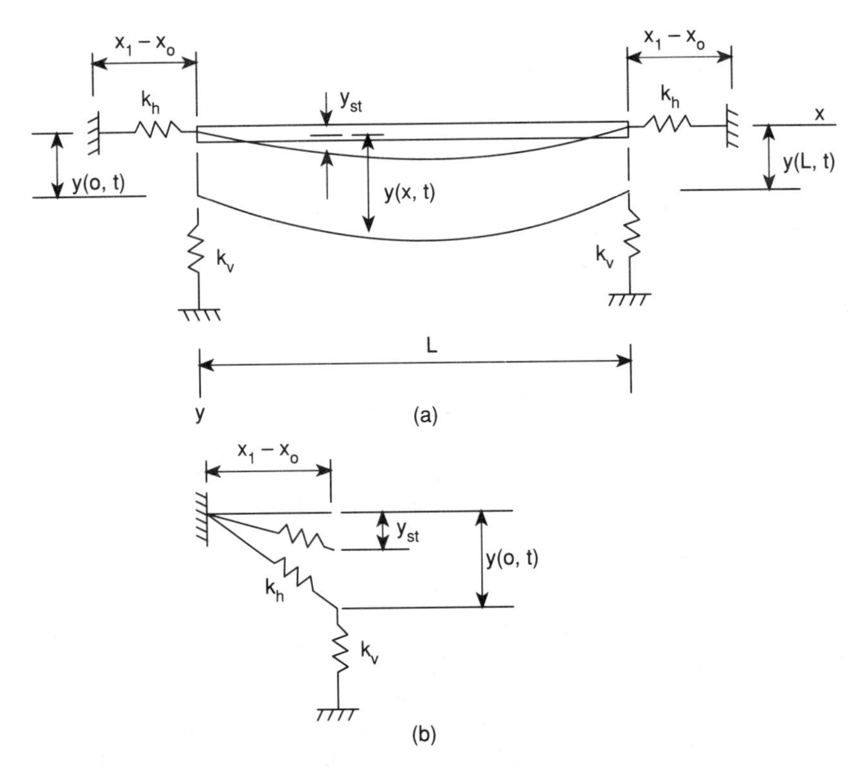

FIGURE 12.6. (a) Deformed configuration of a uniform elastically supported beam; (b) end beam displacements.

the basic differential equation of the nonlinear bending vibration of the member may be based on Timoshenko's[75] beam theory, or it can be derived as in the preceding section. By taking into consideration the horizontal and vertical restraints, as well as the effects of shear and rotatory inertia, the nonlinear differential equation of motion may be written as follows:

$$\begin{aligned} &EI^*(y-y_{st})'''' - \rho\left[\frac{EI^*}{K'G} + I^*\right](\ddot{y})'' + m\ddot{y} \\ &+\left\{k_h x_o - \frac{k_h L}{4}\int_0^L\left[(y')^2 - (y'_{st})^2\right]dx\right\}y'' + \rho^2\left(\frac{I^*}{K'G}\right)\ddddot{y} = 0 \end{aligned} \tag{12.34}$$

where

$$I^* = \frac{I}{\left(1 + \dfrac{k_h x_o}{K'AG}\right)} \tag{12.35}$$

In the above two equations E is Young's modulus of elasticity, ρ is the mass density, A is the cross-sectional area of the member, k_h is the horizontal spring constant, k_v is the vertical spring constant, I is the moment of inertia, G is the shear modulus, K′ is the shear factor, and L is the length of the member. The first and fifth terms in Equation 12.34 incorporate the effects of the vertical and horizontal restraints y_{st} and x_o, respectively, and the shear and rotatory effects are included in the second term of the equation. It should be noted, however, that the displacement y in Equation 12.34 includes the initial transversed static displacement y_{st}.

By introducing the notation

$$\xi = \frac{x}{L} \tag{12.36}$$

$$\beta = \rho L^2\left(\frac{1}{K'G} + \frac{1}{E}\right) \tag{12.37}$$

$$\gamma = \frac{L^4 M}{EI} \tag{12.38}$$

$$M = \rho A \tag{12.39}$$

$$H = \frac{L^4 \rho^2}{EK'G} \tag{12.40}$$

$$R = \frac{k_h x_o L^2}{EI^*} \tag{12.41}$$

$$Z = \frac{k_h L^2}{4EI^*} \tag{12.42}$$

$$I^* = \frac{1}{\left(1 + \dfrac{k_h x_o}{K'AG}\right)} \tag{12.43}$$

Equation 12.34 may be rewritten in a more convenient form as follows:

$$(y - y_{st})_{\xi\xi\xi\xi} - \beta y_{\xi\xi tt} + \gamma y_{tt} + \left\{R - Z\int_0^1 \left[y_\zeta^2 - y_{st}^2\right] d(\xi)\right\} y_{\xi\xi} + Hy_{tttt} = 0 \tag{12.44}$$

The solution of Equation 12.34 or Equation 12.44 may be carried out by assuming in Figure 12.6 that

$$y_{st} = \sum_m a_m \phi_m(\xi) \qquad m = 1, 2, 3, \cdots \tag{12.45}$$

$$y(x,t) = \sum_m b_m(t)\phi_m(\xi) \qquad m = 1, 2, 3, \cdots \tag{12.46}$$

where a_m is a known constant, $b_m(t)$ is an unknown time function, and $\phi_m(\xi)$ represents the mode shapes of vibration. For a symmetric mode shape

$$\phi_m(\xi) = A_m + \sin(m\pi\xi) \qquad m = 1, 3, 5, \cdots \tag{12.47}$$

and for an unsymmetric mode shape

$$\phi_m(\xi) = A_m(1 - 2\xi) + \sin(m\pi\xi) \qquad m = 2, 4, 6, \cdots \tag{12.48}$$

where the constant A_m represents the rigid body motion of the beam and it can be determined by using boundary conditions. i.e., at $\xi = 0$ we have

$$V(0) - k_n x_o \phi_\xi(0) = k_v \phi(0) \tag{12.49}$$

where the shear force V(0) at $\xi = 0$ is given by the expression

$$V(0) = \frac{EI}{\left[1 - \dfrac{\omega^2 \rho I}{MK'G}\right]} \left[-\frac{1}{L^3}\phi_{\xi\xi\xi\xi}(0) - \frac{1}{L}\left(\frac{\rho\omega^2}{K'G} + \frac{\rho I \omega^2}{E} - \frac{k_h x_o}{EI}\right)\phi_\xi(0) \right] \tag{12.50}$$

In the above equation, M is the mass per unit of length of the beam, I is the cross-sectional moment of inertia, and ω is the frequency of vibration in radians per second (rps).

By substituting Equation 12.50 into Equation 12.49 and solving for A_m, since $\phi_m^{(o)} = A_m$, the following expression is obtained for symmetrical vibration modes:

$$A_m = \frac{1}{k_v}\left\{\frac{EI}{\left[1-\frac{\omega^2\rho I}{MK'G}\right]}\left[\left(\frac{m\pi}{L}\right)^3 - \left(\frac{m\pi}{L}\right)\left(\frac{\rho\omega^2}{K'G}+\frac{\rho I\omega^2}{E}-\frac{k_h x_o}{EI}\right)\right] - k_h x_o\left(\frac{m\pi}{L}\right)\right\} \tag{12.51}$$

For unsymmetrical modes of vibration we have

$$A_m = \frac{\frac{EI}{\left[1-\frac{\omega^2\rho I}{MK'G}\right]}\left[-\left(\frac{m\pi}{L}\right)^3 + \left(\frac{\rho\omega^2}{K'G}+\frac{\rho I\omega^2}{E}-\frac{k_h x_o}{EI}\right)\left(\frac{m\pi}{L}\right)\right] + k_h x_o\left(\frac{m\pi}{L}\right)}{\left[\left(\frac{2}{L}\right)\frac{EI}{\left(1-\frac{\omega^2\rho I}{MK'G}\right)} + k_h x_o\left(\frac{2}{L}\right) - k_v\right]} \tag{12.52}$$

By using the orthogonality relation

$$\int_0^1 [\text{Equation}(12-44)]\phi_n(\xi)d\xi = 0 \tag{12.53}$$

and integrating, we obtain

$$\begin{aligned}
&\sum_m b_m\alpha_{mn} - \sum_m a_m\alpha_{mn} - \beta\sum_m b_{m_{tt}}\beta_{mn} + \gamma\sum_m b_{m_{tt}}\alpha^*_{mn} \\
&+\left[R + z\sum_k\sum_p \beta^*_{kp}a_k a_p\right]\sum_m b_m\beta_{mn} \\
&-z\sum_k\sum_p\sum_m \beta^*_{kp}\beta_{mn}b_k b_p b_m + H\sum_m b_{m_{ttt}}\alpha^*_{mn} = 0
\end{aligned} \tag{12.54}$$

where k, p, and m can take values of 1,2,3,..., and

$$\alpha_{mn} = \int_0^1 \phi_{m_{\xi\xi\xi\xi}}(\xi)\phi_n(\xi)d\xi \tag{12.55}$$

$$\alpha^*_{mn} = \int_0^1 \phi_m(\xi)\phi_n(\xi)d\xi \tag{12.56}$$

$$\beta_{mn} = \int_0^1 \phi_{m_{\xi\xi}}(\xi)\phi_n(\xi)d\xi \tag{12.57}$$

$$\beta^*_{kp} = \int_0^1 \phi_{k_\xi}(\xi)\phi_{p_\xi}(\xi)d\xi \tag{12.58}$$

Equation 12.54, however, does not incorporate the axial force effect in the horizontal spring k_h during static deformation of the beam, or during its vibration (see Figure 12.6b). Therefore, Equation 12.54 may be modified as shown below in order to incorporate these effects, i.e.,

$$\begin{aligned}
&\sum_m b_m\alpha_{mn} - \sum_m a_m\alpha_{mn} - \beta\sum_m b_{m_{tt}}\beta_{mn} + \gamma\sum_m b_{m_{tt}}\alpha^*_{mn} \\
&+\left[R + Z\sum_k\sum_p \beta^*_{kp}a_k a_p\right]\sum_m b_m\beta_{mn} - Z\sum_k\sum_p\sum_m \beta^*_{kp}\beta_{mn}b_k b_p b_m \\
&+H\sum b_{m_{tttt}}\alpha^*_{mn} + Z^*\sum_k\sum_p \frac{a_k A_{k_{st}} a_p A_{p_{st}}}{(x_1 - x_o)}\sum_m b_m b_{mn} \\
&-Z^*\sum_k\sum_p \frac{b_k A_k b_p A_p}{(x_1 - x_o)}\sum_m b_m b_{mn} = 0
\end{aligned} \tag{12.59}$$

where

$$Z^* = \frac{k_h L^2}{2EI^*}(x_1 - x_o) \tag{12.60}$$

The quantities $A_{k_{st}}$ and $A_{p_{st}}$ in the 8th term of Equation 12.59 are initial static rigid body displacements of the end spring k_v, and A_k and A_p in the last term of Equation 12.59 are the analogous vibration rigid body displacements. Note that the axial force effects stated earlier are represented by the last two terms of Equation 12.59.

12.3.1 STATIC DEFLECTION AND END FORCE IN SPRING k_h

The static displacement

$$y_{st_m} = \sum_m b_m \phi_m(\xi) \tag{12.61}$$

of the beam with initial deflection a_1, and precompression of the horizontal spring k_h by a displacement x_o, can be determined using Equation 12.59. If m is permitted to take the values m = 1, 2, 3, 4, and 5, we will have five simultaneous cubic equations to solve for the constant b_m. The Newton-Raphson method may be used to determine b_m. On this basis, Equation 12.59 may be written as follows:

$$\begin{aligned}
&\sum_m \alpha_{mn} b_m - a_n \alpha_{1n} - \left[R + Z a_1^2 \beta_{11}^*\right] \sum_m \beta_{mn} b_m \\
&-Z * \sum_m \frac{a_1^2 A_{1_{st}}^2}{(x_1 - x_o)} b_m \beta_{mn} - Z * \sum_k \sum_p \sum_m \frac{b_k b_p A_k A_p}{(x_1 - x_o)} b_m \beta_{mn} \\
&-Z \sum_k \sum_p \sum_m \beta_{kp}^* \beta_{mn} = 0
\end{aligned} \tag{12.62}$$

Thus, by using Equation 12.62 and applying the Newton-Raphson method, b_m (m = 1, 2, 3, 4, 5) can be determined. At the beam center, the static deflection y_{st_m} is

$$y_{st_m} = \sum_m b_m \phi_m\left(\frac{1}{2}\right) \tag{12.63}$$

The actual spring force S in the horizontal spring k_h during static deformation is given by the expression

$$\begin{aligned}
S = {} & k_h x_o + \frac{k_h L}{4} \sum_k \sum_p \beta_{kp}^* a_k a_p - \frac{k_h L}{4} \sum_k \sum_p \beta_{mn} b_m b_n \\
& + \frac{k_h (x_1 - x_o)}{2} \sum_k \sum_p \frac{a_k a_p A_{k_{st}} A_{p_{st}}}{(x_1 - x_o)} \\
& - \frac{k_h (x_1 - x_o)}{2} \sum_k \sum_p \frac{A_k A_p b_k b_p}{(x_1 - x_o)}
\end{aligned} \tag{12.64}$$

The spring force S is a function of x_o. Thus, by increasing x_o, the spring force S increases, and for a given problem there will be a value of x_o which will make the beam buckle. Application of these equations is given in later parts of this chapter.

12.4 VIBRATION ANALYSIS OF THE MEMBER IN FIGURE 12.6

In order to obtain the natural frequencies of vibration of the beam in Figure 12.6, which is under the influence of an initial deflection and an initial compressive spring force due to an x_o displacement, we can assume that the time function $b_m(t)$ is as follows

$$b_m(t) = d_m + d_m^* \cos\omega^* t \tag{12.65}$$

where d_m is a constant that depends on the initial displacement x_o, and it is analogous to the constant d_m that is obtained from the static analysis. The constant d_m^* in Equation 12.65 is associated with the vibration of the member from the static equilibrium position d_m. By substituting Equation 12.65 into Equation 12.59 and carrying out the required manipulations, we obtain the equation

$$\begin{aligned}
&\sum_m \left(d_m + d_m^* \cos\omega^* t\right)\alpha_{mn} - a_1(a_{1n}) + \beta\omega^{*2}\sum_m d_m^* \cos\omega^* t\beta_{mn} \\
&-\omega^{*2}\gamma\sum_m d_m^* \cos\omega^* t a_{mn}^* + \left[R + Za_1^2\beta_{11}^*\right]\sum_m \left(d_m + d_m^* \cos\omega^* t\right)\beta_{mn} \\
&-Z\sum_k\sum_p\sum_m \beta_{kp}^*\beta_{mn}\left(d_k + d_k^* \cos\omega^* t\right)\left(d_p + d_p^* \cos\omega^* t\right) \\
&\left(d_m + d_m^* \cos\omega^* t\right) + Z^*\frac{\left(a_1 A_{1st}\right)^2}{\left(x_1 - x_o\right)}\sum_m \left(d_m + d_m^* \cos\omega^* t\right)\beta_{mn} \\
&-Z^*\sum_k\sum_p\sum_m \frac{A_k A_p}{\left(x_1 - x_o\right)}\beta_{mn}\left(d_k + d_k^* \cos\omega^* t\right)\left(d_p + d_p^* \cos\omega^* t\right) \\
&+\omega^{*4} H\sum_m d_m^* \cos\omega^* t\left(\alpha_{mn}^*\right) = 0
\end{aligned} \tag{12.66}$$

where ω^* is the frequency of vibration in radians per second (rps).

Equation 12.66 satisfies the static d_m, so d_m can be neglected for vibration analysis. Therefore, we retain only the linear terms of d^*_m. On this basis we obtain five homogeneous linear equations in d^*_m, when the first five natural frequencies ω^*_1, ω^*_2, ω^*_3, ω^*_4, and ω^*_5 are to be determined. This is accomplished by using the Bisection method.[75]

Equation 12.66 can be rewritten as

$$\begin{aligned}
&\sum_m d^*_m \alpha_{mn} + \omega^{*2} \beta \sum_m d^*_m \beta_{mn} - \omega^{*2} \gamma \sum_m d^*_m \alpha_{mn} \\
&+ \left[R + Za_1^2 \beta^*_{11}\right] \sum_m d^*_m \beta_{mn} - Z\Delta_n - Z\phi_n - Z\Lambda_n \\
&+ Z * \frac{(a_1 A_{1_{st}})^2}{(x_1 - x_o)} \sum_m d^*_m \beta_{mn} - Z * \Delta^*_n - Z\phi^*_n - Z\Lambda^*_n \\
&+ \omega^{*4} H \sum_m d^*_m \alpha^*_{mn} = 0
\end{aligned} \tag{12.67}$$

where

$$\Delta_n = \sum_k \sum_p \sum_m \beta^*_{kp} \beta_{mn} (d_k d_p d_m) \tag{12.68}$$

$$\phi_n = \sum_k \sum_p \sum_m \beta^*_{kp} \beta_{mn} (d^*_k d_p d_m) \tag{12.69}$$

$$\Lambda_n = \sum_k \sum_p \sum_m \beta^*_{kp} \beta_{mn} (d_k d_p d^*_m) \tag{12.70}$$

$$\Delta^*_n = \sum_k \sum_p \sum_m \frac{A_k A_p}{(x_1 - x_o)} \beta_{mn} (d_k d^*_p d_m) \tag{12.71}$$

$$\phi^*_n = \sum_k \sum_p \sum_m \frac{A_k + A_p}{(x_1 - x_o)} \beta_{mn} (d^*_k d_p d_m) \tag{12.72}$$

$$\Lambda^*_n = \sum_k \sum_p \sum_m \frac{A_k A_p}{(x_1 - x_o)} \beta_{mn} (d_k d_p d^*_m) \tag{12.73}$$

By using combinations of k, p, and m for values ranging from 1 to 5, and performing the required mathematical manipulations for Δ_n, ϕ_n, ϕ_n^*, etc., Equation 12.67 may be put into a matrix form as follows:

$$\begin{bmatrix} A_{11}^* & A_{12}^* & A_{13}^* & A_{14}^* & A_{15}^* \\ A_{21}^* & A_{22}^* & A_{23}^* & A_{24}^* & A_{25}^* \\ A_{31}^* & A_{32}^* & A_{33}^* & A_{34}^* & A_{35}^* \\ A_{41}^* & A_{42}^* & A_{43}^* & A_{44}^* & A_{45}^* \\ A_{51}^* & A_{52}^* & A_{53}^* & A_{54}^* & A_{55}^* \end{bmatrix} \begin{bmatrix} d_1^* \\ d_2^* \\ d_3^* \\ d_4^* \\ d_5^* \end{bmatrix} = 0 \tag{12.74}$$

where

$$\begin{aligned} A_{11}^* &= \alpha_{11}^* + \omega^{*2}\beta_{11} - \omega^*\gamma\alpha_{11} + \left[R + Za_1^2\beta_{11}^*\right]\beta_{11} \\ &-Z\left(\Delta_1^{(1)} + \phi_1^{(1)} + \Lambda_1^{(1)}\right) \\ &+Z^*\frac{\left(a_1A_{1_{st}}\right)^2}{\left(x_1 - x_o\right)}\beta_{11} - Z^*\left(\Delta_1^{*(1)} + \phi_1^{*(1)} + \Lambda_1^{*(1)}\right) + \omega^{*4}H\alpha_{11}^* \end{aligned} \tag{12.75}$$

$$\begin{aligned} A_{12}^* &= \alpha_{21}^* + \omega^{*2}\beta_{21} - \omega^{*2}\gamma\alpha_{21} + \left[R + Za_1^2\beta_{11}^*\right]\beta_{21} \\ &-Z\left(\Delta_1^{(2)} + \phi_1^{(2)} + \Lambda_1^{(2)}\right) \\ &+Z^*\frac{\left(a_1A_{1_{st}}\right)^2}{\left(x_1 - x_o\right)}\beta_{21} - Z^*\left(\Delta_1^{*(2)} + \phi_1^{*(2)} + \Lambda_1^{*(2)}\right) + \omega^{*4}H\alpha_{21}^* \end{aligned} \tag{12.76}$$

$$\begin{aligned} A_{13}^* &= \alpha_{31}^* + \omega^{*2}\beta_{31} - \omega^{*2}\gamma\alpha_{31} + \left[R + Za_1^2\beta_{11}^*\right]\beta_{31} \\ &-Z\left(\Delta_1^{(3)} + \phi_1^{(3)} + \Lambda_1^{(3)}\right) \\ &+Z^*\frac{\left(a_1A_{1_{st}}\right)^2}{\left(x_1 - x_o\right)}\beta_{31} - Z^*\left(\Delta_1^{*(3)} + \phi_1^{*(3)} + \Lambda_1^{*(3)}\right) + \omega^{*4}H\alpha_{31}^* \end{aligned} \tag{12.77}$$

$$\begin{aligned} A_{14}^{*} &= \alpha_{41}^{*} + \omega^{*2} \beta_{41} - \omega^{*2} \gamma\alpha_{41} + \left[R + Za_1^2\beta_{11}^{*}\right]\beta_{41} \\ &-Z\left(\Delta_1^{(4)} + \phi_1^{(4)} + \Lambda_1^{(4)}\right) \\ &+Z * \frac{\left(a_1 A_{1_{st}}\right)^2}{\left(x_1 - x_o\right)} \beta_{41} - Z * \left(\Delta_1^{*(4)} + \phi_1^{*(4)} + \Lambda_1^{*(4)}\right) + \omega^{*4} H\alpha_{41}^{*} \end{aligned} \tag{12.78}$$

$$\begin{aligned} A_{15}^{*} &= \alpha_{51}^{*} + \omega^{*2} \beta_{51} - \omega^{*2} \gamma\alpha_{51} + \left[R + Za_1^2\beta_{11}^{*}\right]\beta_{51} \\ &-Z\left(\Delta_1^{(5)} + \phi_1^{(5)} + \Lambda_1^{(5)}\right) \\ &+Z * \frac{\left(a_1 A_{1_{st}}\right)^2}{\left(x_1 - x_o\right)} \beta_{51} - Z * \left(\Delta_1^{*(5)} + \phi_1^{*(5)} + \Lambda_1^{*(5)}\right) + \omega^{*4} H\alpha_{51}^{*} \end{aligned} \tag{12.79}$$

$$\begin{aligned} A_{21}^{*} &= \alpha_{12}^{*} + \omega^{*2} \beta_{12} - \omega^{*2} \gamma\alpha_{12} + \left[R + Za_1^2\beta_{11}^{*}\right]\beta_{12} \\ &-Z\left(\Delta_2^{(1)} + \phi_2^{(1)} + \Lambda_2^{(1)}\right) \\ &+Z * \frac{\left(a_1 A_{1_{st}}\right)^2}{\left(x_1 - x_o\right)} \beta_{12} - Z * \left(\Delta_2^{*(1)} + \phi_2^{*(1)} + \Lambda_2^{*(1)}\right) + \omega^{*4} H\alpha_{12}^{*} \end{aligned} \tag{12.80}$$

$$\begin{aligned} A_{22}^{*} &= \alpha_{22}^{*} + \omega^{*2} \beta_{22} - \omega^{*2} \gamma\alpha_{22} + \left[R + Za_1^2\beta_{11}^{*}\right]\beta_{22} \\ &-Z\left(\Delta_2^{(2)} + \phi_2^{(2)} + \Lambda_2^{(2)}\right) \\ &+Z * \frac{\left(a_1 A_{1_{st}}\right)^2}{\left(x_1 - x_o\right)} \beta_{22} - Z * \left(\Delta_2^{*(2)} + \phi_2^{*(2)} + \Lambda_2^{*(2)}\right) + \omega^{*4} H\alpha_{22}^{*} \end{aligned} \tag{12.81}$$

$$\begin{aligned} A^*_{23} &= \alpha^*_{32} + \omega^{*2}\beta_{32} - \omega^{*2}\gamma\alpha_{32} + \left[R + Za_1^2\beta^*_{11}\right]\beta_{32} \\ &-Z\left(\Delta_2^{(3)} + \phi_2^{(3)} + \Lambda_2^{(3)}\right) \\ &+Z*\frac{\left(a_1A_{1_{st}}\right)^2}{\left(x_1 - x_o\right)}\beta_{32} - Z*\left(\Delta_2^{*(3)} + \phi_2^{*(3)} + \Lambda_2^{*(3)}\right) + \omega^{*4}H\alpha^*_{32} \end{aligned} \quad (12.82)$$

$$\begin{aligned} A^*_{24} &= \alpha^*_{42} + \omega^{*2}\beta_{42} - \omega^{*2}\gamma\alpha_{42} + \left[R + Za_1^2\beta^*_{11}\right]\beta_{42} \\ &-Z\left(\Delta_2^{(4)} + \phi_2^{(4)} + \Lambda_2^{(4)}\right) \\ &+Z*\frac{\left(a_1A_{1_{st}}\right)^2}{\left(x_1 - x_o\right)}\beta_{42} - Z*\left(\Delta_2^{*(4)} + \phi_2^{*(4)} + \Lambda_2^{*(4)}\right) + \omega^{*4}H\alpha^*_{42} \end{aligned} \quad (12.83)$$

$$\begin{aligned} A^*_{25} &= \alpha^*_{52} + \omega^{*2}\beta_{52} - \omega^{*2}\gamma\alpha_{52} + \left[R + Za_1^2\beta^*_{11}\right]\beta_{52} \\ &-Z\left(\Delta_2^{(5)} + \phi_2^{(5)} + \Lambda_2^{(5)}\right) \\ &+Z*\frac{\left(a_1A_{1_{st}}\right)^2}{\left(x_1 - x_o\right)}\beta_{52} - Z*\left(\Delta_2^{*(5)} + \phi_2^{*(5)} + \Lambda_2^{*(5)}\right) + \omega^{*4}H\alpha^*_{52} \end{aligned} \quad (12.84)$$

$$\begin{aligned} A^*_{31} &= \alpha^*_{13} + \omega^{*2}\beta_{13} - \omega^{*2}\gamma\alpha_{13} + \left[R + Za_1^2\beta^*_{11}\right]\beta_{13} \\ &-Z\left(\Delta_3^{(1)} + \phi_3^{(1)} + \Lambda_3^{(1)}\right) \\ &+Z*\frac{\left(a_1A_{1_{st}}\right)^2}{\left(x_1 - x_o\right)}\beta_{13} - Z*\left(\Delta_3^{*(1)} + \phi_3^{*(1)} + \Lambda_3^{*(1)}\right) + \omega^{*4}H\alpha^*_{13} \end{aligned} \quad (12.85)$$

$$\begin{aligned} A^*_{32} &= \alpha^*_{23} + \omega^{*2}\beta_{23} - \omega^{*2}\gamma\alpha_{23} + \left[R + Za_1^2\beta^*_{11}\right]\beta_{23} \\ &-Z\left(\Delta_3^{(2)} + \phi_3^{(2)} + \Lambda_3^{(2)}\right) \\ &+Z*\frac{\left(a_1A_{1_{st}}\right)^2}{\left(x_1 - x_o\right)}\beta_{23} - Z*\left(\Delta_3^{*(2)} + \phi_3^{*(2)} + \Lambda_3^{*(2)}\right) + \omega^{*4}H\alpha^*_{23} \end{aligned} \quad (12.86)$$

$$
\begin{aligned}
A_{33}^{*} &= \alpha_{33}^{*} + \omega^{*2}\beta_{33} - \omega^{*2}\gamma\alpha_{33} + \left[R + Za_1^2\beta_{11}^{*}\right]\beta_{33} \\
&-Z\left(\Delta_3^{(3)} + \phi_3^{(3)} + \Lambda_3^{(3)}\right) \\
&+Z^{*}\frac{\left(a_1 A_{1_{st}}\right)^2}{\left(x_1 - x_o\right)}\beta_{33} - Z^{*}\left(\Delta_3^{*(3)} + \phi_3^{*(3)} + \Lambda_3^{*(3)}\right) + \omega^{*4} H\alpha_{33}^{*}
\end{aligned}
\tag{12.87}
$$

$$
\begin{aligned}
A_{34}^{*} &= \alpha_{43}^{*} + \omega^{*2}\beta_{43} - \omega^{*2}\gamma\alpha_{43} + \left[R + Za_1^2\beta_{11}^{*}\right]\beta_{43} \\
&-Z\left(\Delta_3^{(4)} + \phi_3^{(4)} + \Lambda_3^{(4)}\right) \\
&+Z^{*}\frac{\left(a_1 A_{1_{st}}\right)^2}{\left(x_1 - x_o\right)}\beta_{43} - Z^{*}\left(\Delta_3^{*(4)} + \phi_3^{*(4)} + \Lambda_3^{*(4)}\right) + \omega^{*4} H\alpha_{43}^{*}
\end{aligned}
\tag{12.88}
$$

$$
\begin{aligned}
A_{35}^{*} &= \alpha_{53}^{*} + \omega^{*2}\beta_{53} - \omega^{*2}\gamma\alpha_{53} + \left[R + Za_1^2\beta_{11}^{*}\right]\beta_{53} \\
&-Z\left(\Delta_3^{(5)} + \phi_3^{(5)} + \Lambda_3^{(5)}\right) \\
&+Z^{*}\frac{\left(a_1 A_{1_{st}}\right)^2}{\left(x_1 - x_o\right)}\beta_{53} - Z^{*}\left(\Delta_3^{*(5)} + \phi_3^{*(5)} + \Lambda_3^{*(5)}\right) + \omega^{*4} H\alpha_{53}^{*}
\end{aligned}
\tag{12.89}
$$

$$
\begin{aligned}
A_{41}^{*} &= \alpha_{14}^{*} + \omega^{*2}\beta_{14} - \omega^{*2}\gamma\alpha_{14} + \left[R + Za_1^2\beta_{11}^{*}\right]\beta_{14} \\
&-Z\left(\Delta_4^{(1)} + \phi_4^{(1)} + \Lambda_4^{(1)}\right) \\
&+Z^{*}\frac{\left(a_1 A_{1_{st}}\right)^2}{\left(x_1 - x_o\right)}\beta_{14} - Z^{*}\left(\Delta_4^{*(1)} + \phi_4^{*(1)} + \Lambda_4^{*(1)}\right) + \omega^{*4} H\alpha_{14}^{*}
\end{aligned}
\tag{12.90}
$$

$$
\begin{aligned}
A_{42}^{*} &= \alpha_{24}^{*} + \omega^{*2}\beta_{24} - \omega^{*2}\gamma\alpha_{24} + \left[R + Za_1^2\beta_{11}^{*}\right]\beta_{24} \\
&-Z\left(\Delta_4^{(2)} + \phi_4^{(2)} + \Lambda_4^{(2)}\right) \\
&+Z^{*}\frac{\left(a_1 A_{1_{st}}\right)^2}{\left(x_1 - x_o\right)}\beta_{24} - Z^{*}\left(\Delta_4^{*(2)} + \phi_4^{*(2)} + \Lambda_4^{*(2)}\right) + \omega^{*4} H\alpha_{24}^{*}
\end{aligned}
\tag{12.91}
$$

$$\begin{aligned} A_{43}^* &= \alpha_{34}^* + \omega^{*2}\beta_{34} - \omega^{*2}\gamma\alpha_{34} + \left[R + Za_1^2\beta_{11}^*\right]\beta_{34} \\ &- Z\left(\Delta_4^{(3)} + \phi_4^{(3)} + \Lambda_4^{(3)}\right) \\ &+ Z^*\frac{\left(a_1 A_{1_{st}}\right)^2}{\left(x_1 - x_o\right)}\beta_{34} - Z^*\left(\Delta_4^{*(3)} + \phi_4^{*(3)} + \Lambda_4^{*(3)}\right) + \omega^{*4} H\alpha_{34}^* \end{aligned} \tag{12.92}$$

$$\begin{aligned} A_{44}^* &= \alpha_{44}^* + \omega^{*2}\beta_{44} - \omega^{*2}\gamma\alpha_{44} + \left[R + Za_1^2\beta_{11}^*\right]\beta_{44} \\ &- Z\left(\Delta_4^{(4)} + \phi_4^{(4)} + \Lambda_4^{(4)}\right) \\ &+ Z^*\frac{\left(a_1 A_{1_{st}}\right)^2}{\left(x_1 - x_o\right)}\beta_{44} - Z^*\left(\Delta_4^{*(4)} + \phi_4^{*(4)} + \Lambda_4^{*(4)}\right) + \omega^{*4} H\alpha_{44}^* \end{aligned} \tag{12.93}$$

$$\begin{aligned} A_{45}^* &= \alpha_{54}^* + \omega^{*2}\beta_{54} - \omega^{*2}\gamma\alpha_{54} + \left[R + Za_1^2\beta_{11}^*\right]\beta_{54} \\ &- Z\left(\Delta_4^{(5)} + \phi_4^{(5)} + \Lambda_4^{(5)}\right) \\ &+ Z^*\frac{\left(a_1 A_{1_{st}}\right)^2}{\left(x_1 - x_o\right)}\beta_{54} - Z^*\left(\Delta_4^{*(5)} + \phi_4^{*(5)} + \Lambda_4^{*(5)}\right) + \omega^{*4} H\alpha_{54}^* \end{aligned} \tag{12.94}$$

$$\begin{aligned} A_{51}^* &= \alpha_{15}^* + \omega^{*2}\beta_{15} - \omega^{*2}\gamma\alpha_{15} + \left[R + Za_1^2\beta_{11}^*\right]\beta_{15} \\ &- Z\left(\Delta_5^{(1)} + \phi_5^{(1)} + \Lambda_5^{(1)}\right) \\ &+ Z^*\frac{\left(a_1 A_{1_{st}}\right)^2}{\left(x_1 - x_o\right)}\beta_{15} - Z^*\left(\Delta_5^{*(1)} + \phi_5^{*(1)} + \Lambda_5^{*(1)}\right) + \omega^{*4} H\alpha_{15}^* \end{aligned} \tag{12.95}$$

$$\begin{aligned} A_{52}^* &= \alpha_{25}^* + \omega^{*2}\beta_{25} - \omega^{*2}\gamma\alpha_{25} + \left[R + Za_1^2\beta_{11}^*\right]\beta_{25} \\ &- Z\left(\Delta_5^{(2)} + \phi_5^{(2)} + \Lambda_5^{(2)}\right) \\ &+ Z^*\frac{\left(a_1 A_{1_{st}}\right)^2}{\left(x_1 - x_o\right)}\beta_{25} - Z^*\left(\Delta_5^{*(2)} + \phi_5^{*(2)} + \Lambda_5^{*(2)}\right) + \omega^{*4} H\alpha_{25}^* \end{aligned} \tag{12.96}$$

$$A_{53}^{*} = \alpha_{35}^{*} + \omega^{*2}\beta_{35} - \omega^{*2}\gamma\alpha_{35} + \left[R + Za_1^2\beta_{11}^{*}\right]\beta_{35}$$
$$-Z\left(\Delta_5^{(3)} + \phi_5^{(3)} + \Lambda_5^{(3)}\right) \tag{12.97}$$
$$+Z^{*}\frac{\left(a_1A_{1_{st}}\right)^2}{\left(x_1 - x_o\right)}\beta_{35} - Z^{*}\left(\Delta_5^{*(3)} + \phi_5^{*(3)} + \Lambda_5^{*(3)}\right) + \omega^{*4}H\alpha_{35}^{*}$$

$$A_{54}^{*} = \alpha_{45}^{*} + \omega^{*2}\beta_{45} - \omega^{*2}\gamma\alpha_{45} + \left[R + Za_1^2\beta_{11}^{*}\right]\beta_{45}$$
$$-Z\left(\Delta_5^{(4)} + \phi_5^{(4)} + \Lambda_5^{(4)}\right) \tag{12.98}$$
$$+Z^{*}\frac{\left(a_1A_{1_{st}}\right)^2}{\left(x_1 - x_o\right)}\beta_{45} - Z^{*}\left(\Delta_5^{*(4)} + \phi_5^{*(4)} + \Lambda_5^{*(4)}\right) + \omega^{*4}H\alpha_{45}^{*}$$

$$A_{55}^{*} = \alpha_{55}^{*} + \omega^{*2}\beta_{55} - \omega^{*2}\gamma\alpha_{55} + \left[R + Za_1^2\beta_{11}^{*}\right]\beta_{55}$$
$$-Z\left(\Delta_5^{(5)} + \phi_5^{(5)} + \Lambda_5^{(5)}\right) \tag{12.99}$$
$$+Z^{*}\frac{\left(a_1A_{1_{st}}\right)^2}{\left(x_1 - x_o\right)}\beta_{55} - Z^{*}\left(\Delta_5^{*(5)} + \phi_5^{*(5)} + \Lambda_5^{*(5)}\right) + \omega^{*4}H\alpha_{55}^{*}$$

In the above equations, the values of $\Delta_1^{(1)}, \Delta_1^{*(1)}, \phi_1^{(1)}, \phi_1^{*(1)}$, etc. may be obtained from Equations 12.68 through 12.73 by expanding the summations for values of k, p, and m from one to five and factoring out d_1^*, d_2^*, d_3^*, d_4^*, and d_5^*. For example, by following this procedure, Equation 12.68 may be written as

$$\Delta_n = \Delta_n^{(1)}d_1^{*} + \Delta_n^{(2)}d_2^{*} + \Delta_n^{(3)}d_3^{*} + \Delta_n^{(4)}d_4^{*} + \Delta_n^{(5)}d_5^{*} \tag{12.100}$$

where, e.g.,

$$\Delta_n^{(1)} = \big[\beta_{11}^*\beta_{1n}(d_1^2) + \beta_{11}^*\beta_{2n}(d_1d_2) + \beta_{11}^*\beta_{3n}(d_1d_3)$$
$$+\beta_{11}^*\beta_{4n}(d_1d_4) + \beta_{11}^*\beta_{5n}(d_1d_5)$$
$$+\beta_{21}^*\beta_{1n}(d_2d_1) + \beta_{21}^*\beta_{2n}(d_2^2) + \beta_{21}^*\beta_{3n}(d_2d_3) + \beta_{21}^*\beta_{4n}(d_2d_4)$$
$$+\beta_{21}^*\beta_{5n}(d_2d_5) + \beta_{31}^*\beta_{1n}(d_3d_1) + \beta_{31}^*\beta_{2n}(d_3d_2) + \beta_{31}^*\beta_{3n}(d_3^2) \qquad (12.101)$$
$$+\beta_{31}^*\beta_{4n}(d_3d_4) + \beta_{31}^*\beta_{5n}(d_3d_5) + \beta_{41}^*\beta_{1n}(d_4d_1) + \beta_{41}^*\beta_{2n}(d_4d_2)$$
$$+\beta_{41}^*\beta_{3n}(d_4d_2) + \beta_{41}^*\beta_{4n}(d_4^2) + \beta_{41}^*\beta_{5n}(d_4d_5) + \beta_{51}^*\beta_{1n}(d_5d_1)$$
$$+\beta_{51}^*\beta_{2n}(d_5d_2) + \beta_{51}^*\beta_{3n}(d_5d_3) + \beta_{51}^*\beta_{4n}(d_5d_4) + \beta_{51}^*\beta_{5n}(d_5^2)\big]$$

The superscript in parenthesis for Δ_n, such as $\Delta_n^{(1)}$, $\Delta_n^{(2)}$, etc. denotes first, second, etc. values of Δ_n. It is not a power of Δ_n.

The frequencies ω_1^*, ω_2^*, ... of the beam can be determined by setting the determinant in Equation 12.74 equal to zero.

12.5 APPLICATIONS

In order to apply the above methodology developed in the preceding section, we consider the elastically supported rectangular beam in Figure 12.6a of length L = 100 in. (2.54 m). The depth h = 8 in. (0.2032 m), width b = 6 in. (0.1524 m), $k_h = 10 \times 10^6$ lb/in. (1751.268×10^6 N/m), $k_v = 400 \times 10^3$ lb/in. ($70{,}050 \times 10^3$ N/m), $E = 30 \times 10^6$ psi (206.8×10^6 kPa), $K' = 0.66667$ and mass M = 0.42222 lb-sec²/in. (73.94 N-sec²/m) per 1 in. (0.0254 m) of length.

By applying the procedure discussed in the preceding section, the variation of the first frequency ω_1^*, in Hz, for various values of x_o and with initial static displacement a_1 is shown in Figure 12.7a. For $a_1 = 0$, which is curve 1 in Figure 12.7a, we note that the first frequency reduces to zero when x_o = 0.756 in. (0.0192 m), and it starts to increase for values of $x_o > 0.756$ in. (0.0192 m). This indicates that the beam buckles at x_o = 0.756 in. (0.192 m). In the same figure, the variation of the first frequency with increasing x_o and for values of a_1 = 0.01, 0.02, and 0.03 is shown by the curves 2, 3, and 4, respectively. Figures 12.7b and 12.7c show the variation of the second and third frequencies of the beam, respectively, with increasing x_o and for values of a_1 = 0, 0.01, 0.02, and 0.03.

In addition, the results in Figure 12.7a indicate that the decreasing pattern of the first frequency with increasing x_o is almost identical for all indicated values of the parameter a_1 when $x_o < 0.3$ in. (0.0076 m), and the plotted curves

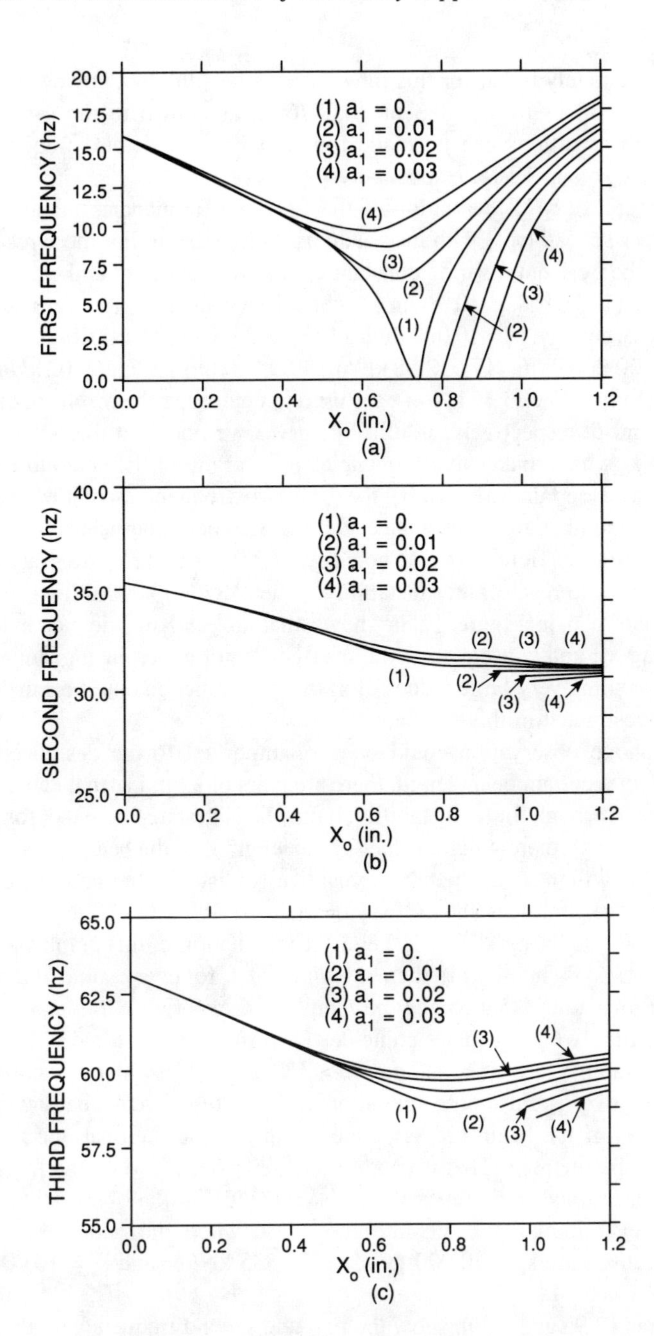

FIGURE 12.7. (a) Variation of first frequency with increasing x_o for various values of a_1 and $k_h = 10 \times 10^6$ lb/in., $k_v = 400,000$ lb/in.; (b) variation of second frequency with x_o for various values of a_1 and $k_h = 10 \times 10^6$ lb/in., $k_v = 400,000$ lb/in.; (c) variation of third frequency with x_o for various values of a_1 and $k_h = 10 \times 10^6$ lb/in., $k_v = 400,000$ lb/in. (1 in. = 0.0254m, 1 lb/in. = 175.1268 N/m).

are approximately linear for this range of x_0. The situation, however, changes rapidly for values of $x_0 > 0.3$ in. (0.0076m), as shown in the same figure. Similar observartions can be made for the second and third frequencies in Figures 12.7b and 12.7c, respectively.

In Figure 12.8a the variation of the first two frequencies ω_1^* and ω_2^*, in radians per second (rps) with increasing x_0, is illustrated. For these results, the value of the horizontal spring constant k_h was kept constant and equal to 10×10^6 lb/in. (1751.268×10^6 N/m), while five values of the vertical spring constant k_v, namely, $k_v = 10{,}000$ lb/in. (1751.268 kN/m), 50,000 lb/in. (8756.335 kN/m), 100,000 lb/in. (17,512.68 kN/m), 150,000 lb/in. (26,269.02 kN/m), and 200,000 lb/in. (35,025.36 kN/m), are used to determine the results in curves 1, 2, 3, 4, and 5, respectively. In all five curves we note that there is a distinct value of x_0 which makes the frequencies ω_1^* and ω_2^* of the beam to coincide. That is, at a certain value of x_0 the first two frequencies of the elastically supported member are at resonance, and the member experiences flutter instability. In the same figure, we also note that a stiffer k_v will require a larger value of x_0 in order to reach flutter instability. When $k_v = 400{,}000$ lb/in. (70,050.72 kN/m), curve 6 in Figure 12.8a shows that ω_1^* and ω_2^* do not meet with increasing x_0, and consequently the member is not experiencing flutter instability. For some very large values of x_0 this phenomenon could occur, but this is not investigated in this problem.

The above observation could be very important to the design engineer, because for a certain beam length there are pairs of k_h and k_v that can make the member go through flutter instability. It may be also stated here that for a given pair of k_h and k_v, there will be values of the length L of the beam where distinct values of x_0 will make the member dynamically unstable. In Figure 12.8b a plot of x_0 vs. the length L is shown for $k_h = 10 \times 10^6$ lb/in. (1751.268×10^6 N/m) and $k_v = 50{,}000$ lb/in. (8756.335 kN/m). Examination of this graph shows that when $0 < L \leq 189$ in. (4.8 m), there is a value of x_0 for every value of the length L within that range that forces the member to experience flutter instability, since the first two frequencies coincide. Therefore, for $0 < L \leq 189$ in. (4.8 m) flutter instability governs, and for $L > 189$ in. (4.8 m) static instability, or buckling, governs. Further examination of the results reveals that the first two frequencies of the beam are associated with the translational and rotational modes of the member. That is, flutter instability occurs when the frequencies of the translational and rotational modes coincide. Similar observations can be made by examining the results in Figure 12.8c, where the values of the pair of k_v and k_h used are $k_h = 50{,}000$ lb/in. (8756.335 kN/m) and $k_v = 10{,}000$ lb/in. (1751.268 kN/m).

Figures 12.9a and 12.9b show the first and second frequency mode shapes, respectively, at the stage where the beam is beginning to buckle. This occurs at $x_0 = 1.00$ in. (0.0254 m). At this point the member may buckle in the downward direction or it may buckle upwards. These two buckling configurations are represented by the curves y_{b1} and y_{b2}, respectively. Since the beam will vibrate either with respect to the static position y_{b1}, or with respect to

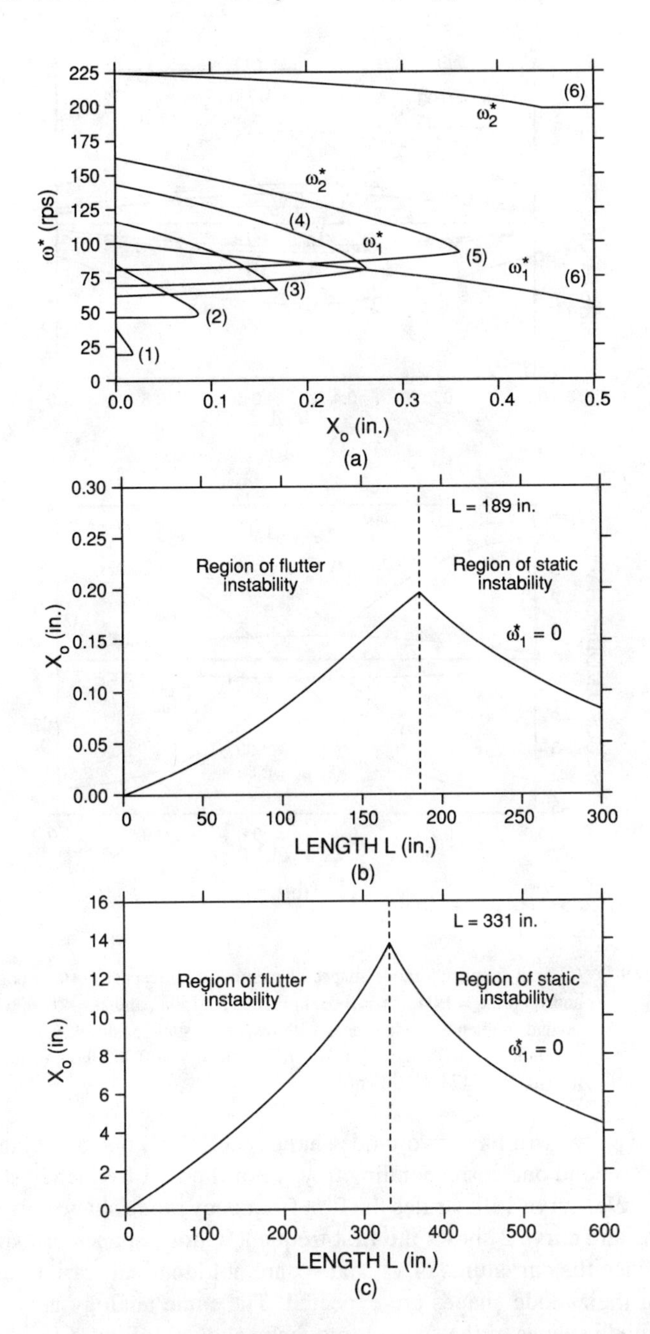

FIGURE 12.8. (a) Curves of flutter instability for various values of k_v and $k_h = 10 \times 10^6$ lb/in.: curve 1 $k_v = 10{,}000$ lb/in., curve 2 $k_v = 50{,}000$ lb/in., curve 3 $k_v = 100{,}000$ lb/in., curve 4 $k_v = 150{,}000$ lb/in., curve 5 $k_v = 200{,}000$ lb/in., and curve 6 $k_v = 400{,}000$ lb/in.; (b) regions of flutter and static instabilities for $k_h = 10 \times 10^6$ lb/in. and $k_v = 50{,}000$ lb/in.; (c) regions of flutter and static instabilities for $k_h = 50{,}000$ lb/in. and $k_v = 10{,}000$ lb/in. (1 in. = 0.0254 m, 1 lb/in. = 175.1268 N/m).

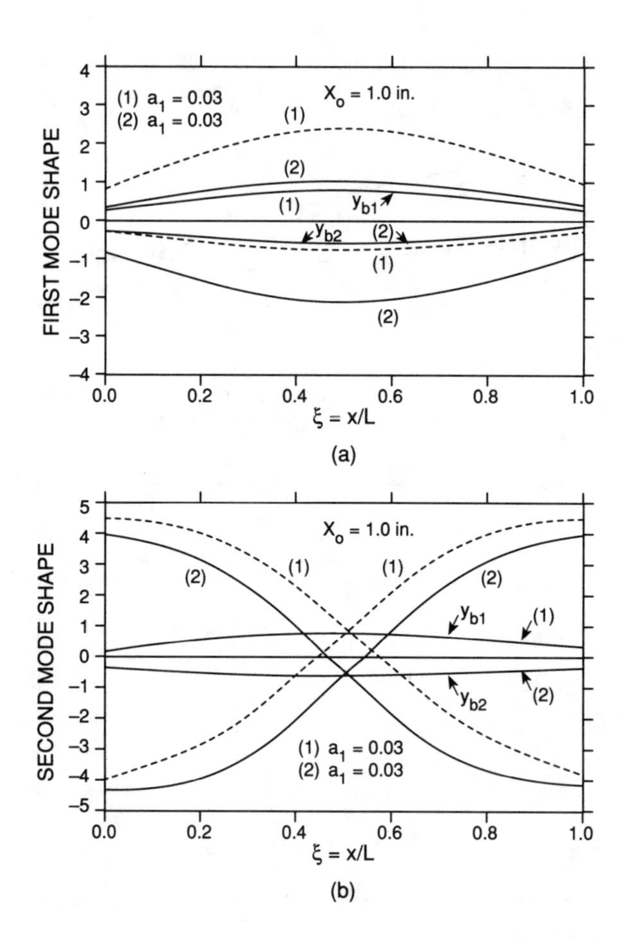

FIGURE 12.9. (a) First frequency mode shapes of member with respect to static positions y_{b1} and y_{b2} for $x_o = 1.0$ in., $a_1 = 0.03$, $k_h = 10 \times 10^6$ lb/in., and $k_v = 400{,}000$ lb/in.; (b) second frequency mode shapes with respect to static positions y_{b1} and y_{b2} for $x_o = 1.0$ in., $a_1 = 0.03$, $k_h = 10 \times 10^6$ lb/in., and $k_v = 400{,}000$ lb/in. (1 in. = 0.0254 m, 1 lb/in. = 175.1268 N/m).

position y_{b2}, we will have two mode shapes each time, one associated with y_{b1} and a second one corresponding to y_{b2}. For the first frequency shown in Figure 12.9a, curve 1 illustrates the first frequency mode shape corresponding to y_{b1} and curve 2 shows the first frequency mode shape corresponding to y_{b2}. Since the curvatures of y_{b1} and y_{b2} are not identical, variations in the shape of their mode shapes are expected. The same analogy may be used for the mode shapes of the second frequency shown in Figure 12.9b. Curve 1 in this figure illustrates the mode shape of the second mode with respect to y_{b1}, and curve 2 dipicts the mode shape of the second mode with respect to y_{b2}.

The variation of the static deflection at the center of the beam with increasing x_o, for values of $a_1 = 0$, 0.01, 0.02, and 0.03 in., is shown in Figure 12.10a.

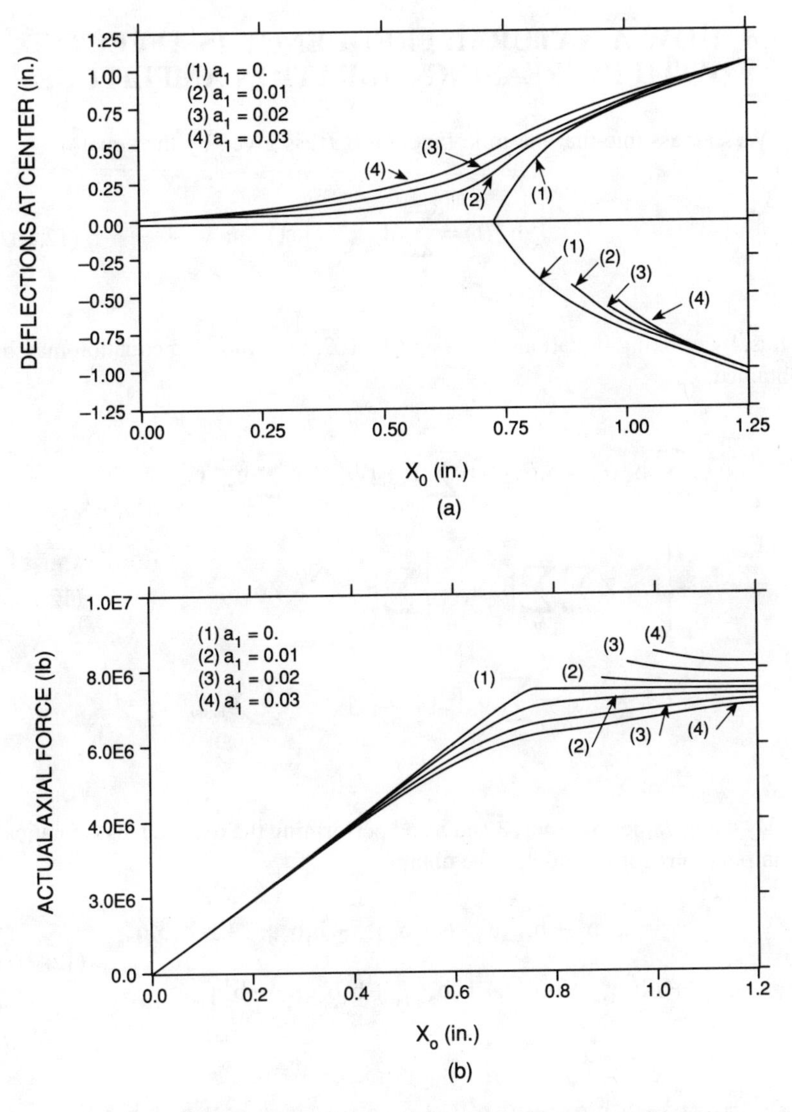

FIGURE 12.10. (a) Variation of static vertical displacement at midspan with increasing x_o for $a_1 = 0, 0.01, 0.02$, and 0.03, $k_h = 10 \times 10^6$ lb/in., $k_v = 400{,}000$ lb/in.; (b) variation of the actual axial force in the k_h spring with increasing x_o for $a_1 = 0, 0.01, 0.02$, and 0.03 and $k_h = 10 \times 10^6$ lb/in., $k_v = 400{,}000$ lb/in. (1 in. = 0.0254 m, 1 lb/in. = 175.1268 N/m).

For the same values of a_1 the variation of the actual axial force at the ends of the member with increasing x_o is shown in Figure 12.10b. In this figure we note that the variation of the axial force with increasing x_o is identical for all values of the parameter a_1 when $x_o < 0.45$ in. (0.0114m), and the indicated curves are linear for this range. The nature of the response changes rapidly when $x_o > 0.45$ in. (0.0114m), as shown in the same figure.

12.6 HOW A NATURAL FREQUENCY IS AFFECTED WITH INCREASING VIBRATION AMPLITUDE

We can assume that the time function $b_m(t)$ is given by the equation

$$b_m(t) = \sum_{j=0}^{3} d^*_{mj} \cos(j\omega t) \tag{12.102}$$

Thus, by utilizing Equations 12.59 and 12.102, the following equation may be obtained.

$$\begin{aligned}
&\sum_m b_m \alpha_m - a_1 \alpha_{1n} - \beta \sum_m b_{m,tt} \beta_{mn} + \alpha \sum_m b_{m,tt} \alpha^*_{mn} \\
&+\left\{ R + Z \sum_k \sum_p \beta^*_{kp} a_k a_p \right\} \sum_m b_m \beta_{mn} - Z\phi_n \\
&+Z * \frac{A^2_{1st} a^2_1}{(x_1 - x_o)} \sum_m b_m \beta_{mn} - Z * \phi^*_n + H \sum_m b_{m,tttt} \alpha^*_{mn} = 0
\end{aligned} \tag{12.103}$$

By using values of m = 1, 3 and 5 and performing the required mathematical manipulations for ϕ_n and ϕ_n^*, we obtain

$$\begin{aligned}
\phi_n &= q_{1n} b_1^3 + b_1^2 b_3 q_{4n} + b_1^2 b_5 q_{5n} + b_1 b_3^2 q_{6n} + b_1 b_3 b_5 q_{7n} \\
&+ b_1 b_5^2 q_{8n} + b_3^3 q_{2n} + b_3^2 b_5 q_{9n} + b_3 b_5^2 q_{10n} + b_5^3 q_{3n}
\end{aligned} \tag{12.104}$$

$$\begin{aligned}
\phi^*_n &= \bar{q}_{1n} b_1^3 + b_1^2 b_3 \bar{q}_{4n} + \bar{q}_{5n} b_1^2 b_5 + \bar{q}_{5n} b_1^2 b_5 + \bar{q}_{6n} b_1 b_3^2 + \bar{q}_{7n} b_1 b_3 b_5 \\
&+ \bar{q}_{8n} b_1 b_5^2 + \bar{q}_{2n} b_3^3 + \bar{q}_{9n} b_3^2 b_5 + \bar{q}_{10n} b_3 b_5^2 + \bar{q}_{3n} b_5^3
\end{aligned} \tag{12.105}$$

where

$$q_{1n} = \beta^*_{11} \beta_{1n}, \quad q_{2n} = \beta^*_{33} \beta_{3n}, \quad q_{3n} = \beta^*_{35} \beta_{5n} \tag{12.106}$$

$$q_{4n} = \beta^*_{11} \beta_{3n} + \beta^*_{13} \beta_{1n} + \beta^*_{31} \beta_{1n} \tag{12.107}$$

$$q_{5n} = \beta_{11}^{*}\beta_{5n} + \beta_{15}^{*}\beta_{1n} + \beta_{51}^{*}\beta_{1n} \tag{12.108}$$

$$q_{6n} = \beta_{13}^{*}\beta_{3n} + \beta_{31}^{*}\beta_{3n} + \beta_{33}^{*}\beta_{1n} \tag{12.109}$$

$$q_{7n} = \beta_{13}^{*}\beta_{3n} + \beta_{15}^{*}\beta_{3n} + \beta_{31}^{*}\beta_{5n} + \beta_{35}^{*}\beta_{1n} + \beta_{51}^{*}\beta_{3n} + \beta_{53}^{*}\beta_{1n} \tag{12.110}$$

$$q_{8n} = \beta_{15}^{*}\beta_{5n} + \beta_{51}^{*}\beta_{5n} + \beta_{55}^{*}\beta_{1n} \tag{12.111}$$

$$q_{9n} = \beta_{33}^{*}\beta_{5n} + \beta_{35}^{*}\beta_{3n} + \beta_{53}^{*}\beta_{3n} \tag{12.112}$$

$$q_{10n} = \beta_{55}^{*}\beta_{5n} + \beta_{53}^{*}\beta_{5n} + \beta_{55}^{*}\beta_{3n} \tag{12.113}$$

$$\bar{q}_{1n} = \bar{\beta}_{11}\beta_{1n}, \quad \bar{q}_{2n} = \bar{\beta}_{33}\beta_{3n}, \quad \bar{q}_{3n} = \bar{\beta}_{35}\beta_{5n} \tag{12.114}$$

$$\bar{q}_{4n} = \beta_{11}\beta_{3n} + \bar{\beta}_{13}\beta_{1n} + \beta_{31}\beta_{1n} \tag{12.115}$$

$$\bar{q}_{5n} = \bar{\beta}_{11}\beta_{5n} + \bar{\beta}_{15}\beta_{1n} + \bar{\beta}_{51}\beta_{1n} \tag{12.116}$$

$$\bar{q}_{6n} = \bar{\beta}_{13}\beta_{3n} + \bar{\beta}_{31}\beta_{3n} + \bar{\beta}_{33}\beta_{1n} \tag{12.117}$$

$$\bar{q}_{7n} = \bar{\beta}_{13}\beta_{3n} + \bar{\beta}_{15}\beta_{3n} + \bar{\beta}_{31}\beta_{5n} + \bar{\beta}_{35}\beta_{1n} + \beta_{51}\beta_{3n} + \bar{\beta}_{53}\beta_{1n} \tag{12.118}$$

$$\bar{q}_{8n} = \bar{\beta}_{15}\beta_{5n} + \bar{\beta}_{51}\beta_{5n} + \bar{\beta}_{55}\beta_{1n} \tag{12.119}$$

$$\bar{q}_{9n} = \bar{\beta}_{33}\beta_{5n} + \bar{\beta}_{35}\beta_{3n} + \bar{\beta}_{53}\beta_{3n} \tag{12.120}$$

$$\bar{q}_{10n} = \bar{\beta}_{55}\beta_{5n} + \bar{\beta}_{53}\beta_{5n} + \bar{\beta}_{55}\beta_{3n} \tag{12.121}$$

and

$$\bar{\beta}_{kp} = \frac{A_k A_p}{(x_1 - x_o)} \tag{12.122}$$

By considering the first term of Equation 12.102 in conjunction with Equations 12.103, 12.104, and 12.105 and applying the harmonic cosine-term balance method, we find

$$\begin{aligned}&\left(d_{10}^{*}\alpha_{1n}+d_{30}^{*}\alpha_{2n}+d_{50}^{*}\alpha_{3n}\right)-a_{1}\alpha_{1n}\\&+\left(R+Za_{1}^{2}\beta_{11}^{*}\right)\left(\beta_{1n}d_{10}^{*}+\beta_{3n}d_{30}^{*}+\beta_{51}d_{50}^{*}\right)\\&+Z*\frac{A_{1}^{2}a_{1}^{2}}{(x_{1}-x_{o})}\left(d_{10}^{*}\beta_{1n}+d_{30}^{*}\beta_{3n}+d_{50}^{*}\beta_{5n}\right)-Z\phi_{n}^{(1)}-Z*\phi_{n}^{*(1)}\end{aligned}\tag{12.123}$$

By considering the second term of Equation 12.102 in conjunction with Equations 12.103, 12.104, and 12.105 and applying the harmonic cosine-term balance method, wc find

$$\begin{aligned}&\left(d_{11}^{*}\alpha_{1n}+d_{31}^{*}\alpha_{2n}+d_{51}^{*}\alpha_{3n}\right)-a_{1}\alpha_{1n}+\omega^{*2}\beta\left(d_{11}^{*}\beta_{1n}+d_{31}^{*}\beta_{3n}+d_{51}^{*}\beta_{5n}\right)\\&-\omega^{*2}\alpha\left(d_{11}^{*}\alpha_{1n}^{*}+d_{31}^{*}\alpha_{3n}^{*}+d_{51}^{*}\alpha_{5n}^{*}\right)+\left(R+Za_{1}^{2}\beta_{11}^{*}\right)\\&\left(\beta_{1n}d_{11}^{*}+\beta_{3n}d_{31}^{*}+\beta_{5n}d_{51}^{*}\right)+Z*\frac{A_{1}^{2}a_{1}^{2}}{(x_{1}-x_{o})}\left(d_{11}^{*}\beta_{1n}+\beta_{3n}d_{31}^{*}+d_{51}^{*}\beta_{5n}\right)\\&+\omega^{*4}H\left(D_{11}^{*}\alpha_{1n}^{*}+d_{31}^{*}\alpha_{3n}^{*}+d_{51}^{*}\alpha_{5n}^{*}\right)-Z\phi_{n}^{2}-Z*\phi_{n}^{*2}=0\end{aligned}\tag{12.124}$$

In a similar manner, by using the third term of the Equation 12.102 in conjunction with Equations 12.103, 12.104, and 12.105 and applying the harmonic cosine-term balance method, we find

$$\begin{aligned}&\left(d_{12}^{*}\alpha_{1n}+d_{32}^{*}\alpha_{2n}+d_{52}^{*}\alpha_{3n}\right)-a_{1}\alpha_{1n}\\&+4\omega^{*2}\beta\left(d_{12}^{*}\beta_{1n}+d_{32}^{*}\beta_{3n}+d_{52}^{*}\beta_{5n}\right)\\&-4\omega^{*2}N\left(d_{12}^{*}\alpha_{1n}^{*}+d_{32}^{*}\alpha_{3n}^{*}+d_{51}^{*}\alpha_{5n}^{*}\right)\\&+\left(R+Za_{1}^{2}\beta_{11}^{*}\right)\left(d_{12}^{*}\beta_{1n}+d_{32}^{*}\beta_{3n}+d_{52}^{*}\beta_{5n}\right)\\&-Z*\left(\frac{A_{1}^{2}a_{1}^{2}}{(x_{1}-x_{o})}\right)\left(d_{12}^{*}\beta_{1n}+d_{32}^{*}\beta_{3n}+d_{52}^{*}\beta_{5n}\right)\\&+16\omega^{*4}H\left(d_{12}^{*}\alpha_{1n}^{*}+d_{32}^{*}\alpha_{1n}^{*}+d_{32}^{*}\alpha_{3n}^{*}+d_{52}^{*}\alpha_{5n}^{*}\right)\\&-Z\phi_{n}^{(3)}-Z*\phi_{n}^{*(3)}=0\end{aligned}\tag{12.125}$$

Finally, by considering in a similar manner the fourth term of Equation 12.102 and applying the harmonic cosine-term balance, we obtain

$$
\begin{aligned}
&\left(d^*_{13}\alpha_{1n} + d^*_{33}\alpha_{2n} + d^*_{53}\alpha_{3n}\right) - a_1\alpha_{1n} + 9\omega^{*2}\,\beta\left(d^*_{13}\beta_{1n} + d^*_{33}\beta_{3n} + d^*_{53}\beta_{5n}\right) \\
&-9\omega^{*2}\,N\left(d^*_{13}\alpha^*_{1n} + d^*_{33}\alpha^*_{3n} + d^*_{53}\alpha^*_{5n}\right) + \left(R + Za_1^2\beta^*_{11}\right) \\
&\left(\beta_{1n}d^*_{13} + \beta_{3n}d^*_{33} + \beta^*_{5n}d^*_{53}\right) - Z^*\left(\frac{A_1^2a_1^2}{(x_1 - x_o)}\right)\left(d^*_{13}\beta_{1n} + \beta_{3n}d^*_{33} + d^*_{53}\beta_{5n}\right) \\
&+81\omega^{*4}\left(d^*_{13}\alpha^*_{1n} + d^*_{33}\alpha^*_{3n} + d^*_{53}\alpha^*_{5n}\right) - Z\phi_n^{(4)} - Z^*\phi_n^{*(4)} = 0
\end{aligned}
\tag{12.126}
$$

In the above equations, the expressions for $\phi_n^{(1)}$, $\phi_n^{*(1)}$, etc. may be obtained from Equations 12.102 through 12.122 by expending the summations for values of n = 1, 3, and 5. For example, by following this procedure the expression for $\phi_n^{(1)}$ may be written as

$$
\begin{aligned}
\phi_n^{(1)} = {} & q_{1n}\left[d^{*3}_{10} + \frac{3}{2}\left(d^{*2}_{11}d^*_{10}\right) + \frac{3}{2}d^{*2}_{11}d_{10} + \frac{3}{2}d^*_{10}d^{*2}_{12} + \frac{3}{2}d^*_{10}d^{*2}_{13}\right. \\
&\left. + \frac{3}{4}d^{*2}_{11}\frac{d^{*3}_{12}}{2}d^*_{11}d^*_{12}d^*_{13}\right] + q_{4n}\left[d^{*2}_{10}d_{30} + d^*_{10}d^*_{11}d^*_{31} + \frac{1}{2}d^{*2}_{11}d^*_{30}\right. \\
& + d^*_{12}d^*_{10}d^*_{32} + \frac{1}{2}d^{*2}_{12}d^*_{30} + d^*_{10}d^*_{13}d^*_{33} + \frac{1}{2}d^{*2}_{13}d^*_{30} + \frac{1}{2}d^*_{11}d^*_{12}d^*_{31} \\
&\left. + \frac{1}{4}d^{*2}_{11}d^*_{32} + \frac{1}{2}d^*_{11}d^*_{12}d^*_{33} + 2d^*_{11}d^*_{13}d^*_{32} + 2d^*_{12}d^*_{13}d^*_{31}\right] \\
& + q_{5n}[- -
\end{aligned}
\tag{12.127}
$$

The superscript in parenthesis for ϕ_n, such as $\phi_n^{(1)}$, $\phi_n^{*(1)}$, etc. denotes first, second, etc. values of ϕ_n.

From Equations 12.123 through 12.126 we form a set of 12 simultaneous cubic equations in d_{mj} which can be solved by using the Newton-Raphson method when a, k_h, k_v, and ω^* are prescribed. The deflection at the midspan of the beam can be determined from the equation

$$(\text{deflection at center})_{rms} = \frac{1}{\sqrt{2}}\left[\left(\sum_{m} d_{mj}\phi_m\left(\frac{1}{2}\right)\right)^2\right]^{1/2} \tag{12.128}$$

$$m = 1,3,5 \qquad j = 0,1,2,3$$

where rms in this equation denotes root mean square.

As an illustration regarding the application of the above methodology we consider the elastically supported beam in Figure 12.6a, which has length L = 100 in. (2.54 m), depth h = 8 in. (0.2032 m), width b = 6 in. (0.1524 m), vertical spring constant k_v = 400,000 lb/in. (70,050.72 kN/m), and horizontal spring constant $k_h = 10 \times 10^6$ lb/in. (1751.268×10^6 N/m). By following the methodology discussed in this section, we determine the variation of the first frequency ω_1^* (in Hz) with increasing midspan deflection. The results are plotted in Figure 12.11 by developing the indicated curves 1 through 7 for values of x_o = –0.4, –0.2, 0, 0.2, 0.4, 0.6, and 0.8, respectively. These curves indicate that ω_1^* increases with increasing midspan deflection. This familly of curves also illustrates that ω_1^* decreases with increasing x_o.

Figure 12.12 illustrates the variation of ω_1^* with increasing midspan amplitude for values of the horizontal spring constant $k_h = 10 \times 10^6$ lb/in., 5×10^6 lb/in., 10^6 lb/in., and 0.5×10^6 lb/in. (1 lb/in. = 175.1268 N/m). The value of the vertical spring constant k_v is kept constant and equal to 400,000 lb/in. (70,050.72 $\times 10^3$N/m). The graphs in Figure 12.12 indicate that ω_1^* increases more rapidly with increasing midspan deflection when the stiffness of the horizontal spring k_h is increased.

PROBLEMS

12.1 The elastically supported beam in Figure 12.1 has a length L = 100 in. (2.54 m), depth h = 8 in. (0.2032 m), width b = 6 in. (0.1524 m), modulus of elasticity $E = 30 \times 10^6$ psi ($206,843 \times 10^6$ Pa), shear coefficient K′ = 2/3, and Poisson ratio ν = 0.25. By following the procedure discussed in Section 12.2 and assuming that k_v = 5000 lb/in. (875.634×10^3 N/m), verify the results obtained in Table 12.2 for the indicated values of the axial compressive load P.

12.2 If k_v in Problem 12.1 is equal to 50,000 lb/in. (8756.34×10^3 N/m) and the axial force $P = 10^6$ lb (4.448×10^6N), determine the length L of the member that makes $\omega_1 = \omega_2$. Also determine the critical length L_{cr} of the member which makes $\omega_1 = 0$.
Answer: L = 60.5 in., $\omega_1 = \omega_2$ = 62.02 rps; L_{cr} = 275 in., $\omega_1 = 0$, ω_2 = 39.14 rps.

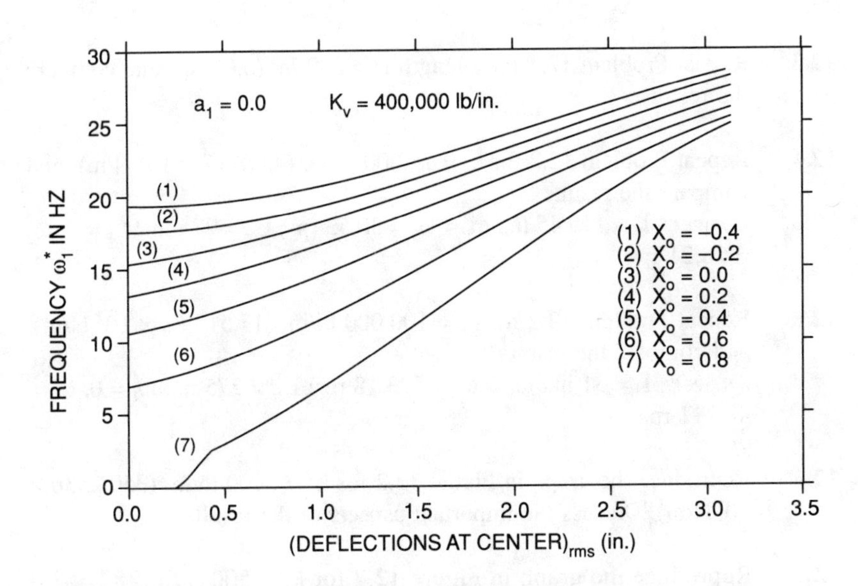

FIGURE 12.11. Variation of $\omega_1{}^*$ with increasing midspan deflection for various values of x_o.

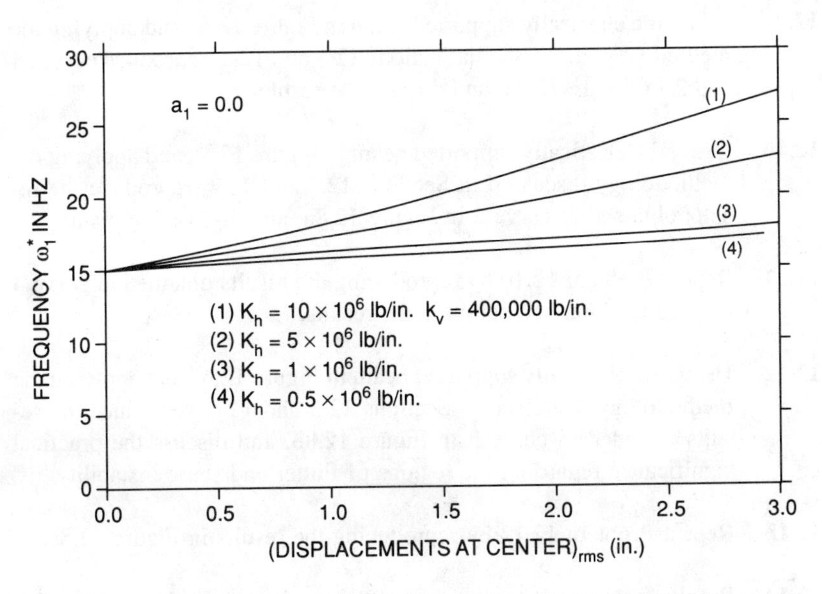

FIGURE 12.12. Variation of $\omega_1{}^*$ with increasing midspan deflection for various values of the horizontal spring constant k_h.

12.3 Repeat Problem 12.1 for a length L = 200 in. (5.08 m), and compare the results.

12.4 Repeat Problem 12.2 for k_v = 25,000 lb/in. (4378.17 × 10^3 N/m), and compare the results.
Answer: L = 115.45 in., $\omega_1 = \omega_2$ = 30.53 rps; L_{cr} = 275 in., ω_1 = 0, ω_2 = 28.21 rps.

12.5 Repeat Problem 12.2 for k_v = 100,000 lb/in. (17,512.68 × 10^3 N/m), and compare the results.
Answer: L = 31 in., $\omega_1 = \omega_2$ = 123.18 rps; L_{cr} = 275 in., ω_1 = 0, ω_2 = 48.211 rps.

12.6 Reproduce the graph in Figure 12.2 for k_v = 2000 lb/in. (350.2536 × 10^3 N/m). Discuss the important aspects of the results.

12.7 Reproduce the graph in Figure 12.2 for k_v = 500 lb/in. (87,563.4 N/m) and discuss the important aspects of the results. Compare the results with the graph obtained in Problem 12.6, and discuss the findings.

12.8 Reproduce the graph obtained in Figure 12.3 and make appropriate comments regarding the indicated regions of instability.

12.9 Using the elastically supported beam in Figure 12.6a and applying the methodology discussed in Sections 12.3 and 12.4, reproduce curves 1 and 2 in Figure 12.7a, and discuss the results.

12.10 Using the elastically supported beam in Figure 12.6a and applying the methodology discussed in Sections 12.3 and 12.4, reproduce the results obtained in curve 3 in Figure 12.8a, and discuss the results.

12.11 Repeat Problem 12.10 by reproducing the results obtained in curve 4 of Figure 12.8a, and compare the results.

12.12 Using the elastically supported beam in Figure 12.6a and applying the methodology discussed in Sections 12.3 and 12.4, reproduce the results obtained in curve 3 in Figure 12.8b, and discuss the practical significance regarding the regions of flutter and static instability.

12.13 Repeat Problem 12.12 by reproducing the results in Figure 12.8c.

12.14 Provide a thorough explanation of the results obtained in Figure 12.9 and discuss their significance to practical engineering problems.

12.15 By following the methodology discussed in Section 12.6, reproduce curves 1 and 2 in Figure 12.10a, and discuss the results.

12.16 By following the methodology discussed in Section 12.6, reproduce curve 6 in Figure 12.11, and discuss the results.

12.17 Repeat Problem 12.16 with $k_v = 200{,}000$ lb/in. ($35{,}025.36 \times 10^3$ N/m), and compare the results.

REFERENCES

1. **Lagrange, J. L.,** Sur la force des ressorts plies, *Mem. Acad. Berlin,* 1770.
2. **Euler, L.,** *Methodus Inveniendi Lineas Curvas,* 1744.
3. **Plana,** *Equation de la courbe fermee par une lame elastique, Mem. R. Soc. Turin,* 1809.
4. **Frisch-Fay, R.,** *Flexible Bars,* Butterworths and Co. Publishers, Ltd., Washington, DC, 1962.
5. **Fertis, D. G. and Afonta, A. O.,** Equivalent systems for large deformation of beams of any stiffness variation, *Eur. J. Mech.* A/Solids, 10(3), 265, 1991.
6. **Fertis, D. G.,** *Dynamics and Vibration of Structures,* John Wiley & Sons, Inc., New York, 1973.
7. **Fertis, D. G.,** *Dynamics and Vibration of Structures,* rev. ed., Robert E. Krieger Publishing Co., Malabar, FL, 1984.
8. **Fertis, D. G. and Lee, C. T.,** Inelastic analysis of flexible bars using simplified nonlinear equivalent systems, *Int. J. Comput. Struct.,* 41(5), 947, 1991.
9. **Fertis, D. G. and Lee, C. T.,** Equivalent systems for inelastic analysis of members subjected to large deformations, in *Proceedings of Plasticity '91 Symposium,* International Symposium on Plasticity, Grenoble, France, August 1991.
10. **Bishoppe, K. E. and Drucker, D. C.,** Large deflection of cantilever beams, *Q. Appl. Math.,* 11, 337, 1953.
11. **Lau, J. H.,** Large deflection of cantilever beams, *J. Eng. Mech., ASCE,* 107(EMI), 259, 1961.
12. **Seide, P.,** Large deflections of a simply supported beam subjected to moment at one end, *J. Appl. Mech., Trans. ASME,* 51(3), 519, 1984.
13. **Wang, T. M., Lee, S. L., and Zienkiewicz, O. C.,** A numerical analysis of large deflections of beams, *J. Mech. Sci.,* 3, 219, 1961.
14. **Ohtsuki, A.,** An analysis of large deflections in a symmetrical three point bending of beam, *Bull. ASME,* 29(253), 1986.
15. **Liebold, R.,** Die Durchbiegung einer Beidseiting fest Eigenspannten, *Z. Angew. Math. Mech.,* 28, 247, 1948.
16. **Prathap, G. and Varadan, T. K.,** Large amplitude free vibration of tapered hinged beams, *AIAA J.,* 16(1), 88.
17. **Hsiao, K. M. et al.,** Nonlinear finite element analysis of elastic frames, *Comput. Struct.,* 26(4), 693, 1987.
18. **Tada, Y. et al.,** Finite element solution to an elastic problem of beams, *Int. J. Numer. Methods Eng.,* 2, 229, 1970.
19. **Yang, T. Y.,** Matrix displacement solution to elastica problems of beams and frames, *Int. J. Solids Struct.,* 9, 829, 1973.
20. **Fertis, D. G. and Afonta, A. O.,** Large deflection of determinate and indeterminate bars of variable stiffness, *J. Eng. Mech., ASCE,* 116(18), 1989.
21. **Fertis, D. G. and Afonta, A. O.,** Free vibration of variable stiffness flexible bars, *J. Comput. Struct.,* 43(3), 445.
22. **Fertis, D. G. and Afonta, A. O.,** Small vibrations of flexible bars by using the finite element method with equivalent uniform stiffness and mass methodology, *Int. J. Sound Vib.,* 161(1), 1993.
23. **Fertis, D. G. and Pallaki, S.,** Pseudolinear and equivalent systems for large deflections of members, *J. Eng. Mech., ASCE,* 115(11), 2440, 1989.
24. **Fertis, D. G. and Lee, C. T.,** Inelastic response of variable stiffness members under cyclic loading, *J. Eng. Mech., ASCE,* 118(7), 1992.
25. **Fertis. D. G. and Lee, C. T.,** Equivalent systems for inelastic analysis of nonprismatic flexible bars with axial restraints, *Int. J. Mech. Sci.,* submitted for publication.
26. **Prathap, G.,** Nonlinear vibration of beams with variable axial restraints, *AIAA J.,* 16(6), 622, 1978.

27. **Mei, C.,** Nonlinear vibration of beams by matrix displacement method, *AIAA J.*, 10(3), 355.
28. **Bhashyam, G. R. and Prathap, G.,** Galerkin finite element method for nonlinear beam vibrations, *J. Sound Vib.*, 72, 191, 1980.
29. **Fertis, D. G. and Lee, C. T.,** Nonlinear vibration of axially restrained elastically supported beams, in *Proceedings of the 1992 Workshop on Dynamics, ASME,* PD-Vol. 44, Dynamics and Vibration, Book No. G00652, 1992, 13–19.
30. **Lau, J. H.,** Large deflection of cantilever beams, *J. Eng. Mech., ASCE,* 107(EM1), 259, 1981.
31. **Gere, J. M. and Timoshenko, S. P.,** *Mechanics of Materials,* 3rd ed., PWS-KENT Publishing Co., Boston, 514.
32. **Fertis, D. G. and Keene, M. E.,** Elastic and inelastic analysis of nonprismatic members, *J. Struct. Eng., ASCE,* 116(2), 1990.
33. **Fertis, D. G. and Taneja, R.,** Equivalent systems for inelastic analysis of prismatic and nonprismatic members, *J. Struct. Eng., ASCE,* 117(2), 473, 1991.
34. **Fertis, D. G., Taneja, R., and Lee, C. T.,** Equivalent systems for inelastic analysis of nonprismatic members, *Comput. Struct.*, 38(1), 31, 1991.
35. **Timoshenko, S.,** *Strength of Materials, Part II,* D. Van Nostrand Company, Inc., Princeton, NJ, 1956, 76, 366.
36. **Eringen, A. C.,** On the non-linear vibration of elastic bars, *Q. Appl. Math.*, 4, 361, 1952.
37. **Timoshenko, S. and Woinosky-Krieger, S.,** *Theory of Plates and Shells,* 2nd ed., McGraw-Hill Book Co., Inc., New York, 1959.
38. **Olsson, R. G.,** Biegung kreisformiger platten von radial veranderlicher Dicke, *Ing. Arch.*, 13, 6, 1937.
39. **Rainville, E. D.,** *Intermediate Differential Equations,* John Wiley & Sons, Inc., New York, 1943.
40. **Conway, H. D.,** Closed-form solutions for circular plates of variable thickness, *J. Appl. Mech.,* 20, 564, 1953.
41. **Conway, H. D.,** The bending of symmetrically loaded circular plates of variable thickness, *J. Appl. Mech.*, 15, 1, 1948.
42. **Favre, H. and Chabloz, E.,** Etude des plaques circulries flechies d'epaisseru linearement variable, *Z. Angew. Math. Phys.,* 1, 317, 1950.
43. **Olsson, R. G.,** Unsymmetrical biegungder kreisring plates von quadrtisch veranderlicher Streifigkeit, *Ing. Arch.,* 10, 14, 1939.
44. **Alwar, R. S. and Nath, Y.,** Application of Chebyskev polynomials to the nonlinear analysis of circular plates, *Int. J. Mech. Sci.,* 18, 589, 1976.
45. **Olsson, R. G.,** Biegung der Rechteckplatte bei linear varanderlicher Biegung Steifigkeif, *Ing. Arch.,* 5, 363, 1934.
46. **Reissner, E.,** Remark on the theory of bending of plates of variable thickness, *J. Math. Phys.,* 5, 363, 1934.
47. **Olsson, R. G.,** Biegung der Rechteckplatte bei linear veranderlicher Strifigkeit und beliebieger Belastung, *Baningenieus,* 22, 10, 1941.
48. **Mukhopadhyay, M.,** A semianalysis solution for rectangular plate bending, *J. Comput. Struct.,* 9, 81, 1977.
49. **Dey, S. S.,** Semi-numerical analysis of rectangular plates in bending, *J. Comput. Struct.,* 14(5–6), 369, 1981.
50. **Ohga , M. and Shigematsu, T.,** Bending analysis of plates with variable thickness by boundary element-transfer matrix method, *J. Comput. Struct.,* 28(5), 635, 1988.
51. **Fertis, D. G. and Mijatov, M. M.,** Equivalent systems for variable thickness plates, *J. Eng. Mech., ASCE,* 115(10), 1989.
52. **Fertis, D. G. and Lee, C. T.,** Equivalent systems for the analysis of rectangular plates of varying thickness, in *Proceedings of the Fifteeth Southeastern Conference on Theoretical and Applied Mechanics,* Atlanta, GA, Vol. 15, 1990, 627.

53. **Fertis, D. G. and Lee, C. T.,** Elastic and Inelastic analysis of variable thickness plates by using equivalent systems, *Int. J. Mech. Struct. Mach.,* 21(2), 201, 1993.
54. **Lin, T. H.,** *Theory of Inelastic Structures,* John Wiley & Sons, Inc., New York, 1986.
55. **Kozma, A. and Fertis, D. G.,** Solution of the deflection of variable thickness plates by the method of the equivalent systems, *Industrial Mathematics,* Vol. 12, Part I, Industrial Mathematics Society, 1962.
56. **Fertis, D. G. and Lee, C. T.,** Inelastic analysis of variable thickness rectangular plates, in Proceedings of the European Joint Conference on Engineering Systems, Design, and Analysis, ASME, Instanbul, Turkey, June 29–July 3, 1992.
57. **Bleich, H. H.,** Response of elastic-plastic structures to transient loads, in Proceedings of the New York Academy of Sciences, Vol. 8, (2nd series) 1955, 135.
58. **Bleich, H. H. and Salvadory, M. G.,** *Impulsive motion of elastic-plastic beams, Trans. ASCE,* 120, 499, 1955.
59. **Baron, M. L., Bleich, H. H., and Widlinger, P.,** Dynamic elastic-plastic analysis of structures, *J. Eng. Mech., ASCE,* 89, 23, 1961.
60. **Berg, G. V. and DaDeppo, D. A.,** Dynamic analysis of elastic-plastic structures, *J. Eng. Mech., ASCE,* 86, 35, 1960.
61. **Lee, E. H. and Symonds, P. S.,** Large plastic deformations of beams under transverse impact, *J. Appl. Mech., ASCE,* 19, 308, 1952.
62. **Toridis, T. and Wen, R. K.,** Inelastic response of beams to moving loads, *J. Eng. Mech., ASCE,* 92, 43, 1966.
63. **Bonc, R.,** Forced vibration of hysteretic systems, in *Proceedings of the IV Conference on Nonlinear Oscillation,* Prague, Czechoslovakia, 1967.
64. **Wen, R. K.,** Lumped elasto-plastic flexibility models, *J. Eng. Mech., ASCE,* 91, 289, 1975.
65. **Iyenger, R. N. and Dash, R. K.,** Study of the random vibration of non-linear systems by the Gaussian closure technique, *J. Eng. Mech., ASCE,* 102, 249, 1976.
66. **Sues, R. H., Mau, S. T., and Wen, Y. K.,** System identification of degrading hysteretic restoric forces, *J. Eng. Mech., ASCE,* 114(5), 833, 1988.
67. **Fertis, D. G. and Lee, C. T.,** Inelastic response of variable stiffness members under cyclic loading, *J. Eng. Mech., ASCE,* 118(7), 1406, 1992.
68. **Kounadis, A. N.,** The existance of regions of divergence instability for nonconservative systems under follower forces, *Int. J. Solid. Struct.,* 19(8), 725, 1983.
69. **Kounadis, A. N.,** Divergence and flutter instability of elastically restraint structures under follower forces, *Int. J. Eng. Sci.,* 19, 553, 1981.
70. **Fertis, D. G. and Lee, C. T.,** Nonlinear vibration and instabilities of elastically supported beams with axial restraints, *Int. J. Sound Vib.,* 162(1), 1993.
71. **Bosela, P. A., Shaker, F. J., and Fertis, D. G.,** Dynamic analysis of space-related linear and nonlinear structures, in *Proceedings of the 15th Southeastern Conference on Theoretical and Applied Mechanics,* SECTAM XV, March 22–25, 1990.
72. **Bosela, P. A., Fertis, D. G., and Shaker, F. J.,** Grounding of space structures, *Int. J. Comput. Struct.,* 45(1), 143, 1992.
73. **Bosela, P. A., Fertis, D. G., and Shaker, F. J.,** A new pre-loaded beam geometric stiffness matrix with full rigid body capabilities, *Int. J. Comput. Struct.,* 45(1), 155, 1992.
74. **Press, W. H., Flannery, B. P., Tecikolsky, S. A.,** and Vetterling, W. T., *Numerical Recipes,* Cambridge University Press, Cambridge, 1989, 243.
75. **Timoshenko, S. P., Young, D. H., and Weaver, W., Jr.,** *Vibration Problems in Engineering,* John Wiley & Sons Co., Inc., New York, 1974.
76. **Bosela, P. A., Shaker, F. J., and Fertis, D. G.,** Dynamic analysis of space-related linear and nonlinear structures, *Int. J. Comput. Struct.,* 44(5), 1145, 1992.
77. **Rao, B. N., Shastry, B. P., and Rao, G. V.,** "Large deflections of Cantilever beams with end rotational load", *ZAMM,* 66, 507, 1986.

INDEX

F

P

R

T

U

V

W

Y